Abiotic Stress Physiology of Horticultural Crops

N.K. Srinivasa Rao • K.S. Shivashankara
R.H. Laxman
Editors

Abiotic Stress Physiology of Horticultural Crops

Springer

Editors
N.K. Srinivasa Rao
Division of Plant Physiology
and Biochemistry
ICAR-Indian Institute of
Horticultural Research
Bengaluru, Karnataka, India

K.S. Shivashankara
Division of Plant Physiology
and Biochemistry
ICAR-Indian Institute of
Horticultural Research
Bengaluru, Karnataka, India

R.H. Laxman
Division of Plant Physiology
and Biochemistry
ICAR-Indian Institute of
Horticultural Research
Bengaluru, Karnataka, India

ISBN 978-81-322-2723-6 ISBN 978-81-322-2725-0 (eBook)
DOI 10.1007/978-81-322-2725-0

Library of Congress Control Number: 2016933032

Printed on acid-free paper

This Springer imprint is published by Springer Nature
The registered company is Springer (India) Pvt. Ltd.

Preface

Abiotic stress factors, mainly salinity, drought, flooding and high temperature, are the main elements which drastically limit the horticultural crop productivity globally. Abiotic stress leads to a series of morphological, physiological, biochemical and molecular changes in plants that adversely affect growth and productivity. Extreme environmental events in the era of global climatic change further aggravate the problem and remarkably restrict the plant growth and development. The mechanisms underlying endurance and adaptation to environmental stress factors have long been the focus of intense research. Plants overcome environmental stresses by the development of tolerance, resistance or avoidance mechanisms. Plant acclimation to environmental stresses is the process to adjust to a gradual change in its environment which allows the plants to maintain performance across a range of adverse environmental conditions. Stress tolerance mechanisms in horticultural crops are gaining attention because most agricultural regions are predicted to experience considerably more extreme environmental fluctuations due to global climate change. It has been estimated that salinity and drought are expected to cause serious salinisation of more than 50 % of all available productive, arable lands by the year 2050. Water availability and water use efficiency are among the important abiotic factors that have had and continue to have a decisive influence on plant evolution. Water stress in its broadest sense encompasses both drought and flooding stress. In-depth knowledge on molecular mechanisms of abiotic stress effects on plants is needed for developing tolerant genotypes. A clear understanding of environmental factors and their interaction with physiological processes is extremely important for improving horticultural practices. Horticultural crops include a wide range of commodities, such as fruits and vegetables that are highly valuable for humanity. They are extensively grown worldwide, and their production can be described as an open and highly complex system affected by many factors, among which we can count weather, soil and cropping system, as well as the interaction between these factors.

Scope of this book includes chapters on tropical and subtropical species written by scientists from different fields of specialisation. The influence of environmental factors, such as temperature, water and salinity on plant physiology and on vegetative and reproductive growth, is comprehensively discussed for each crop. In this book *Abiotic Stress Physiology of*

Horticultural Crops, we present a collection of 19 chapters written by many experts in the field of crop improvement, genetic engineering and abiotic stress tolerance. Various chapters included in this book provide a state-of-the-art account of the information on (1) mechanisms of abiotic stress tolerance responses, and (2) abiotic stress tolerance in various horticulture crops – tomato, onion, capsicum, mango, grapes, banana, litchi, Arid Zone fruit crops, coconut, arecanut, cahew, cocoa, spices, oil palm and tuber crops – which is a resourceful guide suited for scholars and researchers working in the field of crop improvement, genetic engineering and abiotic stress tolerance of horticultural crops.

We, the editors, would like to give special thanks to the authors for their outstanding and timely work in producing such fine chapters. We are highly thankful to Ms. K.C. Pavithra for her valuable help in formatting and incorporating editorial changes in the manuscripts. The editors and contributing authors hope that this book will include a practical update on our knowledge for plant acclimation to environmental stress and lead to new discussions and efforts to the use of various tools for the improvement of horticultural crops for abiotic stress tolerance.

Bengaluru, India

N.K. Srinivasa Rao
K.S. Shivashankara
R.H. Laxman

Contents

Contributors

S.J. Ankegowda ICAR-Indian Institute of Spices Research, Madikeri, Karnataka, India

C.J. Annapoornamma Division of Plant Physiology and Biochemistry, ICAR-Indian Institute of Horticultural Research, Bengaluru, Karnataka, India

K.V. Kasturi Bai ICAR-Central Plantation Crops Research Institute, Kasaragod, Kerala, India

D. Balasimha Central Plantation Crops Research Institute, Karnataka, India

B-404, Mantri Alpyne, Banashankari 5th Stage, Bengaluru, India

ICAR-Central Plantation Crops Research Institute, Kasaragod, Kerala, India

S.K. Behera ICAR-Directorate of Oil Palm Research, Pedavegi, Andhra Pradesh, India

K. Bhanuprakash Section of Seed Science and Technology, ICAR-Indian Institute of Horticultural Research, Bengaluru, Karnataka, India

Geeta Biradar Division of Plant Physiology and Biochemistry, ICAR-Indian Institute of Horticultural Research, Bengaluru, Karnataka, India

Dominic P. D'mello Division of Vegetable Crops, ICAR-Indian Institute of Horticultural Research, Bengaluru, Karnataka, India

A.N. Ganeshamurthy Division of Soil Science and Agricultural Chemistry, ICAR-Indian Institute of Horticultural Research, Bengaluru, Karnataka, India

G.A. Geetha Division of Plant Physiology and Biochemistry, ICAR-Indian Institute of Horticultural Research, Bengaluru, Karnataka, India

Johnson K. George ICAR-Indian Institute of Spices Research, Kozhikode, Kerala, India

Atul Jayapal College of Agriculture, Vellayani, Thiruvananthapuram, Kerala, India

D. Kalaivanan Division of Soil Science and Agricultural Chemistry, ICAR-Indian Institute of Horticultural Research, Bengaluru, Karnataka, India

K.S. Krishnamurthy ICAR-Indian Institute of Spices Research, Kozhikode, Kerala, India

R. Punith Kumar Division of Vegetable Crops, ICAR-Indian Institute of Horticultural Research, Bengaluru, Karnataka, India

Rajesh Kumar ICAR- National Research Centre for Litchi, Muzaffarpur, Bihar, India

S. Naresh Kumar Centre for Environment Science and Climate Resilient Agriculture, ICAR-Indian Agricultural Research Institute, New Delhi, India

R.H. Laxman Division of Plant Physiology and Biochemistry, ICAR-Indian Institute of Horticultural Research, Bengaluru, Karnataka, India

R.K. Mathur ICAR-Directorate of Oil Palm Research, Pedavegi, Andhra Pradesh, India

Vidya S. Murthy Division of Biotechnology, ICAR-Indian Institute of Horticultural Research, Bengaluru, Karnataka, India

V.B. Padmanabhan College of Agriculture, Vellayani, Thiruvananthapuram, Kerala, India

K.C. Pavithra Division of Plant Physiology and Biochemistry, ICAR-Indian Institute of Horticultural Research, Bengaluru, Karnataka, India

V. Rajagopal ICAR-Central Plantation Crops Research Institute, Kasaragod, Kerala, India

N.K. Srinivasa Rao Division of Plant Physiology and Biochemistry, ICAR-Indian Institute of Horticultural Research, Bengaluru, Karnataka, India

Iyyakutty Ravi ICAR-National Research Centre for Banana, Trichy, Tamil Nadu, India

Kundapura V. Ravishankar Division of Biotechnology, ICAR-Indian Institute of Horticultural Research, Bengaluru, Karnataka, India

K. Madhavi Reddy Division of Vegetable Crops, ICAR-Indian Institute of Horticultural Research, Bengaluru, Karnataka, India

A.T. Sadashiva Division of Vegetable Crops, ICAR-Indian Institute of Horticultural Research, Bengaluru, Karnataka, India

Satisha Jogaiah Division of Fruit crops, ICAR-Indian Institute of Horticultural Research, Bengaluru, Karnataka, India

Maryada Sharma Division of Plant Physiology and Biochemistry, ICAR-Indian Institute of Horticultural Research, Bengaluru, Karnataka, India

K.S. Shivashankara Division of Plant Physiology and Biochemistry, ICAR-Indian Institute of Horticultural Research, Bengaluru, Karnataka, India

Aradhana Singh Division of Vegetable Crops, ICAR-Indian Institute of Horticultural Research, Bengaluru, Karnataka, India

V. Sowmya Division of Vegetable Crops, ICAR-Indian Institute of Horticultural Research, Bengaluru, Karnataka, India

K. Suresh ICAR-Directorate of Oil Palm Research, Pedavegi, Andhra Pradesh, India

O. Kumari Swadija College of Agriculture, Vellayani, Thiruvananthapuram, Kerala, India

P. Umadevi ICAR-Indian Institute of Spices Research, Kozhikode, Kerala, India

K.K. Upreti Division of Plant Physiology and Biochemistry, ICAR-Indian Institute of Horticultural Research, Bengaluru, Karnataka, India

M. Mayil Vaganan ICAR-National Research Centre for Banana, Trichy, Tamil Nadu, India

H.S. Yogeesha Section of Seed Science and Technology, ICAR-Indian Institute of Horticultural Research, Bengaluru, Karnataka, India

About the Editors

Dr. N.K. Srinivasa Rao, Principal Scientist (Plant Physiology-Retd.) and Ex-Emeritus Scientist, Indian Institute of Horticultural Research, Bengaluru, has contributed significantly to the understanding of mechanism of abiotic stress tolerance in horticultural crops. He was associated with the release of one important tomato variety Arka Meghali, recommended for cultivation under rainfed conditions, and breeding of tomato and onion for water stress and capsicum, French bean and peas for high-temperature tolerance. Dr. Srinivasa Rao has significantly contributed in identifying genotypes tolerant to abiotic stresses, and the same has been used by the vegetable breeders for breeding varieties for water and high-temperature stress conditions.

Dr. Srinivasa Rao was the Chairman of the group to discuss and prepare a work plan for South Asian countries on the management of heat, moisture and other plant stresses under SAVERNET at the Joint Planning meeting held at the Bangladesh Agricultural Research Council, Dhaka, Bangladesh, from 24 to 27 February 1992. Dr. N. K. Srinivasa Rao was the Principal Investigator of the National Initiative on Climate Resilient Agriculture (NICRA) project. He was instrumental in planning various programmes on horticultural crops under NICRA project. He was the organising Secretary of the National Dialogue on Climate Resilient Horticulture held at IIHR from 28 to 29 January 2012. He was also the PI of ICAR Network Project on Impact, Adaptation and Vulnerability of Indian Agriculture to Climate Change since the inception of the project. He has developed a good laboratory for climate change research at IIHR, Bengaluru. He has devoted his full time in planning and developing these facilities like free air temperature enhancement (FATE), climate-controlled green house (CTGC) and phenomic platform which are of national importance. As a Senior Faculty Member, he had the opportunity of visiting facilities for studies on climate resilient agriculture, CO_2 enrichment and free-air temperature enrichment facilities at the Department of Horticulture, Cornell University and Brookhaven National Laboratory, Long Island, New York. He has published more than 60 scientific papers in national and international journals and he has seven book chapters to his credit. He has edited one book *Climate-Resilient Horticulture: Adaptation and Mitigation Strategies* published by M/S. Springer.

Dr. K.S. Shivashankara Principal Scientist in the Division of Plant Physiology and Biochemistry at Indian Institute of Horticultural Research, Bangalore. He has got more than 20 years of research experience in various fields like fruit aroma, fruit and vegetable antioxidant phytonutrients, antioxidant protection mechanism under stress, mango flowering physiology and ripening and storage disorders of fruits and effect of climate change on fruit and vegetable quality. Dr. Shivashankara has more than 40 research articles in various peer-reviewed national and international journals. He has the experience of working in international laboratories like food engineering lab of NFRI, Tsukuba, Japan, and at Lethbridge Research Centre of Agriculture and Agri-Food Canada at Lethbridge, Canada, in the area of antioxidant phytonutrients and fruit volatile flavours. He has been awarded the Fellow of International College of Nutrition, by the International College of Nutrition, Alberta, Canada. Dr. Shivashankara has been training many researchers and guiding students in the area of fruit and vegetable quality as affected by varieties, storage conditions and environmental factors. He has identified many indigenous fruits with high antioxidant capacity. He was also involved in the evaluation and selection of high antioxidant lines in many vegetables.

Dr. R.H. Laxman Principal Scientist in the Division of Plant Physiology and Biochemistry at ICAR-Indian Institute of Horticultural Research, Bengaluru. He has got more than 21 years of research experience in horticulture crops. He has published about 25 research articles in various peer-reviewed national and international journals and published eight book chapters. He has the experience of working in abiotic stress physiology, production physiology and climate change aspects of important horticultural crops like coconut, banana, mango and tomato.

Part I

Mechanisms of Abiotic Stress Tolerance Responses

Physiological and Morphological Responses of Horticultural Crops to Abiotic Stresses

1

N.K. Srinivasa Rao, R.H. Laxman, and K.S. Shivashankara

Abstract

The crop-environment interaction in horticultural crops is receiving increased attention in the context of changing climatic conditions. Environmental stresses can cause morpho-anatomical, physiological and biochemical changes in crops, resulting in a strong profit reduction. A clear understanding of environmental factors and their interaction with physiological processes is extremely important for improving horticultural practices. Drought, excess moisture, salinity and heat stress are amongst the most important environmental factors influencing crop growth, development and yield processes. A comprehensive understanding of the impact of these stress factors will be critical in evaluating the impact of climate change and climate variability on horticultural crop production. Environmental stresses influence an array of processes including physiology, growth, development, yield and quality of crop. A clear understanding of environmental factors and their interaction with physiological processes is extremely important for improving horticultural practices. This review presents the most recent findings about the effects of the main abiotic environmental factors (water, temperature, salinity) on whole plant physiology of horticultural crops.

1.1 Introduction

Consideration of abiotic stresses in crop species is of vital importance due to the widespread presence of such stresses on agricultural land, the probable increase in their severity and incidence due to global climatic change and other anthropogenic activities and the frequent deleterious effects such stresses have on crop productivity. These effects are the result of processes

N.K.S. Rao (✉) • R.H. Laxman • K.S. Shivashankara
Division of Plant Physiology and Biochemistry, ICAR-Indian Institute of Horticultural Research, Hessaraghatta Lake Post, Bengaluru 560089, Karnataka, India
e-mail: nksraoiihr@gmail.com

N.K.S. Rao et al. (eds.), *Abiotic Stress Physiology of Horticultural Crops*,
DOI 10.1007/978-81-322-2725-0_1

that can be observed at different levels of plant responses, i.e. morphological, physiological and biochemical/molecular changes. At the morphological level, abiotic stress can cause altered shoot, root and leaf growth, as well as developmental changes that result in altered life cycle duration and fewer and/or smaller organs. Physiological processes are also affected, such as photosynthetic rate, transpiration, respiration, partitioning of assimilates to different organs within the plant and mineral uptake. At the cellular level, cell membranes can be damaged, thylakoid structures disorganized, cell size reduced, stomatal guard cell function altered, degree of cellular hydration modified and programmed cell death promoted. And finally, at the biochemical/molecular level, the effects include enzyme inactivation, the production of reactive oxygen species, osmotic damage, changes in primary and secondary metabolite profiles, changed water and ion uptake or movement and altered hormone concentrations.

Adverse environmental conditions such as drought, high soil salinity and temperature extremes are important abiotic stresses causing severe yield loss to agricultural and horticultural crops. Environmental stress is the primary cause of crop losses worldwide, reducing average yields for the major crops by more than 50 % (Bray et al. 2000). Abiotic stresses are often interrelated, either individually or in combination; they cause morphological, physiological, biochemical and molecular changes that adversely affect plant growth and productivity and ultimately yield.

Horticultural crops include a wide range of commodities, such as fruits and vegetables that are highly valuable for humanity. They are extensively grown worldwide, and their production can be described as an open and highly complex system affected by many factors, amongst which we can count weather, soil and cropping system, as well as the interaction between these factors. Given that plant growth and development are directly and indirectly influenced by environmental factors (Schaffer and Andersen 1994), in order to obtain a successful production, it is essential to understand clearly how said factors affect plant physiology (Wien 1997). Being succulent in nature, most of the vegetable crops are sensitive to drought stress, particularly during flowering to seed development stage. Moreover, the legume vegetables, for instance, cowpea, vegetable pea, Indian beans, etc., grown in arid and semiarid regions are generally affected by drought at the reproductive stage. Cullis (1991) opined that a perceptive of how the interaction of physicochemical environment reduces plant development and yield will pave the ways for a combination of breeding methods for plant modification to improve tolerance against environmental stresses.

1.2 Moisture Stress

Moisture stress is one of the greatest environmental factors in reducing yield in the arid and semiarid tropics. Plant experiences drought stress either when the water supply to roots becomes difficult or when the transpiration rate becomes very high. From agricultural point of view, its working definition would be the inadequacy of water availability, including precipitation and soil moisture storage capacity, in quantity and distribution during the life cycle of a crop plant that restricts the expression of full genetic potential of the plant (Sinha 1986). Drought stress modifies photosynthetic rate, relative water content, leaf water potential and stomatal conductance. Ultimately, it destabilizes the membrane structure and permeability, protein structure and function, leading to cell death (Bhardwaj and Yadav 2012). Drought stress is affected by climatic, edaphic and agronomic factors. The susceptibility of plants to drought stress varies in dependence of stress degree, different accompanying stress factors, plant species and their developmental stages (Demirevska et al. 2009). Acclimation of plants to water deficit is the result of different events, which lead to adaptive changes in plant growth and physio-biochemical processes, such as changes in plant structure, growth rate, tissue osmotic potential and antioxidant defences (Duan et al. 2007). It has become imperative to elucidate the responses and adaptation of crops to water deficit and take actions to improve the drought resistance ability of crop plants and to ensure higher crop yields against unfavourable environmental stresses.

Water is fundamental for maintaining normal physiological activity and membrane transport processes (Jones and Tardieu 1998); therefore supplying it adequately is crucial for obtaining maximum productivity of horticultural crops. Further, water plays an important role in horticultural crops, since fruits and vegetables are usually sold on a fresh weight basis and yield is predominantly determined by water content (Marcelis et al. 1998). Drought stress occurs when there is not enough soil water content for successful growth or water supply replenishment (Larcher 2003; Lombardini 2006b). A decline in leaf relative water content (RWC) initially causes stomatal closure, which in turn leads to a decrease in the supply of CO_2 to the mesophyll cells and thus reduces leaf photosynthetic rate. Likewise, drought stress also affects processes such as cell division and expansion, ABA synthesis and sugar accumulation, consequently reducing crop yield (Marsal and Girona 1997; Chartzoulakis et al. 1999; Raviv and Blom 2001; Arquero et al. 2006; Lombardini 2006b). In general, it can be said that horticultural crops require a high water supply through appropriate irrigation schedules. Nevertheless, deficit irrigation can enhance fruit quality by raising dry matter percentage and sugar content (Jones and Tardieu 1998; Spreer et al. 2007). Furthermore, controlled water deficit has been used as a technique to stimulate blossoming in crops such as guava or litchi or to substitute for adequate chilling when temperate crops such as apple are grown in the tropics (Chaikiattiyos et al. 1994). On the other hand, it is important to discuss about flooding, since plant development is affected by either too little or too much water in the root zone. Flooding is produced by storms, over irrigation, poor drainage, high water tables and dam and river overflowing (Rao and Li 2003). As it has been previously mentioned, plants induce a series of physical, chemical and biological processes in response to stress conditions. Under flooding conditions, plants show similar symptoms to those they develop under heat or water stress. Plant responses to waterlogging include increased internal ethylene concentration; low stomatal conductance; decrease in leaf, root and shoot development; changes in osmotic potential and nutrient uptake; and reduced chlorophyll content and photosynthesis (Tamura et al. 1996; Ashraf and Rehman 1999; Rao and Li 2003; Issarakraisila et al. 2007). Flooding also increases the severity of certain diseases, mainly root-rotting fungi (Rao and Li 2003). The decrease of oxygen level in soils affects the bioavailability of nutrients as well as the ability of root systems to uptake and transport water and mineral nutrients (Lizaso et al. 2001). Waterlogging also causes inhibition of N uptake from the soil and reduced leaf concentrations of N, P, K, Ca and Mg in avocado (Schaffer and Andersen 1994) and pea (Rao and Li 2003).

1.2.1 Mechanisms of Drought Resistance

In genetic sense, the mechanisms of drought resistance can be grouped into three categories, viz. drought escape, drought avoidance and drought tolerance. However, crop plants use more than one mechanism at a time to resist drought.

1.2.1.1 Drought Escape

Drought escape is defined as the ability of a plant to complete its life cycle before serious soil and plant water deficits develop. This mechanism involves rapid phenological development (early flowering and early maturity), developmental plasticity (variation in duration of growth period depending on the extent of water deficit) and remobilization of pre-anthesis assimilates to grain.

1.2.1.2 Drought Avoidance

Drought avoidance is the ability of plants to maintain relatively high tissue water potential despite a shortage of soil moisture. Mechanisms for improving water uptake, storing in plant cell and reducing water loss confer drought avoidance. Drought avoidance is performed by maintenance of turgor through increased rooting depth, efficient root system and increased hydraulic conductance and by reduction of water loss through reduced epidermal (stomatal and lenticular) conductance, reduced absorption of radiation by leaf rolling or folding and reduced evaporation surface (leaf area). Plants under drought condition survive by doing a balancing act between maintenance of turgor and reduction of water loss.

1.2.1.3 Drought Tolerance

Drought tolerance is the ability to withstand water deficit with low tissue water potential. The responses of plants to tissue water deficit determine their level of drought tolerance. Drought tolerance is also defined as the ability of a plant to produce its economic product with minimum loss under water deficit environment in relation to the water-constraint-free management (Mitra 2001). The mechanisms of drought tolerance are maintenance of turgor through osmotic adjustment (a process which induces solute accumulation in cell), increase in elasticity in cell and decrease in cell size and desiccation tolerance by protoplasmic resistance.

1.2.2 Plant Responses to Abiotic Stresses

1.2.2.1 Drought

A primary response of plants subjected to drought stress is the growth arrest. Shoot growth inhibition under drought reduces metabolic demands of the plant and mobilizes metabolites for the synthesis of protective compounds required for osmotic adjustment. Root growth arrest enables the root meristem to remain functional and gives rise to rapid root growth when the stress is relieved (Hsiao and Xu 2000). Lateral root inhibition has also been seen to be an adaptive response, which leads to growth promotion of the primary root, enabling extraction of water from the lower layers of soil (Xiong et al. 2006). Growth inhibition can arise due to the loss of cell turgor arising from the lack of water availability to the growing cells. Water availability to cells is low because of poor hydraulic conductance from roots to leaves caused by stomatal closure. Although a decrease in hydraulic conductance decreases the supply of nutrients to the shoot, it also prevents embolism in xylem and could constitute an adaptive response. Osmotic adjustment is another way by which plants cope with drought stress. Synthesis of compatible solutes like polyols and proline under stress prevents the water loss from cells and plays an important role in turgor maintenance (Blum 2005; DaCosta and Huang 2006). Modification of growth priorities as well as reduction in the performance of photosynthetic organs due to stress exposure leads to alterations in carbon partitioning between the source and sink tissues (Roitsch 1999). Hence, carbohydrates that contribute to growth under normal growth conditions are now available for selective growth of roots or for the synthesis of solutes for osmotic adjustment (Lei et al. 2006; Xue et al. 2008).

1.2.2.2 Flooding

Flooding is a major environmental stress that severely limits crop productivity, and it has become a major problem worldwide. More than one third of the world's irrigated area suffers due to flooding, frequently or otherwise. It may be due to heavy rainfall, faulty irrigation, unlevelled land, poor drainage or heavy soil texture. Various morpho-physiological, biochemical and anatomical changes are induced in root system during flooding, for example, reduction in the shoot-root relative growth; formation of thicker adventitious roots or air roots; aerenchyma formation and cuboidal packing of cells for enhancing the longitudinal transport of gases; the major shift in the carbohydrate metabolism of the plants, which move towards lactate; and ethanolic fermentation to provide the required ATP. Oxygen diffusion is 10,000 times slower in waterlogged soil as compared to aerated soil. Continued flooded conditions lead to lack of oxygen in the soil, restricting respiration of the growing roots and other living organisms. Soil chemical properties also change when anaerobic conditions persist for several days. This is followed by loss of chlorophyll of the lower leaves, arrest of crop growth and proliferation of surface root growth with the retreat of water level. Plant transpiration is affected under anaerobic conditions and extended waterlogging results in root death due to inadequate oxygen supply. In a study on tomato, Walter et al. (2004) have demonstrated that tomato had the most vigorous adventitious root growth compared to cucumber, zucchini and bean. Mano and Omori (2007) have reported that dicotyledonous plants (e.g. soybean and tomato) generally form taproot system but develop adventitious roots under flooding conditions. This characteristic allows the root system to obtain oxygen directly from the air because the adventitious roots formed in the soil and even at the soil surface.

Vegetable production occurs in both dry and wet seasons in the tropics. However, production is often limited during the rainy season due to excessive moisture brought about by heavy rain. Most vegetables are highly sensitive to flooding, and genetic variation with respect to this character is limited, particularly in tomato. In general, damage to vegetables by flooding is due to the reduction of oxygen in the root zone which inhibits aerobic processes. Flooded tomato plants accumulate endogenous ethylene that causes damage to the plants (Drew 1979). Low oxygen levels stimulate an increased production of an ethylene precursor, 1-aminocyclopropane-1-carboxylic acid (ACC), in the roots. The rapid development of epinastic growth of leaves is a characteristic response of tomatoes to waterlogged conditions (Kuo and Chen 1980), and the role of ethylene accumulation has been implicated (Kawase 1981). The severity of flooding symptoms increases with rising temperatures; rapid wilting and death of tomato plants are usually observed following a short period of flooding at high temperatures (Kuo et al. 1982). In vegetables waterlogging caused a marked reduction in stomatal conductance of bitter melon. This reduction in stomatal conductance resulted in increased leaf water potential (Liao and Lin 1994). In contrast, Ashraf and Arfan (2005) found no significant correlation between stomatal conductance and water potential of okra plants under waterlogged conditions.

1.3 Alterations in Physiological Mechanisms Under Stress

One of the earliest responses to soil flooding is a reduction in stomatal conductance, inhibition of stomatal aperture. Low O_2 level may also reduce hydraulic conductivity (*Lp*) and consequently decrease root permeability. The decrease in *Lp* may be linked to aquaporins gating by cytosolic pH and also limit water uptake thus in turn leading to internal water deficit (Folzer et al. 2006). Oxygen deficiency generally induces a rapid reduction in the rate of photosynthesis which is generally considered as the result of reduced stomatal aperture, decrease in leaf chlorophyll contents, early leaf senescence and a reduction in leaf area which may also contribute to inhibition of photosynthesis in later stage. Indeed waterlogged soil tends to reduce the translocation of photosynthesis product from 'source' leaf to 'sink' root (Yordanova et al. 2004); as a result, the maintenance of photosynthetic activity and accumulation of soluble sugar in roots are clearly important adaptations to flooding.

Leaf gas exchange parameters have been used to study the photosynthetic capacity of plants during flooding. Except in some flood-tolerant plants that have developed adaptive mechanisms so as to maintain better photosynthetic capacity (Topa and Cheeseman 1992), flooding causes a significant decrease in the capacity for leaf gas exchange in most flooding-susceptible plants. Examples are *Lycopersicon esculentum* (Bradford and Yang 1980) and *Pisum sativum* (Jackson and Kowalewska 1983). Studies showed that in pea plants that were subjected to soil flooding, the foliar ABA content increased, accompanied by closure of stomata (Zhang and Davies 1987). Furthermore, a decrease of stomatal conductance was found to be correlated with an increase of leaf water potential in flooded bitter melon (Liao and Lin 1994). Those observations indicated that stomatal closure results in an increase in leaf water potential. It can be hypothesized that stomatal closure slows the transpiration rate, thus preventing leaf dehydration (Bradford and Hsiao 1982). In normal and flooded plants, stomatal conductance was shown to be correlated with CER (carbon exchange rate) (Liao and Lin 1996; Vu and Yelenosky 1991). These results imply that stomatal aperture is a limiting factor for CER.

The internal CO_2 concentration (*Ci*) of leaves was found to increase linearly with the duration of flooding, despite a reduction of stomatal conductance. The ambient CO_2 concentration was estimated to average 350 μ bars, but less than 300 μ bars were detectable in the leaf intercellular space in leaves of bitter melon plants that had not been subjected to flooding. The *Ci* was found to be 380 μ bars in flooded bitter melon on the sixth day of flooding (Liao and Lin 1994) suggesting that factors such as reduced photosynthetic capacity, a respiratory CO_2 evolution rate exceeding the CO_2 fixation rate and increased stomatal resistance may be involved in *Ci* elevation. Flooding has been reported to cause stomatal closure directly, without affecting the photosynthetic capacity, in *V. ashei* (Davies and Flore 1986a, b) thus decreasing *Ci*. It has been suggested that the stomatal aperture regulates the decline of CER. However, in flooded bitter melon seedlings, *Ci* was observed to increase (Liao and Lin 1996). Other plants, in which *Ci* was found to remain unchanged with flooding, are *C. sinensis* grafted onto *C. jambhiri* and *C. aurantium* rootstocks (Vu and Yelenosky 1991) and bitter melon grafted onto luffa rootstocks (Liao and Lin 1996). These observations suggest that stomatal aperture is not the only limiting factor for CER but is partly responsible for the decrease in the photosynthetic capacity of mesophyll tissue. Avocado is generally considered as a highly flood-sensitive species. High soil moisture levels favour the development of *Phytophthora* root rot. Short periods of waterlogging reduce shoot and root growth (Schaffer and Andersen 1994). Flooding in mango (*Mangifera indica*) determined reductions in net CO_2 assimilation, stomatal conductance and root growth, resulting in increased shoot-root ratios and substomatal CO_2 concentration (Larson et al. 1993). Papaya plants are sensitive to flooding. Waterlogged soils have been reported to cause the falling down of old leaves and chlorosis in the remaining ones as well as the death of these plants after 3 or 4 days (Schaffer and Andersen 1994). Increased ethylene rate and reduced flow of nitrate, hydrogen ions, most protein amino acids, glutamine and abscisic acid to shoots and 40 % yield reduction were observed by Rao and Li (2003) in tomato due to flooding. The translocation of photosynthates from leaf to root decreases in waterlogged soils (Yordanova et al. 2004). Hence, under waterlogged situations sustaining photosynthetic activity and proper translocation of photosynthates to roots is an important adaptation to flooding.

1.3.1 Management Practices to Overcome Flood Damage

Several management practices have been reported to help crops partially or entirely overcome flood damage. For example, the application of nitrogen (N) fertilizers overcomes N deficiency, while natural or synthetic hormones are used to correct hormone imbalances and the addition of fungicides help control soilborne pathogens. Because of reduced root activity, flooding causes a significant decrease in N content and the rate of N accumulation in plants. Plant-available N in soils is also very low because of leaching or run-off. Yellowing of leaves due to loss of chlorophyll from leaves within 2–3 days of waterlogging is probably attributed to N deficiency. Thus, a strategic use of N fertilizer after flooding may alleviate N deficiency and enhance crop recovery from flooding. Growers should apply fertilizers as soon as soils become dry enough for tractor operation. Foliar application of liquid fertilizers is more effective than broadcasting dry fertilizer because of root damage due to flooding. Many kinds of N fertilizers can be used for crops after flooding.

Several fertilizers have been tested for their effectiveness in recovering flood-damaged vegetable crops, and potassium nitrate performed the best, urea the second best and calcium nitrate the third. A regular granular dry fertilizer, such as $10N\text{-}10P_2O_5\text{-}10K_2O$, also can be used for flooded crops, but it is not as effective as foliar and liquid fertilizers. Various plant growth regulators have been associated with alleviation of waterlogging damages, but there is a dearth of information on their effects on waterlogged crops. Spraying shoots with a synthetic cytokinin (6-benzylaminopurine [BAP]) has been reported to reduce flooding damage by improvements in leaf extension and retard premature loss of chlorophyll in older leaves. This was related to BAP compensating for the restricted transport of natural cytokinin from the root system, affecting metabolism of gibberellins and adversely affecting the inhibitory action of abscisic acid on growth.

Flooding increases the severity of diseases. The symptoms of diseased roots are discoloration, rotting of the root and the premature death of the plant. The damage reduces the ability of the root systems to obtain mineral nutrients or perform other functions essential to the shoot. Two common diseases, *Phytophthora* and *Pythium*, cause greatest damage to roots in poorly drained soil. Application of fungicides probably reduces the incidence of disease in waterlogged plants and thereby increases plant tolerance to flooding.

1.4 Environmental Constraints Limiting the Productivity of Horticultural Crops

1.4.1 Vegetables

Environmental stress is the primary cause of crop losses worldwide, reducing average yields for most major crops by more than 50 % (Boyer 1982; Bray et al. 2000). The tropical vegetable production environment is a mixture of conditions that varies with season and region. Climatic changes will influence the severity of environmental stress imposed on vegetable crops. Vegetables, being succulent products by definition, generally consist of greater than 90 % water. Thus, water greatly influences the yield and quality of vegetables; drought conditions drastically reduce vegetable productivity. Drought stress causes an increase of solute concentration in the environment (soil), leading to an osmotic flow of water out of plant cells. This leads to an increase of the solute concentration in plant cells, thereby lowering the water potential and disrupting membranes and cell processes such as photosynthesis. The timing, intensity and duration of drought spells determine the magnitude of the effect of drought.

Moreover, increasing temperatures, reduced irrigation water availability, flooding and salinity are the major limiting factors in sustaining and increasing vegetable productivity. Extreme climatic conditions will also negatively impact soil fertility and increase soil erosion. Thus, additional fertilizer application or improved nutrient-use efficiency of crops will be needed to maintain productivity or harness the potential for enhanced crop growth due to increased atmospheric CO_2. The response of plants to environmental stresses depends on the plant developmental stage and the length and severity of the stress (Bray 2002). Plants may respond similarly to avoid one or more stresses through morphological or biochemical mechanisms (Capiati et al. 2006). Environmental interactions may make the stress response of plants more complex or influence the degree of impact of climate change. The adaptive potential of some plant species reducing water losses was achieved by closing of stomata and reduction in the transpiration rate (Tardieu and Davies 1992). Hence, measurement of transpiration rate is an excellent tool to assess drought-tolerant capacity of crop plants. However, reduction of transpiration rate under drought increases leaf temperature which is deleterious for plants. Abscission of reproductive organs like flower buds and flowers is a major yield-limiting factor in vegetable crops (Kositsup et al. 2009; Wien et al. 1989). The abscission of floral organs during stresses has been associated with the changes in physiological processes (Aloni et al. 1996). In tomato, the abscission of flowers and flower buds and the reduction in photosynthesis were more in susceptible cultivars compared to the tolerant cultivars where the abscission was relatively less (Bhatt et al. 2009).

Fluctuations in temperature occur naturally during plant growth and reproduction. However, extreme variations during hot summers can damage the intermolecular interactions needed for proper growth, thus impairing plant development and fruit set. The increasing threat of climate change is already having a substantial impact on agricultural production worldwide as heat waves cause significant yield losses with great risks for future global food security. Temperature limits the range and production of many crops. In the tropics, high-temperature conditions are often prevalent during the growing season, and, with changing climate, crops in this area will be subjected to increased temperature stress. Analysis of climate trends in tomato-growing locations suggests that temperatures are rising and the severity and frequency of above-optimal temperature episodes will increase in the coming decades (Bell et al. 2000). Vegetative and reproductive processes in tomatoes are strongly modified by temperature alone or in conjunction with other environmental factors (Abdalla and Verkerk 1968). High-temperature stress disrupts the biochemical reactions fundamental for normal cell function in plants. It primarily affects the photosynthetic functions of higher plants (Weis and Berry 1988). High temperatures can cause significant losses in tomato productivity due to reduced fruit set and smaller and lower-quality fruits (Stevens and Rudich 1978). Pre-anthesis temperature stress is associated with developmental changes in the anthers, particularly irregularities in the epidermis and endothecium, lack of opening of the stomium and poor pollen formation (Sato et al. 2002). In pepper, high-temperature exposure at the pre-anthesis stage did not affect pistil or stamen viability, but high postpollination temperatures inhibited fruit set, suggesting that fertilization is sensitive to high-temperature stress (Erickson and Markhart 2002). Hazra et al. (2007) summarized the symptoms causing fruit set failure at high temperatures in tomato; this includes bud drop, abnormal flower development, poor pollen production, dehiscence and viability, ovule abortion and poor viability, reduced carbohydrate availability and other reproductive abnormalities. In tomato, pollen germination and pollen tube growth, ovule viability, stigma and style positions and number of pollen grains retained by the stigma were also seriously affected by high temperature (Foolad 2005). Studies conducted by Wentworth et al. (2006) in common beans showed high-temperature-dependent increases in leaf thickness, palisade development and stomatal density in the adaxial surface of the leaves. In addition, significant inhibition of photosynthesis occurs at temperatures above optimum, resulting in considerable loss of potential productivity (Sage and Kubien 2007).

Salinity is also a serious problem that reduces growth and productivity of vegetable crops in many salt-affected areas. It is estimated that about 20 % of cultivated lands and 33 % of irrigated agricultural lands worldwide are afflicted by high salinity (Foolad 2004). Vegetable production is threatened by increasing soil salinity particularly in irrigated croplands which provide 40 % of the world's food (FAO 2002). Excessive soil salinity reduces productivity of many agricultural crops, including most vegetables which are particularly sensitive throughout the ontogeny of the plant. According to the US Department of Agriculture (USDA), onions are sensitive to saline soils, while cucumbers, eggplants, peppers and tomatoes, amongst the main crops of AVRDC – The World Vegetable Centre, are moderately sensitive. In hot and dry environments, high evapotranspiration results in substantial water loss, thus leaving salt around the plant roots which interferes with the plant's ability to uptake water. Physiologically, salinity imposes an initial water deficit that results from the relatively high solute concentrations in the soil, causes ion-specific stresses resulting from altered K^+/Na^+ ratios and leads to a build-up in Na^+ and Cl^- concentrations that are detrimental to plants (Yamaguchi and Blumwald 2005). Plant sensitivity to salt stress is reflected in loss of turgor, growth reduction, wilting, leaf curling and epinasty, leaf abscission, decreased photosynthesis, respiratory changes, loss of cellular integrity, tissue necrosis and potentially death of the plant

(Jones 1986; Cheeseman 1988). Salinity also affects agriculture in coastal regions which are impacted by low-quality and high-saline irrigation water due to contamination of the groundwater and intrusion of saline water due to natural or man-made events. Salinity fluctuates with season, being generally high in the dry season and low during rainy season when freshwater flushing is prevalent. Furthermore, coastal areas are threatened by specific, saline natural disasters which can make agricultural lands unproductive, such as tsunamis which may inundate low-lying areas with seawater. Although the seawater rapidly recedes, the groundwater contamination and subsequent osmotic stress cause crop losses and affect soil fertility. In the inland areas, traditional water wells are commonly used for irrigation water in many countries. The bedrock deposit contains salts, and the water from these wells are becoming more saline, thus affecting irrigated vegetable production in these areas.

Measures to adapt to these climate change-induced stresses are critical for sustainable tropical vegetable production. Until now, the scientific information on the effect of environmental stresses on vegetables is overwhelmingly on tomato. There is a need to do more research on how other vegetable crops are affected by increased abiotic stresses as a direct potential threat from climate change.

1.4.2 Fruit Crops

Global climate change is having negative effects on flowering and fruit set in tree crops. Freezes and extended periods of high temperature during years have significantly reduced yield and initiated alternate bearing, production of a heavy 'on crop' (high yield) followed by a light 'off crop' (low yield) in citrus, mango, avocado, pistachio and olive. Tree fruit commodity-based industries have endured millions of dollars in lost revenue due to both the immediate effect of the abiotic stress and the prolonged effect of alternate bearing. Further, the high cost and diminishing availability of quality irrigation water and the ever-increasing cost of other inputs (fertilizer, fuel, labour) required in crop production dictate that growers reduce the amount of inputs used or increase yield, including fruit size. New management strategies are needed that solve the production problems resulting from crops being grown under less than optimal (stressful) environmental conditions to increase profitability in order to sustain the tree crop industries of the world. The focus of research on fruit crops in relation to abiotic stresses should be to integrate changes in hormone homeostasis and expression of key genes regulating floral ontogeny and fruit development caused by abiotic stress with changes in whole tree physiology, including alternate bearing, to design corrective strategies to increase productivity and sustain the economic viability of fruit tree productivity. The outcomes of this research will be field-applicable management strategies that protect tree crop productivity from the negative effects of abiotic stress, mitigate alternate bearing and solve production problems to increase yield and grower's income and sustain tree crop industries. It has been observed that vegetative and reproductive growths in trees are differentially sensitive to water stress. Additionally, reproductive growth is differentially sensitive to water stress at different times of the season. It has been reported that mild water stress applied during the intermediate developmental period of slow fruit growth has no effect on crop yields but significantly reduces vegetative growth in peach (Mitchell and Chalmers 1982) and pear (Mitchell et al. 1984). A frequent response of fruit trees to deficit irrigation (DI) is the promotion of flowering. This flowering promotion is often explained in terms of a lesser resource competition with vegetative growth effectively restrained by water deficit in evergreen and deciduous fruit trees (Chaikiattiyos et al. 1994; Behboudian and Mills 1997). This tree response to DI has been successfully exploited to induce out of season blooming and to increase the levels of flowering in many tropical and subtropical fruit crops.

Both high and low temperatures, be they temporary or constant, can induce morpho-anatomical, physiological and biochemical changes in plants, leading to profit reduction

(Higuchi et al. 1998; Wang et al. 2003; Wahid et al. 2007). Heat stress can be a concern in many regions of the tropics and subtropics, since high temperature can cause significant damage such as sunburns on leaves, branches and stems, anticipated leaf senescence and abscission, shoot and root growth inhibition and fruit discoloration and damage (Yamada et al. 1986; Yamada et al. 1996; Higuchi et al. 1998; Almeida and Valle 2007; Wahid et al. 2007). Reproductive processes are also highly affected by heat stress in most plants (Wahid et al. 2007). Through observations in strawberry, Ledesma et al. (2008) found that high-temperature stress negatively affected the number of inflorescences, flowers and fruits and that plant response to high-temperature stress was cultivar dependent. In cherimoya, warm temperatures determined the production of low-viability pollen and therefore of asymmetrical and small fruits containing few seeds (Higuchi et al. 1998). However, it has been observed that pollen viability is reduced in papaya when the temperature drops below 20 °C. This condition can also cause problems of sex change and low-sugar content in fruits (Galán-Saúco and Rodríguez Pastor 2007). In cacao, temperatures above 23 °C seem to accelerate vegetative flushing initiation (Almeida and Valle 2007). Regarding anatomical changes, symptoms observed under heat stress conditions are generally similar to those checked under water stress. Plants had reduced cell size, closure of stomata, curtailed water loss, increased stomatal and trichome densities and greater xylem vessels in both root and shoot (Wahid et al. 2007). In a work conducted by Zhang et al. (2005) in grapes, they found that warm temperatures considerably affected the mesophyll cells, increased plasma membrane permeability, enhanced the loss of grana stacking and determined the swelling of stroma lamellae. Furthermore, an increase in the concentration of abscisic acid (ABA) was observed in grape leaves due to high temperature, suggesting that ABA may be a high-temperature acclimation and heat-tolerance induction factor in this crop (Abass and Rajashekar 1993).

Mango cultivars exposed to different day/night temperatures (15/10, 20/15, 25/20 and 30/25 °C) for 20 weeks showed that vegetative growth increased with increasing temperatures. All cultivars grew vegetatively at 25/20 and 30/25 °C. Like mango, higher temperatures in citrus also enhance vegetative growth, but after certain limit, it retards the shoot elongation. Higher temperature (38/28 °C; day/night temperature) for 10 weeks in sour orange, *Troyer citrange* and Valencia oranges showed that seedlings were with short internodes and leaves were markedly shorter as compared to normal ambient temperature (28/22 °C) (Krishan Kumar et al. 2011). In most fruit crops, generally higher temperature decreased the day interval required for flowering and cooler temperature though required more days for flowering, but the number of flowers produced increased proportionally at this temperature.

1.5 Breeding for Abiotic Stress Tolerance

Vegetable crop plants are herbaceous succulents and much prone to abiotic stresses. Most of them are grown in different agroclimatic situations than their evolutionary regions which make the vegetables more vulnerable to adverse climatic factors and associated losses. In tropical regions, the vegetable production always remains on mercy of environmental condition which varies with season and region. The severity of environmental stress imposed on vegetable crops varies with their genotypes and other crop factors. Climate change-factored rise in temperatures, reduction in irrigation water or drought situation, occurrence of frequent to prolonged flooding, occurrence of acidity or rise in salinity levels and increase in wind velocity are going to be major limiting factors in sustainable vegetable production in tropical islands. These extremities will also affect microbial population in soil and root rhizosphere, soil health and soil fertility and increase soil erosion in tropics which ultimately reduce crop yield. The increase in decomposition rate of organic sources of nutrients and rapid losses of nutrients through leaching or washing out effect are major concerns for vegetable nutrition (Pandey et al. 2009). Abiotic stresses appearing during vegetable production either can be the primary cause for disorders or can influence the susceptibility of a harvest product to such disorders. Thus, appropriate changes are desired in crop plant morphology and physiological processes for increasing their adaptability and productivity in changing climatic situation. The abiotic stress also affects the postharvest life of the vegetables, and therefore, it is appropriate to prioritize amongst the breeding approaches for extending the postharvest life even in abiotic stress situation. This could be through enhancing stress tolerance of edible portion of vegetables through conventional breeding or upregulating the associated genes and pathway by desirable modifications (Taiz and Zeiger 2006).

Such nonconventional breeding efforts are much awaited in tropical crops where climate change-associated factors will certainly reduce the critical time by escalating the speed of vegetable decay. Further, the productivity of vegetables remains low in tropical islands (Olasantan 2007) which could be due to genotypic and environmental factors or their interactions. Bray et al. (2000) reported yield losses of around 50 % in vegetable crop primarily due to environmental stresses. In future, the climate-associated stress events like high temperature, limited soil moisture and salinity stresses will get magnified by climate change impacts. Frequency of extreme events will affect the response of technologies including high-yielding genotypes against soil health degradation or changes in disease and pest equilibriums and reproductive biology with modified microclimate.

1.5.1 Breeding for Water Stress

Water stress is one of the most important abiotic stresses affecting plants where excess water hampers supply of oxygen and essential nutrients to root zone and dilutes the osmotic potential of cell and incites roots to secrete stress-associated hormones, whereas drought stress leads to plant desiccation (Olasantan 2007). Both excess and scarcity of water compromise the plant life and ultimately the yield. Vegetables contain moisture around 90 %, and any deficiency may cause serious damage to their yield and quality. Climate change impacts the availability of drinking and irrigation water and also challenges the vegetable sector in dry months. Mitigation of the impact of flooding on vegetable sector is a major challenge to vegetable growers during rainy season in tropical region where high temperature during sunny hours causes rapid wilting and death of plants. Thus, breeding of genotypes with water stress tolerance for drought period and with high water-use efficiency for drought conditions should be planned through targeted participatory approaches in natural conditions per se. For this, the local genetic resources or wild relatives of respective crops from islands can be explored for waterlogging or moisture stress-tolerant genes. For this, the best strategy is to construct and manage the trait-specific core groups from local genetic resources with regular enrichment. The islands have indigenous tribes and settler communalities which have association with such crop resources, and their traditional knowledge can help in selecting the potential germplasm for targeted breeding. For example, the *Solanum torvum* is a naturally occurring edible plant species and offers opportunity to use in breeding against both excess and deficit of water. The physiological and genetic mechanisms in such plants should be investigated for their differential responses and use them in other crops. A targeted approach for mining good genes and their transfer in target plant using both genomics and transgenics will be more appropriate.

1.5.2 Breeding for Flooding Tolerance

In tropical islands, the high intensity and heavy rains affect crop by severely affecting the physiological and reproductive processes which are associated with limited light availability, high humidity and greater susceptibility to diseases and pests. Thus, breeding strategy should target multi-trait improvement, and for that large-scale germplasm screening followed by sequential or simultaneous breeding approach should be performed. The allele mining for suitable gene sources and transfer of such genes through conventional or nonconventional methods are options for vegetable improvement. However, due attention is required for endemic wild species of vegetable plants which have been found to be tolerant to heavy rains. Thus, breeding for target stress through population improvement can be deployed for mining better alleles along with markers.

1.5.3 Breeding for Temperature Extremes

Vegetables are generally sensitive to environmental extremes, and thus high temperatures and limited soil moisture are the major causes of low yields in the tropics and will be further magnified by climate change. Temperature stresses (35–45 °C) which frequently occur in midday hours in faulty designed polyhouses in tropical islands cause denaturation and destruction of the protoplast leading to cell death. Such damages are observed in tomato, sweet pepper, palak, lettuce and green onion. High day temperature affects anthesis, dehiscence and fruit setting in tomato and capsicum; increases antinutrients and fibre content in leafy vegetables which downgrade their quality; affects movement of pollinating agents; and also changes stigma-pollen interaction. The rise in above-optimal temperature episodes in coming years due to climate change impact will strongly modify the reproductive processes alone or in conjunction with other environmental factors in vegetables. The size, quality and shape of storage organs are greatly affected by day and night temperatures which affect development of storage portion of tuber crops. The breeding programme for heat tolerance finds place in queue to ensure adequate supply of vegetables in coming decades.

1.6 Future Strategies

Understanding the physiological, molecular and biochemical processes in relation to changing climatic factors constitutes the first step towards the development of strategies for designing stress-resistant genotypes. It is important from the crop improvement perspective to identify the specific physiological attributes contributing to the adaptability of crop to the different abiotic factors. This would assist breeders to selectively combine some of these physiological attributes into the high-yielding cultivars required for present-day horticultural production system. There is a need to develop dedicated research programmes aimed at enhancing the tolerance to combinations of different abiotic stresses and particularly those related to drought, waterlogging, high temperature and global change (elevated carbon dioxide and ultraviolet radiation). Tolerance mechanisms for drought and heat may be different; therefore, an integrated approach should be taken for cultivar development. Development of genotypes tolerant to high temperature, moisture stress, salinity and climate proofing through conventional and nonconventional breeding techniques, genomics, biotechnology, etc. is essentially required to meet these challenges.

1.7 Conclusions

The main challenges to improving crop productivity of horticultural crops are the different types of abiotic stresses generally caused by the climate change at regional as well as global levels. Heat, drought, cold and salinity are the major abiotic stresses that induce severe cellular damage in fruit and vegetable crop plants. Waterlogging is one of the major constraints for sustainable horticulture. Its effects are evident on the entire plant as well as cellular levels. High temperatures can affect pollen viability and germination, number of flowers and number of fruits per plant. To make the crop plant survive and yield more under such environmental conditions, it is essential to understand the physiology, biochemistry and biotechnology of the plant. There is a need to understand the crop-environment interactions and develop and use the integrated approach to improve the crop productivity under various types of environmental stresses. Ecophysiological information is a tool that can be used in breeding programmes to obtain improved cultivars. The location-specific breeding approaches are needed for challenging the climate change phenomena in vegetable crops.

References

Abass M, Rajashekar CB (1993) Abscisic-acid accumulation in leaves and cultured cells during heat acclimation in grapes. HortSci 28:50–52

Abdalla AA, Verkerk K (1968) Growth, flowering and fruit-set of the tomato at high temperature. Neth J Agric Sci 16:71–76

Almeida AAF, Valle RR (2007) Ecophysiology of cacao tree. Braz J Plant Physiol 19:425–448

Aloni B, Karni L, Zaidman Z, Schaffer AA (1996) Changes of carbohydrates in pepper (*Capsicum annuum* L.) flowers in relation to their abscission under different shading regimes. Ann Bot 78:163–168

Arquero O, Barranco D, Benlloch M (2006) Potassium starvation increases stomatal conductance in olive trees. HortSci 41:433–436

Ashraf M, Arfan M (2005) Gas exchange characteristics and water relations in two cultivars of *Hibiscus esculentus* under waterlogging. Biol Plant 49:459–462

Ashraf M, Rehman H (1999) Mineral nutrient status of corn in relation to nitrate and long-term waterlogging. J Plant Nutr 22:1253–1268

Behboudian MH, Mills TM (1997) Deficit irrigation in deciduous orchards. Hort Rev 21:105–131

Bell J, Duffy P, Covey C, Sloan L (2000) Comparison of temperature variability in observations and sixteen climate models simulations. Geophys Res Lett 27:261–264

Bhardwaj J, Yadav SK (2012) Genetic mechanisms of drought stress tolerance, implications of transgenic crops for agriculture. agro-eco and strate, for climate change. Sustain Agric Rev 8:213–235

Bhatt RM, Rao NKS, Upreti KK, Shobha HS (2009) Floral abscission and changes in sucrose phosphate synthase and invertase activities in water deficit tomato. Indian J Plant Physiol 14:370–376

Blum A (2005) Drought resistance, water-use efficiency, and yield potential-are they compatible, dissonant, or mutually exclusive? Aust J Agric Res 56:1159–1168

Boyer JS (1982) Plant productivity and environment. Plant Sci 218:443–448

Bradford KJ, Hsiao TC (1982) Physiological responses to moderate water stress. In: Lange OL, Nobel PS, Osmond CB, Ziegler H (eds) Encyclopedia of plant physiology, vol 12b, New series. Springer, New York, pp 263–324

Bradford KJ, Yang SF (1980) Xylem transport of 1-aminocyclopropane-1-carboxylic acid, an ethylene precursor, in waterlogged plants. Plant Physiol 65:322–326

Bray EA (2002) Classification of genes differentially expressed during water-deficit stress in Arabidopsis thaliana: an analysis using microarray and differential gene expression. Ann Bot 89:803–811

Bray EA, Bailey-Serres J, Weretilnyk E (2000) Responses to abiotic stresses. In: Gruissem W, Buchannan B, Jones R (eds) Biochemistry and molecular biology of plants. ASPP, Rockville, pp 1158–1249

Capiati DA, Pais SM, Tellez-Inon MT (2006) Wounding increases salt tolerance in tomato plants: evidence on the participation of calmodulin-like activities in cross-tolerance signalling. J Exp Bot 57:2391–2400

Chaikiattiyos S, Menzel CM, Rasmussen TS (1994) Floral induction in tropical fruit trees: effects of temperature and water supply. J Hortic Sci 69:397–415

Chartzoulakis K, Patakas A, Bosabalidis AM (1999) Changes in water relations, photosynthesis and leaf anatomy induced by intermittent drought in two olive cultivars. Environ Exp Bot 42:113–120

Cheeseman JM (1988) Mechanisms of salinity tolerance in plants. Plant Physiol 87:547–550

Cullis CA (1991) Breeding for resistance for physiological stress. In: Murray DR (ed) Advanced methods in plant breeding and biotechnology. CAB International, Wallingford, pp 340–351

DaCosta M, Huang BR (2006) Osmotic adjustment associated with variation in bentgrass tolerance to drought stress. J Am Soc Hortic Sci 131:338–344

Davies FS, Flore JA (1986a) Short-term flooding effects on gas exchange and quantum yield of rabbiteye blueberry (*Vaccinium ashei* Reade). Plant Physiol 81 (1):289–292

Davies FS, Flore JA (1986b) Gas exchange and flooding stress of high bush and rabbiteye blueberries. J Am Soc Hortic Sci 111:565–571

Demirevska K, Zasheva D, Dimitrov R, Simova-Stoilova-L, Stamenova M, Feller U (2009) Drought stress effects on Rubisco in wheat: changes in the Rubisco large subunit. Acta Physiol Plant 31:1129–1138

Drew MC (1979) Plant responses to anaerobic conditions in soil and solution culture. Curr Adv Plant Sci 36:1–14

Duan B, Yang Y, Lu Y, Korpelainen H, Berninger F, Li C (2007) Interactions between drought stress, ABA and genotypes in *Picea asperata*. J Exp Bot 58:3025–3036

Erickson AN, Markhart AH (2002) Flower developmental stage and organ sensitivity of bell pepper (*Capsicum annuum* L.) to elevated temperature. Plant Cell Environ 25:123–130

FAO (2002) Le the salt of the earth: hazardous for food production. In: Word Food Summit. Five years later, FAO, Rome, Italy, 10–13 June 2012

Folzer H, Dat J, Capelli N, Rieffel D, Badot PM (2006) Response to flooding of sessile oak seedlings (*Quercus petraea*) to flooding: an integrative study. Tree Physiol 26:759–766

Foolad MR (2004) Recent advances in genetics of salt tolerance in tomato. Plant Cell Tissue Organ Cult 76:101–119

Foolad MR (2005) Breeding for abiotic stress tolerances in tomato. In: Ashraf M, Harris PJC (eds) Abiotic stresses: plant resistance through breeding and molecular approaches. The Haworth Press, New York, pp 613–684

Galán-Saúco VG, Rodríguez-Pastor MCR (2007) Greenhouse cultivation of papaya. Acta Horticult 740:191–195

Hazra P, Samsul HA, Sikder D, Peter KV (2007) Breeding tomato (*Lycopersicon Esculentum* Mill) resistant to high temperature stress. Int J Electron Plant Breed 1:31–40

Higuchi H, Utsunomiya N, Sakuratani T (1998) High temperature effects on cherimoya fruit set, growth and development under greenhouse conditions. Sci Hortic 77:23–31

Hsiao TC, Xu LK (2000) Sensitivity of growth of roots versus leaves to water stress: biophysical analysis and relation to water transport. J Exp Bot 51:1596–1616

Issarakraisila M, Ma Q, Turner DW (2007) Photosynthetic and growth responses of juvenile Chinese kale (*Brassica oleracea* var. *alboglabra*) and Caisin (*Brassica rapa* subsp. *parachinensis*) to waterlogging and water deficit. Sci Hortic 111:107–113

Jackson MB, Kowalewska AKB (1983) Positive and negative messages from roots induce foliar desiccation and stomatal closure in flooded pea plants. J Exp Bot 34:493–506

Jones RA (1986) High salt tolerance potential in *Lycopersicon* species during germination. Euphytica 35:575–582

Jones HG, Tardieu F (1998) Modelling water relations of horticultural crops: a review. Sci Hortic 74:21–46

Kawase M (1981) Anatomical and morphological adaptation of plants to water logging. HortSci 16:30–34

Kositsup B, Montpied P, Kasemsap P, Thaler P, Ameglio T, Dreyer E (2009) Photosynthetic capacity and temperature responses of photosynthesis of rubber trees (*Hevea brasiliensis* Müll. Arg.) acclimate to changes in ambient temperatures. Tree Physiol 23:357–365

Kumar K, Rashid R, Bhat JA, Bhat ZA (2011) Effects of high temperature on fruit crops. Elixir Appl Bot 39:4745–4747

Kuo CG, Chen BW (1980) Physical responses of tomato cultivars to flooding. J Am Soc Hortic Sci 105:751–755

Kuo CG, Tsay JS, Chen BW, Lin PY (1982) Screening for flooding tolerance in the genus Lycopersicon. Hortic Sci 17:76–78

Larcher W (2003) Physiological plant ecology, 4th edn. Springer, Berlin

Larson KD, Schaffer B, Davies FS (1993) Physiological, morphological and growth responses of mango trees to flooding. Acta Horticult 342:152–159

Ledesma NA, Nakata M, Sugiyama N (2008) Effect of high temperature stress on the reproductive growth of strawberry cvs. 'Nyoho' and 'Toyonoka'. Sci Hortic 116:186–193

Lei Y, Yin C, Li C (2006) Differences in some morphological, physiological and biochemical responses to drought stress in two contrasting populations of *Populus przewalskii*. Physiol Plant 127:182–191

Liao CT, Lin CH (1994) Effect of flooding stress on photosynthetic activities of *Momordica charantia*. Plant Physiol Biochem 32:1–5

Liao CT, Lin CH (1996) Photosynthetic responses of grafted bitter melon seedlings to flooding stress. Environ Exp Bot 36:167–172

Lizaso JI, Meléndez LM, Ramírez R (2001) Early flooding of two cultivars of tropical maize. II. Nutritional responses. J Plant Nutr 24:997–1011

Lombardini L (2006) Ecophysiology of plants in dry environments. In: D'Odorico P, Porporato A (eds) Dryland ecohydrology. Springer, Berlin, pp 47–66

Mano Y, Omori F (2007) Breeding for flooding tolerant maize using "teosinte" as a germplasm resource. Plant Roots 1:17–21

Marcelis LFM, Heuvenlink E, Goudriaan J (1998) Modeling biomass production and yield of horticultural crops: a review. Sci Hortic 74:83–111

Marsal J, Girona J (1997) Effects of water stress cycles on turgor maintenance processes in pear leaves (*Pyrus communis*). Tree Physiol 17:327–333

Mitchell PD, Chalmers DJ (1982) The effect of reduced water supply on peach tree growth and yield. J Am Soc Hortic Sci 107:853–856

Mitchell PD, Jerie PH, Chalmers DJ (1984) The effects of regulated water deficits on pear tree growth, flowering, fruit growth, and yield. J Am Soc Hortic Sci 109:604–606

Mitra J (2001) Genetics and genetic improvement of drought resistance in crop plants. Curr Sci 80:758–763

Olasantan FO (2007) Vegetable production in tropical africa: status and strategies for sustainable management. J Sustain Agric 30:41–70

Pandey CB, Srivastava RC, Singh RK (2009) Soil nitrogen mineralization and microbial biomass relations; and nitrogen conservation in humid tropics. Soil Sci Soc Am J 73:1142–1149

Rao R, Li YC (2003) Management of flooding effects on growth of vegetable and selected field crops. HortTechnology 13:610–616

Raviv M, Blom TJ (2001) The effect of water availability and quality on photosynthesis and productivity of soilless grown cut roses. Sci Hortic 88:257–276

Roitsch T (1999) Source-sink regulation by sugar and stress. Curr Opin Plant Biol 2:198–206

Sage R, Kubien D (2007) The temperature response of C3 and C4 photosynthesis. Plant Cell Environ 30:1086–1106

Sato S, Peet MM, Thomas JF (2002) Determining critical pre- and post-anthesis periods and physiological processes in *Lycopersicon esculentum* Mill. exposed to moderately elevated temperatures. J Exp Bot 53:1187–1195

Schaffer B, Anderson PC (1994) Handbook of environmental physiology of fruit crops, vol 2, Subtropical and Tropical Crops. CRC Press, Boca Raton, p 310

Sinha SK (1986) Drought resistance in crop plants: a physiological and biochemical analysis. In: Chopra VL, Paroda RS (eds) Approaches for incorporating drought and salinity resistance in crop plants. Oxford/IBH, New Delhi, pp 56–86

Spreer W, Nagle M, Neidhart S, Carle R, Ongprasert S, Müller J (2007) Effect of regulated deficit irrigation and partial rootzone drying on the quality of mango fruits (*Mangifera indica* L., cv. 'Chok Anan'). Agric Water Manag 88:173–180

Stevens MA, Rudich J (1978) Genetic potential for overcoming physiological limitations on adaptability, yield, and quality in tomato. HortSci 13:673–678

Taiz L, Zeiger E (2006) Plant physiology, 4th edn. Sinauer Associates, Sunderland

Tamura F, Tanabe K, Katayama M, Itai A (1996) Effects of flooding on ethanol and ethylene production by pear rootstocks. J Jpn Soc Hortic Sci 65:261–266

Tardieu F, Davies WJ (1992) Stomatal response to abscisic acid is a function of current plant water status. Plant Physiol 98:540–545

Topa MA, Cheeseman JM (1992) Effects of root hypoxia and a low P supply on relative growth, carbon dioxide exchange rates and carbon partitioning in *Pinus serotina* seedlings. Physiol Plant 86:136–144

Vu CV, Yelenosky G (1991) Photosynthetic responses of citrus trees to soil flooding. Physiol Plant 81(1): 7–14

Wahid A, Gelani S, Ashraf M, Foolad MR (2007) Heat tolerance in plants: an overview. Environ Exp Bot 61:199–223

Walter S, Heuberger H, Schnitzler WS (2004) Sensibility of different vegetables of oxygen deficiency and aeration with H_2O_2 in the rhizosphere. Acta Horticult 659:499–508

Wang W, Vinocur B, Altman A (2003) Plant responses to drought, salinity and extreme temperatures: towards genetic engineering for stress tolerance. Planta 218:1–14

Weis E, Berry JA (1988) Plants and high temperature stress. Soc Exp Biol 42:329–346

Wentworth M, Murchie EH, Gray JE, Villegas D, Pastenes C, Pinto M, Horton P (2006) Differential adaptation of two varieties of common bean to abiotic stress. II. Acclimation of photosynthesis. J Exp Bot 57:699–709

Wien HC (1997) The physiology of vegetable crops. CAB International, Wallingford, p 672

Wien HC, Turner AD, Yang SF (1989) Hormonal basis for low light intensity-induced flower bud abscission of pepper. J Am Soc Hortic Sci 114:981–985

Xiong L, Wang RG, Mao G, Koczan JM (2006) Identification of drought tolerance determinants by genetic analysis of root response to drought stress and abscisic acid. Plant Physiol 142:1065–1074

Xue GP, McIntyre CL, Glassop D, Shorter R (2008) Use of expression analysis to dissect alterations in carbohydrate metabolism in wheat leaves during drought stress. Plant Mol Biol 67:197–214

Yamada MH, Yamane T, Hirabayashi T (1986) Studies on cross breeding of Japanese Persimmon (Diospyros Kaki Thumb) 5.Variation of fruit cracking under calyx. Bull Fruit Tree Res Sta E 6:11–20

Yamada M, Hidaka T, Fukamachi H (1996) Heat tolerance in leaves of tropical fruit crops as measured by chlorophyll fluorescence. Sci Hortic 67:39–48

Yamaguchi T, Blumwald E (2005) Developing salt-tolerant crop plants: challenges and opportunities. Trends Plant Sci 10:615–620

Yordanova R, Christov K, Popova L (2004) Antioxidative enzymes in barley plants subjected to soil flooding. Environ Exp Bot 51:91–101

Yordanova R, Christov K, Popova L (2004) Antioxidative enzymes in barley plants subjected to soil flooding. Environ Exp Bot 51:93–101

Zhang J, Davies WJ (1987) ABA in roots and leaves of flooded pea plants. J Exp Bot 38:649–659

Zhang JH, Huang WD, Liu YP, Pan QH (2005) Effects of temperature acclimation pretreatment on the ultrastructure of mesophyll cells in young grape plants (*Vitis vinifera* L. cv. Jingxiu) under cross-temperature stresses. J Integr Plant Biol 47:959–970

Role of Plant Growth Regulators in Abiotic Stress Tolerance

2

K.K. Upreti and Maryada Sharma

Abstract

Endogenous plant growth regulators play an important role in regulating plant responses to abiotic stress by sensitizing growth and developmental processes. While the physiological and molecular mechanisms linked to the role of ABA and cytokinins in stress tolerance are well explained, there is growing interest to elucidate the associations of auxins, ethylene, gibberellins, brassinosteroids, and polyamines in stress tolerance mechanism and also on possible cross talk mechanism among different growth regulators during stress tolerance acquisition. Identification and characterization of the gene regulating synthesis of different endogenous growth regulators and recent progresses on hormonal signaling, mutant research, and physiological actions have provided scope for manipulating their biosynthetic pathways for developing transgenic crop plants with enhanced abiotic stress tolerance. Researches have also provided some leads in exploiting the potential of growth regulators in enhancing the resistance to abiotic stresses of crops.

2.1 Introduction

Plant growth and productivity are adversely affected by various abiotic stresses. Plants are frequently exposed to a plethora of stress conditions such as low temperature, salt, drought, flooding, heat, and oxidative stress. Various anthropogenic activities have accentuated the existing stress factors. All these stress factors prevent them from reaching their full genetic potential and limit the crop productivity. In the event of growing concerns of uncertainties in climatic conditions, the abiotic stresses have become the major threat to agriculture production worldwide (Bray et al. 2000). The plant responses to abiotic stress condition are believed to be complex in nature as these are the reflections of integration of stress effects and responses at various levels of plant organization. To provide tolerance against stresses, plants are equipped with several

K.K. Upreti (✉) • M. Sharma
Division of Plant Physiology and Biochemistry, ICAR-Indian Institute of Horticultural Research, Hessaraghatta Lake Post, Bengaluru 560089, Karnataka, India
e-mail: kku@iihr.ernet.in

N.K.S. Rao et al. (eds.), *Abiotic Stress Physiology of Horticultural Crops*, DOI 10.1007/978-81-322-2725-0_2

inbuilt physiological and biochemical mechanisms occurring at cellular level. An understanding of processes linked to these mechanisms is vital in optimizing the crop growth and productivity under stress conditions.

One of the important and widely discussed aspects in abiotic stress tolerance is the regulatory roles of plant growth regulators (PGR). PGR are chemical substances that profoundly influence the growth and differentiation of plant cells, tissues, and organs and also function as chemical messengers for intercellular communication. Their biosynthesis within plant tissues is not always localized. In general, there are five major classes of PGR, and each is grouped together into one of these classes based on their structural similarities and physiological responses. These include the auxins and gibberellins that stimulate predominantly cell elongation; cytokinins, the purine bases that stimulate cell division; ethylene, the olefinic gaseous molecule that regulates among other plant event fruit ripening; and abscisic acid (ABA), the sesquiterpene that regulates senescence and abscission of plant parts and helps in maintenance of plant water relations. Besides these five classes of naturally existing PGR, the steroid hormones brassinosteroids and polyamines are also reported to exhibit growth-regulating activities in plants and are the topics of extensive researches. While these natural PGR are distinctive both in chemical characteristics and in exhibiting characteristic responses, each of the PGR has the potential to alter almost all the aspects of plant growth and development to confer stress tolerance. It is documented that the induction in stress tolerance is induced by the manipulations in the concentrations of endogenous PGR under stress conditions by helping plants in many ways. Most of the progresses made in the PGR research especially on plant adaptation to abiotic stresses are the outcome of advances made in the precise analysis of PGR employing powerful and reliable physiochemical techniques as well advances made in molecular and genetic approaches. Besides the naturally occurring PGR, a wide range of chemicals with well-defined growth regulatory activities have been synthesized, and several of these have been depicted to have wide applications in improving plant growth and yield and quality of produce. Good progress is also achieved in demonstrating the potential of such PGR in the amelioration of abiotic stress responses in a number of crop plants. In the present chapter, an insight into various physiological and biochemical aspects of PGR in relation to their involvement in abiotic stress is provided.

2.2 Abscisic Acid (ABA)

The ABA is an important chemical signal of plant responses to a range of abiotic stresses, including drought and salinity (Keskin et al. 2010; Verslues and Bray 2006). A dynamic balance between its biosynthesis and degradation, sensitized by developmental and environmental factors, determines the amount of available ABA (Cutler and Krochko 1999). The functions of ABA in plants are multiple. High cellular ABA facilitate modifications in stomatal functioning, root hydraulic conductivity, photosynthesis, biomass allocation between roots and shoots, plant water relations, osmolyte production, and synthesis of stress-responsive proteins and genes to confer stress tolerance (Finkelstein et al. 2002, 2008; Hetherington 2001; Kim et al. 2010a; Hoth et al. 2002; Seki et al. 2002). From the experiments involving radio-labeled ^{18}O, molecular genetic analysis of auxotrophs, and biochemical studies, the pathways for ABA biosynthesis are identified. ABA is biosynthesized in cytosol through the carotenoid biosynthetic pathway (Milborrow 2001). ABA is biosynthesized from the C15 compound, farnesyl pyrophosphate, and a C40 carotenoid, involving isopentenyl pyrophosphate (IPP). IPP is synthesized from mevalonic acid in the cytosol, whereas in plastids, 1-deoxy-D-xylulose-5-phosphate (DXP) from pyruvate and glyceraldehyde-3-phosphate are involved in carotenoid biosynthesis. IPP is converted to a C20 product, geranylgeranyl pyrophosphate, which is further converted to C40 carotenoid, phytoene. The phytoene via series of reaction intermediates like violaxanthin, neoxanthin, and xanthoxin involving cyclization and hydroxylation reactions is converted to ABA (Nambara and Marion-Poll 2005). The plants under stress show inductions in the activities of enzymes associated with the ABA biosynthesis and relative induction in mRNA leading to ABA accumulations. Besides the inhibition of ABA catabolism under stress is also a factor for induced ABA accumulation (Jia and Zhang 1997). The ABA levels in plants are regulated by its degradation irreversibly into its hydroxylated products like phaseic acid (PA) and dihydrophaseic acid (DPA) (Zhou et al. 2004) or reversibly into the physiologically inactive derivative of glucose ester by glucosidase (Boyer and Zeevaart 1982). It is documented that the PA and DPA contents increase parallel to ABA under stress conditions. However, their levels under stress increase even after ABA content reaches plateau. In contrast upon rehydration of plants, the ABA level shows a decrease, but PA or DPA levels either increase or remain unaltered (Zhou et al. 2004).

Drought and salinity induce ABA accumulation in the leaves of many plant species (Benson et al. 1988; Bray 1988; Pekic et al. 1995; Luo et al. 1992; La Rosa et al. 1985, 1987; Jiang and Zhang 2002; Nayyar et al. 2005; Conti et al. 1994; Upreti et al. 1997, 1998, 2000; Upreti and Murti 2004a, 2005; Satisha et al. 2005). The stress release reverses ABA increase and brings back its levels to prestressed levels. The increases in ABA enable plants to restrict their water loss through transpiration following closure of stomata and enhanced plant water status following increased root hydraulic conductivity (Thompson et al. 2007). ABA also participates in the communication between the root and aboveground part, either by stomatal closure or metabolic changes and gene expression (Zhang et al. 2006; Sobeih et al. 2004). However, such regulatory mechanism is more linked to soil moisture content rather to the leaf water status, indicating that ABA acts as chemical signal produced by stressed roots (Davies and Zhang 1991). The sensitivity of stomata to ABA varies in plant species and cultivars and is dependent upon leaf age, climatic factors like temperature and relative humidity, plant nutritional status, ionic status of xylem sap, and leaf water status (Dodd et al. 1996). Such variations in ABA for stomatal response are possibly the consequence of variations in the magnitude of ABA transportation to the active site at guard cell. Tardieu and Simonneau (1998) demonstrated that the xylem ABA concentration and stomatal conductance are linearly inverse related and the slope of relationship varied diurnally. Exogenous application of ABA is effective to increase plant adaptive response to various stress conditions (Marcinska et al. 2013; Javid et al. 2011). However, in some cases, exogenous application of ABA did not increase stress tolerance (Chen and Gusta 1983; Robertson et al. 1987), and this nonresponsive condition to ABA is due to lack of ABA uptake or its degradation by microbes.

The stomata aperture is regulated by turgor potential of its surrounding cells. The guard cell volume is actively responsive to signals produced under stress in order to regulate CO_2 efflux for photosynthesis and transpirational water loss. The ABA increase in guard cells reduces plant water loss through transpiration by promoting stomatal closure (Harris and Outlaw 1991). The influx or efflux of K^+ balanced by flux of anions regulates guard cell volume and this process is ABA regulated (Hetherington and Quatrano 1991). Such results are supported from the studies of MacRobbie (1991); externally applied ABA evokes the efflux of K^+ and anion from guard cells. Besides, efforts are also made toward deciphering electrical responses triggered by ABA in the plasmalemma of the guard cells (Blatt and Thiel 1993; MacRobbie 1991). The cellular electrical changes induced by ABA are the outcome of the depolarization effect which reflects a net influx of cations (Thiel et al. 1992). The depolarization is the driving force for K^+ efflux through outward K^+ channel. Besides, Ca^{+2} also plays an important role in ABA-mediated stomata closure. Ca^{+2} participate as intracellular secondary messenger in mediating the ABA effects on stomatal aperture and/or plasma membrane channel. ABA also evokes alkalization of cytoplasm of guard cells (Irving et al. 1992), which is necessary in the

ABA activation of K^+ channel (Blatt and Armstrong 1993). Wilkinson and Davies (1997) demonstrated pH reduction induced by ABA in sensitizing stomata for closure, as guard cells take up ABA more efficiently at acidic pH (Anderson et al. 1994). Recent studies also depict that the ABA closes stomata by involving signal transduction molecule like H_2O_2 (Luan 2002). Likewise, ABA induces production of nitric oxide in guard cells, the increase which negatively regulates ABA signaling in guard cells (Neill et al. 2008). Furthermore, the ABA is also active in root-to-shoot communication in the plants subjected to stress. The ratio between the growth of root and shoot in a plant is sensitive to abiotic stresses, and there is coordination among them via long distance transport of substrates or signal (Munns and Crammer 1996). Passioura and Stirzaker (1993) opined that the ABA acts as a feed-forward signal from the roots to the aerial plant parts under stress conditions. Jackson (1993) provided evidence for influence of the roots on shoot development via transport of hormones in the xylem. Further Saab et al. (1990) stated that the relationship between ABA and root growth is completely different from that in shoots, as higher ABA levels in roots promote root growth at low water potential (Biddington and Dearman 1982; Watts et al. 1981). However, some investigations also reported inhibition in root expansion by exogenous applications of ABA (Cramer and Jones 1996). ABA increase in roots tends to stimulate the water flow by increasing the root hydraulic conductivity and ion uptake, which causes increase in water potential gradient between soil and roots (Glinka and Reinhold 1971). It also increases water absorbing area of roots and helps the plants transport more water and nutrient under stress situations.

Stress conditions in plants induce metabolic alteration resulting in synthesis and/or accumulation of a wide range of proteins (Pareek et al. 1998; Bray 1988; Bartels et al.1996; Cohen and Bray 1990; Piatkowiski et al. 1990; King et al. 1992; Yokota et al. 2002). The identification and characterization of proteins provide an insight into the complexity of stress response and stress tolerance mechanism (Borkird et al. 1991). Studies have shown the activation of some proteins by drought and salinity stress as well as by ABA. Such information helped in describing ABA involvement in cellular signaling processes in plant-stress interactions (Chandler and Roberstson 1994). Some of the proteins associated with protection of cellular structures are LEA proteins, dehydrin, lipid transfer proteins, desaturase, RAB (responsive to ABA) saturase enzymes, etc. LEA proteins, which are nonenzymatic, and hydrophilic globular proteins are extensively characterized and perform protective functions (Ingram and Bartels 1996; Cushman and Bohnert 2000; Ouvrard et al. 1996; Ismail et al. 1999). Synthesis of ABA is the common dominant factor in the induction of all these proteins and the requirement which is shown through the use of mutants. The ABA-deficient mutant of tomato shows no distinct ABA-responsive proteins when subjected to drought stress compared to the wild type (Bray 1988). ABA treatment to *flacca* resulted in the synthesis of polypeptides similar to the wild type. The stress-responsive proteins have been thought to function in detoxification of cell damage during dehydration (Bartels and Sankar 2005).

A number of genes that respond to stress at transcriptional level have been found to be induced by ABA (Skriver and Mundy 1990; Delasny et al. 1994). However, not all the genes induced by stress are responsive to ABA. There exists ABA-dependent and ABA-independent signal transduction cascade between initial signal of stress and expression for specific gene. Gene expressed during stress helps in protecting cells from stress injury by producing proteins involved in signaling transduction mechanism (Shinozaki and Yamaguchi-Shinozaki 1997). Further, the interaction between transcription factors and their cis-regulatory elements causes expression of stress-inducible genes. Major transcription factor families, which are involved in the regulation of abiotic stress responses, are bZIP, MYB, MYC, NAC, ERF, and DREB/CBF. The ABRE-binding (AREB) proteins or ABRE-binding factors (ABFs) encode bZIP transcription factors among which AREB1/ABF2, AREB2/ABF4, and ABF3 are induced by dehydration, salinity, or ABA treatment and are involved in enhanced drought stress tolerance (Yoshida et al. 2010). Stomatal closure under stress is induced because of the overexpression of GmbZIP1 transgenic plants, leading to enhanced tolerance to stresses. Members of the bZIP family, like ABP9, are associated with enhanced photosynthetic capacity of plants in drought and heat stresses (Zhang et al. 2008a). Similarly, OsABF1 in roots is involved in abiotic stress responses and ABA signaling (Hossain et al. 2010). In tomato, a bZIP transcription factor SIAREB1 participates in abiotic stress by regulating oxidative-stress-related proteins, LEA proteins, and lipid transfer proteins (Orellana et al. 2010).

The ABA controls abiotic stress signaling, regulated by three components: (1) pyrabactin resistance (PYR)/PYR1-like (PYL)/regulatory component of ABA receptor (RCAR), (2) protein phosphatase 2C (PP2C), and (3) SNF1 (sucrose non-fermenting)-related protein kinase 2 (SnRK2) (Mehrotra et al. 2014). PP2C is a negative regulator of ABA signaling and is similar to ABA insensitive 1(ABI1), ABI2 (homolog to ABI1), and AHG3/PP2CA and hypersensitive to ABA1 (HAB1) (Saez et al. 2004; Yoshida et al. 2006). Gene expression and genetic analysis further indicate a dominant role of ABI1, ABI2, and PP2CA in ABA-signaling pathways in plants. Negative regulation of MAPK (MAPK4 and MAPK6) by PP2C and AP2C1 (Schweighofer et al. 2004) associates PP2Cs with cold and drought stress responses. SnRK2 is a positive regulator of ABA signaling and is identified as ABA-activated protein kinase (AAPK) (Li et al. 2000). The FSnRK2 is the key regulator of plant response to abiotic stress. Phosphorylation at the posttranslational level and gene response to changes under stress regulate SnRK2 expression. Salinity and low temperature stress upregulate expression of SnRK2 genes, while high temperature stress downregulate expression of SnRK2 genes (Huai et al. 2008). Further, Nishimura et al. (2010) showed that overexpression of RCAR1/PYL1, PYL5/RCAR8, or PYL8/RCAR3 gives more tolerance toward drought stress in *A. thaliana* and interaction of PYR/PYL/RCAR with ABI1 is ABA dependent.

2.3 Cytokinins

Cytokinins are important growth promotory compounds involved in seed germination, morphogenesis, chloroplast biogenesis, maintenance of assimilate mobilization, fruit and leaf abscission, and in the regulation of stomatal functioning and root-to-shoot communication in the plant experiencing stress. These are synthesized primarily in roots (Chen et al. 1985; Binns 1994), besides the shoot apex and other plant tissues. Cytokinins are N^6-substituted adenosine compounds with branched five-carbon side chain (zeatin and isopentenyl adenine). The cytokinins are biosynthesized following two pathways: de novo biosynthetic pathway (Chen and Melitz 1979) and *t*-RNA pathway (Skoog and Armstrong 1970). Most of the biologically active cytokinins are biosynthesized by de novo biosynthetic pathway. The formation of N^6-(Δ^2-isopentenyl) adenosine-5′-mono phosphate from Δ^2-isopentenyl pyrophosphate and adenosine-5-′-monophosphate through catalytic reactions by isopentenyl transferase (ipt) is the important step in cytokinin biosynthesis (Mok and Mok 2001). In the other pathway, *t*-RNA is degraded and isomerized to cis-zeatin by *cis-trans* isomerase (Mok and Mok 2001). The cytokinin levels are regulated through conversions to riboside derivatives and N- and O-linked glycosides derivatives (Brzobohaty et al. 1994; Murti and Upreti 2000; Zazimalova et al. 1999). These derivatives tend to release free cytokinins for developmental events whenever required, by the action of β-glucosidase present in plants. The cytokinins are also catabolized irreversibly by cytokinin oxidases to inactive forms (Brzobohaty et al. 1994). Plants experiencing abiotic stress tend to decline cytokinin concentration (Naqvi 1999; Pospisilova et al. 2000). Upon stress release, the cytokinin restores their normal levels, and response is fast, and species and cultivar dependent. The cytokinin decline is presumed as the outcome of either reduction in cytokinin biosynthesis or their enhanced degradation or even both. Zhou et al. (2004) reported that the drought stress changes in cytokinins levels in the xylem sap of apple trunk and the changes depended upon drought cycles. During the first cycle of drought and rewatering, cytokinin levels in drought-stressed plants decrease significantly, while in the second, zeatin content declines with no changes in zeatin riboside. During the third cycle, zeatin content did not differ between the treatments. Masia et al. (1994) reported a decrease in cytokinin transport from the root to shoot which occurs during the onset of water stress. Pillay and Beyl (1990) observed reduction in cytokinin concentration in drought-susceptible cultivar of tomato. Upreti et al. (1998) in French bean and (Upreti and Murti 2004a) in onion found profound decline in cytokinins as a result of drought stress with extent of decline dependent upon stress severity, growth stage, and cultivars. Satisha et al. (2005) witnessed a decline in cytokinins in the grape genotypes under soil moisture-deficit conditions. Further, the stress-induced decline in cytokinins depends upon the age of the plant organ, as the young leaves of stressed plants witness greater decline in cytokinins than the old leaves (Upreti and Murti 2004b). Root nodulation is sensitive to drought stress in French bean plants. Upreti and Murti (1999a) reported that the stress-induced decline in root nodulation is linked to decline in cytokinins in roots/nodules. Stoll et al. (2000) in grapevine found that the partial root drying had negative influence on the xylem cytokinin concentration concomitant with distinct increase in xylem-sap pH. Goiocchea et al. (1995) reported decrease in cytokinins in alfalfa under drought, and this was related with accelerated rate of senescence. In desert-grown almond trees, cytokinins showed peak concentrations in the morning and a rapid decrease in the afternoon; these fluctuations preceded daily variation in stomatal conductance (Fusseder et al. 1992). In contrast, Stern et al. (2003) found increased content of zeatin riboside and dihydroxy zeatin riboside in sap with decrease in stem water potential in water-stressed trees of lychee. In grape rootstocks, NaCl salinity caused reduction in root cytokinins content and the rootstocks with high cytokinin contents under salinity maintained high K-Na ratio and root-shoot dry mass ratio (Upreti and Murti 2010). Vankova et al. (2011) in radish reported that the higher stress sensitivity of radish is associated with higher decline in bioactive cytokinin levels, as a consequence of stimulation in cytokinin regulatory enzymes, cytokinin oxidase.

The ipt is an important cytokinin biosynthesis enzyme, and overexpression of it increases cytokinin levels, leading improvements in stress tolerance. Rivero et al. (2007) and Peleg et al. (2011) found overexpression of ipt gene under the control of *senescence-associated receptor kinase* (*SARK*; a maturation- and stress-induced promoter) improves the drought tolerance in different plants. Merewitz et al. (2011) showed that the drought tolerance enhancement associated in ipt-expressed transgenic plants is also linked to maintenance of accumulation of several metabolites, particularly amino acids (proline, γ-aminobutyric acid, alanine, and glycine), carbohydrates (sucrose, fructose, maltose, and ribose), and organic acids, involved in the citric acid cycle. The accumulation of these metabolites could contribute to improved tolerance due to their roles in the stress response pathways such as stress signaling, osmotic adjustment, and respiration for energy production. Similarly, the expression of ipt gene under the control of cold inducible (COR15a) (Belintani et al. 2012) and SAG12 (McCabe et al. 2001) or maize ubiquitin (Hu et al. 2005) promoter induce low temperature tolerance in petunia and chrysanthemum (Khodakovskaya 2005) and lettuce (McCabe et al. 2001). The ipt gene overexpressed plants also trigger tolerance to stress by expression of ROS scavenging enzymes (Gashaw 2014). There are reports relating increased abiotic stress resistance at low cytokinin levels employing mutants lacking the functional cytokinin receptor (Jeon et al. 2010; Kang et al. 2012). Jeon et al. (2010) showed that the *Arabidopsis histidine kinase* (*AHK*) loss-of-function mutants *ahk2*/*ahk3* and *ahk3*/*ahk4* are more resistant to freezing temperatures than the wild type. Similarly, Kang et al. (2012) showed that all *ahk* mutants possess enhanced resistance to dehydration.

Stomata play an important role in the control of water transpiration and gas exchange in plant leaves, and cytokinins are important in keeping stomata open by regulating ABA action (Veselova et al. 2006). Brault and Maldiney (1999) depict that the cytokinins act at plasma membrane in association with other signaling molecules, as cytokinins antagonize many physiological processes like stomatal closure, leaf senescence, and leaf expansion mediated by ABA (Cowan et al. 1999). Such antagonistic relationship between ABA and cytokinins is possible because metabolically cytokinins and ABA share partially the common biosynthetic pathway. This is supported from the observation that the cytokinins increase stomatal aperture and transpiration rate in plants, the responses opposite to that of ABA (Pospisilova et al. 2000). The cytokinins are also expected to override the effects of ABA on stomata (Pospisilova et al. 2000; Blackman and Davies 1983), and thus reduction in cytokinins under stress condition enhances shoot responses to increasing concentrations of ABA. This led to conceptualizing the cytokinins to act as negative signals in plants experiencing stress. Nishiyama et al. (2011) and Wang et al. (2011) also hinted at possibility of cross talk between ABA and cytokinins, as altered ABA sensitivity in plants modifies cytokinins levels and signaling. Further, the mechanism of cytokinin action on guard cell involves membrane hyperpolarization by stimulation of adenylate cyclase activity, leading to increase in intracellular adenosine 3′,5′-cyclic monophosphate content, stimulation of guanylate cyclase activity, or interaction with a calcium calmodulin system (Pospisilova and Dodd 2005; Incoll et al. 1990). Stomatal opening is also regulated by hydraulic as well as chemical signals (Whitehead 1998). Both naturally occurring and synthetic cytokinins increase transpiration rate and increase in stomatal aperture (Incoll et al. 1990; Incoll and Jewer 1987). The stomatal responses of cytokinins are found variable. Blackman and Davies (1983) revealed that zeatin alone did not affect stomatal opening, but partially reverse ABA-induced stomatal closure. In contrast, zeatin riboside or kinetin decreased stomatal opening and had no effect on ABA-induced stomatal closure.

2.4 Gibberellins

Gibberellins are tetracyclic diterpenoid carboxylic acid, which influence growth and various developmental processes, such as elongation, germination, dormancy, flowering, sex expression, enzyme induction, and leaf and fruit senescence. The predominant bioactive forms of gibberellins are GA_1 and GA_4 (Sponsel and Hedden 2004). Gibberellins are biosynthesized from trans-geranylgeranyl diphosphate, formed in plastids via the methyl erythritol phosphate pathway (Kasahara et al. 2002) by the action of terpene cyclases, followed by oxidation by cytochrome P450 monooxygenases and then by soluble 2-oxoglutarate-dependent dioxygenases (Hedden and Thomas 2012; Yamaguchi 2008). The dioxygenases are GA_{20}-oxidase (GA20ox), GA_2-oxidase (GA2ox), and GA_3-oxidase (GA3ox) enzymes, and these help in gibberellin metabolism and function to maintain gibberellin balance during growth. The GA2ox genes are most responsive to developmental events as well as to abiotic stress. Most of the information on the role of gibberellins in stress tolerance mostly came from the results of applications of chemical growth retardants used to control growth of crops and have been shown to enhance drought tolerance (Halevy and Kessler 1963). The primary mechanism by which these chemicals exert influence is through inhibition of the biosynthesis of gibberellins (Rademacher 2000) and application of gibberellin to retardant-treated plants and to GA-deficient mutants (Gilley and Fletcher 1998; Vettakkorumakankav et al. 1999).

There are less number of studies on gibberellin and plant responses to abiotic stress. Radi et al. (2006) showed that the presoaking wheat seeds in gibberellic acid increased the germination potential especially at moderate salinization levels. Between gibberellin levels and the acquisition of stress protection, an intimate relationship exists (Vettakkorumakankav et al. 1999). In sugarcane, supplementing GA_3 as foliar application plays an important role on imparting salt tolerance in terms of enhancing nutrient uptake, as well as the morphological and physiological aspects (Shomeili et al. 2011). Ashraf et al. (2002) showed that GA_3 application increased the nutrient uptake, plant height, leaf area, and yield of wheat under saline conditions. Evidence also shows the involvement of GA_3 in relieving NaCl-induced growth inhibition in rice (Wen et al. 2010). Starck and Kozinska (1980) found that GA_3 causes more absorption of P and Ca^{2+} and less absorption of Na^+, besides adjusting the ion ratios in bean. Bejaoui (1985) concluded that the exogenously applied GA_3 alleviates salt stress due to activation of enzymes which participate in RNA and protein synthesis. Aloni and Pressman (1980) found interaction between salinity and the GA_3 important in petiole elongation, cellular breakdown, and bolting in celery. Maggio et al. (2010) reported that GA_3 treatment in tomato reduced stomatal resistance and enhanced plant water use at low salinity. Likewise, GA_3-priming increases grain yield due to the GA_3-priming-induced modulation of ion uptake and partitioning (within the shoots and roots) as well as hormone homeostasis under saline conditions (Iqbal and Ashraf 2013). Kumar and Singh (1996) witnessed that the seed germination is improved by GA_3 application under saline conditions. The precise mechanisms by which gibberellin is linked to stress tolerance are less understood. One possible mechanism is via its possible interactions with other phytohormones. In this context, the auxin is expected to promote GA biosynthesis (Wolbang et al. 2004). Similarly, gibberellic acid application enhances the catabolism of ABA (Gonai et al. 2004). Application of paclobutrazol, a gibberellin biosynthesis inhibitor, induced stress protection, and application of GA_3 to the dwarf phenotype reversed the inherent stress tolerance. Reversal of the dwarf phenotype by specific gibberellins suggests that the conversion of GA_{20} to GA_1 and GA_9 to GA_4 is compromised, indicating that the modulation of specific gibberellins plays an important role in stress protection. Sankhla et al. (1989) described that the soil drenching treatment of paclobutrazol is important in minimizing water stress injuries in the fruits of ber trees, while Still and Pill (2003) found that foliar applications or seed priming with paclobutrazol is effective in improving drought tolerance in tomato seedlings. Likewise paclobutrazol improves plant water status in apple (Swietlik and Miller 1983), strawberry (Navarro et al. 2007), and peach (George and Nissen 1992) under drought conditions. Upreti and Murti (2000) documented seed priming with mepiquat chloride effective in offering good germination in beans under osmotic stress conditions. Similarly, gibberellins play an important role in submergence tolerance of aquatic plants, and effect is mediated through the regulation of ethylene and ABA biosynthesis. The ethylene production under submergence conditions promotes gibberellin synthesis and inhibits ABA synthesis for plant acclimation by causing elongation growth (Colebrook et al. 2014). The treatment of rice plants with gibberellic acid during submergence promoted elongation growth and compromised survival, indicating that gibberellin negatively impacts tolerance to prolonged submergence (Das et al. 2005). Conversely, treatment of non-tolerant cultivars with gibberellin biosynthesis inhibitor, paclobutrazol, restricts underwater elongation and enhanced submergence survival. Further, gibberellin involvement in submergence is through upregulation of the ethylene response factor (ERF) domain proteins SNORKEL1 and SNORKEL2 in response to elongation-induced ethylene accumulation in submerged plants (Hattori et al. 2009) which directly or indirectly leads to increase in bioactive gibberellin levels. Osmotic stress induces changes in gibberellin metabolism, resulting in the stabilization of DELLAs and earlier onset of endoreduplication. Consequently, this response is absent in mutants with altered gibberellin levels or DELLA activity (Claeys et al. 2012). Zawaski and Busov (2014) following whole-genome microarray, expression, physiological, and transgenic evidences showed that gibberellin catabolism and repressive signaling mediates shoot growth inhibition and physiological adaptation in response to drought. Further drought stress elicits activation of a suite of GA2ox and DELLA encoding genes. The transgenic with upregulated GA 2-oxidase (GA2ox) and DELLA domain proteins showed hypersensitive growth inhibition in response to drought besides displaying greater drought resistance as evident from increase in pigment concentrations and reductions in electrolyte leakage. Comparative transcriptome analysis using whole-genome microarray showed that the GA deficiency and GA insensitivity and drought response share a common region of 684 differentially expressed genes, which suggests that gibberellin metabolism and signaling play a role in plant physiological adaptations in response to alterations in environmental factors. Cold stress directly influences levels of bioactive gibberellins as reflected from increase in expression of three GA2ox genes (Archard et al. 2008). Similarly under salinity stress, six GA2ox genes were shown to be upregulated (Magome et al. 2008). Furthermore, the cold-inducible CBF1/DREB1b protein in *Arabidopsis* imparts freezing tolerance, at least in part by activating the expression of GA2ox genes, which in turn leads to reductions in bioactive gibberellins and suppression of growth (Archard et al. 2008). In *Arabidopsis*, the dwarf and delayed flowering 1 (DDF1) protein, involved in salt stress response, binds to the promoter and activates the GA2ox7 gene (Magome et al. 2008). Archard et al. (2006) showed that mutants with reduced gibberellins content shows enhanced survival under salt stress. Also, the gibberellin signaling participation in mediating growth and stress responses to flooding is shown by Bailey-Serres and Voesenek (2010) based on internodal elongation and plant survival.

2.5 Auxins

Auxins (indole acetic acid, IAA) play essential roles in diverse developmental events like root growth, vascular tissue differentiation, auxiliary bud formation, apical dominance, and flower organ development (Zhao 2010) and respond to abiotic stress. However, little information is available on the relationship between stress and auxins in plants, and the role of auxins in alleviating different stress responses needs better understanding. IAA is mainly synthesized in meristematic tissues through tryptophan-dependent and independent biosynthetic pathways (Zhao et al. 2001). In tryptophan-dependent IAA biosynthetic pathways, indole-3-acetamide (IAM), indole-3-pyruvic acid (IPA), tryptamine (TAM), and indole-3-acetaldoxime (IAOX) are identified as major intermediates. The concentrations and ratios of IAA and IAA derivatives in plant tissues and auxin homeostasis are regulated by its degradation, conjugation to amino acids, and transport. And these processes are sensitive to abiotic stresses in plants. The IAA levels are modified in plant by two possible mechanisms under stress: one by changes in expression of auxin polar transporter gene and the other by inhibitions in polar transport by certain compounds accumulated in response to stress (Potters et al. 2009). Besides, IAA metabolism is also modulated by oxidative degradation of IAA through peroxidases induction under abiotic stress and also reactive oxygen species generated during stress (Kovtun et al. 2000). The salinity stress causes reduction in IAA in rice (Prakash and Prathapasenan 1990). However, gibberellic acid treatment during the salinity stress partly reduced the adverse effect of salinity on IAA levels. Dunlop and Binzel (1996) witnessed significant reduction in the IAA levels of tomato by salinity. Sakhabutdinova et al. (2003) reported a progressive decline in the IAA levels in the root system due to salinity. Afzal et al. (2005) documented exogenous IAA application to seeds prior to sowing alleviates the growth inhibiting effect of salt stress. Likewise, decline in seed germination with increasing salinity levels is improved by IAA or NAA treatments (Gulnaz et al. 1999; Akbari et al. 2007). High temperature affects reproductive development, leading to decline in yield. Upreti et al. (2012) reported that maintenance of high IAA in the reproductive organs of capsicum cultivars is responsible for lower abscission and high temperature stress tolerance. Cold stress and auxin contents are potentially linked (Fukaki et al. 1996; Wyatt et al. 2002), as cold stress inhibits the root gravity response and also auxin. Shibasaki et al. (2009) suggested that auxin concentrations are regulated by changes in auxin transport in plants under cold stress (Shibasaki et al. 2009). Du et al. (2012) documented that the increased tolerance to cold stress is due to the combined effects of IAA and ABA, as ABA or carotenoid-deficient mutants show reduced IAA content and exhibit increased cold resistance (Du et al. 2013). For abiotic stress response, Du et al. (2012) and Du et al. (2013) found that the change in auxin homeostasis could influence ABA synthesis and that the balance of auxin and ABA homeostasis played a crucial role in diverse stress responses. In this regard, Taniguchi et al. (2010) suggested ABA involvement in hydrotropic root responses through modulation in auxin. Kinoshita et al. (2012) illustrated involvement of IAR3 (IAA-Ala Resistant3) in drought tolerance through the production of lateral roots, as IAR3 is effective in producing free auxin by hydrolyzing inactive auxin amino conjugates. Similarly flooding stress negatively affects root development by resisting oxygen supply. Plant tends to develop adventitious roots in response to flooding to alleviate negative effects of flooding stress on plant growth and development. Accumulation of auxins triggers ethylene production, which induces adventitious rooting at the base of the stem (Vidoz et al. 2010). In a study, Muday et al. (2012) reported cross talk between ethylene and auxins in waterlogged plants contributing to the formation adventitious root formation.

Molecular genetic studies made in understanding mechanism of auxin action showed that many auxin-responsive genes are also responsive to cold stress (Jain and Khurana 2009). This is supported from auxin signaling mutants *axr1* and *tir1*, which showed reduced gravity response and responded to cold treatment similar to the wild type (Shibasaki et al. 2009). Likewise, PIN3, an auxin transporter that has been suggested to mediate the early phase of the root gravity response, is inhibited by cold stress (Shibasaki et al. 2009), suggesting that cold stress may affect auxin transport. From transcript analysis, Jain and Khurana (2009) found that many auxin-responsive genes are responsive to cold stress. Du et al. (2012) found that the *OsGH3-2* overexpression decreased free IAA content. Further Du et al. (2013) reported that ABA or carotenoid-deficient mutants have reduced IAA content and exhibit increased cold resistance. PIN proteins associated with auxin transport play an important role in drought tolerance. *OSPIN3T* gene encodes for auxin efflux carrier protein (Zhang et al. 2012) and plays a crucial role in drought tolerance by regulating root and shoot development. Similarly phototropin1, a protein kinase PINOID, is responsible for PIN phosphorylation and is capable of improving drought tolerance at seedling stage (Galen et al. 2007). Besides, several transgenic plants (*PDS-RNAi* transgenic rice, *OsGH3-2* overexpressing rice, *AtYUCCA6* overexpressing *Arabidopsis* plants, and *AtYUCCA6* overexpressing potato plants) with altered IAA level exhibit auxin-related developmental phenotypes together with affected drought stress resistance (Du et al.2012, 2013; Kim et al. 2013). Additionally, endogenous and exogenous auxin positively modulated the expression levels of many abiotic stress-related genes (*RAB18*, *RD22*, *RD29A*, *RD29B*, *DREB2A*, and *DREB2B*) and positively affected reactive oxygen species (ROS) metabolism and underlying antioxidant enzyme activities. Likewise, Cheol Park et al. (2013) showed that the transgenic potato (*Solanum tuberosum* cv. Jowon) overexpressing AtYUC6, members of the YUCCA (YUC) family of flavin-containing monooxygenases, showed high auxin and enhanced drought-tolerant phenotypes. The overexpression of AtYUC6 in potato establishes enhanced drought tolerance through regulated ROS homeostasis. Under salt stress, Zolla et al. (2010) found a reduction in the increase of lateral roots in auxin signaling mutants, axr1, axr4, and tir and the response depended upon the auxin eflux carrier, PIN2. Sun et al. (2008) documented that salt stress alters auxin efflux carrier, besides inhibiting PIN2 expression. This shows that like cold stress, salt stress interferes with root gravitropism, an adaptive response to reduce the damaging effects of salt stress (Galvan-Ampudia and Testerink 2011). Xu et al. (2012) illustrated that the soil alkalinity increases auxin transport activity mediated by PIN2. Jung and Park (2011) suggested for a possible cross talk between salinity and auxin mediated by transcription factor NTM2 via IAA30 gene during seed germination.

2.6 Ethylene

Ethylene is the simplest olefinic gaseous hormone which regulates a wide range of plant developmental processes such as pollination, seed germination, abscission and senescence, flowering, fruit ripening, root formation, and gravitropism. It is biosynthesized by the conversion of methionine to ethylene via the intermediates, S-adenosyl methionine (SAM) and 1-amino cyclopropane-1-carboxylic acid (ACC) by involving enzymes ACC synthase and ACC oxidase (Yang and Hoffman 1984; Kende 1993). Abiotic stress conditions surge ethylene production by inducing ACC synthase and ACC-oxidase activities. Drought stress enhances ethylene in French bean (Upreti et al. 1998), orange (Ben-Yehoshua and Aloni 1974), avocado (Adato and Gazit 1974), *Vicia faba* (El-Beltagy and Hall 1974), and in many other plant species (Narayana et al. 1991; Guinn 1976; Irigoyen et al. 1992). The increase in ethylene under stress is of adaptive significance as it helps plants to cope up stress by reducing water loss through increased senescence of fruits/leaves and reduced growth. The magnitude of ethylene increase under stress depends upon growth stage, stress intensity, and stress duration (Upreti et al. 1998, 2000), and higher stress levels tend to reduce ethylene concentration. In pineapple, drought stress has no effect on flower induction and produced significantly less ethylene and had lower ACC-oxidase activity in leaf and stem tissues than the control plants (Min and Bartholomew 2005). Habben et al. (2014) reported that transgenic field-grown plants with downregulated ACC synthase enzyme activity yield higher under drought stress. Salt stress positively influences ethylene biosynthesis, which helps in promoting salt tolerance by enhancing Na/K homeostasis (Lockhart 2013). Yang et al. (2013) illustrated a key role for ethylene in salt tolerance by relating its ability to retain K^+ rather than decreasing Na in roots and shoots. Likewise Jiang et al. (2013) reported that ethylene overproducing (eto1) mutant under salinity exhibits reduced root Na influx and low root stelar and xylem-sap Na concentrations, leading to restricted root-to-shoot delivery of Na^+, along with high xylem-sap K^+ concentrations. Wang et al. (2009) reported that ethylene alters salt tolerance by interfering with other hormone pathways and NO signaling. It also stimulates H^+-ATPase activity to modulate ion homeostasis and salt tolerance. Cold stress alters ethylene levels in plants and the enhanced ethylene level contributes in cold tolerance (Machacckova et al. 1989; Wang and Adams 1982; Ciardi et al. 1997; Zhao et al. 2014). However, tolerance responses of ethylene are variable and species dependent (Kazan 2013). High temperature stress (35/25 °C) in capsicum increased abscission of reproductive organs, which is due to increase in ethylene concentration by accumulation in ACC and induction in ACC-oxidase activity in flower buds and flowers (Upreti et al. 2012). Shi et al. (2012) reported that in vitro-grown *Arabidopsis* seedlings treated with the ACC and mutant overproducing ethylene, eto1, show reduced freezing tolerance, in contrast to increased freezing tolerance by aminoethoxyvinylglycine (AVG), an ACC biosynthesis inhibitor. Zhao et al. (2009) in tomato suggested positive relationship between ethylene and freezing tolerance from the negative effects of 1-methyl cyclopropene, an ethylene biosynthesis inhibitor on freezing tolerance. Further support for ethylene in cold tolerance is evident from the study of Lockhart (2013) that the ethylene biosynthesis inhibitor 1-methyl cyclopropene (1-MCP) reduces cold tolerance in tomato, whereas ethephon enhances cold tolerance in tomato. Ethylene is also considered important in plant's adaptation to flooding. Under flooding, the lack of oxygen in flooded roots triggers the ACC synthesis, which upon transportation upward in the plants gets oxidized to ethylene to cause nastic movements of the leaves and promote aerenchyma formation (Moore et al. 1998; Colmer 2003). The biochemical mechanism that provokes ethylene biosynthesis under stress is still not clearly understood, and some reports also show variations in ethylene responses. Naylor (1972) suggested greater availability of methionine as a result of increased rate of protein breakdown under stress which leads to the elevated ethylene levels. Beltrano et al. (1997) revealed that the increased production of free radicals under water stress facilitates greater conversion of ACC to ethylene. The increase in ethylene in response to stress is depicted primarily by an increased synthesis of ACC (Yang and Hoffman 1984). Xu and Qi (1993) reported that a slowly developing drought did not promote ethylene or altered ACC levels, while rapidly developing drought enhanced both ethylene and ACC levels. Narayana et al. (1991) also reported more ethylene upon rapid loss of water. Beltrano et al. (1997) observed slight changes in ethylene in leaves under moderate or severe stress conditions. Wright (1980) and Hoffman et al. (1983) showed that ABA interacts with ethylene metabolism by regulating the ACC levels.

Ethylene exerts responses through modulation of gene expression function at transcriptional level by ERF (ethylene response factor) by regulating gene expression under abiotic stress conditions (Zhang et al. 2008b; Hussain et al. 2011). Investigations have suggested the potential of ERF proteins to specifically bind not only to GCC box but also to the DRE/CRT motif, also known as a cis-acting element that responds to cold or osmotic stress (Lee et al. 2004; Wang et al. 2004). DREB proteins are important ERF, widely studied in abiotic stress responses. The members of the DREB1/CBF subfamily are induced in response to cold stress and improve tolerance to freezing (Liu et al. 1998; Kasuga et al.1999). Gilmour et al. (2004) reported that the constitutive expression of *DREB1A* and *DREB1B* induces the expression of cold-regulated genes and increases the freezing tolerance. Similar results are observed for the constitutive action of the protein *DREB2*, under conditions of dehydration and high salinity stresses (Liu et al. 1998; Sakuma et al. 2006). A number of cold-inducible genes, such as LEA proteins and enzymes for sugar metabolism and fatty acid synthesis (Fowler and Thomashow 2002), are also upregulated ectopically expressing *DREB1/CBF* members. Trujillo et al. (2008) found that the *SodERF3* is another ERF that is responsible for improved drought and salt tolerance. Likewise Zhang et al. (2009) illustrated that the transgenic plants overexpressing *GmERF3* exhibited tolerance to high salinity and drought stresses, suggesting its crucial role in both abiotic stresses. Similarly, in ERF-VII genes are involved in the response to submergence and hypoxia (Hattori et al. 2009; Hinz et al. 2010; Licausi et al. 2010). Gibbs et al. (2011) and Licausi et al. (2011) suggested that the constitutive ERF-VII factors, like RAP2.12, act as primary trigger for the oxygen deficiency responses. Xu et al. (2006) reported that the SUB1A promotes a quiescent strategy that allows carbohydrate saving and improves tolerance after flooding stress (Xu et al. 2006). Sl-ERF.B.3 (*Solanum lycopersicum* ethylene response factor B.3) gene encodes for a tomato transcription factor of the ERF (ethylene responsive factor) family, which is induced by cold, heat, and flooding, but is downregulated by salinity and drought (Klay et al. 2014). Xu et al. (2006) found that the *SUB1A-1* allele induces the negative regulation of ethylene, making plants able to survive complete submergence for prolonged periods. Besides providing anoxia tolerance, this allele also provides drought and de-submergence tolerance. Similarly, overexpression of the gene *HRE1* shows increase in anoxia tolerance (Licausi et al. 2010). The ethylene homeostasis during conferring of freezing tolerance seems to be important (Catala et al. 2014), and the 14-3-3 protein, RARE COLD INDUCIBLE 1A (RCI1A), is important in interacting with ACC synthase to modulate freezing tolerance. Hattori et al. (2009) suggested that the ERF transcription factors tends to reduce abscisic acid and gibberellin antagonizing signaling process to support stem elongation and photosynthesis under flooding. Li et al. (2010) showed involvement of TaDi9A, a salt-responsive gene in ethylene signaling. Zhu et al. (2005) reported that the ERF, hos10-1 gene transformed plants, accumulates more Na^+ than wild-type plants but has lower sensitivity to salts, implying salt sensitivity is unrelated to Na^+ accumulation. Archard et al. (2006) found that ACC synthase suppresses the salt sensitivity conferred by NTHK1 in transgenic plants suggesting ethylene is required for counteracting receptor function to improve tolerance.

2.7 Brassinosteroids

Brassinosteroids are polyhydroxylated steroidal phytohormones that are structurally related to animal steroid hormones and possess distinct growth-promoting properties (Bishop and Yokota 2001; Clouse and Sasse 1998). These are considered as phytohormones due to inheriting pleiotropic effects, which influence diverse range of developmental and physiological processes including the promotion of cell elongation and cell division, photomorphogenesis, rhizogenesis, senescence and abscission, xylem differentiation, seed germination, and fruit ripening (Clouse and Sasse 1998; Sasse 1997). Numerous studies indicate brassinosteroid potential in enhancing ability of plant to cope with drought stress, salt stress, and high and low temperature stresses (Fariduddin et al. 2014). The growth regulatory and stress protection capability of brassinosteroids are linked to their action on metabolic processes associated with photosynthesis and nucleic acid and protein biosynthesis (Sasse 1997; Fariduddin et al. 2014). The brassinosteroid biosynthesis is a two-step pathway involving sterol-specific pathway, squalene to campesterol, and the other, brassinosteroid specific pathway with several conversion steps from campesterol to brassinosteroid involving series of hydroxylation, reduction, epimerization, and oxidation reactions (Agarwal and Gehlot 2000). The C-6 oxidation of castasterone is the final step in brassinosteroid synthesis. The brassinosteroids undergo esterification to form 2,3-glucosyl and acyl conjugates at 3-position of its moiety (Asakawa et al. 1996).

Information on mechanism by which brassinosteroids exhibit stress tolerance is lacking. It is suggested that brassinosteroids regulate stress response by a complex sequence of biochemical reactions, such as activation or suppression of key enzymatic reactions, induction of protein synthesis, and the production of various chemical defense compounds (Bajguz and Hayat 2009). Most of the research on the response action of brassinosteroids to stress factors is made employing their exogenous application. Upreti and Murti (2004c) reported that application of epibrassinolide or homobrassinolide prior to drought stress results in increased root nodulation in French bean, and the response is mediated through induction in cytokinin synthesis and nitrogenase activity. Moreover, epibrassinolide is relatively more effective than homobrassinolide in rendering such response. Kagale et al. (2007) reported improved drought tolerance in B *rassica napus* seedlings treated with epibrassinolide, and the improved tolerance is due to reduction in reactive oxygen species, induction in antioxidative enzyme activities, and antioxidant contents (Zhang et al. 2008c; Li et al. 2012). Zhang et al. (2008c) also witnessed improvement in photosynthesis by regulating ribulose-1,5-bisphosphate carboxylase/oxygenase activity and sugar accumulation following brassinosteroid treatment in stressed plants. Rajasekaran and Blake (1999) found delay in stomatal closure induced by drought stress following homobrassinolide treatments. Brassinosteroids are also found effective in modulating salinity stress as evident from improvements in plant tolerance by epibrassinolide treatment. The effect is due to protective action against stress-induced oxidative damage of membrane lipids and induction in antioxidant enzymes (Ozdemir et al. 2004; Ali et al. 2007; Hayat et al. 2010). Similarly, Ding et al. (2012) witnessed improvement in salt tolerance by epibrassinolide in eggplant. Molecular studies reveal redox-sensitive protein NPR1 as possible critical component of brassinosteroid-mediated enhancement in salt tolerance (Divi et al. 2010). Some studies also show induction in high temperature tolerance by brassinosteroids. Singh and Shono (2005) found epibrassinolide-treated tomato plants as more tolerant to high temperature than untreated plants as a result of high accumulation of heat shock proteins and improvement in photosynthetic efficiency. Similarly, epibrassinolide treatment to tomato plants prior to high temperature exposure protects rubisco enzyme and RuBP regeneration under heat stress in order to provide better protection against high temperature stress (Ogweno et al. 2008). Chilling tolerance in plants is influenced by brassinosteroids. The report of Huang et al. (2006) showed that epibrassinolide application upregulated 17 proteins which were downregulated by chilling to confer chilling tolerance. Chilling tolerance is also attributed to increase in membrane permeability (Janeczko et al. 2007), high pigment accumulation, and upregulation in cold-related genes (Kagale et al. 2007) besides high activation of rubisco and expression of photosynthetic genes (Xia et al. 2009). Thus, the physiological responses of brassinosteroids though are variable due to complexities in the molecular mechanism of their action; their potential in improving abiotic stress tolerance has immense utility in managing plant responses to stress. For widening the scope in this area and effectively harnessing the benefits from brassinosteroid research, more investigations are needed on mechanisms by which brassinosteroids confer stress tolerance.

2.8 Polyamines

Polyamines are important growth regulatory polycationic molecules known to be involved in a wide range of developmental events including organogenesis, embryogenesis, floral initiation and development, senescence, fruit development and ripening, and root development (Galston et al. 1997; Kumar et al. 1997). These are biosynthesized by decarboxylation of amino acids, ornithine, or arginine in the reaction catalyzed by enzymes ornithine decarboxylase (ODC) and arginine decarboxylase (ADC), leading to the formation of putrescine, which by subsequent additions of aminopropyl moiety produces spermidine and spermine, respectively. These reactions are catalyzed by enzymes, spermidine synthase, and spermine synthase. The aminopropyl moiety is formed from decarboxylation of SAM employing enzyme SAM decarboxylase. The dynamics of polyamines metabolism is complex due to coexistence of degradation and conjugation pathways of transport and uptake mechanisms (Martin-Tanguy 2001; Federico and Angelini 1991). Molecular studies reveal that the polyamines are involved in signal transduction pathway through effects on calcium fluxes (Thomas et al. 1993) and interaction with certain transcriptional factors (Wang et al. 1999) and protein kinases (Datta et al. 1987). Further, polyamines and ethylene synthesis are co-linked through sharing of common precursor SAM, and thus these tend to inhibit each other's biosynthesis and action (Tiburico et al. 1997). Polyamines are also found to play an important role in conferring tolerance against drought, salinity, flooding, heat stress, and chilling stress in plants (Gill and Tuteja 2010), as evident from polyamine changes under stress alleviation responses of exogenous applied polyamines (Gill and Tuteja 2010) and transgenic studies in plants overexpressing polyamine biosynthetic gene encoding for enzymes, arginine decarboxylase (ADC), ornithine decarboxylase (ODC), *S*-adenosylmethionine decarboxylase (SAMDC), or Spd synthase (SPDS) (Liu et al. 2007). The stress-induced increase in polyamines is a reflection of upward regulation of enzymes associated with their biosynthesis and release from polyamine conjugates (Gupta et al. 2013). In general, plants experiencing abiotic stresses tend to increase polyamine levels, and increase in polyamines helps in the regulation of plant tolerance to stress. The stress-induced increases in polyamine provide tolerance by stabilizing membrane integrity and functionality, altering hormonal balances and inducing antioxidant enzymes. Generally, a polyamine response due to stress depends on species, besides type and concentration of polyamines (Ali et al. 2009). Salt-tolerant barley (Liu et al. 2006) and rice (Krishnamurthy and Bhagwat 1989) cultivars drastically accumulate high levels of spermidine and spermine, with a relative decline in putrescine. Mutlu and Bozcuk (2007) found increase in spermidine content in leaf tissues of sunflower plants, with decrease or no significant changes in other polyamines. Similarly, Zapata et al. (2004) studied effects of salinity on polyamines in varied plant species like *Spinacia oleracea*, *Lactuca sativa*, *Cucumis melo*, *Capsicum annum*, *Brassica oleracea*, *Beta vulgaris*, and *Lycopersicon esculentum* and found that the polyamine titer is greatly altered by salinity with distinct increase in spermidine and spermine contents. Duan et al. (2008) witnessed an increase in enzyme arginine decarboxylase, ornithine decarboxylase, SAM decarboxylase, and diamine oxidase activities, as well as free spermidine and spermine and soluble-conjugated and insoluble-bound putrescine, spermidine, and spermine contents in the roots of cucumber cultivars following salinity stress. Wei et al. (2007) studied the effect of salt stress on polyamine contents in leaves of *Solanum melongena*-grafted plants employing salt-tolerant *Solanum torvum* as rootstock. The increase in the contents of free, soluble-conjugated, and insoluble-bound polyamines, with considerable decline in diamine oxidase and polyamine oxidase activities, was observed. The polyamine increase in grafted plants protects the grafted plants against salt stress by triggering higher activities of antioxidant enzymes like superoxide dismutase, peroxidase, and ascorbate peroxidase and glutathione reductase enzymes. Kim et al. (2010b) reported a decrease in spermidine but an increase in spermine under salinity in Chinese cabbage. Likewise Reggiani et al. (1994) reported an increase in spermidine and spermine but decreased putrescine contents in plants under NaCl stress. Zhao et al. (2003) witnessed salinity dose-dependent increase in polyamines and reduction in bound polyamine fraction and concluded that the plant growth and ratio of bound and free polyamine contents are positively related. Upreti and Murti (2010) found significant alterations in root polyamines in grape rootstocks by salinity stress with tolerant rootstock showing greater increases in spermidine and spermine as well as ABA. The high increase in polyamine helped plants in maintaining high root-shoot biomass ratio and high K-Na ratio in the tolerant rootstock. Similarly, Anjum (2008) reported that citrus rootstock, Cleopatra mandarin, shows better growth and chlorophyll efficiency under salinity stress which accumulated high spermine and carbohydrates but low chloride ions in leaves and roots under salinity. Application of exogenous polyamine is efficient in manipulating the levels of endogenous polyamine during stress, and the stress protection effect is rendered through the maintenance of membrane integrity, regulation of gene expression for the synthesis of osmotically active solutes, reduction in ROS production, and controlling of the accumulation of Na^+ and Cl^- ion in different organs. Anjum (2011) reported that spermidine treatment to NaCl-stressed plants reduced salinity-induced decline in leaf number, chlorophyll content, Fv/Fm, net photosynthetic rate, and N content and reduced Na^+ contents of the plants. Additionally spermidine improved cellular Ca^{2+} and Mg^{2+} contents in salinity-treated plants. In pomegranate plants, putrescine and spermine are effective in reducing the stress-induced decline in growth rate without causing major alteration in Na^+, Cl^-, and K^+ contents of roots and apical and basal leaves (Amry et al. 2011). In addition, the protective role of polyamines against high salt stress is a consequence of altered control of Ca^{2+} allocation through regulating Ca^{2+}-permeable channels, including CAXs (Yamaguchi et al. 2006, 2007). The increase in cytoplasmic Ca^{2+} results in prevention of Na^+/K^+ entry into the cytoplasm, enhancement of Na^+/K^+ influx to the vacuole and, likewise, the suppression of Na^+/K^+ release from the vacuole. Exogenous polyamine partially reversed the NaCl-induced phenotypic and physiological disturbances in citrus. The effect is due to upregulation of expression of polyamine biosynthesis and catabolism genes, regulation of transcript expression and activities of antioxidant enzymes, and restoration of NO-associated genes, such as NR, NADde, NOS-like, and AOX, along with S-nitrosoglutathione reductase and nitrate reductase activities in the salinity-exposed plants (Tanou et al. 2014). In cucumber seedlings, Shu et al. (2012) found that spermidine alleviates salt-induced damage by regulating the levels of endogenous polyamines and improvement of photochemical efficiency under stress conditions. The reduction in Na content in shoots induced by polyamines is be an effective strategy for combating high salinity. Zhao et al. (2007) found that the polyamines combat deleterious effects of salinity in barley by altering Na and K balance to improve K^+/Na^+ homeostasis, restricting Na^+ influx into roots and by preventing K^+ loss from shoots. Lakra et al. (2006) also demonstrated increase shoot K allocation by salinity. In pistachio seedlings, Kamiab et al. (2014) found spermidine and spermine treatments were effective in improving salinity tolerance by inducing superoxide dismutase and catalase activities and by decreasing the hydrogen peroxide (H_2O_2), thus balancing ions toward lower Na-K ratio. Similarly, drought stress leads to accumulation of free or conjugated polyamines in many plant species, indicating that polyamine biosynthesis plays an important role in plant response to stress (Liu et al. 2007). Upreti and Murti (2005) reported cultivar difference in polyamine content changes in French bean cultivars under drought stress conditions. Moreover, the stress response on individual polyamine varied with stress duration. The putrescine which increased initially with stress declined under severe stress regimes. In contrast, spermidine levels consistently declined and spermine levels progressively increased with stress. The spermine level under stress was related with ABA and stress tolerance of cultivars. Differential response of drought stress on changes in individual polyamines is also shown by Turner and Stewart (1986). Exogenous polyamine applications have been tried in providing evidence for its role in counteracting stress. The polyamine treatments increased endogenous polyamine levels in plants under stress (Tiburico et al. 1997) and also reversed stress-induced changes in growth and cellular injuries. Spermine application is found effective in improving net photosynthesis rate and water use efficiency in wheat leaves experiencing drought (Farooq et al. 2009). In tomato, spermidine is responsive in improving drought resistance by increasing gas exchange parameters and reducing internal CO_2 concentration by preventing stomatal closure and stimulating CO_2 uptake (Zhang et al. 2010). Evidences indicate the role of polyamine in the modulation of stomata aperture, an effect similar to ABA, by targeting KAT1-like inward K^+ channel in guard cells (Liu et al. 2000). Polyamines are also implicated in plant performance in flooded soil. The accumulation in putrescine in flooded roots stimulated PM ATPase activity, which helps in cell homeostasis and nutrient acquisition (Bertini et al. 1997). Yiu et al. (2009) witnessed alleviation of flood stress in onion by putrescine through reduction in superoxide radicals and H_2O_2 (Yiu et al. 2009). Polyamines are also involved in the regulation of intracellular homeostasis under flooding (Reggiani et al. 1993). The anoxic condition under flooding tends to decrease polyamines in the absence of K^+, but supplementation of K^+ under such conditions reduces negative effect of anoxia on polyamines. Jia et al. (2010) reported that application of spermidine to cucumber roots enhances ATP production and alleviates flooding responses by enhancing aerobic respiration and declining fermentation metabolism. Shi et al. (2009) also observed that the putrescine application is efficient in the alleviation of stress-induced reduction of gas exchange variables of cucumber subjected to root-zone hypoxia by enhancing nitrate reductase activities. Likewise high temperature stress alters polyamine balance, which helps in providing thermotolerance to plant by stabilizing membrane structural integrity and functionality (Edreva et al. 1998). Such effects are due to the polycationic nature of polyamines which facilitates their strong binding to nucleic acids, proteins, and membranes (Childs et al. 2003). Thermotolerance by polyamines is also possible by regulating heat stress-induced inhibition of photosynthetic efficiency. Upreti et al. (2012) reported that high accumulation in spermidine and spermine in flower buds and flowers of capsicum cultivars is an important attribute of thermotolerance as found by lower decline in abscission of floral parts. The increased polyamine also downregulates ethylene production leading to reduction in abscission of reproductive organs under high temperature. Murkowski (2001) found that the spermidine application improves high temperature tolerance in tomato plants, by lowering thermal damage to the pigment-protein complex structure. Huang et al. (1991) witnessed high temperature-induced accumulation of putrescine at filling stage of rice which helps in thermotolerance by regulating the stress-induced decline in photosynthetic capacity, chlorophyll content, and RuBPC activity. Similarly, exogenous spermine is effective in alleviating heat-induced damage to the photosynthetic apparatus of cucumber by shielding protein complexes in thylakoid against heat damage (Li et al. 2003). Genetic modifications of the polyamine biosynthetic pathway are useful to establish the function of polyamines in plant responses to abiotic stress. The investigations on gene expression associated with polyamines under stress have been made, and reports indicate the presence of complicated transcriptional profiling (Gonzalez de Mejia et al. 2003). The mRNA of some polyamine biosynthetic genes is rapidly induced immediately after stress in some species, and in others it was induced when stress was exerted for a certain period, indicating that the polyamine genes are differentially regulated under stress (Malamberg et al. 1998). The elevated putrescine levels as a result of overexpressing ADC2 induce drought tolerance, which is related to reduction of water loss by the induction of stomata closure (Alcazar et al. 2010). The EMS mutants of *Arabidopsis thaliana* spe1-1 and spe2-1 displaying reduced ADC activity are deficient in polyamine accumulation after acclimation to high NaCl concentrations and exhibit sensitivity to salt stress (Kasinathan and Wingler 2004). The mutant, adc2-1, showing diminished putrescine content is more sensitive to salt stress, whereas exogenous addition of putrescine protects salt-induced injury in such mutant (Urano et al. 2004). Similarly, spermine-deficient mutants are sensitive to salt, while the addition of spermine suppresses the salt sensitivity, suggesting a protective role of this polyamine to high salinity (Yamaguchi et al. 2006). Alcazar et al. (2010) opined that upregulation of polyamine biosynthetic genes and accumulation of polyamines under stress are ABA-dependent responses as ABA modulates polyamine metabolism at the transcription level by upregulating the expression of ADC2, SPDS1, and SPMS genes under stress conditions. Transcript profiling also revealed that cold enhances the expression of ADC1, ADC2, and SAMDC2 genes (Cuevas et al. 2009).

2.9 Conclusions and Future Perspectives

Plant hormones are vital components of plant growth and development under abiotic stresses. The stress conditions alter their levels which help in plant adaptation through their responses on stomatal functioning, plant water balance, nutrient allocations, and source-sink transitions, besides maintaining antioxidant status. There is either an increase or decrease in the endogenous PGR levels in plants under stress conditions, and responses are cultivars, stress duration, and stress intensity dependent. While stressed plants invariably showed an increase in ABA and decrease in cytokinins, the gibberellins, auxin, ethylene, and polyamine levels show variable responses to abiotic stress factors. Significant progresses have been made over the last few years in understanding processes regulating the biosynthesis and metabolism of naturally present PGR and their associated roles in signaling mechanisms. Besides, PGR often alter gene expression by inducing or preventing the degradation of transcriptional regulators. Use of mutants with modified hormone biosynthesis pathways is helpful in unfolding mechanism of actions associated with different PGR under stress conditions. The progresses on molecular aspects of hormonal physiology have led identification of genes associated with biosynthesis of different PGR and genes encoding their receptors. And information on stress-induced manipulations in genes has been vital in establishing the role of PGR in adaptation of plant to abiotic stresses. The efforts are also made in manipulating genes for balancing plant-stress responses and in interaction of PGR and other cellular metabolites following stress acquisition. However, most of the studies on hormonal role in stress tolerance are carried out in isolation. As the PGR are interrelated at cellular level synergistic or antagonistic cross talk, and there is overlapping of various stress factors at cellular level at plant level, the mutual interactions and communications between PGR through involvement of unique set of genes are vital to plant responses to different stresses. This aspect needs due emphasis in defining involvement of PGR in plant adaptation to abiotic stresses. The success in elucidating roles of PGR in stress tolerance both at cellular and molecular levels helped in showing positive effects of exogenous application of PGR and their synthetic substitutes and of compounds capable in modifying PGR metabolism in improving stress tolerance in wide range of crop species. However, commercial benefits of such results are yet to be established because of gaps in understanding the physiological basis of their actions. Bacteria and mycorrhiza are efficient in producing PGR and have potential to manipulate their endogenous levels for eliciting growth responses in plants and inducing stress tolerance. Efforts are needed to exploit their benefits as alternate tool in the management of abiotic stresses. Further, the involvement of polyamines and brassinosteroids in the regulation of plant growth and development and stress tolerance in plants is well understood. However, more research is needed in unraveling the mechanism of their stress-protective roles, especially from the point of their interactions and interrelations with other PGR as well as with stress-responsive genes.

References

Adato I, Gazit S (1974) Water-deficit, ethylene production, and ripening in avocado fruits. Plant Physiol 53:45–46

Afzal I, Basra S, Iqbal A (2005) The effect of seed soaking with plant growth regulators on seedling vigor of wheat under salinity stress. J Stress Physiol Biochem 1:6–14

Agarwal S, Gehlot HS (2000) An update on brassinosteroids. In: Hemantranjan A (ed) Advances in plant physiology. Scientific Publishers (India), Jodhpur, pp 149–178

Akbari G, Sanavy S, Yousefzadeh S (2007) Effect of auxin and salt stress (NaCl) on seed germination of wheat cultivars (*Triticum aestivum* L.). Pak J Biol Sci 10:2557–2561

Alcazar R, Altabella T, Marco F, Bortolotti C, Reymond M, Koncz C, Carrasco P, Tiburcio AF (2010) Polyamines: molecules with regulatory functions in plant abiotic stress tolerance. Planta 231:1237–49

Ali B, Hayat S, Ahmad A (2007) 28-homobrassinolide ameliorates the saline stress in *Cicer arietinum* L. Environ Exp Bot 59:217–223

Ali RM, Abbas HM, Kamal RK (2009) The effects of treatment with polyamines on dry matter and some metabolites in salinity-stressed chamomile and sweet marjoram seedlings. Plant Soil Environ 55:477–483

Aloni B, Pressman E (1980) Interaction with salinity of GA_3-induced leaf elongation, petiole pithiness and bolting in celery. Sci Hortic 13:135–142

Amry E, Mirzaei M, Moradi M, Zare K (2011) The effects of spermidine and putrescine polyamines on growth of pomegranate (*Punica granatum* L. cv 'Rabbab') in salinity circumstance. Int J Plant Physiol Biochem 3:43–49

Anderson BE, Ward JM, Schoeder JI (1994) Evidence for extracellular reception site for abscisic acid in *Commelina* guard cell. Plant Physiol 104:1177–1183

Anjum MA (2008) Effect of NaCl concentrations in irrigation water on growth and polyamine metabolism in two citrus rootstocks with different levels of salinity tolerance. Acta Physiol Plant 30:43–52

Anjum MA (2011) Effect of exogenously applied spermidine on growth and physiology of citrus rootstock Troyer citrange under saline conditions. Turk J Agric 35:43–53

Archard P, Cheng H, De Grauwe L, Decat J, Schoutteten H, Moritz T, Van Der Straeten D, Peng J, Harberd NP (2006) Integration of plant responses to environmentally activated phytohormonal signals. Science 311:91–94

Archard P, Gong F, Cheminant S, Alioua M, Hedden P, Genschik P (2008) The cold inducible CBF1 factor-dependent signaling pathway modulates the accumulation of the growth-repressing DELLA proteins via its effect on gibberellin metabolism. Plant Cell 20:2117–2129

Asakawa S, Abe H, Nishikawa N, Natsume M, Koshioka M (1996) Purification and identification of new acyl-conjugated teasterones in lily pollen. Biosci Biotechnol Biochem 60:1416–1420

Ashraf M, Karim F, Rasul E (2002) Interactive effects of gibberellic acid (GA_3) and salt stress on growth, ion accumulation and photosynthetic capacity of two spring wheat (*Triticum aestivum* L.) cultivars differing in salt tolerance. Plant Growth Regul 36:49–59

Bailey-Serres J, Voesenek LA (2010) Life in the balance: a signaling network controlling survival of flooding. Curr Opin Plant Biol 13:489–494

Bajguz A, Hayat S (2009) Effects of brassinosteroids on the plant responses to environmental stresses. Plant Physiol Biochem 47:1–8

Bartels D, Sankar R (2005) Drought and salt tolerance in plants. Crit Rev Plant Sci 24:23–58

Bartels D, Furni A, Ingram J, Salamani F (1996) Responses of plants to dehydration stress: a molecular analysis. Plant Growth Regul 20:111–118

Bejaoui M (1985) Interaction between NaCl and some phytohormones on soybean growth. J Plant Physiol 120:95–110

Belintani G, Guerzoni S, Moreira P, Vieira E (2012) Improving low-temperature tolerance in sugarcane by expressing the ipt gene under a cold inducible promoter. Biol Plant 56(1):71–77

Beltrano J, Montaldi E, Bartoli C, Carbone A (1997) Emission of water-stress ethylene in wheat (*Triticum aestivum* L.) ears: effects of rewatering. Plant Growth Regul 21:121–126

Benson RJ, Boyer JS, Mullet JE (1988) Water deficit-induced changes in abscisic acid, growth, polysomes and translatable RNA in soybean hypocotyls. Plant Physiol 88:289–294

Ben-Yehoshua S, Aloni B (1974) Effect of water stress on ethylene production by detached leaves of Valencia orange (*Citrus sinensis* Osbeck). Plant Physiol 53:863–865

Bertini A, Brambilla I, Mapellis S, Reggiani R (1997) Elongation growth in the absence of oxygen: the rice coleoptile. Russ J Plant Physiol 44:543–547

Biddington NL, Dearman AS (1982) The effect of abscisic acid on root growth and shoot growth of cauliflower plants. Plant Growth Regul 1:15–24

Binns A (1994) Cytokinin accumulation and action: biochemical, genetic and molecular approaches. Annu Rev Plant Physiol Plant Mol Biol 45:173–196

Bishop GJ, Yokota T (2001) Plants steroid hormones, brassinosteroids: current highlights of molecular aspects on their synthesis/metabolism, transport, perception and response. Plant Cell Physiol 42:114–120

Blackman PG, Davies WJ (1983) The effects of cytokinins and ABA on stomatal behavior of maize and commelina. J Exp Bot 34:1619–1626

Blatt MR, Armstrong F (1993) K^+ channels of stomatal guard cells: abscisic acid evoked control of the outward rectifier mediated by cytoplasmic pH. Planta 191:543–567

Blatt MR, Thiel G (1993) Hormonal control of ion channel gating. Annu Rev Plant Physiol Plant Mol Biol 44:543–567

Borkird C, Simoens C, Villaroel R, Van Montagu M (1991) Gene expression associated with water stress – adaptation of rice cells and identification of two rice genes as *hsp 70* and ubiquitin. Physiol Plant 82:449–457

Boyer GL, Zeevaart JA (1982) Isolation and quantization of beta-d-glucopyranosyl abscisate from leaves of xanthium and spinach. Plant Physiol 70:227–231

Brault M, Maldiney R (1999) Mechanisms of cytokinin action. Plant Physiol Biochem 37:403–412

Bray EA (1988) Drought- and ABA-induced changes in polypeptide and mRNA accumulation in tomato leaves. Plant Physiol 88:1210–1214

Bray EA, Bailey-Serres J, Weretilnyk E (2000) Responses to abiotic stresses. In: Buchannan B, Gruissem W, Jones R (eds) Biochemistry and molecular biology of plants. American Society of Plant Biology, Rockville, MD, pp 1158–1203

Brzobohaty B, Moore I, Palme K (1994) Cytokinin metabolism: implications for regulation of plant growth and development. Plant Mol Biol 26:1483–1497

Catala R, López-Cobollo R, Mar Castellano M, Angosto T, Alonso JM, Ecker JR, Salinas J (2014) The *Arabidopsis 14*-3-3 protein RARE COLD INDUCIBLE 1A links low-temperature response and ethylene biosynthesis to regulate freezing tolerance and cold acclimation. Plant Cell 26:3326–3342

Chandler PM, Roberstson M (1994) Gene expression regulated by abscisic acid and its relation to stress tolerance. Ann Rev Physiol Plant Mol Biol 45:113–141

Chen TH, Gusta LV (1983) Abscisic acid induced freezing resistance in cultured plant cells. Plant Physiol 73:71–75

Chen CM, Melitz D (1979) Cytokinin biosynthesis in a cell free system from cytokinin autotrophic tobacco tissue cultures. FEBS Lett 107:15–20

Chen CC, Ertl JR, Lesiner SM, Chang C (1985) Localisation of cytokinin biosynthetic sites in pea plants and carrot roots. Plant Physiol 78:510–513

Cheol Park H, Cha JY, Yun DJ (2013) Roles of YUCCAs in auxin biosynthesis and drought stress responses in plants. Plant Signal Behav 8, e24495

Childs AC, Mehta DJG, Erner EW (2003) Polyamine-dependent gene expression. Cell Mol Life Sci 60:1394–1406

Ciardi JA et al (1997) Increased ethylene synthesis enhances chilling tolerance in tomato. Physiol Plant 101:333–340

Claeys H, Skirycz A, Maleux K, Inzé D (2012) DELLA signaling mediates stress-induced cell differentiation in Arabidopsis leaves through modulation of anaphase-promoting complex/cyclosome activity. Plant Physiol 159:739–747

Clouse SD, Sasse JM (1998) Brassinosteroids: essential regulators of plant growth and development. Annu Rev Plant Physiol Plant Mol Biol 49:427–445

Cohen A, Bray EA (1990) Characterization of the RNAs that accumulate in wilted tomato leaves in response to elevated levels of endogenous abscisic acid. Planta 182:27–33

Colebrook EH, Thomas SG, Phillips AL, Hedden P (2014) The role of gibberellin signalling in plant responses to abiotic stress. J Exp Biol 217:67–75

Colmer TD (2003) Long-distance transport of gases in plants: a perspective on internal aeration and radial oxygen loss from roots. Plant Cell Environ 26:17–36

Conti S, Landi P, Sanguineti MC, Stefanelli S, Tuberosa R (1994) Genetic and environmental effects of abscisic acid accumulation in leaves of field grown maize. Euphytica 78:81–89

Cowan AK, Cairns ALP, Bartels-Rahm B (1999) Regulation of abscisic acid metabolism: towards a metabolic basis of abscisic acid-cytokinin antagonism. J Exp Bot 50:595–603

Cramer GR, Jones RL (1996) Osmotic stress and abscisic acid reduce calcium activities in roots of *Arabidopsis thaliana*. Plant Cell Environ 19:1291–1298

Cuevas JC, Lopez-Cobollo R, Alcazar R, Zarza X, Koncz C, Altabella T, Salinas J, Tiburcio AF, Ferrando A (2009) Putrescine as a signal to modulate the indispensable ABA increase under cold stress. Plant Signal Behav 4:219–222

Cushman J, Bohnert HJ (2000) Genomic approaches to plant stress tolerance. Curr Opin Plant Biol 3:117–124

Cutler AJ, Krochko JE (1999) Formation and breakdown of ABA. Trends Plant Sci 4:472–478

Das KK, Sarkar RK, Ismail AM (2005) Elongation ability and non-structural carbohydrate levels in relation to submergence tolerance in rice. Plant Sci 168:131–136

Datta N, Schell MB, Roux SJ (1987) Stimulation of a nuclear NH kinase from pea plumules and its role in the phosphorylation of a nuclear Polypeptide. Plant Physiol 84:1397–1401

Davies WJ, Zhang J (1991) Root signals and the development of plants growing in drying soil. Ann Rev Plant Physiol Mol Biol 42:55–76

Delasny M, Gaubier P, Hull G, Saez-Vasquez J, Gallois P, Raynal M, Cooke R, Grellet F (1994) Nuclear genes expressed during seed desiccation: relationship with responses to stress. In: Basra AS (ed), Stress-induced gene expression in plants. Chapter 2, Harwood Academic Publication, Reading, UK, pp 25–59

Ding HD, Zhu XH, Zhu ZW, Yang SJ, Zha DS, Wu XX (2012) Amelioration of salt-induced oxidative stress in eggplant by application of 24-epibrassinolide. Biol Plant 56:767–770

Divi UK, Rahman T, Krishna P (2010) Brassinosteroids-mediated stress tolerance in Arabidopsis shows interactions with abscisic acid, ethylene and salicylic acid pathways. BMC Plant Biol 10:151

Dodd IC, Stikic R, Davies WJ (1996) Chemical regulation of gas exchange and growth of plants in drying soil in the field. J Exp Bot 47:1475–1490

Du H, Wu N, Fu J, Wang S, Li X, Xiao J, Xiong L (2012) A GH3 family member, OsGH3-2, modulates auxin and abscisic acid levels and differentially affects drought and cold tolerance in rice. J Expt Bot 63:6467–6480

Du H, Wu N, Chang Y, Li X, Xiao J, Xiong L (2013) Carotenoid deficiency impairs ABA and IAA biosynthesis and differentially affects drought and cold tolerance in rice. Plant Mol Biol 83:475–488

Duan J, Guo Li S, Kang Y (2008) Exogenous spermidine affects polyamine metabolism in salinity-stressed *Cucumis sativus* roots and enhances short-term salinity tolerance. J Plant Physiol 165:1620–1635

Dunlop JR, Binzel ML (1996) NaCl reduces indole acetic acid levels in the roots of tomato plants independent of stress induced abscisic acid. Plant Physiol 112:379–384

El-Beltagy AS, Hall MA (1974) Effect of water stress upon endogenous ethylene levels in *Vicia faba*. New Phytol 73:47–60

Edreva A, Yordanov I, Kardjieva R, Gesheva E (1998) Heat shock responses of bean plants: involvement of free radicals, antioxidants and free radical/active oxygen scavenging systems. Biol Plant 41:185–191

Fariduddin Q, Yusuf M, Ahmad I, Ahmad A (2014) Brassinosteroids and their role in response of plants to abiotic stresses. Biol Plant 58:9–17

Farooq M, Wahid A, Lee DJ (2009) Exogenously applied polyamines increase drought tolerance of rice by improving leaf water status, photosynthesis and membrane properties. Acta Physiol Plant 31:937–945

Federico R, Angelini R (1991) Polyamines catabolism in plants. In: Slocum RD, Flores HE (eds) Biochemistry and physiology of plants. CRC Press, Boca Raton, FL, 41–56

Finkelstein R, Gampala SSI, Rock CD (2002) Abscisic acid signaling in seeds and seedling. Plant Cell 14:515–545

Finkelstein R, Reeves W, Ariizumi T, Steber C (2008) Molecular aspects of seed dormancy. Ann Rev Plant Biol 59:387–415

Fowler S, Thomashow MF (2002) Arabidopsis transcriptome profiling indicates that multiple regulatory pathways are activated during cold acclimation in addition to the CBF cold response pathway. Plant Cell 14:1675–1690

Fukaki H, Fujisawa H, Tasaka M (1996) Gravitropic response of inflorescence stems in Arabidopsis thaliana. Plant Physiol 110:933–943

Fusseder A, Wartinger A, Hartung W, Schulze ED, Heilmeier H (1992) Cytokinins in the xylem sap of desert grown almonds (*Prunus dulcis*) trees: daily courses and their possible interactions with abscisic acid and leaf conductance. New Physiol 122:45–52

Galen C, Rabenold JJ, Liscum E (2007) Functional ecology of a blue light photoreceptor: effects of phototropin-1 on root growth enhance drought tolerance in *Arabidopsis thaliana*. New Phytol 173:91–99

Galston AN, Kaur-Sawhney R, Altabella T, Tiburico AF (1997) Plant polyamines in reproductive activity and response to abiotic stress. Bot Acta 110:197–207

Galvan-Ampudia CS, Testerink C (2011) Salt stress signals shape the plant root. Curr Opin Plant Biol 14:296–302

Gashaw A (2014) Manipulation of endogenous cytokinin with ipt gene confers tolerance to multiple abiotic stresses. Int J Agron Plant Prod 5:44–51

George AP, Nissen RJ (1992) Effects of water stress, nitrogen and paclobutrazol on flowering, yield and fruit quality of the low-chill peach cultivar Flordaprince. Sci Hortic 49:197–209

Gibbs DJ, Lee SC, Isa NM, Gramuglia S, Fukao T, Bassel GW, Correia CS, Corbineau F, Theodoulou FL, Bailey-Serres J, Holdsworth MJ (2011) Homeostatic response to hypoxia is regulated by the N-end rule pathway in plants. Nature 479:415–418

Gill SS, Tuteja N (2010) Polyamines and abiotic stress tolerance in plants. Plant Signal Behav 5:26–33

Gilley A, Fletcher RA (1998) Gibberellin antagonizes paclobutrazol-induced stress protection in wheat seedlings. Plant Physiol 103:200–207

Gilmour SJ, Fowler SG, Thomashow MF (2004) Arabidopsis transcriptional activators CBF1, CBF2, and CBF3 have matching functional activities. Plant Mol Biol 54:767–781

Glinka Z, Reinhold L (1971) Abscisic acid raises the permeability of plant cell to water. Plant Physiol 48:103–105

Goiocchea N, Dolczal K, Antolin MC, Strnad M, Sanchez-Diaz M (1995) Influence of mycorrhizae and *Rhizobium* on cytokinin content in drought-stressed alfalfa. J Exp Bot 46:1543–1549

Gonai T, Kawahara S, Tougou M, Satoh S, Hashiba T, Hirai N, Kawaide H, Kamiya Y, Yoshioka T (2004) Abscisic acid in the thermo inhibition of lettuce seed germination and enhancement of its catabolism by gibberellin. J Exp Bot 55:111–118

Gonzalez de Mejia E, Martinez Resendiz V, Castano Tostado E, Loarca Pina G (2003) Effect of drought on polyamine metabolism, yield, protein content and in vitro protein digestibility in tepary (Phaseolus acutifolius) and common (Phaseolus vulgaris) bean seeds. J Sci Food Agric 83:1022–1030

Guinn G (1976) Water deficit and ethylene evolution by young cotton bolls. Plant Physiol 57:403–405

Gulnaz A, Iqbal J, Azam F (1999) Seed treatment with growth regulators and crop productivity. II. Response of critical growth stages of wheat (*Triticum aestivum* L.) under salinity stress. Cereal Res 27:419–426

Gupta K, Dey A, Gupta B (2013) Plant polyamines in abiotic stress responses. Acta Physiol Plant 35:2015–2036

Habben JE, Bao X, Bate NJ, DeBruin JL, Dolan D, Hasegawa D, Helentjaris TG, Lafitte RH, Lovan N, Mo H, Reimann K, Schussler JR (2014) Transgenic alteration of ethylene biosynthesis increases grain yield in maize under field drought-stress conditions. Plant Biotechnol J 12:685–693

Halevy AH, Kessler B (1963) Increased tolerance of bean plants to soil drought by means of growth-retarding substances. Nature 197:310–311

Harris MJ, Outlaw WH Jr (1991) Rapid adjustment of guard-cell abscisic acid level to current leaf water status. Plant Physiol 95:171–173

Hattori Y, Nagai K, Furukawa S, Song XJ, Kawano R, Sakakibara H, Matsumoto T, Yoshimura A, Kitano H, Matsuoka M, Mori H, Ashikari M (2009) The ethylene response factors SNORKEL1 and SNORKEL2 allow rice to adapt to deep water. Nature 460:1026–1030

Hayat Q, Hayat Iran M, Ahmad A (2010) Effect of exogenous salicylic acid under changing environment: a review. Environ Exp Bot 68:14–25

Hedden P, Thomas GS (2012) Gibberellin biosynthesis and its regulation. Biochem J 444:11–25

Hetherington AM (2001) Guard cell signaling. Cell 107:711–714

Hetherington AM, Quatrano RS (1991) Mechanisms of action of abscisic acid at the cellular level. New Phytol 119:9–32

Hinz M, Wilson IW, Yang J, Buerstenbinder K, Llewellyn D, Dennis ES, Sauter M, Dolferus R (2010) Arabidopsis RAP2.2: an ethylene response transcription factor that is important for hypoxia survival. Plant Physiol 153:757–772

Hoffman NE, Fu JR, Yang SF (1983) Identification and metabolism of 1-(malonylamino) cyclopropane-1-carboxylic acid in germinating peanut seeds. Plant Physiol 71:197–199

Hossain MA, Lee Y, Cho JI, Ahn CH, Lee SK, Jeon JS, Kang H, Lee C-H, An G, Park PB (2010) The bZIP transcription factor OsABF1 is an ABA responsive element binding factor that enhances abiotic stress signaling in rice. Plant Mol Biol 72:557–560

Hoth S, Morgante M, Sanchez JP, Hamfley MK, Tingey SV, Chua NH (2002) Genome wide gene expression profiling in *Arabidopsis thaliana* reveals new targets of abscisic acid and largely impaired gene regulation in the abi t-1 mutant. Cell Sci 115:489–490

Hu X, Neill ST, Tang Z, Cai W (2005) Nitric oxide mediates gravitropic bending in soybean roots. Plant Physiol 137:663–670

Huai J, Wang M, Zheng J, Dong ZL, Zhao J, Wang G (2008) Cloning and characterization of the SnRK2 gene family from Zea mays. Plant Cell Rep 12:1861–1870

Huang YJ, Luo YF, Huang XZ, Rao ZM, Liu YB (1991) Varietal difference of heat tolerance at grain filling stage and its relationship to photosynthetic characteristics and endogenous polyamine of flag leaf in rice. China J Rice Sci 13:205–210

Huang LF, Zheng JH, Zhang YY, Hu WH, Mao WH, Zhou YH, Yu JQ (2006) Diurnal variations in gas exchange, chlorophyll fluorescence quenching and light allocation in soybean leaves: the cause for midday depression in CO_2 assimilation. Sci Hortic 110:214–218

Hussain SS, Kayani MA, Amjad M (2011) Transcription factors as tools to engineer enhanced drought tolerance in plants. Biotechnol Prog 27:297–306

Incoll LD, Jewer PC (1987) Cytokinins and the water relations of whole plants. In: Cytokinins – plant hormones in search of a role. Monograph 14. British Plant Growth Regulator Group, pp 85–97

Incoll LD, Ray JP, Jewer PC (1990) Do cytokinins act as root to shoot signals? In: Davies WJ, Jeffcoat B (eds) Importance of root to shoot communication in the response to environmental stress. British Society for Plant Growth Regulators, Bristol, pp 185–197

Ingram J, Bartels D (1996) The molecular basis of dehydration tolerance in plants. Annu Rev Plant Physiol Plant Mol Biol 47:377–403

Iqbal M, Ashraf M (2013) Gibberellic acid mediated induction of salt tolerance in wheat plants: growth, ionic partitioning, photosynthesis, yield and hormonal homeostasis. Environ Exp Bot. 86:76–85

Irigoyen JJ, Emerich DW, Sánchez-Díaz M (1992) Alfalfa leaf senescence induced by drought stress: photosynthesis, hydrogen peroxide metabolism, lipid peroxidation and ethylene evolution. Physiol Plant 84:67–72

Irving HR, Gehring CA, Parish RW (1992) Changes in cytosolic pH and calcium of guard cells precedes stomatal movement. Proc Natl Acad Sci U S A 89:1790–1794

Ismail AM, Hall AE, Close TJ (1999) Purification and partial characterization of a dehydrin involved in chilling tolerance during seedling emergence of cowpea. Plant Physiol 120:237–244

Jackson MB (1993) Are plant hormones involved in root to shoot communication? Adv Bot Res 19:104–187

Jain M, Khurana JP (2009) Transcript profiling reveals diverse roles of auxin-responsive genes during reproductive development and abiotic stress in rice. FEBS J 276:3148–3162

Janeczko A, Gullner G, Skoczowski A, Dubert F, Barna B (2007) Effects of brassinosteroid infiltration prior to cold treatment on ion leakage and pigment contents in rape leaves. Biol Plant 51:355–358

Javid MG, Sorooshzadeh A, Moradi F, Modarres Sanavy SAM, Allahdadi I (2011) The role of phytohormones in alleviating salt stress in crop plants. Aust J Crop Sci 5:726–734

Jeon J, Kim NY, Kim S, Kang NY, Novák O, Ku SJ, Cho C, Lee DJ, Lee EJ, Strnad M, Kim J (2010) A subset of cytokinin two-component signaling system plays a role in cold temperature stress response in Arabidopsis. J Biol Chem 285:23371–23386

Jia YX, Sun J, Guo SR, Li J, Hu XH, Wang SP (2010) Effect of applied spermidine on growth and respiratory metabolism in roots of cucumber (*Cucumis sativa*) seedlings under hypoxia. Russ J Plant Physiol 57:648–655

Jia W, Zhang J (1997) Comparison of exportation and metabolism of xylem-delivered ABA in maize leaves at different water status and xylem sap pH. Plant Growth Regul 21:43–49

Jiang M, Zhang (2002) Water stress-induced abscisic acid accumulation triggers the increased generation of reactive oxygen species and up-regulates the activities of antioxidant enzymes in maize leaves. J Exp Bot 53:2401–2410

Jiang C, Belfield EJ, Cao Y, Smith JA, Harberd NP (2013) An Arabidopsis soil-salinity-tolerance mutation confers ethylene-mediated enhancement of sodium/potassium homeostasis. Plant Cell 25:3535–3552

Jung JH, Park CM (2011) Auxin modulation of salt stress signaling in Arabidopsis seed germination. Plant Signal Behav 6(8):1198–1200

Kagale S, Divi UK, Krochko JE, Keller WA, Krishna P (2007) Brassinosteroids confers tolerance in *Arabidopsis thaliana* and *Brassica napus* to a range of abiotic stresses. Planta 225:353–364

Kamiab F, Talaie A, Khezri M, Javanshah A (2014) Exogenous application of free polyamines enhance salt tolerance of pistachio (*Pistacia vera* L.) seedlings. Plant Growth Regul 72:257–268

Kang NY, Cho C, Kim NY, Kim J (2012) Cytokinin receptor-dependent and receptor-independent pathways in the dehydration response of *Arabidopsis thaliana*. J Plant Physiol 169:1382–1391

Kasahara H, Hanada A, Kuzuyama T, Takagi M, Kamiya Y, Yamaguchi S (2002) Contribution of the mevalonate and methylerythritol phosphate pathways to the biosynthesis of gibberellins in Arabidopsis. J Biol Chem 277:45188–45194

Kasinathan V, Wingler A (2004) Effect of reduced arginine decarboxylase activity on salt tolerance and on polyamine formation during salt stress in *Arabidopsis thaliana*. Physiol Plant 121:101–107

Kasuga M, Liu Q, Miura S, Yamaguchi-Shinozaki K, Shinozaki K (1999) Improving plant drought, salt, and freezing tolerance by gene transfer of a single stress-inducible transcription factor. Nat Biotechnol 17:287–291

Kazan K (2013) Auxin and the integration of environmental signals into plant root development. Ann Bot 112:1655–1665

Kende H (1993) Ethylene biosynthesis. Annu Rev Plant Physiol Plant Mol Biol 44:283–307

Keskin BC, Saikaya AT, Yuksel B, Memoh AR (2010) Abscisic acid regulated gene expression in bread wheat. Aust J Crop Sci 4:617–625

Khodakovskaya M (2005) Effects of cor15a-IPT gene expression on leaf senescence in transgenic Petunia x hybrida and Dendranthema x grandiflorum. J Exp Bot 56:1165–1175

Kim TH, Bohmer M, Nishilmura N, Schorder JI (2010a) Guard cell signal transduction network: advances in understanding abscisic acid, CO_2 and Ca^{+2} signalling. Ann Rev Plant Biol 61:561–569

Kim JS, Shim S, Kim MJ (2010b) Physiological response of Chinese cabbage to salt stress. Kor J Hortic Sci Technol 28:343–352

Kim JI, Baek D, Park HC, Chun HJ, Oh DH, Lee MK, Cha JY, Kim WY, Kim MC, Chung WS, Bohnert HJ, Lee SY, Bressan RA, Lee SW, Yun DJ (2013) Overexpression of Arabidopsis YUCCA6 in potato results in high-auxin developmental phenotypes and enhanced resistance to water deficit. Mol Plant 6:337–349

King SW, Joshi CP, Nguyen HT (1992) DNA sequence of an ABA-responsive gene (rab 15) from water-stressed wheat roots. Plant Mol Biol 18:119–121

Kinoshita N, Wang H, Kasahara H, Liu J, MacPherson C, Machida Y, Kamiya Y, Matthew A, Hannah MA, Chua N (2012) IAA-Ala Resistant3, an evolutionarily conserved target of miR167, mediates arabidopsis root architecture changes during high osmotic stress. Plant Cell 24:3590–3602

Klay I, Pirrello L, Riahi A, Bernadac A, Cherif M, Bouzayen S (2014) Ethylene response factor Sl-ERF. B.3 is responsive to abiotic stresses and mediates salt and cold stress response regulation in tomato. Sci World J. doi:10.1155/2014/167681

Kovtun Y, Chiu WL, Tena G, Sheen J (2000) Functional analysis of oxidative stress-activated mitogen-activated protein kinase cascade in plants. Proc Natl Acad Sci U S A 97:2940–2945

Krishnamurthy R, Bhagwat KA (1989) Polyamines as modulators of salt tolerance in rice cultivars. Plant Physiol 91:500–504

Kumar B, Singh B (1996) Effect of plant hormones on growth and yield of wheat irrigated with saline water. Ann Agron Res 17:209–212

Kumar A, Altabella T, Taylor MA, Tiburcio AF (1997) Recent advances in polyamine research. Trends Plant Sci 2:124–130

La Rosa PC, Handa AK, Hasegawa PM, Bressan RA (1985) Hormonal regulation of protein synthesis associated with salt tolerance in plant cells. Plant Physiol 79:138–142

La Rosa PC, Hasigawa PM, Rhodes D, Clithero JM, Watad AEA, Bressan R (1987) Abscisic acid stimulated osmotic adjustment and its involvement in adaptation of tobacco cells to NaCl. Plant Physiol 85:174–181

Lakra N, Mishra SN, Singh DB, Tomar PC (2006) Exogenous putrescine effect on cation concentration in leaf of *Brassica juncea* seedlings subjected to Cd and Pb along with salinity stress. J Environ Biol 27:263–269

Lee J, Hong S, Lee D, Choi S, Woo TK (2004) The ethylene-responsive factor like protein 1 (CaERFLP1) of hot pepper (*Capsicum annuum L.*) interacts in vitro with both GCC and DRE/CRT sequences with different binding affinities: possible biological roles of CaERFLP1 in response to pathogen infection and high salinity conditions in transgenic tobacco plants. Plant Mol Biol 55:61–81

Li J, Wang XQ, Watson MB, Assmann SM (2000) Regulation of abscisic acid-induced stomatal closure and anion channels by guard cell AAPK kinase. Science 287:300–310

Li ZJ, Nada K, Tachibana S (2003) High-temperature-induced alteration of ABA and polyamine contents in leaves and its implication in thermal acclimation of photosynthesis in cucumber (*Cucumis sativus* L.). J Japan Soc Hortic Sci 72:393–401

Li S, Xu C, Yang Y, Xia G (2010) Functional analysis of TaDi19A, a salt-responsive gene in wheat. Plant Cell Environ 33:117–129

Li YH, Liu YJ, Xu XL, Jin M, An LZ, Zhang H (2012) Effect of 24-epibrassinolide on drought stress-induced changes in *Chorispora bungeana*. Biol Plant 56:192–196

Licausi F, Dongen JT, Giuntoli B, Novi G, Santaniello A, Geigenberger P, Perata P (2010) HRE1 and HRE2, two hypoxia-inducible ethylene response factors, affect anaerobic responses in *Arabidopsis thaliana*. Plant J 62:302–315

Licausi F, Kosmacz M, Weits DA, Giuntoli B, Giorgi FM, Voesenek LA, Perata P, van Dongen JT (2011) Oxygen sensing in plants is mediated by an N-end rule pathway for protein destabilization. Nature 479:419–22

Liu Q, Kasuga M, Sakuma Y, Abe H, Miura S, Goda H, Shimada Y, Yoshida S, Shinozaki K, Yamaguchi-Shinozaki K (1998) Two transcription factors, DREB1 and DREB2, with an EREBP/AP2 DNA binding domain separate two cellular signal transduction pathways in drought- and low-temperature-responsive gene expression, respectively, in Arabidopsis. Plant Cell 10:391–406

Liu K, Fu H, Bei Q, Luan S (2000) Inward potassium channel in guard cells as a target for polyamines regulation of stomatal movements. Plant Physiol 124:1315–1326

Liu J, Yu BJ, Liu YL (2006) Effects of spermidine and spermine levels on salt tolerance associated with tonoplast H^+-TPase and H^+-PPase activities in barley roots. Plant Growth Regul 49:119–126

Liu J-H, Kitashiba H, Wang J, Ban Y, Moriguchi T (2007) Polyamines and their ability to provide environmental stress tolerance to plants. Plant Biotechnol 24:117–126

Lockhart J (2013) Salt of the earth: ethylene promotes salt tolerance by enhancing Na/K homeostasis. Plant Cell 25:3150

Luan S (2002) Signaling drought in guard cell. Plant Cell Environ 5:229–237

Luo M, Liu JH, Mahapatra S, Hiu RD, Mahapatra SS (1992) Characterization of a gene family encoding abscisic acid- and environmental stress-inducible proteins of alfalfa. J Biol Chem 267:432–436

Machacckova I, Hanisova A, Krekule J (1989) Levels of ethylene, ACC, MACC, ABA and proline as indicators of cold hardening and frost resistance in winter wheat. Physiol Plant 76:603–607

MacRobbie EAC (1991) Effect of ABA on ion transport and stomatal regulation. In: Davies WJ, Jones HG (eds) Abscisic acid: physiology and biochemistry, Environ Plant Biol Series BIOS. Scientific Publishers, Oxford, 63–79

Maggio A, Barbieri G, De Raimondi G, Pascale S (2010) Contrasting effects of GA_3 treatments on tomato plants exposed to increasing salinity. J Plant Growth Regul 29:63–72

Magome H, Yamaguchi S, Hanada A, Kamiya Y, Oda K (2008) The DDF1 transcriptional activator upregulates expression of a gibberellin-deactivating gene, GA2ox7, under high-salinity stress in Arabidopsis. Plant J 56:613–626

Malamberg RL, Waston MB, Galton GL, Yu W (1998) Molecular genetic analyses of plant polyamines. Crit Rev Plant Sci 17:199–224

Marcinska I, Czyczyło-Mysza E, Skrzypek M, Grzesiak F, Janowiak M, Filek M, Dziurka K, Dziurka P, Waligórski K, Juzoń K, Cyganek I, Grzesiak S (2013) Alleviation of osmotic stress effects by exogenous application of salicylic or abscisic acid on wheat seedlings. Int J Mol Sci 14:13171–13193

Martin-Tanguy J (2001) Metabolism and function of polyamines in plants: recent development (new approaches). Plant Growth Regul 34:135–148

Masia A, Pitacco A, Braggio L, Giulivo C (1994) Hormonal response to partial root drying of the root system of *Helianthus annuus*. J Exp Bot 45:69–76

McCabe S, Garratt C, Schepers F, Jordi M, Stoopen M, Davelaar E, van Rhijn JH, Power JB, Davey MR (2001) Effects of PSAG12-IPT gene expression on development and senescence in transgenic lettuce. Plant Physiol 127:505–516

Mehrotra R, Bhalothia P, Bansal P, Basantani MK, Bharti V, Mehrotra S (2014) Abscisic acid and abiotic stress tolerance-different tiers of regulation. J Plant Physiol 171:486–496

Merewitz EB, Gianfagna T, Huang B (2011) Protein accumulation in leaves and roots associated with improved drought tolerance in creeping bentgrass expressing an ipt gene for cytokinin synthesis. J Exp Bot 62:5311–5333

Milborrow BV (2001) The pathway of biosynthesis of abscisic acid in vascular plants: a review of the present state of knowledge of ABA biosynthesis. J Exp Bot 52:1145–1164

Min XJ, Bartholomew DP (2005) Effects of flooding and drought on ethylene metabolism, titratable acidity and fruiting of pineapple. Acta Hortic 666:135–148

Mok DW, Mok MC (2001) Cytokinin metabolism and action. Annu Rev Plant Physiol Plant Mol Biol 52:89–118

Moore R, Clark WD, Voodpich DS (1998) Botany. WBC/McGraw-Hill, New York

Muday GK, Rahman A, Binder BM (2012) Auxin and ethylene: collaborators or competitors? Trends Plant Sci 17:181–195

Munns R, Crammer GR (1996) Is coordination of leaf and root growth mediated by abscisic acid? Opinion Plant Soil 185:33–49

Murkowski A (2001) Heat stress and spermidine: effect on chlorophyll fluorescence in tomato plants. Biol Plant 44:53–57

Murti GSR, Upreti KK (2000) Plant hormones. In: Hemantaranjan A (ed) Advances in plant physiology, vol 3. Scientific Publishers, Jodhpur, pp 109–148

Mutlu F, Bozcuk S (2007) Salinity-induced changes of free and bound polyamine levels in sunflower (*Helianthus annuus* L.) roots differing in salt tolerance. Pak J Bot 39:1097–1102

Nambara E, Marion-Poll A (2005) Abscisic acid biosynthesis and catabolism. Ann Rev Plant Biol 56:165–185

Naqvi SSM (1999) Plant hormones and stress phenomenon. In: Pessarakli M (ed) Handbook of plant and crop stress. Marcel-Dekker, New York, pp 709–730

Narayana IS, Lalonde, Saini HS (1991) Water-stress-induced ethylene production in wheat: a fact or artifact? Plant Physiol 96(2):406–410

Navarro A, Sanchez-Blanco MJ, Banon S (2007) Influence of paclobutrazol on water consumption and plant performance of *Arbutus unedo* seedlings. Sci Hortic 111:133–139

Naylor AW (1972) Water deficits and nitrogen metabolism. In: Kozlowski TT (ed) Water deficits and plant growth, vol 3. Academic, New York, pp 241–254

Nayyar H, Kaur S, Smita SKJ, Dhir KK, Bains T (2005) Water-stress induced injury to reproductive phase in chick pea: Evaluation of stress sensitivity in wild and cultivated species in relation to abscisic acid and polyamines. Agron Crop Sci 191:450–457

Neill S, Barros R, Bright J, Desikan R, Hancock J, Harrison J, Morris P, Ribeiro D, Wilson I (2008) Nitric oxide, stomatal closure, and abiotic stress. J Exp Bot 59:165–176

Nishimura N, Sarkeshik A, Nito K, Park S, Wang A, Carvalho PC, Lee S, Caddell DF, Cutler SR, Chory J, Yates JR, Schroeder JI (2010) PYR/PYL/RCAR family members are major in-vivo ABI1 protein phosphatase 2C-interacting proteins in Arabidopsis. Plant J 61:290–300

Nishiyama R, Watanabe Y, Fujita Y, Le DT, Kojima M, Werner T, Vankova R, Yamaguchi-Shinozaki K, Shinozaki K, Kakimoto T, Sakakibara H, Schmulling T, Tran LS (2011) Analysis of cytokinin mutants and regulation of cytokinin metabolic genes reveals important regulatory roles of cytokinins in drought, salt and abscisic acid responses, and abscisic acid biosynthesis. Plant Cell 23:2169–2183

Ogweno JO, Song XS, Shi K, Hu WH, Mao WH, Zhou YH, Yu JQ, Nogues S (2008) Brassinosteroids alleviate heat-induced inhibition of photosynthesis by increasing carboxylation efficiency and enhancing antioxidant systems in *Lycopersicon esculentum*. J Plant Growth Regul 27:49–57

Orellana S, Yanez M, Espinoza A, Verdugo I, Gonzalez E, Lara SR, Casaretto JA (2010) The transcription factor SlAREB1 confers drought, salt stress tolerance and regulates biotic and abiotic stress-related genes in tomato. Plant Cell Environ 33:2191–2200

Ouvrard O, Cellier F, Ferrare K, Tousch D, Lamaze T, Dupuis JM, Casse Delbart F (1996) Identification and expression of water stress and abscisic acid-regulated genes in a drought tolerant sunflower genotype. Plant Mol Biol 31:819–829

Ozdemir F, Bor M, Demiral T, Turkan I (2004) Effects of 24-epibrassinolide on seed germination, seedling growth, lipid peroxidation, proline content and antioxidative system of rice (*Oryza sativa* L.) under salinity stress. Plant Growth Regul 42:203–211

Pareek A, Singla SL, Grover A (1998) Protein alterations associated with salinity, desiccation, high and low temperature stresses and abscisic acid application in seedlings of Pusa 169, a high yielding rice (*Oryza sativa* L.) cultivar. Curr Sci 75:1023–1035

Passioura JB, Stirzaker RJ (1993) Feed forward responses of plants to physically inhospitable soil. Crop Science Society of America, Madison, WI, USA, pp 715–719

Pekic S, Stikic R, Tomljanovic L, Andjelkovis V, Quarrie SA (1995) Characterization of maize lines differing in leaf abscisic content in the field Abscisic acid physiology. Ann Bot 75:67–73

Peleg Z, Reguera M, Tumimbang E, Walia H, Blumwald E (2011) Cytokinin-mediated source/sink modifications improve drought tolerance and increase grain yield in rice under water-stress. Plant Biotechnol J 9:747–758

Piatkowiski D, Schneider K, Salamani F, Bartles D (1990) Characterization of five abscisic-acid-responsive cDNA clones isolated from the desiccation-tolerant plant *Craterostigma plantagineum* and their relation to other water stress genes. Plant Physiol 94:1682–1688

Pillay I, Beyl C (1990) Early responses of drought resistant and susceptible tomato plants subjected to water stress. Plant Growth Regul 9:213–219

Pospisilova J, Dodd IC (2005) Role of plant growth regulators in stomatal limitation to photosynthesis during water stress. In: Pessarakli M (ed) Hand book of photosynthesis, IInd edn. Taylor and Francis, New York, pp 811–825

Pospisilova J, Synkova H, Rulcova J (2000) Cytokinin and water stress. Biol Plant 43:321–328

Potters G, Pasternak TP, Guisez Y, Jansen MAK (2009) Different stresses, similar morphogenic responses: integrating a plethora of pathways. Plant Cell Environ 32:158–169

Prakash L, Prathapasenan G (1990) NaCl and gibberellic acid induced changes in the content of auxin, the activity of cellulose and pectin lyase during leaf growth in rice (Oryza sativa). Ann Bot 365:251–257

Rademacher W (2000) Growth retardants: effects on gibberellin biosynthesis and other metabolic pathways. Annu Rev Plant Physiol Plant Mol Biol 51:501–531

Radi A, Lange T, Niki T, Koshioka M, Lange MJ (2006) Ectopic expression of pumpkin gibberellin oxidases alters gibberellins biosynthesis and development of transgenic Arabidopsis plants. Plant Physiol 140:528–536

Rajasekaran LR, Blake TJ (1999) New plant growth regulators protect photosynthesis and enhance growth under drought of jack pine seedlings. J Plant Growth Regul 18:175–181

Reggiani R, Aurisano N, Mattana M, Bertini A (1993) Effect of K^+ ions on polyamine levels in wheat seedlings under anoxia. J Plant Physiol 142:94–98

Reggiani R, Spartak B, Bertini A (1994) Changes in polyamines metabolism in seedlings of three wheat (*Triticum aestivum* L.) cultivars differing in salt sensitivity. Plant Sci 102:121–126

Rivero M, Kojima M, Gepstein A, Sakakibara H, Mittler R, Gepstein S, Blumwald E (2007) Delayed leaf senescence induces extreme drought tolerance in a flowering plant. Proc Natl Acad Sci U S A 104:19631–19636

Robertson AJ, Gusta LV, Reany MJT, Ischigawa M (1987) Protein synthesis in bromegrass (*Bromus inermis* Leyss) cultured cells during the induction of frost tolerance by abscisic acid or low temperature. Plant Physiol 84:1331–1336

Saab IN, Sharp RE, Pritchard J, Voetberg GS (1990) Increased endogenous abscisic acid maintains primary root growth and inhibits shoot growth of maize seedlings at low water potentials. Plant Physiol 93:1329–1336

Saez A, Apostolova N, Gonzalez-Guzman M, Gonzalez-Garcia MP, Nicolas C, Lorenzo O, Rodriguez PL (2004) Gain-of-function and loss-of-function phenotypes of the protein phosphatase 2C HAB1 reveal its role as a negative regulator of abscisic acid signalling. Plant J 37:354–369

Sakhabutdinova AR, Fatkhutdinova DR, Bezrukova MV and Shakirova FM (2003) Salicylic acid prevents the damaging action of stress factors on wheat plants. Bulg J Plant Physiol (Special issue 2013) 314–319

Sakuma Y, Maruyama K, Osakabe Y, Qin F, Seki M, Shinozaki K, Yamaguchi-Shinozaki K (2006) Functional analysis of an *Arabidopsis* transcription factor, DREB2A, involved in drought-responsive gene expression. Plant Cell 18:1292–1309

Sankhla N, Sankhla D, Upadhayaya A, Davis TD (1989) Amelioration of drought and high temperature injury in fruits of ber by paclobutrazol. Acta Hortic 239:197–202

Sasse JM (1997) Recent progress in brassinosteriod research. Physiol Plant 100:696–701

Satisha J, Prakash GS, Murti GSR, Upreti KK (2005) Response of grape genotypes to water deficit: root, shoot growth and endogenous hormones. Indian J Plant Physiol 10:225–230

Schweighofer A, Hirt H, Meskiene I (2004) Plant PP2C phosphatases: emerging functions in stress signaling. Trends Plant Sci 9:236–240

Seki M, Ishida J, Narusaka M, Fujita M, Nanjo T, Umezawa T, Kamiya A, Nakajima M, Enju A, Sakurai T, Satou M, Akiyama K, Yamaguchi-

Shinozaki K, Carninci P, Kawai J, Hayashizaki Y, Shinozaki K (2002) Monitoring the expression pattern of ca 7000. Arabidopsis genes under ABA treatment using full length cDNA microarray. Funct Integr Genomics 2:282–291

Shi K, Gu M, Yu HJ, Jiang YP, Zhou YH, Yu J (2009) Physiological mechanism of putrescine enhancement of root-zone hypoxia tolerance in cucumber plants. Sci Agric Sin 42:1854–1858

Shi Y, Tian S, Hou L, Huang X, Zhang X, Guo H, Yang S (2012) Ethylene signaling negatively regulates freezing tolerance by repressing expression of CBF and type-A ARR genes in Arabidopsis. Plant Cell 24:2578–2595

Shibasaki K, Uemura M, Tsurumi S, Rahman A (2009) Auxin response in Arabidopsis under cold stress: underlying molecular mechanisms. Plant Cell 21:3823–3838. doi:10.1105/tpc.109.069906

Shinozaki K, Yamaguchi-Shinozaki K (1997) Gene expression and signal transduction in water-stress response. Plant Physiol 115:327–334

Shomeili M, Nabipour M, Meskarbashee M, Memari HR (2011) Effects of gibberellic acid on sugarcane plants exposed to salinity under a hydroponic system. African J Plant Sci 5:609–616

Shu S, Yuan LY, Guo SR, Sun J, Liu CJ (2012) Effect of exogenous spermidine on photosynthesis, xanthophyll cycle and endogenous polyamines in cucumber seedlings exposed to salinity. Afr J Biol 11:6064–6074

Singh I, Shono M (2005) Physiological and molecular effects of 24-epibrassinolide, a brassinosteroid on thermotolerance of tomato. Plant Growth Regul 47:111–119

Skoog F, Armstrong DJ (1970) Cytokinins. Ann Rev Plant Physiol 21:359–384

Skriver K, Mundy J (1990) Gene expression in response to abscisic acid and osmotic stress. Plant Cell 2:503–512

Sobeih WY, Dodd IC, Bacon MA, Grierson D, Davies WJ (2004) Long-distance signals regulating stomatal conductance and leaf growth in tomato (*Lycopersicon esculentum*) plants subjected to partial root-zone drying. J Exp Bot 55:2353–2363

Sponsel VM, Hedden P (2004) Gibberellin biosynthesis and inactivation. In: Davies PJ (ed) Plant hormones: biosynthesis, signal transduction, action. Kluwer Academic Publishers, Dordrecht, pp 63–94

Starck Z, Kozinska M (1980) Effect of phytohormones on absorption and distribution of ions in the salt stressed bean plants. Acta Soc Bot Pol 49:111–125

Stern RA, Naor A, Bar N, Gozits S, Bravdo BA (2003) Xylem sap zeatin-riboside and dihydrozeatin riboside levels in relation to plant and soil water status and flowering in 'Mauritius' lychee. Sci Hortic 98:285–291

Still JR, Pill WG (2003) Germination, emergence and seedlings growth of tomato and impatiens in response to seed treatment with paclobutrazol. Hortic Sci 38:1201–1204

Stoll M, Loveys B, Dry P (2000) Hormonal changes induced by partial root-zone drying of irrigated grapevine. J Exp Bot 51:1627–1634

Sun F, Zhang W, Hu H, Li B, Wang Y, Zhao Y, Li K, Liu M, Li X (2008) Salt modulates gravity signaling pathway to regulate growth direction of primary roots in Arabidopsis. Plant Physiol 146:178–188

Swietlik D, Miller SS (1983) The effect of paclobutrazol on growth and response to water stress of apple seedlings. J Am Soc Hortic Sci 108:1076–1080

Taniguchi YY, Taniguchi M, Tsuge T, Oka A, Aoyama T (2010) Involvement of *Arabidopsis thaliana* phospholipase Dzeta 2 in root hydrotropism through the suppression of root gravitropism. Planta 231:491–497

Tanou G, Ziogas V, Belghazi M, Christou A, Filippou P, Job D, Fotopoulos V, Molassiotis A (2014) Polyamines reprogram oxidative and nitrosative status and the proteome of citrus plants exposed to salinity stress. Plant Cell Environ 37(4):864–885

Tardieu F, Simonneau T (1998) Variability among species of stomatal control under fluctuating soil water status and evaporative demand: Modelling isohydric and anisohydric behaviours. J Exp Bot 49:419–432

Thiel G, MacRobbie EAC, Blatt MR (1992) Membrane transport in stomatal guard cells: the importance of voltage control. J Membr Biol 126:1–18

Thomas T, Gunnia UB, Yurkow EJ, Seibold JR, Thomas TJ (1993) Inhibition of calcium signalling in murine spilenocytes by polyamines: differential effects of CD4 and CD8 T-cells. Biochem J 291:375–381

Thompson AJ, Barry JA, Mulholland J, McKee JMT, Howard W, Hilton HW, Horridge JS, Farquhar GD, Smeeton RC, Smillie IRA, Black CR, Taylor IB (2007) Overproduction of abscisic acid in tomato increases transpiration efficiency and root hydraulic conductivity and influences leaf expansion. Plant Physiol 143:1905–1917

Tiburico AF, Altabelllana T, Borrell Masgrau C (1997) Polyamines metabolism and its regulation. Physiol Plant 100:664–674

Trujillo LE, Sotolongo M, Menendez C, Ochogavia ME, Coll Y, Hernandez I, Borras-Hidalgo O, Thomma BP, Vera P, Hernandez L (2008) SodERF3, a novel sugarcane ethylene responsive factor (ERF), enhances salt and drought tolerance when overexpressed in tobacco plants. Plant Cell Physiol 49:512–525

Turner LB, Stewart GR (1986) The effect of water stress upon polyamine levels in barley (*Hordeum vulgare* L.). J Expt Bot 37:170–177

Upreti KK, Murti GSR (1999) Water stress induced changes in root nodulation and cytokinin levels in French Bean. Ind J Plant Physiol 26:1–4

Upreti KK, Murti GSR (2004a) Changes in electrolyte leakage, chlorophyll concentration, endogenous hormones and bulb weight in onion in response to water stress. Trop Agric 81:127–132

Upreti KK, Murti GSR (2004b) Leaf growth and endogenous hormones under water stress and stress recovery in French bean. J Plant Biol 31:61–64

Upreti KK, Murti GSR (2004c) Effects of brassinosteroids on growth, nodulation, phytohormone content and nitrogenase activity in French bean under water stress). Biol Plant 48:407–411

Upreti KK, Murti GSR (2005) Water stress induced changes in common polyamines and abscisic acid in French bean. Ind J Plant Physiol 10:145–150

Upreti KK, Murti GSR (2010) Response of grape rootstocks to salinity: changes in root growth, polyamines and abscisic acid. Biol Plant 54:730–734

Upreti KK, Murti GSR, Bhatt RM (1997) Response of French bean varieties to water deficits: changes in morphological characters and yield attributes. Indian J Hortic 54:66–74

Upreti KK, Murti GSR, Bhatt RM (1998) Response of French bean cultivars to water deficits: changes in endogenous hormones, proline and chlorophyll. Biol Plant 40:381–388

Upreti KK, Murti GSR, Bhatt RM (2000) Response of pea cultivars to water stress: changes in morpho-physiological characters, endogenous hormones and yield. Veg Sci 27:57–61

Upreti KK, Srinivasa Rao NK, Jayaram HL (2012) Floral abscission in capsicum under high temperature: role of endogenous hormones and polyamines. Indian J Plant Physiol 17:207–214

Urano K, Yoshiba Y, Nanjo T, Ito T, Yamaguchi-Shinozaki K, Shinozaki K (2004) Arabidopsis stress-inducible gene for arginine decarboxylase AtADC2 is required for accumulation of putrescine in salt tolerance. Biochem Biophys Res Commun 313:369–375

Vankova R, Gaudinova A, Dobrev P, Malbeck J, Haisel D, Motyka V (2011) Comparison of salinity and drought stress effects on abscisic acid metabolites activity of cytokinin oxidase/dehydrogenase and chlorophyll levels in radish and tabacco. Eco Quest 14:99–100

Verslues PE, Bray EA (2006) Role of abscisic acid (ABA) and *Arabidopsis thaliana* ABA-insensitive loci in low water potential induced ABA and Proline accumulation. J Exp Bot 57:201–212

Veselova RG, Farkhutdinov D, Veselov S, Kudoyarova GR (2006) Role of cytokinins in the regulation of stomatal conductance of wheat seedlings under conditions of rapidly changing local temperature. Russ J Plant Physiol 53:756–761

Vettakkorumakankav NN, Falk D, Saxena P, Fletcher RA (1999) A crucial role for gibberellins in stress protection of plants. Plant Cell Physiol 40:542–548

Vidoz ML, Loreti E, Mensuali A, Alpi A, Perata P (2010) Hormonal interplay during adventitious root formation flooded tomato plants. Plant J 63:551–562

Wang CY, Adams DO (1982) Chilling-induced ethylene production in cucumbers (*Cucumis sativus* L.). Plant Physiol 69(1982):424–427

Wang Y, Devereux W, Stewart TM, Casero RA (1999) Cloning and characterization of polyamine-modulated factor-1, a transcriptional cofactor that regulators the transcription of the spermidine/spermine N1-acetyl transferase gene. J Biol Chem 274:22095–22101

Wang WZ, Huang Q, Chen Q, Zhang Z, Zhang H, Wu Y, Huang D, Huang R (2004) Ectopic overexpression of tomato JERF3 in tobacco activates down stream gene expression and enhances salt tolerance. Plant Mol Biol 55:183–192

Wang Y, Bao ZL, Zhu Y, Hua J (2009) Analysis of temperature modulation of plant defense against biotrophic microbes. Mol Plant Microbe Interact 22:498–506

Wang Y, Li L, Ye T, Zhao S, Liu Z, Feng YQ, Wu Y (2011) Cytokinin antagonizes ABA suppression to seed germination of Arabidopsis by downregulating ABI5 expression. Plant J 68:249–261

Watts S, Rodriguez JL, Evans SE, Davies WJ (1981) Root and shoot growth of plants treated with abscisic acid. Ann Bot 47:595–602

Wei GP, Zhu YL, Liu ZL, Yang LF, Zhang GW (2007) Growth and ionic distribution of grafted eggplant seedlings with NaCl stress. Acta Bot Boreal-Occident Sin 27:1172–1178

Wen F, Zhang Z, Bai T, Xu Q, Pan Y (2010) Proteomics reveals the effects of gibberellic acid (GA_3) on salt-stressed rice (*Oryza sativa* L.) shoots. Plant Sci 178:170–175

Whitehead D (1998) Regulation of stomatal conductance and transpiration in forest canopies. Tree Physiol 18:633–644

Wilkinson S, Davies WJ (1997) Xylem sap pH increase: a drought signal received at the apoplastic face of the guard cell that involves the suppression of saturable abscisic acid uptake by the epidermal symplast. Plant Physiol 113:559–573

Wolbang CM, Chandler PM, Smith JJ, Ross JJ (2004) Auxin from the developing inflorescence is required for the biosynthesis of active gibberellins in barley stems. Plant Physiol 134:769–776

Wright STC (1980) The effect of plant growth regulator treatments on the level of ethylene emanating from excised turgid and wilted leaves. Planta 148:381–388

Wyatt SE, Rashotte AM, Shipp MJ, Robertson D, Muday GK (2002) Mutations in the gravity persistence signal loci in Arabidopsis disrupt the perception and/or signal transduction of gravitropic stimuli. Plant Physiol 130:1426–435

Xia XJ, Huang LF, Zhou YH, Mao WH, Shi K, Wu JX, Asami T, Chen Z, Yu JQ (2009) Brassinosteroids promote photosynthesis and growth by enhancing activation of Rubisco and expression of photosynthetic genes in *Cucumis sativus*. Planta 230:1185–1196

Xu CC, Qi Z (1993) Effect of drought on lipoxygenase activity, ethylene and ethane formation 21 in leaves of soybeans plants. Acta Bot Sin 35:31–37

Xu K, Xu X, Fukao T, Canlas P, Maghirang-Rodriguez R, Heuer S, Ismail AM, Bailey-Serres J, Ronald PC, Mackill DJ (2006) Sub1A is an ethyleneresponse-factor-like gene that confers submergence tolerance to rice. Nature 442:705–708

Xu W, Jia L, Baluška F, Ding G, Shi W, Ye N, Zhang J (2012) PIN2 is required for the adaptation of

Arabidopsis roots to alkaline stress by modulating proton secretion. J Exp Bot 63:6105–6114

Yamaguchi S (2008) Gibberellin metabolism and its regulation. Annu Rev Plant Physiol Plant Mol Biol 59:225–251

Yamaguchi K, Takahashi Y, Berberich T, Imai A, Miyazaki A, Takahashi T, Michael A, Kusano T (2006) The polyamine spermine protects against high salt stress in *Arabidopsis thaliana*. FEBS Lett 580:6783–6788

Yamaguchi K, Takahashi Y, Berberich T, Imai A, Takahashi T, Michael AJ, Kusano T (2007) A protective role for the polyamine spermine against drought stress in Arabidopsis. Biochem Biophys Res Commun 352:486–490

Yang SF, Hoffman NE (1984) Ethylene biosynthesis and its regulation in higher plants. Ann Rev Plant Physiol Mol Biol 35:155–189

Yiu JC, Juang D, Fang CW, Liu L, Wu I (2009) Exogenous putrescine reduces flooding-induced oxidative damage by increasing the antioxidant properties of Welsh onion. Sci Hortic 120:306–314

Yokota A, Kawasaki S, Iwano M, Nakamura C, Miyake C, Akashi K (2002) Citrullin and Drip-1 protein (ArgE Homologue) in drought tolerance of wild water melon. Ann Bot 89:825–832

Yoshida R, Umezawa T, Mizoguchi T, Takahashi S, Takahashi F, Shinozaki K (2006) The regulatory domain of SRK2E/OST1/SnRK2.6 interacts with ABI1 and integrates ABA and osmotic stress signals controlling stomatal closure in Arabidopsis. J Biol Chem 281:5310–5320

Yoshida T, Fujita Y, Sayama H, Kidokoro S, Maruyama K, Mizoi J, Shinozaki K, Yamaguchi-Shinozaki K (2010) AREB1, AREB2and ABF3 are master transcription factors that cooperatively regulate ABRE-dependent ABA signaling involved in drought stress tolerance and require ABA for full activation. Plant J 61:672–680

Zapata PJ, Serrano M, Pretel MT, Amoros A, Botella MA (2004) Polyamines and ethylene changes during germination of different plant species under salinity. Plant Sci 167:781–788

Zawaski CV, Busov B (2014) Roles of gibberellin catabolism and signaling in growth and physiological response to drought and short-day photoperiods in Populus trees. PLoS One 9:1–12

Zazimalova E, Kaminek M, Brezinova A, Motyka V (1999) Control of cytokinin biosynthesis and metabolism. In: Hooykaas PJJ, Hall MA, Libbenga KR (eds) Biochemistry and molecular biology of plant hormones. Elsevier, Amersterdan, pp 141–160

Zhang J, Jia W, Yang J, Ismail AM (2006) Role of ABA in integrating plant responses to drought and salt stresses. Field Crops Res 97:111–119

Zhang GM, Chen X, Chen X, Xu Z, Guan S, Li LC, Li A, Guo J, Mao L, Ma Y (2008a) Phylogeny, gene structures, and expression patterns of the ERF gene family in soybean (*Glycine max* L.). J Exp Bot 59:4095–4107

Zhang M, Zhai Z, Tian X, Duan L, Li Z (2008b) Brassinolide alleviated the adverse effect of water deficits on photosynthesis and the antioxidant of soybean (*Glycine max* L.). Plant Growth Regul 56:257–264

Zhang X, Wollenweber B, Jiang D, Liu F, Zhao J (2008c) Water deficit and heat shock effects on photosynthesis of a transgenic *Arabidopsis thaliana* constitutively expressing ABP9, a bZIP transcription factor. J Exp Bot 59:839–840

Zhang G, Chen M, Li L, Xu Z, Chen X, Guo J, Ma Y (2009) Overexpression of the soybean GmERF3 gene, an AP2/ERF type transcription factor for increased tolerances to salt, drought and diseases in transgenic tobacco. J Exp Bot 60:3781–3796

Zhang Q, Li J, Zhang W, Yan S, Wang R, Zhao J, Li Y, Qi Z, Sun Z, Zhu Z (2012) The putative auxin efflux carrier OsPIN3t is involved in the drought stress response and drought tolerance. Plant J 72:805–816

Zhao Y (2010) Auxin biosynthesis and its role in plant development. Ann Rev Plant Biol 61:49–64

Zhao Y, Christensen SK, Fankhauser C, Cashman JR, Cohen JD, Weigel D, Chory J (2001) A role for flavin monooxygenase-like enzymes in auxin biosynthesis. Science 291:306–309

Zhao FG, Sun C, Liu YL, Zhang HW (2003) Relationship between polyamine metabolism in roots and salt tolerance of barley seedlings. Acta Bot Sin 45:295–300

Zhao F, Song CP, He J, Zhu H (2007) Polyamines improve K^+/Na^+ homeostasis in barley seedlings by regulating root Ion channel activities. Plant Physiol 145:1061–1072

Zhao D, Shen L, Fan B, Yu B, Zheng Y, Lv S, Sheng J (2009) Ethylene and cold participate in the regulation of *LeCBF1* gene expression in postharvest tomato fruits. FEBS Lett 583:3329–3334

Zhao M et al (2014) Cold acclimation-induced freezing tolerance of *Medicago truncatula* seedlings is negatively regulated by ethylene. Physiol Plant 152:115–129

Zhou R, Cutler AJ, Ambrose SJ, Galka MM, Nelson KM, Squires TM, Loewen MK, Ashok S, Jadhav AS, Ross ARS, Taylor DC, Suzanne R, Abrams SR (2004) A new abscisic acid catabolic pathway. Plant Physiol 134:361–369

Zhu JH, Verslues PE, Zheng XW, Lee B, Zhan X, Manabe Y, Sokolchik I, Zhu Y, Dong C, Zhu J, Hasegawa PM, Bressan RA (2005) HOS10 encodes an R2R3-type MYB transcription factor essential for cold acclimation in plant. Proc Natl Acad Sci U S A 102:9966–9971

Zolla G, Heimer YM, Barak S (2010) Mild salinity stimulates a stress-induced morphogenic response in *Arabidopsis thaliana* roots. J Expt Bot 61:211–224

Antioxidant Protection Mechanism During Abiotic Stresses

3

K.S. Shivashankara, K.C. Pavithra, and G.A. Geetha

Abstract

Abiotic stresses are the major constraints for plant growth and productivity. Adverse effects of abiotic stresses are primarily by the production of various reactive oxygen species (ROS). Reactive oxygen species are broadly categorized into superoxides ($O_2^{\bullet-}$), hydroxyl radicals ($OH^{\bullet}$), alkoxyl radicals ($RO^{\bullet}$), peroxyl radicals ($ROO^{\bullet}$), singlet oxygen ($^1O_2$), lipid hydroperoxides (ROOH), hydrogen peroxides (H_2O_2), and excited carbonyls (RO^*). In plants the major sites of ROS production are peroxisomes and chloroplast. Other sites of ROS production are apoplast, cell membrane, endoplasm reticulum, and cell wall. Production of ROS in excess results in lipid peroxidation of membranes, protein degradation, pigment bleaching, and damage to nucleic acids ultimately resulting in cell death. Plants are protected against oxidative stress by enzymatic and non-enzymatic antioxidant mechanisms. Enzymatic antioxidant system include enzymes such as superoxide dismutase (SOD), peroxidase, monodehydroascorbate reductase (MDHAR), dehydroascorbate reductase (DHAR), catalase (CAT), glutathione reductase (GR), ascorbate peroxidase (APX), glutathione peroxidase (GPX), guaiacol peroxidase (GPOX), glutathione S-transferase (GST), and polyphenol oxidase (PPO), whereas non-enzymatic antioxidants include ascorbic acid (AA), α-tocopherol, reduced glutathione (GSH), carotenoids, phenolic acids, anthocyanins, flavonoids, and proline. In addition to creating oxidative stress, ROS also act as signalling molecules leading to the expression of many genes involved in stress tolerance, control of cell cycle, cell elongation,

K.S. Shivashankara (✉) • K.C. Pavithra • G.A. Geetha
Division of Plant Physiology and Biochemistry,
ICAR-Indian Institute of Horticultural Research,
Hessaraghatta Lake Post, Bengaluru 560089,
Karnataka, India
e-mail: shiva@iihr.ernet.in; shivaiihr@yahoo.com

N.K.S. Rao et al. (eds.), *Abiotic Stress Physiology of Horticultural Crops*,
DOI 10.1007/978-81-322-2725-0_3

osmolyte accumulation, and other systemic responses. The present review covers the biochemistry of ROS synthesis, their scavenging mechanisms, types of ROS, effect of abiotic stresses on their production, and their impact on horticultural crops.

3.1 Introduction

An inevitable by-product of oxidation reactions in living beings is the spillover of electrons to oxygen leading to the production of reactive oxygen species. The production of ROS has been estimated to be around 1 % of the oxygen consumed. Abiotic stress conditions cause extensive losses to agriculture production worldwide. All the crop plants and their wild relatives are quite sensitive to abiotic stress conditions (Boscaiu et al. 2008). Abiotic stresses cause changes at the morphological to molecular levels that adversely affect plant growth, productivity, and yield. Drought, high temperature, flooding, chilling, salinity, high light, and heavy metals are the major abiotic stresses that affect plants and also crop yield (Bita and Gerats 2013). Various stresses have crosstalks through osmotic stress and ion homeostasis in the cell. This is brought about mainly by the changes in expressions of a group of genes that affect growth rates and productivity. Therefore, to understand the stress response mechanisms in plants, it is very important to identify altered gene expression in response to abiotic stresses.

The major abiotic stresses such as drought, salinity, flooding, and low and high temperatures cause oxidative stresses in plant cells due to the over production of ROS. Under normal conditions there is a fine balance between the production and scavenging of ROS in plants, and this is disturbed by the abiotic and biotic stresses. The excess production of ROS results in membrane lipid peroxidation, discoloration of pigments, protein inactivation, and damage to DNA (Yordanov et al. 2000; Apel and Hirt 2004) in plant cell, which ultimately leads to cell death. A better understanding of the chemistry of ROS production and its removal is needed in the development of abiotic stress-tolerant genotypes.

3.2 Reactive Oxygen Species (ROS)

Reactive oxygen species are the derivatives of oxygen containing one or more unpaired electrons which are highly reactive and also unstable (Mittler 2002). ROS are the natural consequence of the electron spillover to oxygen in aerobic metabolism, and plants have developed defence mechanisms to deal with them in normal conditions, controlling the formation and removal rates.

3.2.1 Types of ROS

ROS in plants include singlet oxygen ($^1O_2$), superoxide radical ($O_2^{\bullet -}$), hydrogen peroxide (H_2O_2), and hydroxyl radical ($OH^\bullet$). There are other ROS in the form of alkoxyl radical ($RO^\bullet$), peroxyl radical ($ROO^\bullet$), organic hydroperoxide (ROOH), and excited carbonyl (RO^*).

3.2.1.1 Singlet Oxygen ($^1O_2$)

Singlet oxygen is generated by the reaction of triplet chlorophyll (Chl) with O_2 in the light-harvesting antenna system of chloroplast:

$$\text{Chl} + \text{light} \rightarrow {}^3\text{Chl}\,(\text{triplet chlorophyll})$$

$$^3\text{Chl} + {}^3\text{O} \rightarrow \text{Chl} + {}^1\text{O}_2$$

Abiotic stresses reduce the intra- and intercellular CO_2 concentration by affecting stomatal conductance in plants. Reduced CO_2 concentration increases the probability of electrons spilling over to oxygen leading to the production of $^1O_2$ which has a half-life of about 3 μs (Hatz et al. 2007; Das and Roychoudhury 2014). Singlet oxygen triggers the upregulation of many genes involved in the protection against photo-oxidative stress (Krieger-Liszkay et al. 2008). Carotenoid content in chloroplast is also triggered by singlet oxygen.

3.2.1.2 Superoxide Radical ($O_2^{\bullet-}$)

Superoxide ($O_2^{\bullet-}$) does not cause large damages due to relatively shorter half-life of 2–4 μs, but can lead to the formation of highly reactive free radicals like hydroxyl radicals and singlet oxygen which cause extensive damage to the membranes (Halliwell 2006; Das and Roychoudhury 2014). The major sites of superoxide production in plants are peroxisomes, reaction centres of photosystems in chloroplasts, and mitochondria (Rhoads et al. 2006; Moller et al. 2007). Superoxides are scavenged by the enzyme superoxide dismutase by converting them into hydrogen peroxide. Superoxides are also produced in the plasma membrane by NADPH oxidase (Moller et al. 2007) by the following reaction:

$$NADPH + 2O_2NADP^+ + 2O_2^{\bullet-} + H^+$$

3.2.1.3 Hydrogen Peroxide (H_2O_2)

Hydrogen peroxide is not considered as a free radical but an oxidant and is involved in the production of many reactive oxygen species. H_2O_2 is formed when $O_2^{\bullet-}$ undergoes both reduction and protonation. It can occur both non-enzymatically under low pH conditions and by the enzyme superoxide dismutase (SOD) (Das and Roychoudhury 2014):

$$2O_2^{\bullet-} + 2H^+ \rightarrow H_2O_2 + O_2$$

Electron transport chain of chloroplast and mitochondria are the major sites of H_2O_2 production in plant cells. In addition to this, it is also produced in ER and in peroxisomes during photorespiration and β-oxidation of fatty acids, by NADPH oxidase and xanthine oxidase enzymes. It has a longer half-life (1 ms) compared to other ROS members; due to this, it can traverse longer distances and cross plant cell membranes (Das and Roychoudhury 2014).

3.2.1.4 Hydroxyl Radical ($OH^\bullet$)

Hydroxyl radicals ($OH^\bullet$) are the most reactive and toxic among the ROS in biological system. It is generated at neutral pH by the reaction between H_2O_2 and $O_2^{\bullet-}$ catalyzed by transition metals like Fe (Fe^{2+}, Fe^{3+}) (Das and Roychoudhury 2014). The reaction is referred to as Heber-Weiss reaction.

3.2.1.4.1 Fenton's Reaction

Ferrous ion is oxidized by hydrogen peroxide to produce hydroxyl radicals which are the most reactive free radicals. Ferric ion is converted back to ferrous by another hydrogen peroxide molecule leading to the formation of peroxyl radical:

$$Fe^{2+} + H_2O_2 \rightarrow Fe^{3+} + OH^\bullet + OH^-$$
$$Fe^{3+} + H_2O_2 \rightarrow Fe^{2+} + OOH^\bullet + H^+$$

Due to their high reactivity, hydroxyl radicals cause extensive damages to cellular components by lipid peroxidation, protein damage, and membrane destruction. Unlike superoxides, hydroxyl radicals cannot be eliminated by an enzymatic reaction. It has a very short half-life of 1 μs (Das and Roychoudhury 2014). Therefore, the only way to avoid the damages by hydroxyl radicals is by preventing its formation.

3.2.1.5 Peroxyl ($ROO^\bullet$) and Alkoxyl ($RO^\bullet$) Radicals

$ROO^\bullet$ and $RO^\bullet$ are moderately strong oxidants. They can be generated from organic hydroperoxide (ROOH) decomposition induced by heat or radiation and by ROOH reaction with transition metal ions and other oxidants capable of abstracting hydrogen. $ROO^\bullet$ are also important intermediates in processes involving carbon-centred radicals, which react rapidly with O_2. In addition, biomolecule-derived $ROO^\bullet$ and $RO^\bullet$ can be generated from the oxidation of lipids, proteins, and nucleic acids (Augusto and Miyamoto 2011):

$$ROOH + Fe^{2+} \rightarrow RO^{\bullet} + OH^{-} + Fe^{3+}$$
$$ROOH + Fe^{3+} \rightarrow ROO^{\bullet} + H^{+} + Fe^{2+}$$

The reactivity of $ROO^{\bullet}$ and $RO^{\bullet}$ is influenced by the substituents at the α-carbon. An electron-withdrawing group increases the reactivity (for instance, chloroperoxyl radical ($CCl_3OO^{\bullet}$)), whereas electron-donating group decreases it (for instance, phenoxyl radicals). Aromatic $ROO^{\bullet}$ and $RO^{\bullet}$ tend to be less reactive because of unpaired electron delocalization. The reactions of $ROO^{\bullet}$ and $RO^{\bullet}$ with biomolecules often involve hydrogen abstraction, which is facilitated in compounds containing weakly bound hydrogens. This is the case for lipids, thiols, and several chain-breaking antioxidants. Lipids are particularly susceptible to hydrogen abstraction, and this reaction is the rate-limiting step in the propagation of lipid peroxidation chain reactions. $ROO^{\bullet}$ formed on aromatic rings and those with α-carbon linked to hydroxy or amino groups can decompose to liberate $O_2^{\bullet-}$ or $HO_2^{\bullet}$. This type of reaction has been reported for amino acids, such as lysine. $ROO^{\bullet}$ can also react with another $ROO^{\bullet}$ by the Russell mechanism, generating a ketone, an alcohol, and $^1O_2$.

3.2.1.6 Organic Hydroperoxides (ROOH)

Organic hydroperoxides (ROOH) are produced upon oxidation of biomolecules, including lipids, proteins, and DNA. Lipid hydroperoxides can be formed enzymatically during lipoxygenase, cyclooxygenase, cytochrome P_{450}, and heme-peroxidase turnover. Moreover, a great number of ROOH are generated non-enzymatically by the oxidation of biomolecules mediated by free radicals and $^1O_2$. ROOH are relatively stable and can participate in reactions that change their toxicity. Normally, cells contain enzymes that reduce ROOH to their corresponding alcohols, decreasing their reactivity and toxicity:

$$\underset{\text{(Lipid)}}{ROO^{\bullet} + RH} \rightarrow ROOH + \underset{\text{(Lipidalkylradical)}}{R^{\bullet}}$$

In the presence of transition metals like iron and copper, organic hydroperoxides decompose to form alkoxyl radicals. Hydroperoxides are responsible for initiation of lipid peroxidation and also form α, β-unsaturated aldehydes by their decomposition that can damage biological macromolecules like DNA (Zhu et al. 2007):

$$ROOH + Me^{(n-1)+} \rightarrow RO^{\bullet} + OH^{-} + Me^{n+}$$

where Me represents a transition metal, such as iron or copper. ROOH can react with reduced and oxidized states of free transition metal ions to form alkoxyl radicals. In the first case, $RO^{\bullet}$ is produced, whereas in the second case, $ROO^{\bullet}$ is produced. Alkoxyl radicals further initiate the chain of oxidation process and produce other reactive molecules like singlet oxygen, ketones, epoxides, electrophilic aldehydes, and the excited carbonyl species.

3.2.1.7 Excited Carbonyl (RO^*)

The occurrence of excited carbonyl is formed by the action of singlet oxygen on unsaturated fatty acids. This reaction forms an intermediate called dioxetane which has luminescence properties. Dioxetane undergoes cleavage to form excited carbonyls that can damage lipids and DNA.

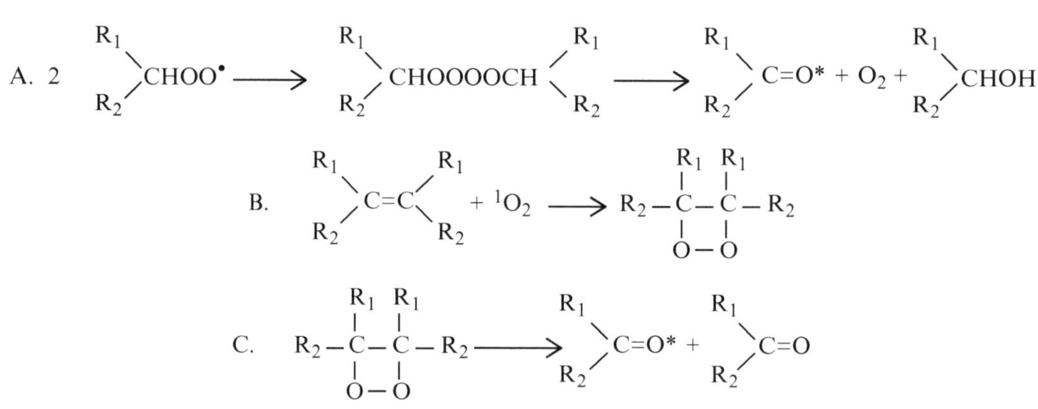

(A) Disproportionation of *sec*-peroxyl radicals generates triplet carbonyl compound. (B) and (C) carbonyl emissions originate from a dioxetane breakdown; this emission involves the participation of $^1O_2$ by its reaction with electron-rich olefins (B) to give up a dioxetane.

3.3 Generation of Reactive Oxygen Species

During photosynthesis, singlet oxygen ($^1O_2$) is produced in the chloroplast thylakoids where the energy is transferred to triplet oxygen ($^3O_2$) by the triplet chlorophyll (^{3}Chl). Triplet chlorophyll is produced due to insufficient energy dissipation by the excited chlorophylls. Superoxide radical ($O_2^{\bullet -}$) is produced when electrons spill over to oxygen from the electron transport chains of chloroplast and mitochondria, and it is also produced by the NADPH oxidase. Hydrogen peroxide (H_2O_2) and hydroxyl radicals ($OH^{\bullet}$) are produced from superoxides ($O_2^{\bullet -}$)(Fig. 3.1). Hydrogen peroxide inactivates enzymes by oxidizing the thiol groups. H_2O_2 can also be considered as a secondary messenger for triggering the defence reactions or cell death because of its long half-life and permeability through the membranes. On the other hand, hydroxyl radicals are highly toxic to the cells and non-enzymatically destroy all the biological molecules like DNA, protein, and lipids leading to cell death.

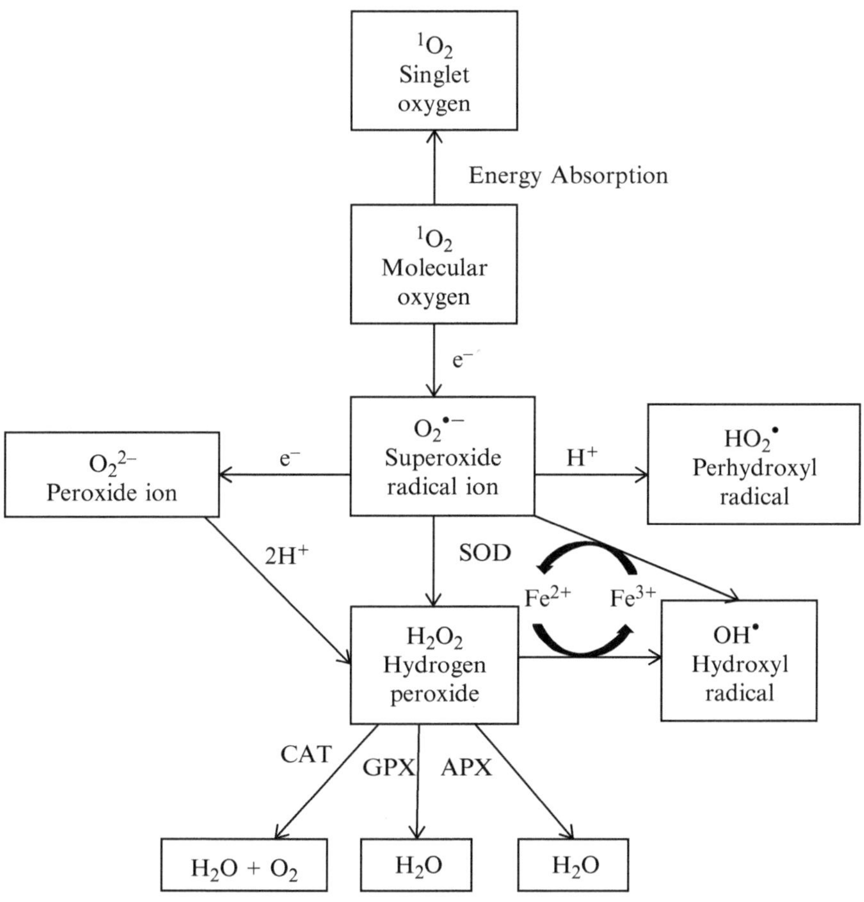

Fig. 3.1 Schematic representation of generation of reative oxygen species (ROS)

3.3.1 Generation of ROS by NADPH Oxidase

The primary ROS superoxide is formed by single-electron reduction of molecular oxygen. The reaction is catalyzed by NADPH oxidase with NADPH as the electron donor:

$$O_2 + e^- \rightarrow O_2^{\bullet -} 2O_2 + \text{NADPH} \rightarrow 2O_2^{\bullet -} + \text{NADP}^+ + H^+$$

NADPH oxidase is a membrane-bound multi-component complex which transfers electrons from NADPH to O_2 via FAD and haem groups (Bolwell and Wojtaszek 1997).

3.3.2 Generation of ROS by Glycolate Oxidase

In leaves of C_3 plants, the consumption of oxygen by RuBisCO during daytime prevents the photooxidation of photosystems by acting as an additional sink for electron of light reactions. However, the glycolate synthesized by the oxygenation reactions is transported to peroxisomes where glycolate oxidase oxidizes glycolate and releases H_2O_2. This is the major source of H_2O_2 produced during photosynthesis (Apel and Hirt 2004):

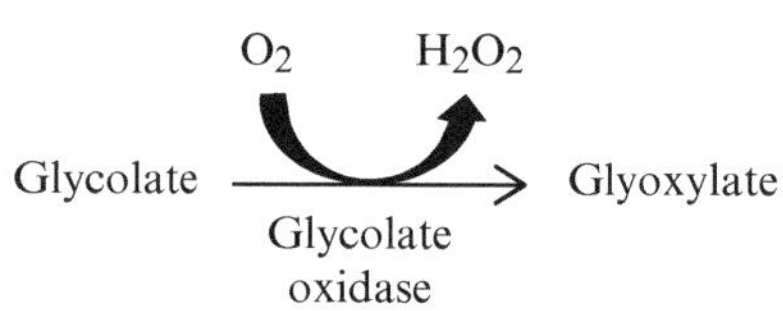

3.3.3 Generation of ROS by Xanthine Oxidase

Xanthine oxidase (XOD) is responsible for the production of superoxide radicals in peroxisomes. Superoxides react with nitric oxide produced in the peroxisomes by nitric oxide synthase and forms peroxynitrite. Peroxynitrites are involved in the nitration of proteins, and in the peroxisome, it is involved in the conversion of xanthine dehydrogenase (XDH) into the superoxide-generating XOD (Corpas et al. 2001):

$$\text{Hypoxanthine} + 2O_2 + H_2O \overset{XO}{\rightarrow} \text{Xanthine} + 2O_2^{\bullet -} + 2H^+ + H^+$$

3.3.4 Generation of ROS by Lipoxygenase

Lipoxygenase catalyzes the oxidation of polyunsaturated fatty acids with *cis,cis*-1,4-pentadiene units leading to the formation of respective hydroperoxides (Foyer and Noctor 2009). There are three types of lipoxygenases, namely, 5-, 12-, and 15-lipoxygenase, based on the carbon number oxidized. 12-LOX oxidizes the unsaturated fatty acid and generates fatty acid radical. This radical reacts with oxygen and generates 12-hydroperoxyeicosatetraenoic acid (12-HPETE). Finally, glutathione peroxidase catalyzes the formation of 12(S)-hydroxy-(5Z,8Z,10E,14Z)-eicosatetraenoic acid (12-HETE) from 12-HPETE by using GSH which is converted to GSSG.

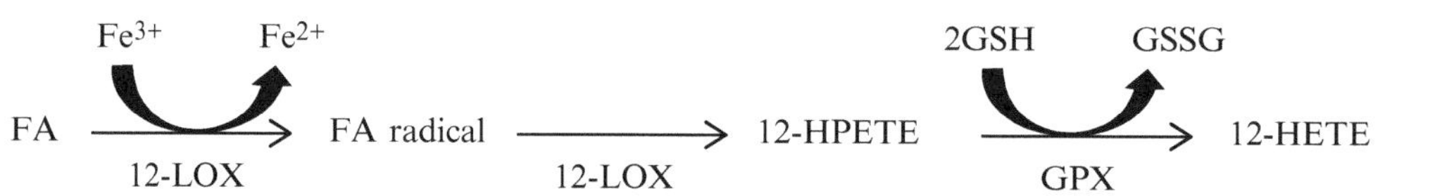

3.4 Sites of ROS Generation in the Cell

ROS are continuously produced in the cells both under normal and stressed conditions. The active sites of ROS generation are most of the cell organelles like chloroplasts, peroxisomes, mitochondria, membranes, and cell walls. In the chloroplast ROS is produced in reaction centres of thylakoids. The reduced availability of CO_2 during stress and the reduced fixation coupled with over-reduction of the electron transport chain are the main causes of ROS production in the chloroplast. The triplet state of chlorophyll in the light-harvesting complex and the electron transport chain are the major sites of ROS production in chloroplasts ($O_2^{\bullet -}$, $^1O_2$, and H_2O_2) (Gill and Tuteja 2010) during daytime. During daytime the major process of H_2O_2 production in plants is the photorespiration where the oxidation of glycolate by glycolate oxidase in peroxisomes accounts for the majority of H_2O_2 production. Superoxides are also produced in peroxisomes (Sharma et al. 2012). Peroxidation of polyunsaturated fatty acids by cell wall-localised lipoxygenases is another major source of ROS like $OH^{\bullet}$, $O_2^{\bullet -}$, H_2O_2, and $^1O_2$ (Das and Roychoudhury 2014). ROS are also produced by the electron-transporting oxidoreductases commonly present in plasma membranes. NADPH oxidase present in plasma membranes produces superoxides by using cytoplasmic NADPH in the presence of O_2 which is followed by the dismutation to form H_2O_2 (Sharma et al. 2012). Therefore, the important site of H_2O_2 production is the apoplast. NADPH-dependent production of superoxides also takes place in ER using CYP450. Hydrogen peroxide when produced in excess under stress affects the Calvin cycle enzymes (Fig. 3.2 and Table 3.1).

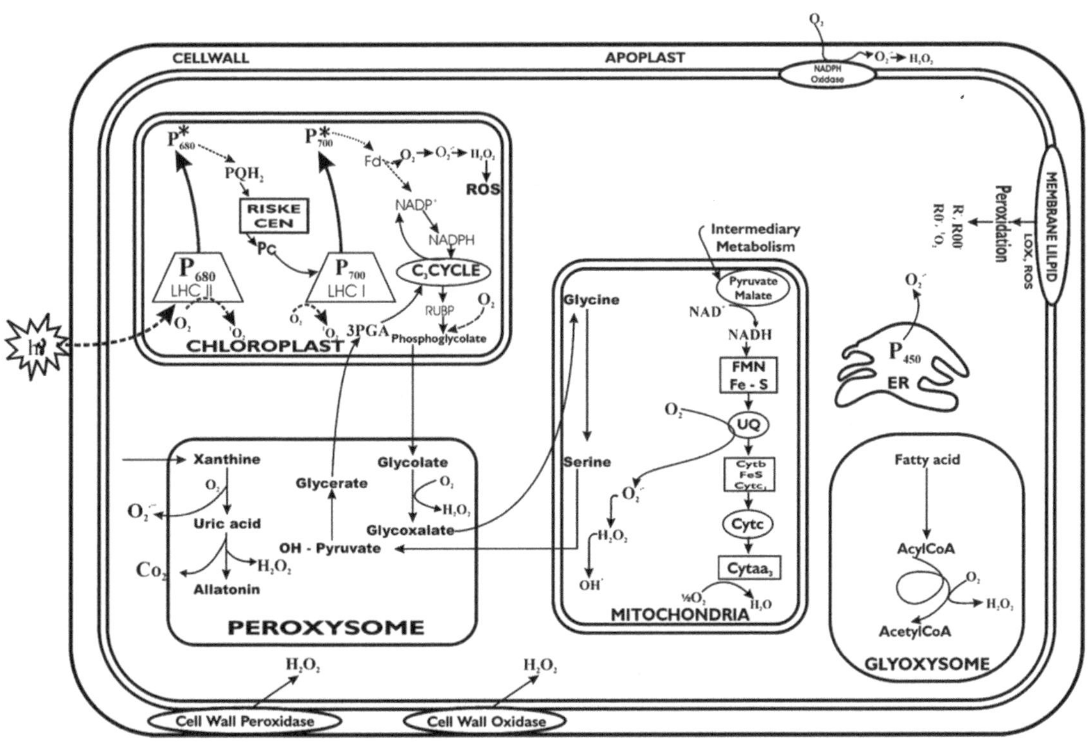

Fig. 3.2 Schematic representation of sources of reative oxygen species (ROS) (Bhattacharjee 2005)

Table 3.1 The list of reactive oxygen species and their sources with mode of action

ROS	Half-life ($t_{1/2}$)	Mode of action	Sources	Mechanism of production
Singlet oxygen ($^1O_2$)	1–4 μs	Oxidizes protein, DNA, and polyunsaturated fatty acids	Chloroplasts	Excited chlorophyll
Superoxide ($O_2^{\bullet -}$)	1–4 μs	Attacks double bonds of iron-sulphur (Fe-S) clusters of proteins and unsaturated fatty acids. Strongly reacts with nitric oxide (NO) to form peroxynitrite ($ONOO^-$)	Plasma membrane	NADPH oxidase
			Chloroplasts	Photosystem I: ETC
			Mitochondria	
			Cell wall	Ferredoxin, 2Fe-2S, and 4Fe-4S clusters
			Peroxisomes	Aconitase, 1-galactono-γ-lactone dehydrogenase
			Endoplasmic reticulum	Xanthine oxidase
				NAD(P)H-dependent electron transport involving Cyt P_{450}
Hydroxyl radical ($OH^{\bullet}$)	1 μs	Extremely reactive with like protein, lipids, and DNA	Cell wall	Decomposition of O_3
			Chloroplasts	Fenton reaction
Hydrogen peroxide (H_2O_2)	1 ms	Oxidizes proteins and reacts with $O_2^{\bullet -}$ in a Fe-catalyzed reaction to form $OH^{\bullet}$	Chloroplasts	Photosynthesis ET and PSI or II
			Mitochondria	Mitochondrial form of SOD
			Cell wall	Peroxidases, Mn^{2+}, and NADH
			Plasma membrane	SOD
			Peroxisomes	Glycolate oxidase, fatty acid β-oxidation
			Apoplast	Amine oxidase, oxalate oxidase

3.5 Abiotic Stress and ROS Interaction

All aerobic organisms produce reactive oxygen species as a by-product of the oxidative metabolism but kept in balance by the antioxidative mechanism of the cells. In addition to the deleterious effects, ROS also have a role in cell signalling processes (Foyer and Noctor 2003); therefore, the concentration of ROS needs to be regulated (Mittler 2002). The major sources of ROS are chloroplast reaction centres, mitochondrial electron transport complexes, peroxisomes, and also the enzymes like NADPH oxidases, amine oxidases, and cell wall peroxidases (Mittler 2002). Under abiotic stress conditions, production of ROS can exceed the scavenging capability of the cell (Gill and Tuteja 2010). The raise in the level of ROS can induce the stronger defence mechanisms in plants in order to overcome the oxidative stresses. However, the excess ROS has the potential of damaging the biomolecules leading to death of the cells (Boguszewska and Zagdańska 2012). One of the major sites of ROS production during stress is chloroplast and is mainly due to stomatal closure-induced CO_2 deficiency. Lack of CO_2 inside chloroplast results in the spilling over of electrons to O_2 leading to the overproduction of singlet oxygen. Overproduction of singlet oxygen bleaches the chlorophyll molecules by oxidizing carotenoids. Yellowing of leaves during abiotic stresses is mainly due to the oxidation of carotenoids and chlorophylls by singlet oxygen species.

Lipid peroxidation (LPO) takes place when ROS molecules exceed the threshold levels in the cells and membranes. Lipid peroxidation leads to loss of membrane integrity resulting in leakage of substances from the cells and inactivation of membrane-bound receptors, protein channels, and enzymes (Gill and Tuteja 2010). In addition to the peroxidation of lipids, ROS modifies proteins by oxidation of a number of amino acids leading to the production of protein carbonyls (Boguszewska and Zagdańska 2012). Some of the amino acids that form free carbonyl groups upon oxidation by ROS are proline, lysine, arginine, threonine, histidine, and tryptophan resulting in inactivation of protein cross-linkages (Moller et al. 2007; Foyer and Noctor 2009). Further ROS like hydroxyl radicals also damage DNA molecules by oxidizing purine, pyrimidine and deoxyribose backbone. Singlet oxygen ($^1O_2$) damages guanine, and H_2O_2 and $O_2^{\bullet-}$ do not react with DNA molecules (Boguszewska and Zagdańska 2012).

3.6 Antioxidant Defence Systems

During stress conditions, plants adapt a defence system involving antioxidants to protect cells from the negative effects of ROS (Noctor and Foyer 1998). The system acts as a scavenger and comprizes two types of antioxidant components, namely, enzymatic and non-enzymatic antioxidants.

3.6.1 Enzymatic Antioxidants

The enzymatic antioxidants include many enzymes operating in different organelles. The most important enzymes are superoxide dismutase (SOD), ascorbate peroxidase (APX), glutathione peroxidase (GPX) catalase (CAT), monodehydroascorbate reductase (MDHAR), dehydroascorbate reductase (DHAR), guaiacol peroxidase (GPOX), glutathione reductase (GR), glutathione S-transferase (GST), and polyphenol oxidase (PPO).

3.6.1.1 Superoxide Dismutase (SOD) (E.C.1.15.1.1)

SOD is the first line of defence against reactive oxygen species attack in plants (McCord and Fridovich 1969; Boguszewska and Zagdańska 2012). SOD belongs to the family of metalloenzymes. The SOD catalyzes the removal of $O^{\bullet}{}_2^-$ by dismutating it into O_2 and H_2O_2. By removing superoxides, SOD reduces the risk of $OH^{\bullet}$ formation by the Haber-Weiss reaction. SODs are classified into three isozymes based on the associated metal ions like Mn-SOD is localized in the mitochondria, Fe-SOD is localized in the chloroplasts, and Cu/Zn-SOD is localized in the cytosol, peroxisomes, and chloroplasts (Mittler 2002; Das and Roychoudhury 2014):

$$O_2^{\bullet-} + O_2^{\bullet-} + 2H^+ \xrightarrow{SOD} 2H_2O_2 + O_2$$

3.6.1.2 Catalase (CAT) (E.C.1.11.1.6)

One of the first antioxidant enzymes to be discovered and characterized is catalase (Mhamdi et al. 2010; Boguszewska and Zagdańska 2012). CAT converts H_2O_2 to H_2O and O_2 and is a haem-containing enzyme. The enzyme has high affinity for H_2O_2, but lesser specificity for organic peroxides (R-O-O-R) with highest turnover rates for H_2O_2. One molecule of CAT can scavenge about six million molecules of H_2O_2 per minute (Boguszewska and Zagdańska 2012). It is the main scavenger of excess ROS under abiotic stresses (Mittler 2002) and is essential for removing peroxisomal H_2O_2 produced by photorespiration (Noctor et al. 2000). Catalase activity decreases under conditions that suppress photorespiration, such as elevated CO_2.

$$2H_2O_2 \xrightarrow{CAT} O_2 + 2H_2O$$

3.6.1.3 Ascorbate Peroxidase (APX) (E.C.1.1.11.1)

APX is associated with the ascorbate-glutathione cycle. It removes H_2O_2 in the cytosol and in the chloroplast using ascorbate as a reducing agent and releases water and dehydroascorbic acid (Das and Roychoudhury 2014):

$$H_2O_2 + AA \xrightarrow{APX} 2H_2O + DHA$$

APX is a more efficient scavenger of H_2O_2 than CAT due to its higher affinity for H_2O_2 (Das and Roychoudhury 2014). Better redox state of the cells can be maintained by having higher level of ascorbates which is necessary for scavenging the ROS molecules. There are five different isoforms of APX based on the localization: thylakoid APX (tAPX), glyoxysome membrane APX (gmAPX), chloroplast stromal soluble form of APX (sAPX), and cytosolic form of APX (cAPX) (Boguszewska and Zagdańska 2012).

3.6.1.4 Monodehydroascorbate Reductase (MDHAR) (E.C.1.6.5.4)

MDHAR is an enzyme involved in the glutathione-ascorbate cycle and reduces monodehydroascorbate to ascorbic acid. It is a FAD-containing enzyme that uses NADPH as an electron donor for reducing the oxidized ascorbate. It is present as isozymes in chloroplast, cytosol, glyoxysome, mitochondria, and peroxisome. MDHAR has high specificity for monodehydroascorbate (MDHA). It is the only known enzyme to use an organic radical (MDA) as a substrate and is also capable of reducing phenoxyl radicals:

$$MDHA + NADPH \xrightarrow{MDHAR} AA + NADP^+$$

3.6.1.5 Dehydroascorbate Reductase (DHAR) (E.C.1.8.5.1)

DHAR reduces dehydroascorbic acid to ascorbic acid using reduced glutathione (GSH) as the reducing substrate; therefore, it is very important for maintaining ascorbic acid pool in reduced form. It is critical thus in maintaining the redox state of the plant cell (Chen and Gallie 2006). DHAR is found abundantly in seeds, roots, and both green and etiolated shoots (Das and Roychoudhury 2014):

$$DHA + 2GSH \xrightarrow{DHAR} AA + GSSG$$

Dehydroascorbate is reduced to ascorbate by DHAR in the presence of glutathione which is important for ascorbate recycling (Boguszewska and Zagdańska 2012).

3.6.1.6 Glutathione Reductase (GR) (E.C.1.6.4.2)

Glutathione reductase is a flavoprotein localized mainly in the chloroplasts, mitochondria, and cytosol. It is involved in regenerating ascorbates via ascorbate-glutathione cycle (Gill and Tuteja 2010; Boguszewska and Zagdańska 2012). The reduced glutathione maintains redox equilibrium with the oxidized form of glutathione (GSSG) in the cell. It reduces the disulphide (GSSG) form to sulphydryl (GSH) form and sustains the reduced status of glutathione (Reddy and Raghavendra 2006; Boguszewska and Zagdańska 2012):

$$GSSG + NADPH \xrightarrow{GR} 2GSH + NADP^+$$

3.6.1.7 Guaiacol Peroxidase (GPOX) (E.C.1.11.1.7)

GPOX is also involved in scavenging excess cellular H_2O_2 and localized in the cytoplasm and apoplast. However, the major role of GPOX is in the synthesis of lignin upon maturation or infection of tissues. GPOX has a role in defence against biotic stress by degrading indoleacetic acid (IAA) and utilizing H_2O_2 in the process (Das and Roychoudhury 2014). Guaiacol peroxidase oxidizes mainly guaiacol or pyrogallol in addition to a large number of organic compounds such as phenols, aromatic amines, and hydroquinones, etc. (Boguszewska and Zagdańska 2012).

3.6.1.8 Glutathione Peroxidase (GPX) (E.C.1.11.1.9)

Glutathione peroxidase reduces organic hydroperoxides or H_2O_2 using GSH as an electron donor and forms GSSG. The enzyme is non-specific for hydroperoxides. It is localized mainly in the cytosol and mitochondria:

$$ROOH/H_2O_2 + GSH \xrightarrow{GPX} ROH/H_2O + GSSG$$

3.6.1.9 Glutathione S-Transferase (GST) (E.C.2.5.1.18)

GST is also localized in the cytosol and mitochondria and catalyzes almost similar reactions as glutathione peroxidase. The only difference is the transfer reactions where GSH is converted into its sulphenic acid form. This spontaneously forms a disulphide with another GSH molecule leading to the formation of GSSG.

$$\text{R-O-OH} + \text{GSH} \xrightarrow{\text{GST}} \text{R-OH} + \text{GS-OH} \xrightarrow{\text{GSH}} \text{GS-SG}$$

3.6.1.10 Polyphenol Oxidase (PPO) (E.C.1.14.18.1)

The PPO is involved in oxidation of phenolic compounds. Its activity is usually seen when the tissue is damaged. The enzyme is present in the plastids and then comes in contact with the phenolic compounds present in the vacuole leading to the formation of oxidized phenols. It is a copper-containing enzyme and catalyzes two types of reactions in the presence of oxygen, the hydroxylation of monophenols called monophenolase activity and the oxidation of ortho-phenols called diphenolase activity into ortho-quinones. This reaction is followed by non-enzymatic polymerization of the quinones giving rise to melanins, pigments of high molecular mass and dark colour.

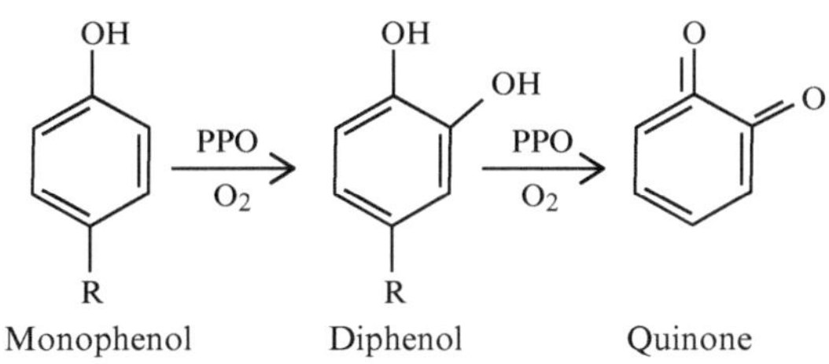

3.6.2 Non-enzymatic Antioxidants

The non-enzymatic antioxidants include ascorbic acid, glutathione, carotenoids, proline, phenolic acids, flavonoids, and α-tocopherol.

3.6.2.1 Ascorbic Acid (AA)

Ascorbic acid is the most abundant water-soluble antioxidant in plants and protects the cells from ROS damages. Ascorbic acid is reported in almost all plant cell types, their organelles, and also in apoplasts (Smirnoff and Wheeler 2000; Das and Roychoudhury 2014). Ascorbic acid in addition to its role in maintaining the redox state of cells can also directly react with H_2O_2, $O_2^{\bullet -}$, and $OH^{\bullet}$ radicals and lipid hydroperoxidases. Ascorbic acid is also important for regenerating tocopherol from tocopheroxyl radicals which gives protection to membranes. The primary role of AA is to protect metabolic processes against H_2O_2 (Boguszewska and Zagdańska 2012).

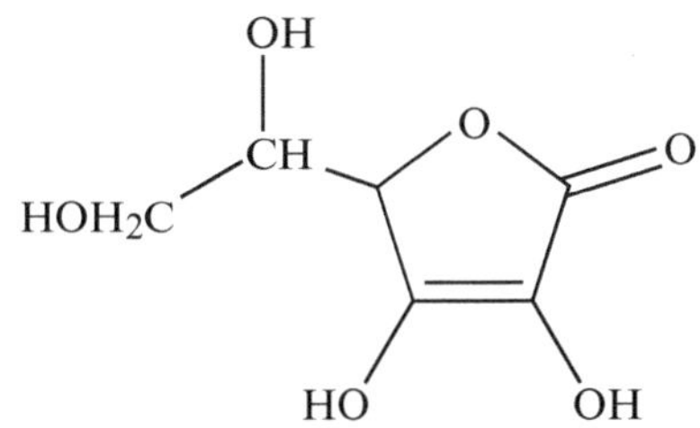

Structure of ascorbic acid (AA)

3.6.2.2 Reduced Glutathione (GSH)

Glutathione is abundantly found in almost all the cell organelles. It is a low-molecular-weight thiol tripeptide (γ-glutamyl-cysteinyl-glycine). It is involved in a wide range of processes like cell differentiation, cell growth, senescence and cell death, regulation of sulphate transport, conjugation of metabolites, and regulation of enzymatic activity (Das and Roychoudhury 2014). GSH scavenges H_2O_2, $^1O_2$, $OH^{\bullet}$, and $O^{\bullet}{}_2^{-}$ and protects the different biomolecules by forming adducts (glutathiolated) or by reducing them in the presence of ROS or organic free radicals and generating GSSG as a by-product. GSH also plays a vital role in regenerating AA to yield GSSG (Das and Roychoudhury 2014). The ratio of reduced to oxidized glutathione during the degradation of H_2O_2 influences the signalling pathways for the initiation of tolerance mechanisms against ROS (Foyer and Noctor 2003).

The ascorbate-glutathione cycle (Fig. 3.3) is the major mechanism involved in the detoxification of cells including the scavenging of ROS. This cycle removes hydrogen peroxides produced in the cell by the action of SOD. Some of the major enzymes involved in the ROS detoxification reactions are ascorbate peroxidase (APX), glutathione peroxidase (GPX), monodehydroascorbate reductase (MDHAR), and dehydroascorbate reductase (DHAR). The cycle regenerates reduced form of ascorbate and glutathione and maintains the pool cells in a reduced state.

3.6.2.3 α-Tocopherol

Tocopherols are membrane-bound antioxidants having the potential for scavenging ROS and lipid radicals. There are four types of tocopherols namely α-, β-, γ-, and δ- in plants. Out of these, α- tocopherol was found to have the highest antioxidative activity due to the presence of three methyl groups in its molecular structure. It is estimated that one molecule of α-tocopherol can scavenge up to 120 $^{1}O_2$ molecules by resonance energy transfer.

3.6.2.4 Carotenoids

Carotenoids are the lipophilic antioxidants located mainly in the plastids of both

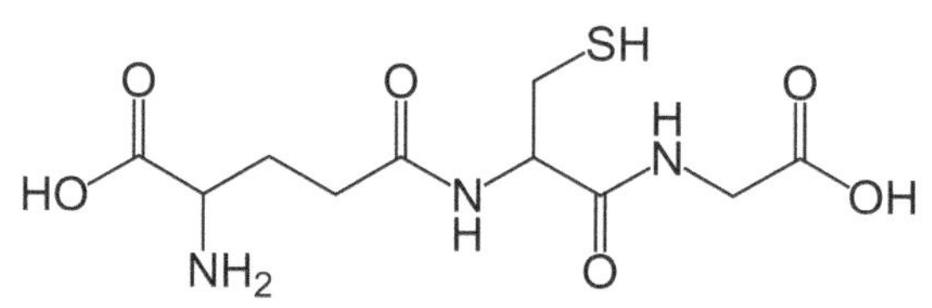

Structure of reduced glutathione

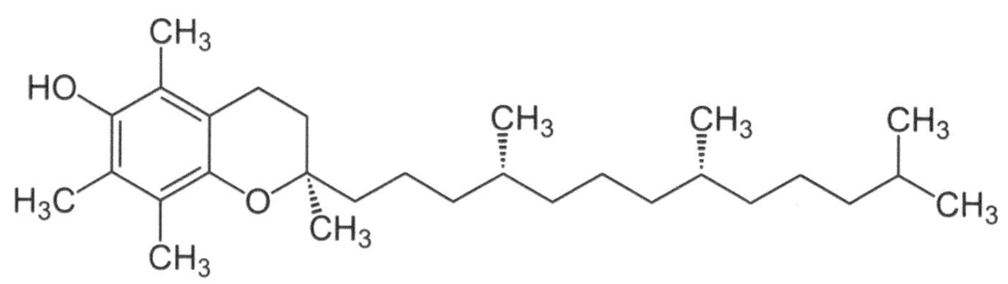

Structure of α-Tocopherol

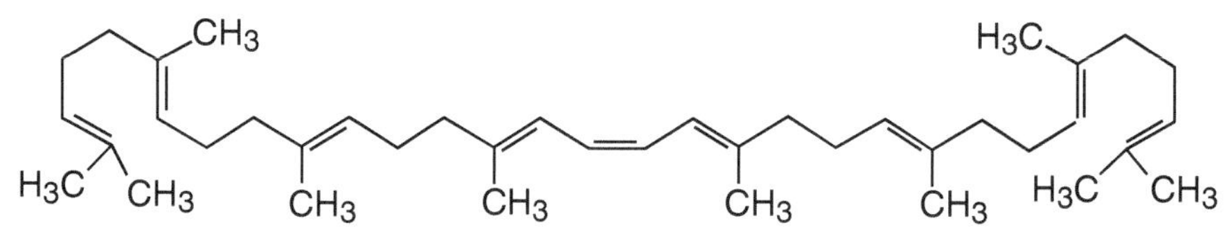

Structure of lycopene

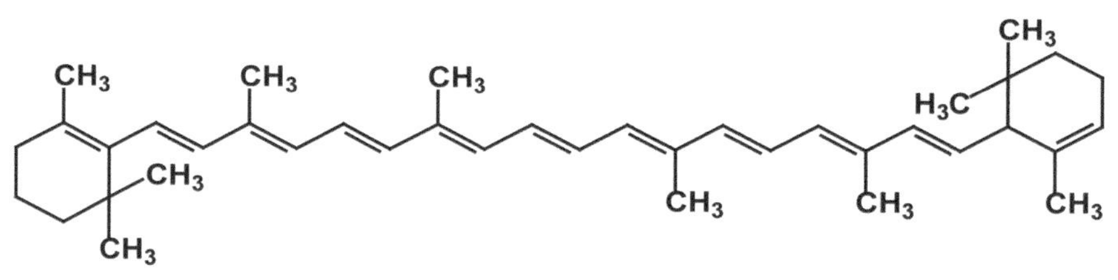

Structure of β-carotene

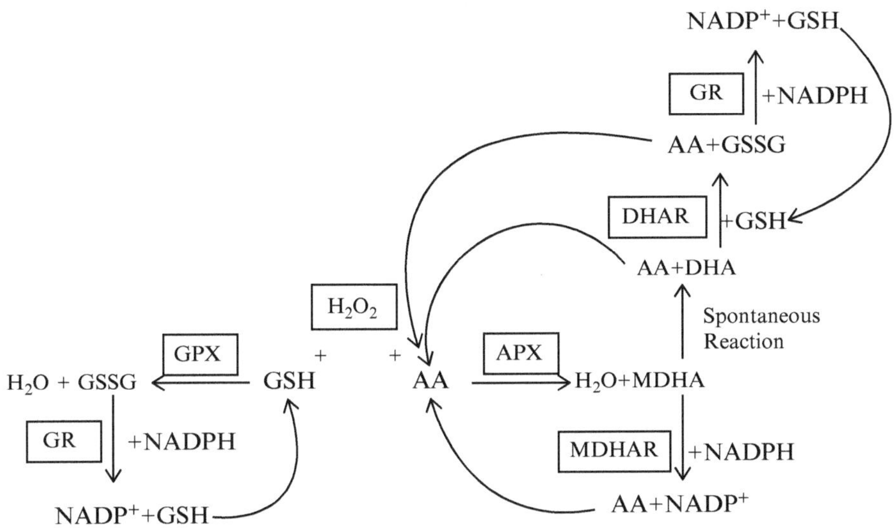

Fig. 3.3 The ascorbate-glutathione cycle. (**a**) Non-enzymatic compounds: *AA* ascorbic acid (ascorbate), *GSH* glutathione (reduced), *DHA* dehydroascorbate, *GSSG* glutathione (oxidized), *MDHA* monodehydroascorbate; (**b**) Enzymes: *APX* ascorbate peroxidase, *GR* glutathione reductase, *GPX* glutathione peroxidase, *DHAR* dehydroascorbate reductase, *MDHAR* monodehydroascorbate reductase

photosynthetic and non-photosynthetic plant tissues (Das and Roychoudhury 2014). Carotenoids protect chlorophylls from the harmful singlet oxygen and also transfer energy from the shorter wavelengths (400 and 550 nm) to the light-harvesting complex. They are important for the photosystem (PS I) assembly and the stability of light-harvesting complex proteins as well as thylakoid membrane stabilization.

3.6.2.5 Flavonoids

Flavonoids are widely found in leaves and floral organs. They are grouped into anthocyanins, flavones, flavonols, flavanols, flavones, and isoflavones (Das and Roychoudhury 2014). Flavonoids have been considered as a secondary ROS scavenging system in plants (Fini et al. 2011). They also scavenge $^1O_2$ and protect the chloroplast especially the outer envelope of chloroplasts (Agatia et al. 2012).

List of different types of flavonoids

Flavonoids	Compounds
Flavanones	Hesperetin, naringenin, eriodictyol
Anthocyanins	Delphinidin, cyanidin, malvidin, pelargonidin, peonidin, petunidin
Flavonols	Quercetin, kaempferol, myricetin, isorhamnetin, quercetagetin, rutin
Flavanols	(+)-catechin, (+)-gallocatechin, (−)-epigallocatechin 3-gallate
Isoflavones	Genistein, daidzein, glycitein
Flavones	Luteolin, apigenin, tangeretin

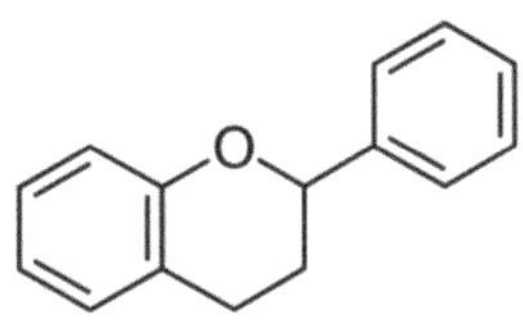

Basic structure of flavonoids

3.6.2.6 Proline

Proline is an antioxidant α-amino acid synthesized from glutamic acid via pyrroline 5-carboxylate pathway. It is also an inhibitor of programmed cell death. It removes $OH^\bullet$ and $^1O_2$ (Das and Roychoudhury 2014). During stress, proline accumulates in plants in large amounts which is mainly due to enhanced synthesis (Verbruggen and Hermans 2008) (Table 3.2).

Table 3.2 Cellular functions and locations of enzymatic and non-enzymatic antioxidants

Enzymatic antioxidants	Reaction catalyzed	Subcellular location
Superoxide dismutase (SOD)	$O_2^{\bullet}$-+$O_2^{\bullet}$-+2H+$\rightarrow$ $2H_2O_2$+O_2	Peroxisomes, mitochondria, cytosol, and chloroplast
Catalase (CAT)	$2H_2O_2 \rightarrow O_2+2H_2O$	Peroxisome and mitochondria
Ascorbate peroxidase (APX)	H_2O_2+AA $\rightarrow$ $2H_2O$+DHA	Peroxisomes, mitochondria, cytosol, and chloroplast
Monodehydroascorbate reductase (MDHAR)	MDHA+NADPH $\rightarrow$ AA+NADP+	Mitochondria, cytoplasm, and chloroplast
Dehydroascorbate reductase (DHAR)	DHA+2GSH $\rightarrow$ AA+GSSG	Mitochondria, cytoplasm, and chloroplast
Glutathione reductase (GR)	GSSG+NADPH $\rightarrow$ 2GSH+NADP+	Mitochondria, cytoplasm, and chloroplast
Guaiacol peroxidase (GPOX)	H_2O_2+GSH $\rightarrow$ H_2O+GSSG	Mitochondria, cytoplasm, chloroplast, and ER
Glutathione peroxidase (GPX)	ROOH/H_2O_2+GSH $\rightarrow$ ROH/H_2O +GSSG	Mitochondria, cytosol
Glutathione S-transferase (GST)	R-O-OH+GSH $\rightarrow$ R-OH+GS-OH $\rightarrow$ GS-SG	Mitochondria, cytosol
Polyphenol oxidase (PPO)	Monophenol $\xrightarrow{O_2}$ Diphenol $\xrightarrow{O_2}$ Quinone	Chloroplast
Non-enzymatic antioxidants	**Function**	**Subcellular location**
Ascorbic acid (AA) and reduced glutathione (GSH)	Need enzymes for scavenging ROS	Cytosol, chloroplast, mitochondria, peroxisome, and apoplast
α-Tocopherol	Detoxifies products of membrane LPO	Membrane
Carotenoids	Protect the photosynthetic apparatus by quenching a triplet sensitizer (Chl), $^1O_2$, and other harmful free radicals	Chloroplasts
Flavonoids	Direct scavengers of H_2O_2 and $^1O_2$ and $OH^{\bullet}$	Vacuole
Proline	Scavenger of $OH^{\bullet}$ and prevent damages due to LPO	Mitochondria, cytosol, and chloroplast

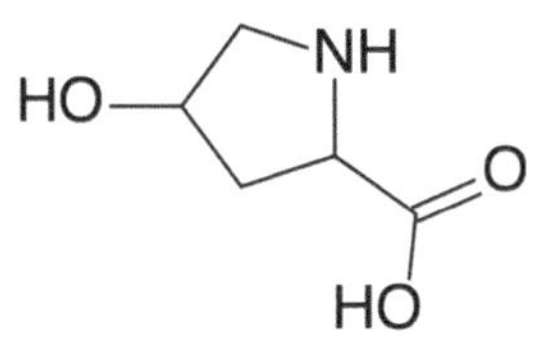

Structure of proline

3.7 Impact of Abiotic Stress on Antioxidant Defence System

Abiotic stresses like water deficit, excess moisture, soil and water salinity, and high and low temperature affect almost all the metabolism of plants. Water deficit and flooding are considered to be the major stresses that severely affect the plant growth. Stomatal closure causes a decrease in internal CO_2 concentration and thus results in a decline in photosynthesis by the diminished availability of CO_2 for carbon fixation. Reduction of CO_2 concentration increases the amount of harmful ROS within the leaf because of ongoing light reaction, which leads to leaf senescence and even death of the plant (Sairam et al. 1997). In addition to the drought and flooding, temperature stresses like heat, cold, and freezing are the other important abiotic factors which severely reduce the crop yields (Boyer 1982). In addition to the production of ROS, heat and cold stress also affect the membrane fluidity thus altering the membrane-related processes. ROS and membrane fluidity together affect the membrane-bound functional and structural proteins. ROS also activates the signalling pathway leading to the production of stress proteins, accumulation of compatible solutes, and upregulation of antioxidants.

3.7.1 Drought

Increased peroxidase activity and higher phenol content were recorded in apple in response to drought stress (Bolat et al. 2014). Increased proline concentration along with phenolic compounds was also observed in banana leaves during stress (Ismail et al. 2004). In some studies increased levels of ascorbic acid were also observed along with phenols and H_2O_2 during stress (El-Enany et al. 2013). Water stress increased the activities of POD, PPO along with increased membrane damage, and reduced the relative water content (Krishnamurthy et al. 2000). However, the activities of other antioxidant enzymes CAT and SOD were found to decrease. When stress was imposed on tomato plants, SOD activity was increased in all the growth stages. It was reported that SOD enhanced water stress tolerance in plants (Sivakumar 2014). Stronger correlation was observed between the activities of antioxidant enzymes POD, GR, and GST in the shoot with leaf water content after 1 week of water withdrawal in onion (Csiszár et al. 2007). Increased activities of antioxidant enzymes like CAT and SOD were observed in water-stressed mulberry plants (Chaitanya et al. 2002). Water stress increased LPO, CAT, SOD, APX, membrane injury index, H_2O_2, and $OH^{\cdot}$ in leaves of stressed bean plants (Zlatev et al. 2006). In most of the water stress studies, the activities of antioxidant enzymes increased, along with the increased membrane injuries and increased levels of ROS. Tolerant genotypes show lesser membrane damages, increased activities of antioxidant enzymes, and lower levels of ROS compared to the susceptible ones.

3.7.2 Salinity

Salt-stressed fava bean plants treated with stigmasterol had an increased membrane stability index, photosynthetic pigment, and carbohydrate contents compared with salt-stressed untreated plants (Hassanein et al. 2012). Salinity stress (50, 100, or 200 mM) increased electrolyte leakage and decreased activities of SOD, CAT, POD, APX, and GR. Sitosterol enhanced membrane stability and improved the chemical constituents and the enzyme activity in capsicum plants (Abu-Muriefah 2015). Activities of many antioxidant enzymes did not change when the NaCl concentrations were low (70 mol m^{-1}) but at higher concentrations (110 ± 130 mol m^{-1}) enhanced the activity of cytosolic Cu/Zn-SOD I and chloroplast Cu/Zn-SOD II in pea plants (Hernandez et al. 1999). Increased activities of APX and SOD were observed in the leaves of pea varieties EC 33866 and Puget at all the concentrations of NaCl. Increase in the activities of GR and DHAR were observed at 150 mM of NaCl; however at 200 mM NaCl activities declined (Ahmad et al. 2008), indicating that the antioxidant enzymes increase at moderate stress but at severe stress the response is negative. Anastasia et al. (2013) also studied moderate salt stress on tomato plants and observed an enhancement in lycopene and content of other potentially antioxidant compounds in fruits.

3.7.3 Chilling Stress

Increased activities of APX, GPX, GR, DHAR, and MDHAR along with higher MDA and H_2O_2, ascorbate, and glutathione while decreased activities of SOD and CAT were observed during chilling stress in capsicum (Guo et al. 2012). Chilling stress on cucumber plants also enhanced the activities of SOD, APX, GR, and POD while it decreased the CAT activity (Lee and Lee 2011). In strawberry plants also increased activities of POD, SOD, CAT, APX, DHAR, and GR enzymes and $O_2^{\bullet-}$ and H_2O_2 content were observed during chilling stress (Luo et al. 2011). In addition to this, increased dehydroascorbate and reduced glutathione content were also noticed. However, a reduction in MDA during chilling stress was reported by Christou et al. (2014).

3.7.4 High Temperature

High temperature (40. 8°C) increased the MDA and H_2O_2 concentrations in apple leaves. The content of total ascorbate, reduced ascorbic acid (AA), and total glutathione (GSH) was highest at 2 h in the high-temperature treatment, followed by a continuous decline with further increases in treatment duration (Ma et al. 2008). Tomato plants, when subjected to high-temperature stress, antioxidant enzymes like SOD and POD were increased (Rivero et al. 2001; Ogweno et al. 2009; Laxman et al. 2014). Watermelon plants were grown for 30 days at different temperatures (15, 25, and 35 °C). An increase in the accumulation of soluble phenolics and activity of PAL, POD, and PPO enzymes was observed at high temperatures (Rivero et al. 2001). Lower MDA and higher activities of POD, SOD, and APX under heat stress were observed in thermotolerant genotype of mung bean genotype NM 19-19 compared to the susceptible genotype NM 20-21, indicating the importance of antioxidant enzymes under heat stress (Mansoor and Naqvi 2013). Total soluble proteins decreased while total amino acids and proline content increased under high-temperature (40 °C) treatment (Chaitanya et al. 2001).

3.7.5 Flooding

Waterlogging decreased the activities of antioxidant enzymes like SOD, CAT, APX, and GR in mung bean (Ahmed et al. 2002). Waterlogging decreased the chlorophyll content and membrane stability index in the leaves but increased the activities of SOD, APX, GR, and CAT in pigeon pea (Kumutha et al. 2009). The activity of APX in roots significantly increased during the period of continuous waterlogging. However, the activities of CAT, SOD, and GR and the contents of AA, GSSG, and α-tocopherol in the roots were unaffected by waterlogging in tomato (Lin et al. 2004).

The antioxidant response of horticulture crops for various abiotic stresses is summarized Table 3.3.

3.8 Abiotic Stress and ROS: Transgenic Studies

It is known that most of the damages by the oxidative stress during abiotic stresses is due to the overproduction of ROS and reduced defence mechanisms. However, tolerant genotypes can show better defence mechanisms or lower production of ROS under stress. The defence mechanisms could be enzymatic or non-enzymatic or both. The timely expression of these processes is important for the tolerance to build up. There are several transgenic studies showing that the overexpression of antioxidant enzymes can give stress tolerance in plants.

Glutathione peroxidase genes (ApGPX2 and AcGPX2) were found to give tolerance against oxidative stress, drought, and salinity in *Arabidopsis* plants (Gaber et al. 2006). Ascorbate peroxidase genes have been shown to enhance the tolerance to chilling and salinity in tomato (APX gene) and for high light and drought in *Arabidopsis* (APX2 gene) (Kornyeyev et al. 2003; Rossel et al. 2006). From these studies it is evident that ascorbate and glutathione cycle enzymes are involved in giving tolerance against multiple stresses, due to their ability to scavenge ROS. Transfer of APX3 gene to tobacco enhanced the protection against oxidative stress further proving the importance of ascorbate in antioxidative processes (Wang et al. 1999). Downstream genes of the ascorbate cycle (DHAR) are also capable of providing tolerance to plants against salinity in *Arabidopsis* plants (Ushimaru et al. 2006). Salt tolerance in tobacco was enhanced by transferring the katE gene from *E. coli* which showed increased hydrogen peroxide scavenging (Al-Taweel et al. 2007). However, the overexpression of GST gene in *Arabidopsis* and cotton plants did not change any antioxidant activity (Light et al. 2005) indicating that the effect may be dependent on the type of stresses and plants. Transgenic tobacco overexpressing SOD in the chloroplast, mitochondria, and cytosol was generated (Bowler et al. 1991), and these have been shown to enhance tolerance to oxidative stress induced by methyl viologen (MV) in

Table 3.3 Response of enzymatic and non-enzymatic antioxidants under abiotic stresses in different crop plants

Stress	Crop	Response	References
Drought	Apple	POD ↑ and phenol content↑	Bolat et al. (2014)
Drought	Banana	Proline concentration ↑	Ismail et al. (2004)
Drought	Black pepper	CAT and SOD ↓ while POD and PPO↑	Krishnamurthy et al. (2000)
Drought	Cowpea	Phenolic compounds, H_2O_2 and ascorbic acids ↑	El-Enany et al. (2013)
Drought	Tomato	MDA, SOD, CAT activity↑	Sivakumar (2014)
Drought	Onion	POD, GR and GST↑ ↑severe	Csiszár et al. (2007)
Drought	Mulberry	CAT and SOD↑	Chaitanya et al. 2002
Drought	Beans	Lipid peroxidation, CAT, SOD, APX, H_2O_2, and $OH^{\cdot}$↑	Zlatev et al. (2006)
Drought	Grapes	H_2O_2↑ and carotenoids ↓	Zonouri et al. (2014)
Salinity	Beans	CAT, ascorbic acid, POD, and reduced glutathione↑	Hassanein et al. (2012)
Salinity	Capsicum	Proline↑ and SOD, CAT, POD, APX, GR ↓	Abu-Muriefah (2015)
Salinity	Pea	APX, MDHAR, GR, DHAR, and MDA↑	Ahmad et al. (2008)
Salinity	Strawberry	Ascorbate and glutathione content↑	Christou et al. (2013)
Salinity	Tomato	Lycopene ↑	Anastasia et al. (2013)
Chilling	Capsicum	MDA, H_2O_2, GR, GPX↑ and SOD, CAT ↓	Guo et al. (2012)
Chilling	Cucumber	SOD, APX, GR, POD↑ and CAT	Lee and Lee (2011)
Chilling	Strawberry	POD, SOD, CAT, APX, DHAR, GR, $O_2^{\cdot -}$, H_2O_2 ↑	Luo et al. (2011)
Chilling	Strawberry	MDA, H_2O_2, and NO ↓	Christou et al. (2014)
High temperature	Apple	MDA and H_2O_2↑	Ma et al. (2008)
High temperature	Tomato, watermelon	Soluble phenolics, PAL, POD, PPO activity↑	Rivero et al. (2001)
High temperature	Tomato	SOD, POX, and GR↑	Laxman et al. (2014)
High temperature	Tomato	APX, SOD, GPOX, MDA↑, and CAT↓	Ogweno et al. 2009
High temperature	Mung bean	MDA content↓ and POD, SOD, APX↑	Mansoor and Naqvi (2013)
High temperature	Mulberry	Total soluble proteins ↓ and total amino acids, proline↑	Chaitanya et al. (2001)
Flooding	Mung bean	SOD, CAT, APX, and GR ↓	Ahmed et al. (2002)
Flooding	Pigeon pea	SOD, APX, GR, and CAT↑	Kumutha et al. (2009)
Flooding	Tomato	APX↑ and CAT, SOD, GR, AA, GSSG, tocopherols unaffected	Lin et al. (2004)

↑, increased; ↓, decreased

leaf disc assays. Transgenic alfalfa (*Medicago sativa*) plants cv. RA3 overexpressing Mn-SOD in the chloroplasts showed reduced injury from water stress, as determined by chlorophyll fluorescence, electrolyte leakage, and regrowth (McKersie et al. 1996) while Mn-SOD transferred to alfalfa plants showed tolerance to freezing stress (McKersie et al. 1993). Overexpression of chloroplast Cu/Zn-SOD showed a dramatic improvement in the photosynthetic performance under chilling stress conditions in transgenic tobacco (Gupta et al. 1993) and potato plants (Perl et al. 1993). Cu/Zn-SOD gene transferred to tobacco cultivars enhanced the tolerance to salt, water, and PEG stresses (Badawi et al. 2004a, b) while Tepperman and Dunsmuir (1990) transferred Cu/Zn-SOD gene to tobacco and tomato plants, which did not show any significant differences during stress condition.

In *Arabidopsis* overexpression of Mn-SOD or DHAR was found to enhance salt tolerance (Wang et al. 2004; Ushimaru et al. 2006); in tobacco also DHAR overexpression enhanced salt tolerance (Eltayeb et al. 2007). Exogenous application of ascorbic acid helped in fast recovery from the salt stress in tomato seedlings (Shalata and Neumann 2001). APX overexpression in the chloroplasts of *Nicotiana tabacum* also reduced the toxic effects of H_2O_2 and generated drought tolerance (Badawi et al. 2004a). Better photosynthesis, growth, and higher chlorophyll content were observed when ascorbic acid was applied to wheat cultivars during drought stress (Malik and Ashraf 2012). Decreased oxidative damage in terms of lipid peroxidation and electrolyte leakage along with a reduction in H_2O_2 was observed when *Arabidopsis* VTE1 gene was overexpressed in tobacco plants and subjected to drought conditions. Transgenic plants also exhibited increased chlorophyll content compared to wild type (Liu et al. 2008).

Reduced oxidative damage, better recovery, and growth after drought stress were seen in many plants where antioxidative enzymes like GST, GPX, Cu/Zn-SOD, Mn-SOD, and APX were overexpressed (Gupta et al. 1993; Van Camp et al. 1996; Yu et al. 1999; Roxas et al. 2000; Lu et al. 2010) indicating the importance of scavenging the ROS during stress. Enhanced proline accumulation in transgenic petunia plants using *Arabidopsis* P5CS gene increased the drought tolerance by scavenging ROS (Yamada et al. 2005). Overexpression of signal transduction genes (MAPKKK) also increased the performance of plants during drought stress (Shou et al. 2004).

The effect of antioxidant genes on the abiotic stress tolerance in various plants is summarized in Table 3.4.

3.9 Future Strategies

A lot of work has been done in understanding the abiotic stress responses in plants and identifying the traits related to the tolerance. Physiological and molecular analyses have played a major role in understanding the abiotic stress responses in various plants. Detailed studies on the mechanism tolerance to various abiotic stresses at physiological, biochemical, and molecular level are a prerequisite in improving crop productivity. The anticipated global environmental changes in the near future may further aggravate the abiotic stress-related plant responses. This necessitates intensive research efforts so as to better understand and manage crops from adverse effect of abiotic stresses. Salinity and drought stress are increasing drastically at the global level. Identification of proper traits for overcoming these stresses in wild-type relatives of crops and their incorporation into the cultivated species are need of the hour. Creating transgenic plants overexpressing certain genes of antioxidant enzymes or signal transduction mechanism or transcription factors also helps in overcoming the adverse effects of these stresses. In addition to the tolerance, more emphasis needs to be given for better acquisition of water, selective avoidance of salt ions, and more conservation of water leading to more productivity per unit of water utilized in horticultural crops. These mechanisms not only reduce the ROS production but also increase productivity.

3.10 Conclusions

Horticultural crop plants face various abiotic stresses of seasonal occurrence which influence plant growth and productivity. These abiotic stresses increase the generation of ROS that are harmful to plant cells at high concentrations. These ROS cause damage to the metabolites such as proteins, lipids, and nucleic acids which ultimately leads to cell death. Normally plants possess specific mechanisms to detoxify the ROS through the activation of antioxidant defence system. During the past few decades, genetic engineering approach has given encouraging results in terms of improving tolerance to abiotic stresses by enhanced activities of enzymatic and non-enzymatic antioxidants. However, more efforts are needed in this direction especially in horticultural crops to cope up with current environmental challenges.

Table 3.4 Overexpression of ROS scavenging enzymatic antioxidants and their effects in transgenic plants for abiotic stress tolerance

Gene	Source	Effect	References
AA	Tomato	Salt stress tolerance	Shalata and Neumann (2001)
AA	Wheat	Higher chlorophyll contents under drought stress	Malik and Ashraf (2012)
ApGPX2 and AcGPX2	*Arabidopsis*	Oxidative stress, drought, and salt resistance	Gaber et al. (2006)
cAPX	Tobacco	Reduced the toxic effects of H_2O_2 under drought tolerance	Badawi et al. (2004a)
APX	Cotton	Chilling and salt tolerance	Kornyeyev et al. (2003)
Mn-SOD	*Arabidopsis*	Salt tolerance	Ushimaru et al. (2006)
APX2	*Arabidopsis*	High light- and drought-tolerant mutant	Rossel et al. (2006)
APX3	Tobacco	Increased protection against oxidative stress	Wang et al. (1999)
P5CS	Petunia plants	Resistance to drought	Yamada et al. (2005)
VTE1	Tobacco	Decreased LPO, H_2O_2, increased chlorophyll under drought	Liu et al. (2008)
AtMDAR1	Tobacco	Ozone, salt, and polyethylene glycol tolerance	Eltayeb et al. (2007)
Cu/Zn-SOD	Potato, tobacco	Improvement in the photosynthesis under chilling and heat stress	Perl et al. (1993); Gupta et al. (1993)
Cu/Zn-SOD	Tobacco	Enhanced tolerance to salt, water, and PEG stresses	Badawi et al. (2004b)
Cu/Zn-SOD	Tobacco, tomato	No protection seen against superoxide toxicity	Tepperman and Dunsmuir (1990)
Cu/Zn-SOD, APX	Sweet potato	Enhanced tolerance and recovery from drought stress	Lu et al. (2010)
DHAR	*Arabidopsis*	Salt tolerance	Wang et al. (2004)
katE	Tobacco	Salt tolerance by hydrogen peroxide scavenging	Al-Taweel et al. (2007)
GPX	Tobacco	Chilling and salt resistance	Yoshimuru et al. (2004)
GST	Cotton	No antioxidant activity under salt stress	Light et al. (2005)
GST, GPX	Tobacco	Reduced oxidative damage in the stressed transgenic seedlings	Roxas et al. (2000)
Mn/Fe-SOD	Alfalfa	Increased photosynthesis	Rubio et al. (2002)
Mn-SOD	Alfalfa	Tolerance to freezing stress; water deficit; winter survival	McKersie et al. (1993, 1996, 1999)
Mn-SOD	Canola	Aluminium tolerance	Basu et al. (2001)
Mn-SOD	Tobacco	Reduced cellular damage under oxidative stress	Bowler et al. (1991)
Mn-SOD	Tobacco	Tolerance to manganese deficiency	Yu et al. (1999)
Nt107 GST	Tobacco	Sustained growth under cold and salinity stress	Roxas et al. (1997)
NtPox GST	*Arabidopsis*	Protect against aluminium toxicity and oxidative stress	Ezaki et al. (2000)
NPK1	Maize	Enhanced drought tolerance	Shou et al. (2004)

References

Abu-Muriefah SS (2015) Effect of sitosterol on growth, metabolism and protein pattern of pepper (*Capsicum Annuum* L.) plants grown under salt stress conditions. Int J Agric Crop Sci 8(2):94–106

Agatia G, Azzarellob E, Pollastrib S, Tattinic M (2012) Flavonoids as antioxidants in plants: location and functional significance. Plant Sci 196:67–76

Ahmad P, John R, Sarwat M, Umar S (2008) Responses of proline, lipid peroxidation and antioxidative enzymes in two varieties of *Pisum sativum* L. under salt stress. Int J Plant Prod 2:353–366

Ahmed S, Higuchi H, Nawata E, Sakuratani T (2002) Jpn J Trop Agric 46:166–174

Al-Taweel K, Iwaki T, Yabuta Y, Shigeoka S, Murata N, Wadano A (2007) A bacterial transgene for catalase protects translation of D1 protein during exposure of salt-stressed tobacco leaves to strong light. Plant Physiol 145:258–265

Anastasia E, Giannakoula, Ilias IF (2013) The effect of water stress and salinity on growth and physiology of tomato (*Lycopersicon esculentum* Mill.). Arch Biol Sci 65(2):611–620

Apel K, Hirt H (2004) Reactive oxygen species: metabolism, oxidative stress, and signal transduction. Annu Rev Plant Biol 55:373–399

Augusto O, Miyamoto S (2011) Oxygen radicals and related species. In: Pantopoulos K, Schipper HM (eds) Principles of free radical biomedicine. Nova Science Publishers, Brazil, pp 1–23

Badawi GH, Kawano N, Yamauchi Y, Shimada E, Sasaki R, Kubo A, Tanaka K (2004a) Overexpression of ascorbate peroxidase in tobacco chloroplasts enhances the tolerance to salt stress and water deficit. Physiol Plant 121:231–238

Badawi GH, Yamauchi Y, Shimada E, Sasaki R, Kawano N, Tanaka K (2004b) Enhanced tolerance to salt stress and water deficit by overexpressing superoxide dismutase in tobacco (*Nicotiana tabacum*) chloroplasts. Plant Sci 166:919–928

Basu U, Good AG, Taylor GJ (2001) Transgenic Brassica *napus* plants overexpressing aluminium-induced mitochondrial manganese superoxide dismutase cDNA are resistant to aluminium. Plant Cell Environ 24:1269–1278

Bhattacharjee S (2005) Reactive oxygen species and oxidative burst: roles in stress, senescence and signal transduction in plants. Curr Sci 89(7): 1113–1121

Bita CE, Gerats T (2013) Plant tolerance to high temperature in a changing environment: scientific fundamentals and production of heat stress-tolerant crops. Front Plant Sci 4:273–290

Boguszewska D and Zagdańska B (2012) ROS as signaling molecules and enzymes of plant response to unfavorable environmental conditions. In: Lushchak Volodymyr Lushchak and Semchyshyn HM (eds) In oxidative stress – molecular mechanisms and biological effects, 341-375, ISBN: 978-953-51-0554-1, Published by InTech Janeza Trdine 9, 51000 Rijeka, Croatia

Bolat I, Dikilitas M, Ercisli S, Ikinci A, Tonkaz T (2014) The effect of water stress on some morphological, physiological, and biochemical characteristics and bud success on apple and quince rootstocks. Sci World J 2014:1–8

Bolwell GP, Wojtaszek P (1997) Mechanisms for the generation of reactive oxygen species in plant defence – a broad perspective. Physiol Mol Plant Path 51:347–366

Boscaiu M, Lull C, Lidon A, Bautista I, Donat P, Mayoral O, Vicente O (2008) Plant response to abiotic stress in their natural habitats. Bull UASVM, Horti 65:53–58

Bowler C, Slooten L, Vandenbranden S, De Rycke R, Botterman J, Sybesma C, Van Montagu M, Inze D (1991) Manganese superoxide dismutase can reduce cellular damage mediated by oxygen radicals in transgenic plants. EMBO J 10(7):1723–1732

Boyer JS (1982) Plant productivity and environment. Sci 218:443–448

Chaitanya KV, Sundar D, Ramachandra RA (2001) Mulberry leaf metabolism under high temperature stress. Biol Plant 44(3):379–384

Chaitanya KV, Sundar D, Masilamani S, Reddy AR (2002) Variation in heat stress-induced antioxidant enzyme activities among three mulberry cultivars. Plant Growth Regul 36:175–180

Chen Z, Gallie DR (2006) Dehydroascorbate reductase affects leaf growth, development, and function. Plant Physiol 142:775–787

Christou A, Manganaris GA, Papadopoulos I, Fotopoulos V (2013) Hydrogen sulfide induces systemic tolerance to salinity and non-ionic osmotic stress in strawberry plants through modification of reactive species biosynthesis and transcriptional regulation of multiple defence pathways. J Exp Bot 64(7):1953–1966

Christou A, Filippou P, Manganaris GA, Fotopoulos V (2014) Sodium hydrosulfide induces systemic thermotolerance to strawberry plants through transcriptional regulation of heat shock proteins and aquaporin. BMC Plant Biol 14:111

Corpas FJ, Barroso JB, del Río LA (2001) Peroxisomes as a source of reactive oxygen species and nitric oxide signal molecules in plant cells. Trends Plant Sci 6 (4):145–150

Csiszár J, Lantos E, Tari I, Madoşă E, Wodala B, Vashegyi A, HorváthF PA, Szabó M, Bartha B, Gallé A, Lazăr A, Coradini G, Staicu M, Postelnicu S, Mihacea S, Nedelea G, Erdei L (2007) Antioxidant enzyme activities in Allium species and their cultivars under water stress. Plant Soil Environ 53(12):517–523

Das K, Roychoudhury A (2014) Reactive oxygen species (ROS) and response of antioxidants as ROS-scavengers during environmental stress in plants. Front Environ Sci 2:53

El-Enany AE, AL-Anazi AD, Dief N, Al-Taisan WA (2013) Role of antioxidant enzymes in amelioration of water deficit and waterlogging stresses on *Vignasinensis* plants. J Biol Earth Sci 3(1):B44–B53

Eltayeb AE, Kawano N, Badawi GH, Kaminaka H, Sanekata T, Shibahara T, Inanaga S, Tanaka K (2007) Overexpression of monodehydroascorbate reductase in transgenic tobacco confers enhanced tolerance to ozone, salt and polyethylene glycol stresses. Planta 225:1255–1264

Ezaki B, Gardner RC, Ezaki Y, Matsumoto H (2000) Expression of aluminum-induced genes in transgenic *Arabidopsis* plants can ameliorate aluminum stress and/or oxidative stress. Plant Physiol 122:657–665

Fini A, Brunetti C, Ferdinando MD, Ferrini F, Tattini M (2011) Stress-induced flavonoid biosynthesis and the antioxidant machinery of plants. Plant Signal Behav 6 (5):709–711

Foyer CH, Noctor G (2003) Redox sensing and signalling associated with reactive oxygen in chloroplasts, peroxisomes and mitochondria. Physiol Plant 119:355–364

Foyer CH, Noctor G (2009) Redox regulation in photosynthetic organisms: signaling, acclimation, and practical implications. Antioxid Redox Signal 11 (4):861–905

Gaber A, Yoshimura K, Yamamoto T, Yabut Y, Takeda T, Miyasaka H, Nakano Y, Shigeoka S (2006) Glutathione peroxidase-like protein of *Synechocystis* PCC 6803 confers tolerance to oxidative and environmental stresses in transgenic *Arabidopsis*. Physiol Plant 128:251–262

Gill SS, Tuteja N (2010) Reactive oxygen species and antioxidant machinery in abiotic stress tolerance in crop plants. Plant Physiol Biochem 48:909–930

Guo WL, Chen RG, Gong ZH, Yin YX, Ahmedand SS, He YM (2012) Exogenous abscisic acid increases antioxidant enzymes and related gene expression in pepper (*Capsicum annuum*) leaves subjected to chilling stress. Genet Mol Res 11(4):4063–4080

Gupta AS, Heinen JL, Holaday AS, Burke JJ, Allen RD (1993) Increased resistance to oxidative stress in transgenic plants that over-express chloroplastic Cu/Zn superoxide dismutase. Proc Natl Acad Sci 90:1629–1633

Halliwell B (2006) Reactive species and antioxidants. Redox biology is a fundamental theme of aerobic life. Plant Physiol 141:312–322

Hassanein RA, Hashem HA, Khalil RR (2012) Stigmasterol treatment increases salt stress tolerance of faba bean plants by enhancing antioxidant systems. Plant Omics J 5(5):476–485

Hatz S, Lambert JD, Ogilby PR (2007) Measuring the lifetime of singlet oxygen in a single cell: addressing the issue of cell viability. Photochem Photobiol Sci 6:1106–1116

Hernandez JA, Campillo A, Jimenez A, Alarcon JJ, Sevilla F (1999) Response of antioxidant systems and leaf water relations to NaCl stress in pea plants. New Phytol 141:241–251

Ismail MR, Yusoff MK, Maziah M (2004) Growth, water relations, stomatal conductance and proline concentration in water stressed banana (Musa spp.) plants. Asian J Plant Sci 3(6):709–713

Kornyeyev D, Logan BA, Allen RD, Holaday AS (2003) Effect of chloroplastic overproduction of ascorbate peroxidase on photosynthesis and photoprotection in cotton leaves subjected to low temperature photo inhibition. Plant Sci 165:1033–1041

Krieger-Liszkay A, Fufezan C, Trebst A (2008) Singlet oxygen production in photosystem II and related protection mechanism. Photosynth Res 98:551–564

Krishnamurthy KS, Ankegowda SJ, Saji KV (2000) Water stress effects on membrane damage and activities of catalase, peroxidase and superoxide dismutase enzymes in black pepper (*Piper nigrum* L.). J Plant Biol 27(1):39–42

Kumutha D, Ezhilmathi K, Sairam RK, Srivastava GC, Deshmukhand PS, Meena RC (2009) Waterlogging induced oxidative stress and antioxidant activity in pigeon pea genotypes. Biol Plant 53(1):75–84

Laxman RH, Rao NKS, Biradar G, Sunoj VSJ, Shivashankara KS, Pavithra CB, Dhanyalakshmi KH, Manasa KM, Bhatt RM, Sadashiva AT, Christopher MG (2014) Antioxidant enzymes activity and physiological response of tomato (*Lycopersicon esculentum* M.) genotypes under mild temperature stress. Ind J Plant Physiol 19(2):161–164

Lee DH, Lee CB (2011) Chilling stress-induced changes of antioxidant enzymes in the leaves of cucumber: in gel enzyme activity assays. Plant Sci 159(1):75–85

Light GG, Mahan JR, Roxas VP, Allen RD (2005) Transgenic cotton (*Gossypium hirsutum* L.) seedlings expressing a tobacco glutathione S-transferase fail to provide improved stress tolerance. Planta 222:346–354

Lin KR, Weng C, Lo H, Chen J (2004) Study of the root antioxidative system of tomatoes and eggplants under waterlogged conditions. Plant Sci 167:355–365

Liu X, Hua X, Guo J, Qi D, Wang L, Liu Z, Jin Z, Chen S, Liu G (2008) Enhanced tolerance to drought stress in transgenic tobacco plants overexpressing *VTE1* for increased tocopherol production from *Arabidopsis thaliana*. Biotechnol Lett 30:1275–1280

Lu Y, Deng X, Kwak S (2010) Over expression of Cu/Zn superoxide dismutase (Cu/Zn SOD) and ascorbate peroxidase (APX) in transgenic sweet potato enhances tolerance and recovery from drought stress. Afr J Biotechnol 9(49):8378–8391

Luo Y, Tang H, Zhang Y (2011) Production of reactive oxygen species and antioxidant metabolism about strawberry leaves to low temperature. J Agric Sci 3 (2):89–96

Ma Y, Ma FJ, Zhang J, Li M, Wang Y, Liang D (2008) Effects of high temperature on activities and gene expression of enzymes involved in ascorbate-glutathione cycle in apple leaves. Plant Sci 175 (6):761–766

Malik S, Ashraf M (2012) Exogenous application of ascorbic acid stimulates growth and photosynthesis of wheat (*Triticum aestivum* L.) under drought. Soil Environ 31:72–77

Mansoor S, Naqvi FN (2013) Effect of heat stress on lipid peroxidation and antioxidant enzymes in mung bean (*Vigna radiata* L) seedlings. Afr J Biotechnol 12 (21):3196–3203

McCord JM, Fridovich I (1969) Superoxide dismutase. An enzymic function for erythrocuprein (Hemocuprein). J Biol Chem 244(22):6049–6065

McKersie BD, Chen Y, de Beus M, Bowley SR, Bowler C, Inze D, D'Halluin K, Boterman J (1993) Superoxide dismutase enhances tolerance of freezing stress in transgenic alfalfa (*Medicago sativa* L.). Plant Physiol 103:1155–1163

McKersie BD, Bowley SR, Harjanto E, Leprince O (1996) Water deficit tolerance and field performance of transgenic alfalfa overexpressing superoxide dismutase. Plant Physiol 111:1177–1181

McKersie BD, Bowley SR, Jones KS (1999) Winter survival of transgenic alfalfa overexpressing superoxide dismutase. Plant Physiol 199:839–847

Mhamdi A, Queval G, Chaouch S, Vanderauwera S, Breusegem FV, Noctor G (2010) Catalase function in plants: a focus on Arabidopsis mutants as stress-mimic models. J Exp Bot 61(15):4197–4220

Mittler R (2002) Oxidative stress, antioxidants and stress tolerance. Trends Plant Sci 7(9):405–410

Moller IM, Jensen PE, Hansson A (2007) Oxidative modifications to cellular components in plants. Annu Rev Plant Biol 58:459–481

Noctor G, Foyer CH (1998) Ascorbate and glutathione: keeping active oxygen under control. Annu Rev Plant Physiol Plant Mol Biol 49:249–279

Noctor G, Veljovic-Jovanovic S, Foyer CH (2000) Peroxide processing in photosynthesis: antioxidant coupling and redox signalling. Philos Trans R Soc Lond B 355:1465–1475

Ogweno JO, Zhou YH, Yu JQ (2009) Changes in activities of antioxidant enzymes and photosynthesis in detached leaves of tomato after exposure to different temperatures. Afr J Hortic Sci 2:124–137

Perl A, Perl-Treves R, Galili S, Aviv D, Shalgi E, Malkin S, Galun E (1993) Enhanced oxidative stress defense in transgenic potato expressing tomato Cu, Zn superoxide dismutases. Theor Appl Genet 85:568–576

Reddy AR, Raghavendra AS (2006) Photooxidative stress. In: Madhava Rao KV, Raghavendra AS, Reddy KJ (eds) Physiology and molecular biology of stress tolerance in plants. Springer, Dordrecht, pp 157–186

Rhoads DM, Umbach AL, Subbaiah CC, Siedow JN (2006) Mitochondrial reactive oxygen species. Contribution to oxidative stress and interorganellar signaling. Plant Physiol 141:357–366

Rivero RM, Ruiz JM, García PC, López-Lefebre LR, Sánchez E, Romero L (2001) Resistance to cold and heat stress: accumulation of phenolic compounds in tomato and watermelon plants. Plant Sci 160 (2):315–321

Rossel JB, Walter PB, Hendrickson L, Chow WS, Poole A, Mullineaux PM, Pogson BJ (2006) A mutation affecting ascorbate peroxidase 2 gene expression reveals a link between responses to high light and drought tolerance. Plant Cell Environ 29:269–281

Roxas VP, Smith RK, Allen ER, Allen RD (1997) Overexpression of glutathione S-transferase/ glutathione peroxidase enhances the growth of transgenic tobacco seedlings during stress. Nat Biotechnol 15:988–991

Roxas VP, Lodhi SA, Garrett DK, Mahan JR, Allen RD (2000) Stress tolerance in transgenic tobacco seedlings that overexpress glutathione S-transferase/ glutathione peroxidase. Plant Cell Physiol 41:1229–1234

Rubio MC, González EM, Minchin FR, Webb KJ, Arrese-Igor C, Ramos J, Becana M (2002) Effects of water stress on antioxidant enzymes of leaves and nodules of transgenic alfalfa overexpressing superoxide dismutases. Physiol Plant 115:531–540

Sairam RK, Deshmukh PS, Shukla DS (1997) Tolerance of drought and temperature stress in relation to increased antioxidant enzyme activity in wheat. J Agron Crop Sci 178:171–177

Shalata A, Neumann PM (2001) Exogenous ascorbic acid (vitamin C) increases resistance to salt stress and reduces lipid peroxidation. J Exp Bot 52:2207–2211

Sharma P, Jha AB, Dubey RS, Pessarakli M (2012) Reactive oxygen species, oxidative damage, and antioxidative defense mechanism in plants under stressful conditions. J Bot 2012:1–26

Shou H, Bordallo P, Wang K (2004) Expression of the *Nicotiana* protein kinase (NPK1) enhanced drought tolerance in transgenic maize. J Exp Bot 55:1013–1019

Sivakumar R (2014) Effect of drought on plant water status, gas exchange and yield parameters in contrasting genotypes of tomato (*Solanum lycopersicum*). Am Int J Res Form Appl Nat Sci 8 (1):57–62

Smirnoff N, Wheeler GL (2000) Ascorbic acid in plants: biosynthesis and function. Crit Rev Biochem Mol Biol 35(4):291–314

Tepperman JM, Dunsmuir P (1990) Transformed plants with elevated levels of chloroplast SOD are not more resistant to superoxide toxicity. Plant Mol Biol 14:501–511

Ushimaru T, Nakagawa T, Fujioka Y, Daicho K, Naito M, Yamauchi Y, Nonaka H, Amako K, Yamawaki K, Murata N (2006) Transgenic *Arabidopsis* plants expressing the rice dehydroascorbate reductase gene are resistant to salt stress. J Plant Physiol 163:1179–1184

Van Camp W, Capiau K, Van Montagu M, Inze D, Slooten L (1996) Enhancement of oxidative stress tolerance in transgenic tobacco plants overproducing Fe-superoxide dismutase in chloroplasts. Plant Physiol 112:1703–1714

Verbruggen N, Hermans C (2008) Proline accumulation in plants: a review. Amino Acids. doi:10.1007/s00726-008-0061-6

Wang J, Zhang H, Allen RD (1999) Overexpression of an *Arabidopsis* peroxisomal ascorbate peroxidase gene in tobacco increases protection against oxidative stress. Plant Cell Physiol 40(7):725–732

Wang Y, Ying Y, Chen J, Wang X (2004) Transgenic *Arabidopsis* overexpressing Mn-SOD enhanced salt-tolerance. Plant Sci 167:671–677

Yamada M, Morishita H, Urano K, Shiozaki N, Yamaguchi-Shinozaki K, Shinozaki K, Yoshiba Y (2005) Effects of free proline accumulation in petunias under drought stress. J Exp Bot 56:1975–1981

Yordanov I, Velikova V, Tsone V (2000) Plant response to drought, acclimation and stress tolerance. Photosynthetica 30:171–186

Yoshimuru K, Miyao K, Gaber A, Takeda T, Kanaboshi H, Miyasaka H, Shigeoka S (2004) Enhancement of stress tolerance in transgenic tobacco plants over expressing *Chlamydomonas* glutathione peroxidase in chloroplasts or cytosol. Plant J 37:21–33

Yu Q, Osborne LD, Rengel Z (1999) Increased tolerance to Mn deficiency in transgenic tobacco overproducing superoxide dismutase. Ann Bot 84:543–547

Zhu BZ, Zhao HT, Kalyanaraman B, Liu J, Shan GQ, Du YG, Frei B (2007) Mechanism of metal-independent decomposition of organic hydroperoxides and formation of alkoxyl radicals by halogenated quinones. Proc Natl Acad Sci 104(10):3698–3702

Zlatev ZS, Lidon FC, Ramalho JC, Yordanov IT (2006) Comparison of resistance to drought of three bean cultivars. Biol Plant 50(3):389–394

Zonouri M, Javadi T, Ghaderi N, Saba MK (2014) Effect of foliar spraying of ascorbic acid on chlorophyll a, chlorophyll b, total chlorophyll, carotenoids, hydrogen peroxide, leaf temperature and leaf relative water content under drought stress in grapes. Bull Environ Pharmacol Life Sci 3:178–184

Molecular Mechanisms of Heat Shock Proteins and Thermotolerance in Plants

4

Vidya S. Murthy and Kundapura V. Ravishankar

Abstract

Plants are sessile organisms which are regularly exposed to many stresses due to changing environmental factors. The majority of the stresses induce the production of a group of proteins called heat shock proteins (Hsps). These proteins are grouped into different classes according to their molecular weight: Hsp100, Hsp90, Hsp70, Hsp60, small heat shock proteins (sHsps), and ubiquitins. Hsps act as molecular chaperones and regulate the protein folding, accumulation and transportation, and removal of damaged proteins. In case of temperature stress signaling pathway, the plasma membrane carries out the primary role with the help of other secondary messengers, such as Ca^{2+} ions, hydrogen peroxide (H_2O_2), and nitric oxide (NO). Other components, such as mitogen-activated protein kinases (MAPK) and calmodulins, regulate the activation of heat shock transcription factors (HSFs) and thus the synthesis of Hsps. In this chapter, we discuss the significance of Hsps in thermotolerance and also the role of different classes of Hsps and their interactions with other stress-induced components.

4.1 Introduction

Heat stress is often defined as the increase in temperature beyond a threshold level for a specific time period, sufficient to cause the irreversible change to plant growth and development. However, heat stress is a complex phenomenon, and it is a function of intensity, duration, and rate of increase in temperature. Heat tolerance is generally defined as the ability of the plant to grow and produce an economic yield under high temperature stress (Peet and Willits 1998).

Heat stress is a detrimental factor in crop production worldwide due to increasing ambient temperatures (Hall 2001). Different global circulation models predict that greenhouse gasses will gradually increase the world's average ambient temperature. According to the Intergovernmental Panel on Climate Change (IPCC), the global

V.S. Murthy • K.V. Ravishankar (✉)
Division of Biotechnology, ICAR-Indian Institute of Horticultural Research, Hessaraghatta Lake Post, Bengaluru 560089, Karnataka, India
e-mail: kvravi@iihr.ernet.in

N.K.S. Rao et al. (eds.), *Abiotic Stress Physiology of Horticultural Crops*,
DOI 10.1007/978-81-322-2725-0_4

mean temperature will rise by 0.3 °C per decade (Jones et al. 1999) reaching to approximately 1 and 3 °C above the present value by years 2025 and 2100, respectively, leading to global warming.

Plants respond to high temperature (HT) with variation in the intensity of temperature, time, and development stage. At extreme HT, damage or death of the cellular content may occur within minutes, which may lead to a catastrophic disintegration of the cellular organization (Ahuja et al. 2010). Heat stress affects many aspects of plant processes like development, germination, growth, reproduction, and yield. Under heat stress stability of various proteins, RNA species and cytoskeleton structures, membranes, and the efficiency of enzymatic reactions in the cell are seen to be altered. This finally leads to the breakdown of major physiological processes in the plants, resulting into metabolic imbalance (Hasanuzzaman et al. 2013). Like other organisms, plants have both an inherent ability to survive at high temperatures (basal thermotolerance) and the ability to acquire thermotolerance (Kumar et al. 2007). Acquired thermotolerance may be induced by either exposure to short but sublethal high temperatures (Pumisutapon et al. 2012) or by a gradual temperature increase to lethally high levels, as experienced under natural conditions (Larkindale and Vierling 2008). In natural environments, plants frequently experience gradual daily temperature range fluctuations and have become acclimated to gradual increases to high temperatures that would otherwise be lethal. Acquired thermotolerance may, therefore, reflect the natural mechanism that contributes to thermotolerance in plants (Fig. 4.1). Thermotolerance is thus facilitated by synthesis of a distinctive class of proteins called heat shock proteins (Hsps).

4.2 Evolution/Discovery of Heat Shock Proteins (Hsps)

The synthesis of heat shock proteins (Hsps) has been shown to increase by heat stress. First it was studied in soybean seedlings, where they were grown at 28–40 °C and production of proteins changed dramatically with synthesis of new set of proteins called Hsps (Key et al. 1981). The discovery of Hsps was first seen in *Drosophila* and later in a wide range of organisms like *Escherichia coli* and *Saccharomyces cerevisiae* where the responses were seen to be highly similar in terms of homologous regions and molecular mechanisms (Ritossa 1962). Different Hsp classes are found in various cellular compartments, semiautonomous organelles, mitochondria, and chloroplasts. It was also proven that the production of Hsps is essential and is the most fundamental response to temperature change.

4.3 Classification of Hsps

Heat shock proteins and other stress-associated proteins are known to protect plant cells against deleterious effects. They are classified based on the size, site of synthesis, and function. The size of Hsps range from~15 to 110 kDa and are divided into different groups based on their size and function. Six structurally conserved distinct classes are: Hsp100, Hsp90, Hsp70, Hsp60 (chaperonins), approximately 17–30 kDa molecular weight referred to as small Hsps (sHsps), and ubiquitin (8.5 kDa) (Lindquist 1986; Waters et al. 1996; Helm et al. 1997; Bakau and Horwich 1998).

4.3.1 Hsp90 Family

These proteins vary from 82 to 96 kDa in size. These families of protein are mostly found to be abundant after the "stress/shock." Hsp90 functions as an ATP-dependent chaperone that binds and prevents aggregation and it is also capable of performing alone or in consonance with other proteins forming hetero-complexes like the cytoplasmic chaperone hetero-complex (CCH) (Reddy et al. 1999). The Hsp90 family found in *Arabidopsis* has seven members: one found in cytosol, one in each organelle (chloroplast and mitochondria), one in the endoplasmic

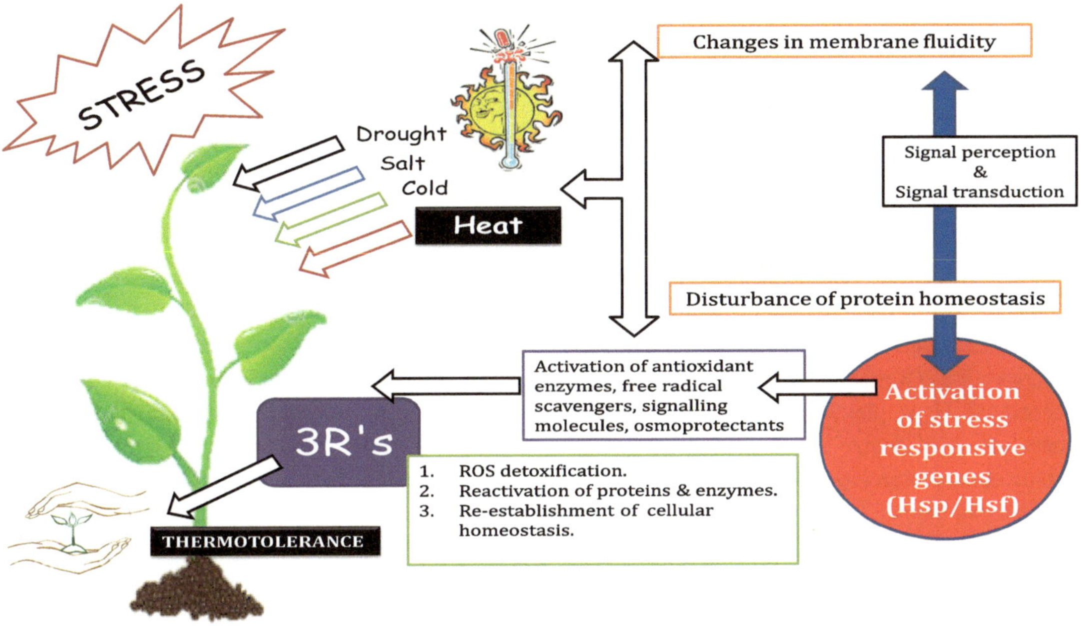

Fig. 4.1 Schematic model for temperature sensing in plants: this figure summarizes the major forms of abiotic stresses and focuses on heat stress and its independent mode of action. Heat stress leads to activation of stress-responsive mechanisms like ROS detoxification and cellular homeostasis and activation of Hsp leading to thermotolerance

reticulum (ER), and three very similar ones in the endomembrane system (Krishna and Gloor 2001). Hsp90 inhibitors like geldanamycin and radicicol have been shown to enhance thermotolerance by induction of a heat-shock-like response in *Arabidopsis* and *Physcomitrella* (Yamada et al. 2007; Saidi et al. 2009). A decrease in the cytosolic Hsp90 activity was seen in *Arabidopsis* during initial heat shock, which suggests that Hsp90 may be an inhibitor of HSFs in the absence of heat stress (Helm et al. 1997). Hence, it is hypothesized that, under heat stress, Hsp transcription would take place because the initial inactivation of Hsp90 would allow the release of HSFs (Yamada et al. 2007; Saidi et al. 2011).

4.3.2 Hsp70 Family

The Hsp70 family of proteins is highly ubiquitous and consists of the most temperature-sensitive and highly conserved Hsps. The Hsp70 family consists of ATP-binding proteins which show 60–80 % base similarity among eukaryotic cells. Both under normal and stress conditions, Hsp70 has essential functions in assisting refolding of nonnatives and in preventing aggregation. Hsp70 is also involved in the import of proteins, translocation, and facilitation of the proteolytic degradation of unstable proteins by targeting the proteins to lysosomes or proteasomes. The *Arabidopsis* genome contains Hsp70 family encoding members of at least 18 genes, of which 4 belongs to Hsp110/SSE subfamily and 14 to the DnaK subfamily (Wang et al. 2004). It has been demonstrated that members of Hsp70 chaperones are expressed in response to various abiotic stress conditions such as heat, cold, and drought, as well as to chemical and other stresses (Kregel 2002). A study on *A. thaliana* indicates the requirement of Hsp70 for the differentiation of germinating seeds, and its tolerance to heat was found in the stroma of chloroplast (Mizzen and Welch 1988; Li et al. 2011).

4.3.3 Hsp60 Family

The members of Hsp60 were the first proteins to be named as molecular chaperones that were supported by both biochemical and genetic studies in *E. coli*. The Hsp homologous are found both in chloroplast and mitochondria. Plant chloroplast chaperones had 50 % amino acid identity with two distinct polypeptides, Cpn60a and Cpn60b. The chloroplast homologue was originally identified as the ribulose bisphosphate carboxylase (Rubisco) binding protein. Mitochondrial Hsps play an important role during heat stress of maize seedlings. These proteins contribute a larger portion of the total protein content in maize seedlings than the 4-day-old plants. Hsp60 is known to be associated with nucleus-encoded folding, imported proteins, and perfect assembly of proteins inside the mitochondria (Vierling 1991).

Studies show that a mutation in *Arabidopsis* Cpn60a resulted in defective development of the plant embryo and seedling (De Klerk and Pumisutapon 2008). While in the case of Cpn60b (antisense), transgenic tobacco plants exhibited drastic phenotypic changes, including delayed flowering, slow growth, stunting, and leaf chlorosis (Wang et al. 2004).

4.3.4 Low Molecular Weight (LMW) Hsp

LMW Hsp sizes range from 15 to 42 kDa. The LMW Hsps belong to four multi-gene families, viz., two gene families encode cytoplasmic proteins, one gene family encodes chloroplast-localized protein, and one gene family in endomembrane proteins. Some studies suggest that these small Hsps act as molecular chaperones that prevent thermal aggregation of proteins (Diamant et al. 2001; Vierling 1991). The LMW Hsp was first identified in chloroplasts, in *Arabidopsis* it is encoded by a single gene (Vierling 1991), and in soybean and pea it is encoded by a small gene family (Liu et al. 2006a, b). The expression of DcHsp17.7 (*Daucus carota* L. class CI) gene in carrot which was genetically modified showed greater thermotolerance under DcHsp17.7-CI-overexpressing plants. Furthermore, plants overexpressing DcHsp17.7-CI in an antisense orientation were observed to be comparatively susceptible to heat stress (Sun et al. 2001).

4.3.5 Hsp100

The Hsp100/Clp class of chaperones are members of the ATPase superfamily with diverse functional properties of reactivation of aggregated proteins. Hsp100 family proteins have been reported in many plant species, such as *Arabidopsis*, tobacco, soybean, rice, maize, and wheat. The expression of Hsp100/Clp protein is developmentally regulated and is induced by heat stress. In a recent study on rice, Hsp100 protein production was correlated with the disappearance of protein granules in yeast cells (Wang et al. 2006; Sarkar et al. 2009; Sengupta and Majumder 2009). It was also observed that, after the stress period, the dissolution of electron-dense granules takes place by Hsp101, indicating the of Hsp100 in the recovery phase (Wang et al. 2004).

4.3.6 Ubiquitin

Ubiquitins are highly conserved proteins which contain 76 amino acids. The levels of ubiquitin increase due to increase in heat stress with the increased demand for removal of damaged proteins. In the studies conducted with the mammalian cell strain ts85, which is temperature sensitive, the cell line has a defective ubiquitin activating enzyme E1. At higher temperatures there is an increase in the production of ubiquitin and other Hsps like Hsp70 which suggests that inactivation and overloading of damaged proteins lead to the induction of heat shock proteins (Finley et al. 1984). Heat stress has been shown to result in a decrease in monomer pool size and accumulation of high-molecular-mass conjugates of ubiquitin in plant tissue. In potato, two heat shock-inducible genes were identified using polyubiquitin gene-specific probes (Belknap and Garbarino 1996). Similarly, in both maize and sunflower, transcripts from two polyubiquitin genes were observed to increase with temperature, while the mRNA levels remained unchanged (Liu et al. 1995) (Table 4.1).

4.4 Functions of Hsps

Heat shock proteins (Hsps) are molecular chaperones, a class of proteins which are involved in protein folding, assembly, translocation, and degradation in a broad array of normal cellular processes (Vierling 1991; Pratt 2001; Wang et al. 2004). Hsps are ubiquitous in the cell and are also being found in all subcellular locations. A few cytoprotective functions have been associated with Hsps and, in particular, the Hsp70 family. This Hsp70 family preferably includes the folding of proteins, intracellular compartments, refolding of misfolded proteins, protein structure maintenance, transmembrane protein translocation into various cellular compartments, prevention of aggregation of proteins, and finally degradation of unstable proteins (Deshaies et al. 1988; Palleros et al. 1991).

Further Hsps also play a role in apoptosis. Some proteins that were predominantly antiapoptotic were Hsp27, Hsp70, and Hsp90, whereas Hsp60 is proapoptotic. An increase in the role of Hsps in the apoptotic signaling pathway at multiple points was also observed (Kregel 2002). Hsps have a more general folding function, which associates them by activating and/or transferring the protein to their place of function (Pratt 2001). In plants some important protein targets were identified, such as R proteins that initiate the hypersensitive response (Hubert et al. 2003; Takahashi et al. 2003a, b) and clavata proteins that regulate the shoot and floral meristem (Ishiguro et al. 2002).

4.5 What Is Thermotolerance?

Abiotic stress can have an adverse impact on plant growth, development, reproduction, and yield. Temperature is one of the crucial factors that affect a wide range of plants. So the ability to survive temperature stress that would cause deleterious effects can be slightly overcome by exposing the plants to lower temperature or non lethal temperature; this process of acclimatizing them to temperature stress is termed as acquired thermotolerance (Lin et al. 1984; Lindquist and Craig 1988; Hahn and Li 1990; Srikanthbabu et al. 2002).

Plants which grow under natural climatic conditions might experience lethal stress because of the non-acquaintance to heat stress conditions; conversely, if plants have the ability to withstand the lethal temperature upon exposures to sublethal stress, they have what is referred to as "induction stress" tolerance (Kumar et al. 1999; Piterkovaa et al. 2013). Plants reshape their metabolism by different ways in response to high temperature, particularly by acquiring compatible solutes that help in organizing proteins and cellular structures, maintaining turgor pressure by osmotic adjustment, and reestablishing the cellular redox balance and homeostasis by modifying the ROS scavenging system. Heat stress tolerance leads to the alterations in the gene expression at molecular levels.

These include genes that are regulatory in function and are responsible for the expression of osmoprotectants, transporters, detoxifying enzymes, and stress-associated proteins (SAP) that gradually leads to the development of heat tolerance in the form of acclimation or to adaptation (Huang and Xu 2008) (Fig. 4.2).

Studies on *Arabidopsis* mutants have shown that phytohormones, such as ABA, ROS, and SA pathways, were also seen to be increased at different stages of development and maintenance of acquired thermotolerance apart from heat shock proteins (Hsp32 and Hsp101) (Larkindale and Huang 2005; Charng et al. 2006; Panchuk et al. 2002), while others, cytokinin (CK), auxin (AUXs), and gibberellic acids (GAs), decreased that ultimately cause premature plant senescence (Talanova et al. 2003; Larkindale and Huang 2004; Larkindale et al. 2005). For example, the abscission caused by higher ABA and ethylene levels and reduced levels and transport of

Table 4.1 Heat shock proteins, their families, and proposed functions

Family	Function	References
Hsp100	Dissociation and degradation of aggregated proteins (ATP dependent)	Swindell et al. (2007)
Hsp90	Co-regulator, helps in signal transduction and manages protein folding (ATP dependent)	Hubert et al. (2003)
Hsp70	Primary stabilization of newly formed proteins (ATP-dependent protein)	Palleros et al. (1991)
Hsp60	Folding machinery (ATP dependent)	Kulz (2003)
Hsp20 or small Hsp (sHsp)	Formation of high-molecular-weight oligomeric complexes that serve as cellular matrix for stabilization of unfolded proteins	Morrow and Tanguay (2012); Saidi et al. (2011); Sun et al. (2002)
	Hsp100, Hsp70, and Hsp40 are needed for its release	
Ubiquitin	Removal of damaged proteins for degradation	Ferguson et al. (1990)

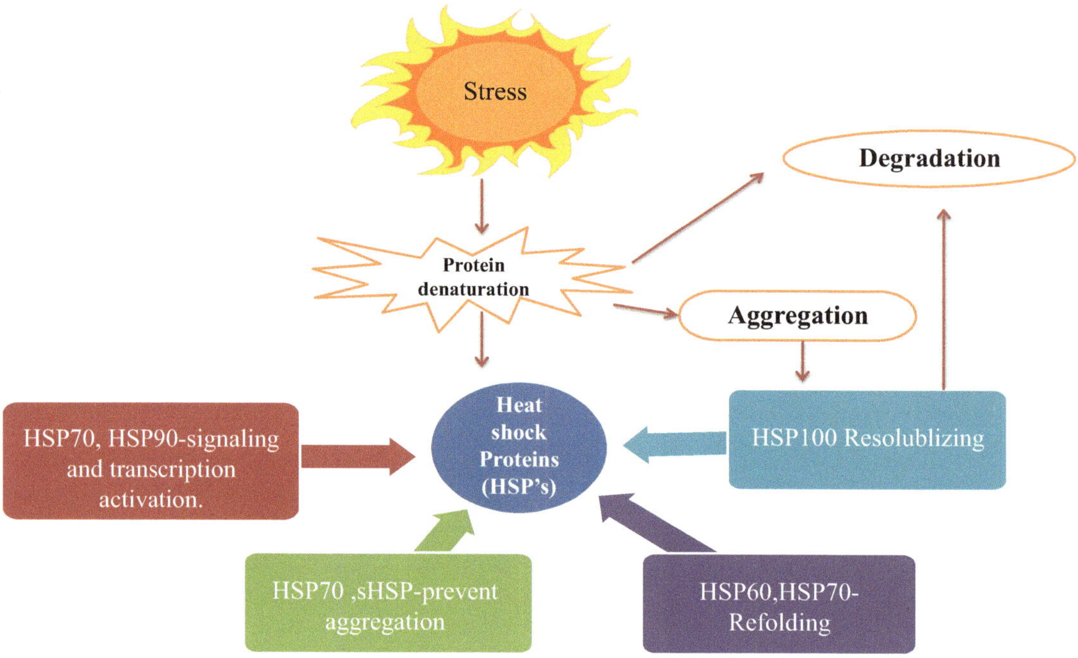

Fig. 4.2 Heat shock proteins: the heat shock protein (Hsp) plays a vital role in terms of the response to abiotic stress. Heat stress often leads to denaturation of structural and functional proteins. Thus, survival of cells under stress depends on maintenance of conformational changes and prevention of protein folding and aggregation. Some members of the Hsp/chaperone families [e.g., small Hsp (sHsp) and Hsp70] help in stabilizing protein conformation, and Hsp60, Hsp70, and Hsp90 prevent aggregation of misfolded proteins. The Hsp100/Clp can be resolubilized followed by refolding. Hsp70 and Hsp90 participate in signal transduction and transcription and lead to the synthesis of HSFs and ROS/antioxidants

AUXs to reproductive organs leads to a vital effect under heat stress (Binder and Patterson 2009).

A transient peak in ABA levels was reported in response to HS in pea plants (Liu et al. 2006a, b) and during recovery of creeping bentgrass from HS treatments. The ABA signaling mutants, *abi1* and *abi2* (insensitive), showed reduced survival after heat stress; inversely, accumulation of Hsps was not affected in these mutants of *Arabidopsis* (Larkindale et al. 2005). The key factors in the acquisition of thermotolerance are the adaptive mechanisms that protect cells from harmful effects under heat stress.

The heat shock response (HSR) is defined as a transient reprogramming of the gene expression of a conserved response of cells and organisms to elevated temperatures. The induction temperature for the HSR is normally 5–10 °C above the normal optimum growth temperature for that particular species. The features of this response include induction of Hsps synthesis and subsequent acquisition of a higher level of thermotolerance (Huang and Xu 2008).

In wheat seeds the accumulation of Hsp58, Hsp60, Hsp46, Hsp40, and Hsp14 induced by mild heat pretreatments during the initial hours conferred no acquired thermotolerance (Abernethy et al. 1989). However, Hsp101 and Hsp17.6 induced by water stress in cotyledons of cotton seedling did not contribute to thermotolerance (Bruke 2001; Burke and O'Mahony 2001). Plants synthesize sHsps during the response to heat shock, and thermotolerance was correlated with the accumulation of H25 (Guo et al. 2007). In addition, a gene of rice, OsHsp16.9, conferred thermotolerance to *Escherichia coli* (Yeh et al. 1997), and the altered expression of a gene for carrot, Hsp17.7, correlated with the thermotolerance of transformed carrot plants. These observations indicate the sHsps' role in thermotolerance (Sanmiya et al. 2004). In the case of young grape plants, heat acclimation was studied in leaves where Hsp70 and sHsp17.6 were predominantly distributed in the chloroplast, nucleus, cytoplasm, and mesophyll cells (Zhang et al. 2008).

4.6 Can Plants Sense the Temperature and Activate the Signaling Pathways?

The relay of the signal, perception of stress, and switching on the mechanism of adaption are important steps under plant stress tolerance. Various stimuli to heat stress involve signaling via the redox response system. Signaling molecules like ROS, Ca^{2+}, and other plant hormones activate genomic reprogramming via signal cascades (Joyce et al. 2003; Suzuki and Mittler 2006). Generation and later reactions due to activated oxygen species (AOS) including singlet oxygen ($^1O_2$), superoxide radical ($O_2^{\bullet-}$), hydrogen peroxide (H_2O_2), and hydroxyl radical ($OH^{\bullet}$) are results of cellular injury due to heat stress (Liu and Huang 2000). Autocatalytic peroxidation of membrane lipids and pigments takes place due to AOS, which finally leads to altered membrane permeability and later loss of its functions (Xu et al. 2006) (Fig. 4.3).

A model that describes the integration of temperature fluctuation and regulation under heat shock responses was demonstrated to trigger membrane fluidity which activates sensing of temperature stress and influencing gene expression. Macromolecules in the cells, such as protein complexes, membranes, and nucleic acid polymers, together "perceive" the heat shock. The increased kinetic movement of these macromolecules results in reversible physical changes, such as transcended membrane fluidity, partial melting of DNA, RNA strands, protein subunit dissociation, and exposure of hydrophobic cores (Mittler et al. 2012).

The rigidification of thylakoid membranes alters the gene expression profiles of Hsps, suggesting that the thylakoid membranes are the location for sensing variation in temperature (Horvath et al. 1998). The possibility of the thylakoid membrane acting as a heat sensor is physiologically crucial, since it is highly susceptible to increased temperature, owing to unsaturation, and the presence of photosystems, which are insubstantial to temperature changes (Sung et al. 2003). As a response to temperature stress signaling, cytosolic Ca^{2+} concentration drastically increased (Larkindale and Knight 2002). This increase might lead to the acquisition of stress tolerance by acquisition of high temperature-induced signals by MAPK. MAPK cascades participate in signal transduction pathways. They function ubiquitously in many responses to external signals (Kaur and Gupta 2005). A heat-shock-activated MAPK (HAMK) has been identified, the activation of which triggers the apparent opposite changes in membrane fluidity coupled with cytoskeletal remodeling (Sangwan and Dhindsa 2002).

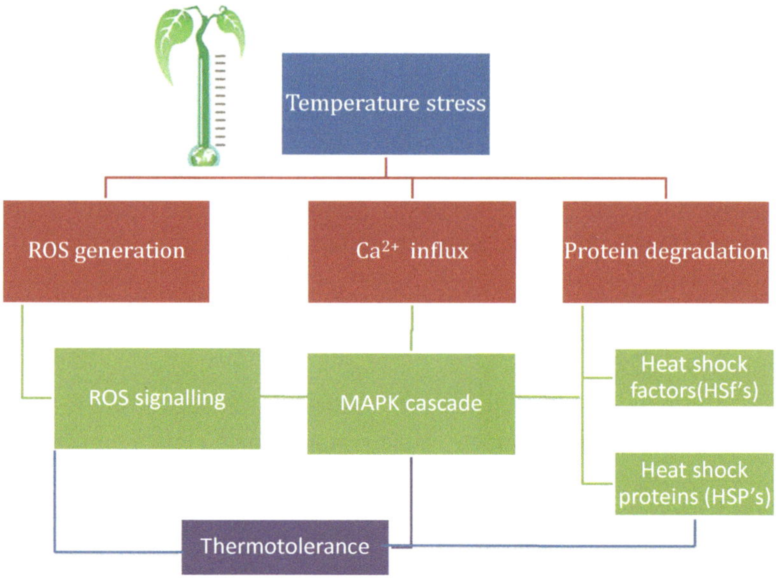

Fig. 4.3 Heat stress tolerance mechanisms in plants: ROS and Ca^{2+} play an important role as transducers (messengers) in mitogen-activated protein kinases (MAPK) cascades. They are centrally positioned in signaling pathways. Activation of these channels leads to expression of heat shock proteins and ultimately imparting thermotolerance (Rejeb et al. 2014)

Ca^{2+} influx and the action of Ca-dependent protein kinases (CDPK) have been closely correlated with the expression of Hsps (Sangwan and Dhindsa 2002). However, a recent study suggested that Ca^{2+} is not required for synthesis of Hsps, despite the fact that heat stress may alleviate induction of Ca^{2+} and calmodulin (CaM) uptake-related genes (Gong et al. 1997; Wahid et al. 2007).

Ca^{2+} is a signaling molecule which binds to CaM and activates a cascade of signals which in turn regulates many heat shock proteins and genes (Liu et al. 2003). It was also observed that under heat stress, Ca^{2+} accumulation might decrease the risk of heat injury, relatively increasing the antioxidant activity (Gong et al. 1997), maintenance of turgor pressure in guard cells (Webb et al. 1996), and equip cell survival. However, excess release of Ca^{2+} into cytosol might have cytotoxic effects on plants (Wang and Li 1999). Specific groups of potential signaling molecules like SA, ABA, $CaCl_2$, H_2O_2, and ACC might induce tolerance of plants to heat stress by reducing oxidative damage (Larkindale and Huang 2004; Wahid et al. 2007). H_2O_2 also plays key role in plant cell signaling under stress conditions (Dat et al. 2000). A MAPK protein phosphorylation cascade is shown to be activated by H_2O_2. Methyl-SA plays a major signaling role under heat stress (Llusia et al. 2005).

In summary, sensing of high temperature and induction of cascade signaling are important adaptive mechanisms in coping with adversaries where ROS, hormones, and ethylene have been identified for the perception of heat stress cues (Singh et al. 2012).

4.7 Transcription Factors and Genes Involved in Thermotolerance

HSFs are transcription regulators involved in thermotolerance. HSFs play an important role in the heat shock response by mediating the activation of many heat shock responsive genes and heat shock proteins. All these HSFs have a conserved domain near the N-terminus called DNA-binding domain (DBD). Many transgenic studies have shown that overexpression of HSFs imparts thermotolerance under heat stress and also in other abiotic stresses. For example, overexpression of *HsfA*1 in tomato conferred thermotolerance at a high temperature of 45 °C for 1 h followed by 51 °C for 1 h (Mishra et al. 2002).

In the case of plant transcriptomics, approximately 5 % upregulation greater than twofold in response to high temperature stress was observed. Often greatly induced by heat stress response, chaperones conquer only a minor part (Larkindale and Vierling 2008; Saidi et al. 2011). Most of these transcripts represented during stress responses are those genes that are involved in primary and secondary metabolism, translation, transcription, regulation, and responses to environmental stresses and in processes such as calcium and phytohormone signaling, sugar, lipid signaling, or protein phosphorylation (Saidi et al. 2011).

Studies on *Arabidopsis* show the absence of Hsp101, the mitochondrial transcription termination factor-related protein that enhances thermotolerance via mitochondrial oxidative damage control (Kim et al. 2012). Chloroplast protein synthesis transcription factor (elongation factor EF-Tu) accumulation is necessary for plant tolerance to high temperature stress (Ristic et al. 2008).

Other important genes such as late embryogenesis abundant (LEA) proteins and dehydrins are also involved in protection from heat, drought, and other abiotic stress (Peet and Willits 1998).

4.8 Role of Hsps in Different Environmental Stresses

Water deficit or drought, heat stress, chilling and freezing, salinity, and oxygen deficiency are major stress factors affecting plant growth and development (Radin and Boyer 1982; Kregel 2002). Several studies suggest that plants exhibit a phenomenon called cross-tolerance (Levitt 1980; Crawford 1989) where the exposure of plants to one stress can lead to tolerance to several other stresses. Under drought stress the accumulation of dehydrins and Hsps represent two very different classes of stress proteins. The functions of dehydrin in plants are not well understood, yet synthesis of transcripts of dehydrins was observed in sunflower plants under drought tolerance (Cellier et al. 1998) and drought-tolerant wheat (Labhilili et al. 1995).

R genes which are involved in plant disease resistance mediate protection to a variety of pathogens, including fungi, bacteria, nematodes, viruses, insects, and oomycetes (Bentem et al. 2005). In plants, Hsp90 is essential for resistance mediated by several R proteins, nucleotide-binding leucine-rich repeat (NB-ARC-LRR) class (Hubert et al. 2003; Lu et al. 2003; Takahashi et al. 2003a, b). In tomato I-2 is a CC-NB-ARC-LRR protein that induces resistance to race 2 *Fusarium oxysporum* (Ori et al. 1997; Simons et al. 1998). By using recombinant proteins, they performed in vitro protein–protein interaction analysis in which they were able to show a direct interaction between I-2 and PP5 and between I-2 and Hsp90. This suggests that R proteins require a steroid hormone-like receptors and Hsp90/PP5 complex for their folding and functioning of resistance (Bentem et al. 2005). Genetic improvement of proteins involved in osmotic adjustments, ROS detoxification, photosynthetic reactions, and protein biosynthesis has shown positive results in developing transgenic plants with thermotolerance (Awasthi et al. 2015).

4.9 Genome-Wide Survey of Hsps/HSFs in Plants

In recent years, many whole genome sequences of plants are available. This gives an opportunity to study the evolution, role, and diversity of HSFs and Hsps. In rice, there are 25 HSFs, 29 sHsps, 26 Hsp70, 9 Hsp90, and 10 Hsp100 family genes that were identified (Guo et al. 2008). It was also shown that Hsps and HSFs are important components in the heat shock regulatory network (Hu et al. 2009; Lee et al. 2010). In carrot the genome-wide survey resulted in identification of 35 DcHSFs. They were divided into three classes based on structure and phylogenetic relationship. Many plant HSFs have been reported, viz., 21 in *Arabidopsis thaliana*, 25 in *Oryza sativa*, 52 in *Brassica rapa*, 28 in *Populus trichocarpa*, 16 in *Medicago truncatula*, 25 in *Zea mays*, and 26 in *Glycine max*, respectively (Huang et al. 2015).

4.10 Conclusion

Over the years, plants have evolved a wide range of mechanisms to deal with biotic and abiotic stresses. Like in other living systems, plants exposed to high temperature stress induce a response known as the heat shock response. The various physiological processes in response to heat lead to the protection of cells against damage and restoration of cellular functions thus leading to acquired thermotolerance. One of the characteristics of heat shock response is the accumulation of a specific group of proteins called heat shock proteins (Hsps). In addition to high temperature, other stresses like salinity, drought, and pathogen attack are known to increase Hsp levels in plants. Based on molecular weight, Hsps are classified as small Hsp (sHsp), Hsp60, Hsp70, Hsp90, and Hsp100. Hsps are major stress-responsive proteins involved in distinct processes like protein folding as molecular chaperones and also participate in the import of proteins and translocation of unstable proteins for degradation to lysosomes or proteasomes. Hsp expression in plants is regulated at the transcriptional level by heat shock factors, hormones, and intracellular components. Whole genome sequencing of many crop plants has helped to understand their evolution, distribution, and regulation of expression in a wide variety of stresses. This has considerably improved our knowledge on the role and functions of Hsps in plants. In recent years, efforts have been made to overexpress many Hsps and heat shock factors by genetic engineering of crop plants to improve thermotolerance.

References

Abernethy RH, Thiel DS, Peterson NS, Helm K (1989) Thermotolerance is developmentally dependent in germinating wheat seed. Plant Physiol 89:569–576

Ahuja I, DeVos RC, Bones AM, Hall RD (2010) Plant molecular stress responses face climate change. Trends Plant Sci 15:664–674

Awasthi R, Bhandari K, Nayyar N (2015) Temperature stress and redox homeostasis in agricultural crops. Front Environ Sci 3:11

Bakau B, Horwich AL (1998) The Hsp70 and Hsp60 chaperone machines. Cell 92:351–366

Belknap WR, Garbarino JE (1996) The role of ubiquitin in plant senescence and stress responses. Trends Plant Sci 1(10):331–335

Binder BM, Patterson SE (2009) Ethylene-dependent and -independent regulation of abscission. Stewart Postharvest Rev 5:1–10. doi:10.2212/spr.2009.1.1

Burke JJ (2001) Identification of genetic diversity and mutations in higher plant acquired thermotolerance. Physiol Plant 112:167–170

Burke JJ, O'Mahony PJ (2001) Protective role in acquired thermotolerance of developmentally regulated heat shock proteins in cotton seeds. J Cotton Sci 5:174–183

Cellier F, Coneje´ ro G, Breitler JC, Casse F (1998) Molecular and physiological responses to water deficit in drought-tolerant and drought-sensitive sunflower lines (*Helianthus Annuus* L.): accumulation of dehydrin transcripts correlates. Plant Physiol 116:319–328

Charng Y, Liu H, Liu N, Hsu F, Ko S (2006) Arabidopsis Hsa32, a novel heat shock protein, is essential for acquired thermotolerance during long recovery after acclimation. Plant Physiol 140:1297–1305

Crawford RMM (1989) Studies in plant survival. Blackwell Scientific Publications, Oxford

Dat J, Vandenbeele S, Vranova E, Van Montagu M, Inze D, Van Breusegem F (2000) Dual action of the active oxygen species during plant stress responses. Cell Mol Life Sci 57:779–795

De Klerk GJ, Pumisutapon P (2008) Protection of in vitro grown Arabidopsis seedlings against abiotic stresses. Plant Cell Tissue Org 95:149–154

De la Fuente van Bentem S, Vossen JH, de Vries KJ, van Wees S, Tameling WI, Dekker HL, de Koster CG, Haring MA, Takken FL, Cornelissen BJ (2005) Heat shock protein 90 and its co-chaperone protein phosphatase 5 interact with distinct regions of the tomato I-2 disease resistance protein. Plant J 43(2):284–298

Deshaies RJ, Koch BD, Werner-Washburne M, Craig EA, Schekman R (1988) A subfamily of stress proteins facilitates translocation of secretory and mitochondria precursor polypeptides. Nature 332:800–805

Diamant S, Eliahu N, Rosenthal D, Goloubinoff P (2001) Chemical chaperones regulate molecular chaperones in vitro and in cells under combined salt and heat stresses. J Biol Chem 276:39586–39591

Ferguson DL, Guikema JA, Paulsen GM (1990) Ubiquitin pool modulation and protein degradation in wheat roots during high temperature stress. Plant Physiol 92:740–746

Finley D, Crechanover A, Varshavsky A (1984) Thermolability of ubiquitin-activating enzyme from the mammalian cell cycle mutant ts85. Cell 37:43–55

Gong M, Chen SN, Song YQ, Li ZG (1997) Effect of calcium and calmodulin on intrinsic heat tolerance in relation to antioxidant systems in maize seedlings. Aust J Plant Physiol 24:371–379

Guo SJ, Zhou HY, Zhang XS, Li XG, Meng QW (2007) Overexpression of CaHSP26 in transgenic tobacco alleviates photoinhibition of PSII and PSI during chilling stress under low irradiance. J Plant Physiol 164 (2):126–136

Guo J, Wu J, Ji Q, Wang C, Luo L, Yuan Y, Wang Y, Wang J (2008) Genome-wide analysis of heat shock transcription factor families in rice and *Arabidopsis*. J Genet Genomics 35:105–118

Hahn GM, Li GC (1990) Thermotolerance, thermoresistance, and thermosensitization. Stress Proteins Biol Med. doi:10.1101/087969337.19.79

Hall AE (2001) Crop responses to environment. CRC Press LLC, Boca Raton

Hasanuzzaman M, Nahar K, Alam M, Roychowdhury R, Fujita M (2013) Physiological, biochemical, and molecular mechanisms of heat stress tolerance in plants. Int J Mol Sci 14:9643–9684

Helm KW, Lee GJ, Vierling E (1997) Expression and native structure of cytosolic class II small heat-shock proteins. Plant Physiol 114:1477–1485

Horvath G, Arellano JB, Droppa M, Baron M (1998) Alterations in photosystem II electron transport as revealed by thermoluminescence of Cu-poisoned chloroplasts. Photosynth Res 57:175–181

Hu W, Hu G, Han B (2009) Genome-wide survey and expression profiling of heat shock proteins and heat shock factors revealed overlapped and stress specific response under abiotic stresses in rice. Plant Sci 176:583–590

Huang B, Xu C (2008) Identification and characterization of proteins associated with plant tolerance to heat stress. J Integr Plant Biol 50:1230–1237

Huang Y, Li MY, Wang F, Xu ZS, Hang W, Wang GL, Ma J, Xiong AS (2015) Heat shock factors in carrot: genome-wide identification, classification, and expression profiles response to abiotic stress. Mol Biol Rep 42:893–905

Hubert DA, Tornero P, Belkhadir Y, Krishna P, Takahashi A, Shirasu K, Dangl JL (2003) Cytosolic HSP90 associates with and modulates the Arabidopsis RPM1 disease resistance protein. EMBO J 22:5679–5689

Ishiguro S, Watanabe Y, Ito N, Nonaka H, Takeda N, Sakai T, Kanaya H, Okada K (2002) SHEPHERD is the *Arabidopsis* GRP94 responsible for the formation of functional CLAVATA proteins. EMBO J 21:898–908

Jones PD, New M, Parker DE, Mortin S, Rigor IG (1999) Surface area temperature and its change over the past 150 years. Rev Geophys 37:173–199

Joyce SM, Cassells AC, Mohan JS (2003) Stress and aberrant phenotypes in vitro culture. Plant Cell Tissue Org 74:103–121

Kaur N, Gupta AK (2005) Signal transduction pathways under abiotic stresses in plants. Curr Sci 88:1771–1780

Key J, Lin C, Chen Y (1981) Heat shock proteins of higher plants. Proc Natl Acad Sci U S A 78:3526–3530

Kim M, Lee U, Small I, des Francs-Small CC, Vierling E (2012) Mutations in an *Arabidopsis* mitochondrial transcription termination factor–related protein enhance thermotolerance in the absence of the major molecular chaperone HSP101. Plant Cell 24:3349–3365

Kregel KC (2002) Heat shock proteins: modifying factors in physiological stress responses and acquired thermotolerance. J Appl Physiol 92:2177–2186

Krishna P, Gloor G (2001) The Hsp90 family of proteins in Arabidopsis thaliana. Cell Stress Chaperones 6(3):238

Kulz D (2003) Evolution of the cellular stress proteome: from monophyletic origin to ubiquitous function. J Exp Biol 206:3119–3124

Kumar G, Krishnaprasad BT, Savitha M, Gopalakrishna R, Mukhopadhyay K, Ramamohan G, Udayakumar M (1999) Enhanced expression of heat-shock proteins in thermo-tolerant lines of sunflower and their progenies selected on the basis of temperature-induction response. Theor Appl Genet 99(1):359–367

Kumar SM, Kumar G, Srikanthbabu V, Udayakumar M (2007) Assessment of variability in acquired thermotolerance: potential option to study genotypic response and the relevance of stress genes. J Plant Physiol 164(2):111–125

Labhilili M, Joudrier P, Gautier MF (1995) Characterization of cDNAs encoding *Triticum durum* dehydrins and their expression patterns in cultivars that differ in drought tolerance. Plant Sci 112:219–230

Larkindale J, Huang B (2004) Thermotolerance and antioxidant systems in *Agrostis stolonifera*: involvement of salicylic acid, abscisic acid, calcium, hydrogen peroxide, and ethylene. J Plant Physiol 161:405–413

Larkindale J, Huang B (2005) Effects of abscisic acid, salicylic acid, ethylene and hydrogen peroxide in thermotolerance and recovery for creeping bentgrass. Plant Growth Regul 47:17–28

Larkindale J, Knight MR (2002) Protection against heat stress-induced oxidative damage in *Arabidopsis* involves calcium, abscisic acid, ethylene and salicylic acid. Plant Physiol 128:682–695

Larkindale J, Vierling E (2008) Core genome responses involved in acclimation to high temperature. Plant Physiol 146:748–761

Larkindale J, Hall JD, Knight MR, Vierling E (2005) Heat stress phenotypes of Arabidopsis mutants implicate multiple signaling pathways in the acquisition of thermotolerance. Plant Physiol 138:882–897

Lee J, Song H, Han CT, Lim YP, Chung SM, Hur Y (2010) Expression characteristics of heat shock protein genes in two comparable inbred lines of Chinese cabbage. Chiifu Kenshin Genes Genomics 32 (3):247–257

Levitt J (1980) Responses of plants to environmental stresses. Acad Press 1:496

Li S, Li F, Wang J, Zhang W, Meng Q, Chen TH, Murata N, Yang X (2011) Glycinebetaine enhances the tolerance of tomato plants to high temperature during germination of seeds and growth of seedlings. Plant Cell Environ 34(11):1931–1943

Lin C, Roberts JK, Key J (1984) Acquisition of thermotolerance in soybean seedlings. Plant Physiol 74:152–160

Lindquist S (1986) The heat-shock response. Annu Rev Biochem 55:1151–1191

Lindquist S, Craig EA (1988) The heat-shock proteins. Annu Rev Genet 22:631–677

Liu X, Huang B (2000) Heat stress injury in relation to membrane lipid peroxidation in creeping bentgrass. Crop Sci 40:503–510

Liu Y, Steussy CN, Goebl MG (1995) Intragenic suppression among CDC34 mutations defines a class of ubiquitin-conjugating catalytic domains. Mol Cell Biol 15:5635–5644

Liu HT, Li B, Shang ZL, Li XZ, Mu RL, Sun DY, Zhou RG (2003) Calmodulin is involved in heat shock signal transduction in wheat. Plant Physiol 132:1186–1195

Liu HT, Liu YY, Pan QH, Yang HR, Zhan JC, Huang WD (2006a) Novel interrelationship between salicylic acid, abscisic acid, and PIP2-specific phospholipase C in heat acclimation-induced thermotolerance in pea leaves. J Exp Bot 57:3337–3347

Liu N, Ko S, Yeh KC, Charng Y (2006b) Isolation and characterization of tomato Hsa32 encoding a novel heat-shock protein. Plant Sci 170:976–985

Llusia J, Penuelas J, Munne-Bosch S (2005) Sustained accumulation of methyl salicylate alters antioxidant protection and reduces tolerance of holm oak to heat stress. Physiol Plant 124:353–361

Lu R, Malcuit I, Moffett P, Ruiz MT, Peart J, Wu AJ, Rathjen JP, Bendahmane A, Day L, Baulcombe DC (2003) High throughput virus-induced gene silencing implicates heat shock protein 90 in plant disease resistance. EMBO J 22:5690–5699

Mishra SK, Tripp J, Winkelhaus S, Tschiersch B, Theres K, Nover L, Scharf KD (2002) In the complex family of heat stress transcription factors, HsfA1 has a unique role as master regulator of thermotolerance in tomato. Genes Dev 16(12):1555–1567

Mittler R, Finka A, Goloubinoff P (2012) How do plants feel the heat? Trends Biochem Sci 37(3): 118–25

Mizzen LA, Welch WJ (1988) Characterization of the thermotolerant cell. I. Effects on protein synthesis activity and the regulation of heat-shock protein 70 expression. J Cell Biol 106:1105–1116

Morrow G, Tanguay RM (2012) Small heat shock protein expression and functions during development. Int J Biochem Cell Biol 44:1613–1621

Ori N, Eshed Y, Paran I, Presting G, Aviv D, Tanksley S, Zamir D, Fluhr R (1997) The I2C family from the wilt disease resistance locus I2 belongs to the nucleotide binding, leucine-rich repeat super family of plant resistance genes. Plant Cell 9:521–532

Palleros DR, Welch WJ, Fink AL (1991) Interaction of Hsp70 with unfolded proteins: effects of temperature and nucleotides on the kinetics of binding. Proc Natl Acad Sci U S A 88:5719–5723

Panchuk II, Volkov RA, Schoffl F (2002) Heat stress and heat shock transcription factor-dependent expression and activity of ascorbate peroxidase in *Arabidopsis*. Plant Physiol 129:838–853

Peet MM, Willits DH (1998) The effect of night temperature on greenhouse grown tomato yields in warm climate. Agric For Meteorol 92:191–202

Piterkovaa J, Luhovaa L, Mieslerovab B, Lebedab A, Petrivalsky M (2013) Nitric oxide and reactive oxygen species regulate the accumulation of heat shock proteins in tomato leaves in response to heat shock and pathogen infection. Plant Sci 207:57–65

Pratt WB (2001) Hsp90-binding immunophilins in plants: the protein movers. Trends Plant Sci 6:54–58

Pumisutapon, Visser RGF, de Klerk GJ (2012) Moderate abiotic stresses increase rhizome growth and outgrowth of axillary buds in Alstroemeria cultured in vitro. Plant Cell Tissue Organ Cult 110:395–400

Radin JW, Boyer JS (1982) Control of leaf expansion by nitrogen nutrition in sunflower plants. Plant Physiol 69:771–775

Reddy VS, Safadi F, Zielinski RE, Reddy ASN (1999) Interaction of a kinesin-like protein with calmodulin isoforms from *Arabidopsis*. J Biol Chem 274:31727–31733

Rejeb IB, Pastor V, Mauch-Mani B (2014) Plant responses to simultaneous biotic and abiotic stress: molecular mechanisms. Plants 3(4):458–475

Ristic Z, Bukovnik U, Momcilovic I, Fu J, Vara Prasad PV (2008) Heat-induced accumulation of chloroplast protein synthesis elongation factor, EF-Tu, in winter wheat. J Plant Physiol 165(2):192–202

Ritossa F (1962) A new puffing pattern induced by temperature shock and DNP in *Drosophila*. Experientia 18:571–573

Saidi Y, Finka A, Muriset M, Bromberg Z, Weiss YG, Maathuis FJ, Goloubinoff P (2009) The heat shock response in moss plants is regulated by specific calcium-permeable channels in the plasma membrane. Plant Cell 21:2829–2843

Saidi Y, Finka A, Goloubinoff P (2011) Heat perception and signalling in plants: a tortuous path to thermotolerance. New Phytol 190:556–565

Sangwan V, Dhindsa RS (2002) In vivo and in vitro activation of temperature responsive plant map kinases. FEBS Lett 531:561–564

Sanmiya K, Suzuki K, Egawa Y, Shono M (2004) Mitochondrial small heat-shock protein enhances thermotolerance in tobacco plants. FEBS Lett 557 (1–3):265–268

Sarkar NK, Kim Y, Grover A (2009) Rice sHsp genes: genomic organization and expression profiling under stress and development. BMC Genomics 10:393

Sengupta S, Majumder AL (2009) Insight into the salt tolerance factors of a wild halophytic rice, Porteresia coarctata: a physiological and proteomic approach. Planta 229(4):911–929

Simons G, Groenendijk J, Wijbrandi J, Reijans M, Groenen J, Diergaarde P, der Lee TV, Bleeker M, Onstenk J, de Both M, Haring M, Mes J, Cornelissen B, Zabeau M, Vos P (1998) Dissection of the Fusarium *I2* gene cluster in tomato reveals six homologs and one active gene copy. Plant Cell 10:1055–1068

Singh K, Chugh V, Sahi GK, Chhuneja P (2012) Wheat: mechanisms and genetic means for improving heat tolerance. In: Improving crop resistance to abiotic stress, vols 1&2. Wiley-VCH Verlag GmbH & Co. KGaA, Weinheim

Srikanthbabu V, Ganeshkumar, Krishnaprasada BT, Gopalakrishnaa R, Savitha M, Udayakumar M (2002) Identification of pea genotypes with enhanced thermotolerance using temperature induction response technique (TIR). J Plant Physiol 159(5):535–545

Sun W, Bernard C, van de Cotte B, van Montagu M, Verbruggen N (2001) At-HSP17.6A, encoding a small heat-shock protein in Arabidopsis, can enhance osmotolerance upon overexpression. Plant J 27:407–415

Sun W, Van Montagu M, Verbruggen N (2002) Small heat shock proteins and stress tolerance in plants. Biochim Biophys Acta 1577(1):1–9

Sung DY, Kaplan F, Lee KJ, Guy CL (2003) Acquired tolerance to temperature extremes. Trends Plant Sci 8:179–187

Suzuki N, Mittler R (2006) Reactive oxygen species and temperature stresses: a delicate balance between signaling and destruction. Physiol Plant 126:45–51

Swindell WR, Huebner M, Weber AP (2007) Transcriptional profiling of *Arabidopsis* heat shock proteins and transcription factors reveals extensive overlap between heat and non-heat stress response pathways. BMC Genomics 8:125

Takahashi A, Casais C, Ichimura K, Shirasu K (2003a) HSP90 interacts with RAR1 and SGT1, and is essential for RPS2-mediated disease resistance in Arabidopsis. Proc Natl Acad Sci 100:11777–11782

Takahashi Y, Berberich T, Miyazaki A, Seo S, Ohashi Y, Kusano T (2003b) Spermine signalling in tobacco: activation of mitogen-activated protein kinases by spermine is mediated through mitochondrial dysfunction. Plant J 36(6):820–829

Talanova VV, Akimova TV, Titov AF (2003) Effect of whole plant and local heating on the ABA content in cucumber seedling leaves and roots and on their heat tolerance. Russ J Plant Physl 50(1):90–94

Vierling E (1991) The roles of heat shock proteins in plants. Annu Rev Plant Physiol Plant Mol Biol 42:579–620

Wahid A, Gelani S, Ashraf M, Foolad MR (2007) Heat tolerance in plants: an overview. Environ Exp Bot 61:199–223

Wang JB, Li RQ (1999) Changes of Ca^{2+} distribution in mesophyll cells of pepper under heat stress. Acta Hortic Sin 26:57–58

Wang W, Vinocur B, Shoseyov O, Altman A (2004) Role of plant heat-shock proteins and molecular chaperones in the abiotic stress response. Trends Plant Sci 9:244–252

Wang L, Chena S, Kong W, Lia S, Archbold DD (2006) Salicylic acid pretreatment alleviates chilling injury and affects the antioxidant system and heat shock proteins of peaches during cold storage. Postharvest Biol Technol 41(3):244–251

Waters ER, Lee GJ, Vierling E (1996) Evolution, structure and function of the small heat shock proteins in plants. J Exp Bot 47:325–338

Webb AAR, Mcainsh MR, Taylor JE, Hetherington AM (1996) Calcium ions as intercellular second messengers in higher plants. Adv Bot Res 22:45–96

Xu S, Li J, Zhang X, Wei H, Cui L (2006) Effects of heat acclimation pretreatment on changes of membrane lipid peroxidation, antioxidant metabolites, and ultrastructure of chloroplasts in two cool-season turfgrass species under heat stress. Environ Exp Bot 56:274–285

Yamada K, Fukao Y, Hayashi M, Fukazawa M, Suzuki I, Nishimura M (2007) Cytosolic HSP90 regulates the heat shock response that is responsible for heat acclimation in *Arabidopsis thaliana*. J Biol Chem 282:37794–37804

Yeh CH, Chang PL, Yeh KW, Lin WC, Chen YM, Lin CY (1997) Expression of a gene encoding a 16.9-kDa heat-shock protein, OsHsp16.9, in *Escherichia coli* enhances thermotolerance. Proc Natl Acad Sci U S A 94:10967–10972

Zhang JH, Wang LJ, Pan QH, Wang YZ, Zhan JC, Huang WD (2008) Accumulation and subcellular localization of heat shock proteins in young grape leaves during cross-adaptation to temperature stresses. Sci Hortic 117(3):231–240

Mechanisms of Heavy Metal Toxicity in Plants

5

D. Kalaivanan and A.N. Ganeshamurthy

Abstract

Pollution of the environment with the toxic heavy metals has become one of the major causes for worry for human health in both emerging and advanced countries. Metal contamination issues are becoming more and more common in India and elsewhere, with many documented cases of metal toxicity in mining industries, foundries, smelters, coal-burning power plants, and agriculture. As land application becomes one of the foremost waste utilization and disposal practices, soil is increasingly being seen as a major source of metal(loid)s reaching food chain, largely through plant uptake and animal transfer. Heavy metal buildup in soils is of concern in agricultural production due to the adverse effects on food safety and marketability, crop growth due to phytotoxicity, and environmental health of soil organisms. Metal toxicity has high impact and relevance to plants, and consequently, it affects the ecosystem, where the plants form an integral component. A few metals, including copper, manganese, iron, cobalt, zinc, and chromium, are, however, essential to plant metabolism in trace quantities. It is only when metals are present in bioavailable forms and at excessive levels; they have the potential to turn out to be toxic to plants through formation of complex compounds within the cell. Plants growing in metal-contaminated sites exhibit altered metabolism, growth reduction, lower biomass production, and metal accumulation. Various physiological and biochemical processes in plants are affected by metal toxicities. The present-day investigations into toxicity and tolerance in metal-stressed plants are prompted by the growing metal pollution in the environment. This article details the range of heavy metals, toxicity for plants, and mechanisms of plants to cope with metal toxicity.

D. Kalaivanan • A.N. Ganeshamurthy (✉)
Division of Soil Science and Agricultural Chemistry,
ICAR-Indian Institute of Horticultural Research,
Hessaraghatta Lake Post, Bengaluru 560089,
Karnataka, India
e-mail: angmurhty@iihr.ernet.in

N.K.S. Rao et al. (eds.), *Abiotic Stress Physiology of Horticultural Crops*,
DOI 10.1007/978-81-322-2725-0_5

5.1 Introduction

Currently, the available literature focuses mainly on different features of heavy metal buildup both in the ecosystem and in the food chain. Reports are also available on the occurrence of high concentration of these metals in crop plants, and on consumption of the produce or the plants, animals and human beings are also affected (Gopal et al. 2003). Being sessile organisms, plants are continuously exposed during their life cycle to adverse environmental conditions that detrimentally affect growth, development, or productivity. The occurrence of toxic compounds, such as heavy metals (HMs), is one important factor that can cause damage to plants by altering most important plant physiological and metabolic processes. Heavy metals are important environmental pollutants, and their toxicity is a problem of increasing significance for ecological, evolutionary, nutritional, and environmental reasons. The term "heavy metals" refers to any metallic element that has a relatively high density and is toxic or poisonous even at low concentration (Lenntech Water Treatment and Air Purification 2004). "Heavy metals" in a general collective term, which applies to the group of metals and metalloids with atomic density greater than 4 g/cm^3, or five times or more, are greater than water (Hawkes 1997). However, chemical properties of the heavy metals are the most influencing factors compared to their density. Heavy metals include lead (Pb), cadmium (Cd), nickel (Ni), cobalt (Co), zinc (Zn), chromium (Cr), iron (Fe), arsenic (As), silver (Ag), and the platinum group elements (Nagajyoti et al. 2010).

Heavy metals are mostly found in dispersed form in rock formations. Industrialization and urbanization have increased the anthropogenic contribution of heavy metals in the environment. Unlike organic pollutants, heavy metals once introduced into the surroundings cannot be biodegraded. They persist indefinitely and cause pollution of air, water, and soils. Heavy metals have the largest availability in soil and aquatic ecosystems and to a relatively smaller proportion in the atmosphere as particulate or vapors. Heavy metal toxicity in plants varies with plant species, specific metal, concentration, chemical form and soil composition, and pH, as many heavy metals are considered to be essential for plant growth. Some of these heavy metals like Cu and Zn either serve as cofactor and activators of enzyme reactions, e.g., informing enzymes/substrate metal complex (Mildvan 1970) or exert a catalytic property such as prosthetic group in metalloproteins. These essential trace metal nutrients take part in redox reactions, electron transfer, and structural functions in nucleic acid metabolism. Some of the heavy metals such as Cd, Hg, and As are strongly poisonous to metal-sensitive enzymes, resulting in growth inhibition and death of organisms.

5.2 Nature and Classification of Heavy Metals

No organic life can develop and persist without the involvement of metal ions. Current research has revealed that life is as much inorganic as organic. Commonly, the term trace element is rather loosely used in current literature to designate the elements which occur in small concentrations in natural biological systems. The growing public concern over the deteriorating quality of the environment has led to a widespread usage when referring to trace elements. Thus, for all practical purposes, other terms such as "trace metals," "trace inorganics," "heavy metals," "microelements," and "micronutrients" have been treated as synonyms with the term trace elements (Nagajyoti et al. 2010). The elementary constituents of plant, animal, and human life may be classified as major and trace elements, the latter group comprising both essential and nonessential elements (including toxic elements).

Some of the heavy metals such as Fe, Cu, and Zn are essential for plants and animals (Wintz et al. 2002). The availability of heavy metals in medium varies, and metals such as Cu, Zn, Fe, Mn, Mo, Ni, and Co are essential micronutrients

(Reeves and Baker 2000), whose uptake in excess to the plant requirements result in toxic effects (Monni et al. 2000; Blaylock and Huang 2000). They are also called as trace elements due to their presence in trace (10 $mgkg^{-1}$, or mgL^{-1}) or in ultra-trace (1 μgkg^{-1}, or μgL^{-1}) quantities in the environmental matrices. The essential heavy metals (Cu, Zn, Fe, Mn, and Mo) play biochemical and physiological functions in plants and animals. Two major functions of essential heavy metals are the following: (a) participation in redox reaction (Fe, Cu Cr, and Co) and (b) direct participation, being an integral part of several enzymes (Zn, Cd, Ni, Al, etc.).

An alternative classification of metals based on their coordination chemistry categorizes heavy metals as class B metals that come under nonessential trace elements, which are highly toxic elements such as Hg, Ag, Pb, and Ni (Nieboer and Richardson 1980). Some of these heavy metals are bioaccumulative, and they neither break down in the environment nor are easily metabolized. Such metals accumulate in the ecological food chain through uptake at primary producer level and then through consumption at consumer levels. Plants are stationary, and roots of a plant are the primary contact site for heavy metal ions in soil. In aquatic systems, whole plant body is exposed to these ions. Heavy metals are also absorbed directly to the leaves due to particles deposited on the foliar surfaces.

5.3 Mechanism of Heavy Metal Toxicity in Plants

Oxygen-free radicals are produced when molecular oxygen accepts electrons from other molecules, and many intracellular reactions reduce oxygen to superoxide (O_2^-) or hydrogen peroxide (H_2O_2). Even though these molecules are not very reactive, they can form hydroxyl radicals (−OH), which are possibly responsible for most of the oxidative damage in biological systems (Halliwell and Cutteridge 1990). The one-electron reduction of molecular oxygen to the superoxide radical is thermodynamically unfavorable (Illan et al. 1976), but can nevertheless take place by interaction with another paramagnetic center. Transition metals such as iron and copper (M) have frequently unpaired electrons and are, therefore, very good catalysts of oxygen reduction, following the reaction:

$$M^n + O_2 \rightarrow M^{n+1} + O_2^-$$

In aqueous solutions at neutral pH, O_2^- can generate H_2O_2, which can subsequently decompose to produce OH by the Haber–Weiss reaction, copper or iron (M) being again involved as follows:

$$M^{n+1} + O_2^- \rightarrow M^n + O_2$$
$$M^n + H_2O_2 \rightarrow M^{n+1} + OH^- + OH$$

These reactions are summarized as

$$O_2 + H_2O_2 \rightarrow O_2 + OH^- + OH$$

When iron is the transition metal in the Haber–Weiss reaction, it is called the Fenton reaction. Hydroxyl radicals (OH) produced by the Haber–Weiss reaction can oxidize biological molecules, leading to major cellular damages and ultimately to cell death. Hydroxyl radicals can be produced in close proximity to DNA, adding or removing H atoms to DNA bases or the DNA backbone, respectively (Pryor 1988). Their action could be responsible for 10^4–10^5 DNA base modifications per cell per day (Ames et al. 1991). Fe^{2+} ions either complexed to a phosphate residue or coordinated with ring nitrogens or free in solution are involved in these hydroxyl radical-mediated DNA alterations (Luo et al. 1994). Metal ions, through Haber–Weiss reaction, also play an important role in the oxidative modifications of free amino acids and proteins, and this aspect has been reviewed by Stadtman (1993). Histidine, arginine, lysine, proline, methionine, and cysteine residues are the most common sites of oxidation in proteins, and their major oxidation products have been identified. Frequently, only one amino acid residue in a given protein is modified by oxidation. These modifications correspond to site-specific processes, amino acid residues at metal-binding sites being specific targets. A major consequence of oxygen-free radical damage to proteins is to target them for degradation by proteases (Roseman and Levine 1987). Release of Fe^{2+} from [4Fe-4S] clusters of some dehydratases such as aconitases is also another important aspect of protein oxidation (Goldstein and Czapski 1986). Finally, oxygen and transition metals (in particular iron) are implicated in lipid peroxidation, and biological membranes rich in polyunsaturated fatty acids are extremely susceptible to these reactions. Ascorbic acid is known to quench lipid peroxy and alkoxy radicals in the aqueous phase. However, ascorbate can also function as a prooxidant by reducing Fe^{3+} or Cu^{2+}, allowing the metal-catalyzed transition of lipid hydroperoxides to radical species (Scholz et al. 1990).

5.3.1 Genotoxicity

Metal binding to the cell nucleus causes promutagenic damage including DNA base modifications, inter- and intramolecular cross-linkage of DNA and proteins, DNA strand breaks, rearrangements, and depurination. Chemical reactions driving this damage, and the resulting mutations, are characteristics of an oxidative DNA attack (Kasprzak 1995). Metal-mediated production of reactive oxygen species in the DNA vicinity generates principally the promutagenic adduct 8–oxoG (7, 8-dihydro-8-oxoguanine) that could miss pair with adenine in the absence of DNA repair, resulting in C to T transversion mutations (Cunningham 1997). Although oxidative damage explains most of the mechanisms involved in metal-mediated carcinogenicity and acute toxicity, other pathways also have to be considered. DNA methylation aberration has been shown to play an important role in tumorigenesis (Zingg and Jones 1997). Cell treatment with carcinogenic nickel can cause chromatin condensation through competition with magnesium ions. This condensed chromatin is hypermethylated, leading to silencing of putative anti-oncogenic gene expression, thus driving treated cells to a carcinogenic state (Lee et al. 1995). Most of the cellular and molecular aspects of metal toxicity in plants are unknown, even though deleterious effects on crop production have long been recognized. Preliminary observations on putative metal genotoxic effects in plant are scarce. At the organ level, this symptom is common to numerous metals. Nickel accumulation in maize root apex reduces meristem mitotic activity, and this could be due to the lack of integrity of root meristems (L'Huillier et al. 1996). Concentration and time-dependent cadmium, copper, and nickel clastogenic effects were observed in *Helianthus annuus* (Chakravarty and Srivastava 1992). Taken together, these observations suggest that genotoxic effects could be in part responsible for metal phytotoxicity, deserving more work to elucidate the underlying mechanisms.

5.3.2 Heavy Metals on Antioxidant Activity

Heavy metal toxicity is reported to increase the activity of enzymes such as glucose-6-phosphate dehydrogenase and peroxidase in the leaf of plants grown in polluted soil (Van Assche and Clijsters 1987). Metal ions play an important role in the antioxidant network, as these are essential cofactors of most antioxidant enzymes. For example, all isoforms of superoxide dismutase (SOD) contain bound heavy metal ions. Cu and Zn constitute the cofactor of the Cu/Zn-SOD associated with chloroplast; glyoxysomes contain Mn-SOD. Fe-SOD has been found in the chloroplast of some plants. Metals are involved in the direct or indirect generation of free radicals (FR) and reactive oxygen species (ROS) in the following ways: (1) direct transfer of electron in single-electron reduction, (2) disturbance of metabolic pathways resulting in an increase in the rate of FR and ROS formation, (3) inactivation and downregulation of the enzymes of the antioxidative defense system, and (4) depletion of low molecular weight antioxidants (Aust et al. 1985). The ROS produced in leaf cells are removed by complex enzyme catalase (CAT), ascorbate peroxidases (APX), glutathione peroxidases (GPX), superoxide dismutase (SOD), and glutathione reductases (GR) of antioxidant systems. Heavy metals induced oxidative damage in senescing oat leaf cells (Luna et al. 1994), primary leaves of mung bean (Weckex and Clijsters 1997), and in wheat leaves (Panda and Patra 2000). Proline is known to play a role in the detoxification of active oxygen in *Brassica juncea* and *Cajanus cajan* under heavy metal stress (Alia Prasad and Pardha Saradhi 1995). Accumulation of proline has been observed in various plant species subjected to heavy metal stress (Shah and Dubey 1998). It is reported that antioxidant enzyme activities increase under Zn stress in *Brassica juncea* (Prasad and Hagmeyer 1999). Verma and Dubey (2003) have reported that with increase in the level of Pb treatment, antioxidant enzymes such as guaiacol peroxidase, SOD, APX, and GR activities increased compared to control in rice plants. It has been further suggested that SOD and GR play a pivotal role in combining oxidative stress in rice plants and Pb toxicity. Thus, it is evident from the several research reports that judicious use and presence of heavy metals have toxic effects on plants, animals, and other living organisms and affect the same after certain limits. Therefore, it is well needed to intensify the research programs for better understanding of heavy metal toxicity on plants and allied areas to maintain the ecological harmony of the globe.

5.4 Mechanisms of Plants to Cope with Metal Toxicity

Like all other organisms, plants have evolved different mechanisms to maintain physiological concentrations of essential metal ions and to minimize exposure to nonessential heavy metals. Plant tolerance to a particular heavy metal is governed by an interrelated network of physiological and molecular mechanisms, and understanding of these mechanisms and their genetic basis is an important aspect to developing plants as agents of phytoremediation. Plants have both constitutive and adaptive mechanisms to withstand excess heavy metals (Meharg 1994). Physiological, biochemical, and molecular approaches continue to be employed to identify the underlying mechanisms of heavy metal accumulation, tolerance, and adaptive mechanisms to cope with heavy metal stress. Some mechanisms are ubiquitous because they are also required for general metal homeostasis, and they minimize the damage caused by high concentrations of heavy metals in plants by detoxification, thereby conferring tolerance to heavy metal stress. Other mechanisms target individual metal ions (indeed some plants have more than one mechanism to prevent the accumulation of specific metals), and these processes may involve the exclusion of particular metals from the intracellular environment or the sequestration of toxic ions within compartments to isolate them from sensitive cellular components. As a first line of defense, many plants exposed to toxic concentrations of metal ions attempt to prevent or reduce uptake into root cells by restricting metal ions to the apoplast, binding them to the cell wall or to cellular exudates or by inhibiting long-distance transport (Manara 2012). If this fails, metals already in the cell are addressed using a range of storage and detoxification strategies, including immobilization, synthesis of specific heavy metal transporters, chelation, trafficking, and sequestration of heavy metals by particular ligands (phytochelatins and metallothioneins). When these options are exhausted, plants activate oxidative stress defense mechanisms contrasting the effects of ROS and MG (such as upregulation of antioxidant and glyoxalase system), and the synthesis of stress-related proteins and signaling molecules, such as heat shock proteins, hormones, the biosynthesis of Pro, polyamines, and signaling molecule such as salicylic acid and nitric oxide (Hossain et al. 2012).

5.5 Effects of Toxic Heavy Metals on Plants

Plants, like all living organisms, are often sensitive both to the deficiency and to the excess availability of some heavy metal ions as essential micronutrient, while the same at higher concentrations and even more ions such as cadmium (Cd), mercury (Hg), arsenic (As) are strongly poisonous to the metabolic activities (Foy 1978). Contamination of agricultural soil by heavy metals has become a critical environmental concern due to their potential adverse ecological effects. Such toxic elements are considered as soil pollutants due to their widespread occurrence and their acute and chronic toxic effect on plants grown from such soils (Reeves and Baker 2000).

5.5.1 Zinc Toxicity in Plants

Zinc (Zn) is an essential micronutrient that affects several metabolic processes of plants and has a long biological half-life. Concentrations of Zn found in contaminated soils frequently exceed to those required as nutrients and may cause phytotoxicity. Zn concentrations in the range of 150–300 mg/kg have been measured in polluted soils (Warne et al. 2008). High levels of Zn in soil inhibit many plant metabolic functions, result in retarded growth, and cause senescence. The phytotoxicity of Zn and Cd is indicated by decrease in growth and development, metabolism, and an induction of oxidative damage in various plant species such as *Phaseolus vulgaris* (Cakmak and Marshner 1993) and *Brassica juncea* (Prasad and Hagmeyer 1999). Cd and Zn have reported to cause alternation in catalytic efficiency of enzymes in *Phaseolus vulgaris* (Somasekharaiah et al. 1992) and pea plants (Romero-Puertas et al. 2004). Zinc toxicity in plants limited the growth of both root and shoot (Ebbs and Kochian 1997; Fontes and Cox 1998). Zinc toxicity also causes chlorosis in the younger leaves, which can extend to older leaves after prolonged exposure to high soil Zn levels (Ebbs and Kochian 1997). The chlorosis may arise partly from an induced iron (Fe) deficiency as hydrated Zn^{2+} and Fe^{2+} ions have similar radii (Marschner 1986). Excess Zn can also give rise to manganese (Mn) and copper (Cu) deficiencies in plant shoots. Such deficiencies have been attributed to a hindered transfer of these micronutrients from root to shoot. This hindrance is based on the fact that the Fe and Mn concentrations in plants grown in Zn-rich media are greater in the root than in the shoot (Ebbs and Kochian 1997). Another typical effect of Zn toxicity is the appearance of a purplish-red color in leaves, which is ascribed to phosphorus (P) deficiency (Lee et al. 1996).

5.5.2 Cadmium Toxicity in Plants

Cadmium is a nonessential element that adversely affects plant growth and development. It is considered as one of the significant pollutants due to its high toxicity and more solubility in water. The monitoring limit of cadmium (Cd) in agricultural soil is 100 mg/kg soil (Salt et al. 1995). Concentration of cadmium varies extensively among plant species and varieties within species. Cadmium absorption and translocation varied with cultivars of amaranthus (Varalakshmi and Ganeshamurthy 2009) and radish (Varalakshmi and Ganeshamurthy 2013). The variation in Cd uptake among amaranthus cultivars may be due to genotypic variations in cultivars to absorb or translocate specific toxic metals. Cadmium can alter the uptake of mineral nutrients by plants through its effects on availability of nutrients from the soil. Plants grown in soil containing high levels of Cd show visible symptoms of injury reflected in terms of chlorosis, leaf rolls, growth inhibition, browning of root tips, and finally death (Mohanpuria et al. 2007; Guo et al. 2008). The inhibition of root Fe(III) reductase induced by Cd led to Fe(II) deficiency, and it seriously affected photosynthesis (Alcantara et al. 1994). In general, cadmium has been shown to interfere with the uptake, transport, and use of several elements, viz., Ca, Mg, P, and K and water by plants (Das et al. 1997). Cd also reduced the absorption of nitrate and its transport from roots to shoots, by inhibiting the nitrate reductase activity in the shoots (Hernandez et al. 1996). Appreciable inhibition of the nitrate reductase activity was also found in plants of *Silene cucubalus* (Mathys

1975). Nitrogen fixation and primary ammonia assimilation decreased in nodules of soybean plants during Cd treatments (Balestrasse et al. 2003). Metal toxicity can affect the plasma membrane permeability, causing a reduction in water content; in particular, Cd has been reported to interact with the water balance (Costa and Morel 1994). Cadmium treatments have been shown to reduce ATPase activity of the plasma membrane fraction of wheat and sunflower roots (Fodor et al. 1995). Cadmium produces alterations in the functionality of membranes by inducing lipid peroxidation (Fodor et al. 1995) and disturbances in chloroplast metabolism by inhibiting chlorophyll biosynthesis and reducing the activity of enzymes involved in CO_2 fixation (De Filippis and Ziegler 1993).

5.5.3 Copper Toxicity in Plants

Copper (Cu) is considered as a micronutrient for plants (Thomas et al. 1998) and plays an important role in CO_2 assimilation and ATP synthesis. Cu is also an essential component of various proteins like plastocyanin of photosynthetic system and cytochrome oxidase of respiratory electron transport chain (Demirevska-Kepova et al. 2004). But enhanced industrial and mining activities have contributed to the increasing occurrence of Cu in ecosystems. Mining activities generate a large amount of waste rocks and tailings, which get deposited at the surface. Excess of Cu in soil plays a cytotoxic role, induces stress, and causes injury to plants. This leads to plant growth retardation and leaf chlorosis (Lewis et al. 2001). Interveinal chlorosis is a common initial symptom of copper toxicity. The chlorosis often takes the form of cream or white spots or lesions. With increasing exposure, leaf tips and margins can become necrotic. In acute Cu toxicity, leaves may become wilted before eventually becoming necrotic (Foy et al. 1995). Exposure of plants to excess Cu generates oxidative stress and ROS (Stadtman and Oliver 1991). Oxidative stress causes disturbance of metabolic pathways and damage to macromolecules (Hegedus et al. 2001). Copper toxicity affected the growth of *Alyssum montanum* (Ouzounidou 1994) and Cd of cucumber (Moreno-Caselles et al. 2000) and *Brassica juncea* (Singh and Tewari 2003). Copper and Cd in combination have affected adversely the germination, seedling length, and number of lateral roots in *Solanum melongena* (Neelima and Reddy 2002). Citrus seedlings exposed to excess Cu produce few new roots and have a thickened tap root (Zhu and Alva 1993). Thickening of root apices was also evident in *Pinus* seedlings (Arduini et al. 1995). In *Betula papyrifera* (paper birch) and *Lonicera tatarica* (honeysuckle) seedlings high, Cu concentrations have been shown to inhibit the production of root hairs (Patterson and Olson 1983).

5.5.4 Mercury Toxicity in Plants

Mercury poisoning has become a problem of current interest as a result of environmental pollution on a global scale. Natural emissions of mercury form two-thirds of the input; manmade releases form about one-third. Significant amounts of mercury may be added to agricultural land with sludge, fertilizers, lime, fungicides, and manures. The most important sources of contaminating agricultural soil have been the use of organic mercurials as a seed coat dressing to prevent fungal diseases in seeds (Patra and Sharma 2000). The large input of mercury (Hg) into the arable lands has resulted in the widespread occurrence of mercury contamination in the entire food chain. Mercury is a unique metal due to its existence in different forms, e.g., HgS, Hg^{2+}, Hg^0, and methyl-Hg. However, in agricultural soil, ionic form (Hg^{2+}) is predominant (Han et al. 2006). Mercury released to the soil mainly remains in solid phase through adsorption onto sulfides, clay particles, and organic matters. Increasing evidence has shown that Hg^{2+} can readily accumulate in higher and aquatic plants (Kamal et al. 2004; Israr et al. 2006). High level of Hg^{2+} is strongly phytotoxic to plant cells. Toxic level of Hg^{2+} can induce visible injuries and physiological disorders in plants (Zhou et al. 2007). For example, Hg^{2+} can bind to water channel proteins, thus inducing leaf stomata to close and physical obstruction of water flow in plants (Zhang and Tyerman 1999). High level of Hg^{2+} interferes the mitochondrial activity and induces oxidative stress by triggering the generation of ROS. This leads to the disruption of biomembrane lipids and cellular metabolism in plants (Cargnelutti et al. 2006). Mercury affects both light and dark reactions of photosynthesis. Replacement of the central atom of chlorophyll, magnesium, by mercury in vivo, prevents photosynthetic light harvesting in the affected chlorophyll molecules, resulting in a breakdown of photosynthesis (Patra and Sharma 2000).

5.5.5 Chromium Toxicity in Plants

Chromium is found in all phases of the environment including air, water, and soil. Chromium compounds are highly toxic to plants and are injurious to their growth and development. Although some crops are not affected by low Cr ($3.8–10^{-4}$ μM) concentrations (Huffman and Allaway 1973a, b), Cr is toxic to higher plants at 100 μg kg^{-1} dry weight (Davies et al. 2002). Since seed germination is the first physiological process affected by Cr, the ability of a seed to germinate in a medium containing Cr would be indicative of its level of tolerance to this metal (Peralta et al. 2001). Seed germination of the weed *Echinochloa colona* was reduced to 25 % with 200 μM Cr (Rout et al. 2000). High levels (500 ppm) of hexavalent Cr in soil reduced germination up to 48 % in the bush bean *Phaseolus vulgaris* (Parr and Taylor 1982). Peralta et al. (2001) found that 40 ppm of Cr (VI) reduced by 23 % the ability of seeds of Lucerne (*Medicago sativa* cv. Malone) to germinate and grow in the contaminated medium. Reductions of 32–57 % in sugarcane bud germination were observed with 20 and 80 ppm Cr, respectively (Jain et al. 2000). The reduced germination of seeds under Cr stress could be a depressive effect of Cr on the activity of amylases and on the subsequent transport of sugars to the embryo axes (Zeid 2001). Protease activity, on the other hand, increases with the Cr treatment, which could also contribute to the reduction in germination of Cr-treated seeds (Zeid 2001). Decrease in root growth is a well-documented effect due to heavy metals in trees and crops (Tang et al. 2001). Prasad et al. (2001) reported that the order of metal toxicity to new root primordia in *Salix viminalis* is Cd > Cr > Pb, whereas root length was more affected by Cr than by other heavy metals studied. Chromium stress is one of the important factors that affect photosynthesis in terms of CO_2 fixation, electron transport, photophosphorylation, and enzyme activities (Nagajyoti et al. 2010). In higher plants and trees, the effect of Cr on photosynthesis is well documented (Van Assche and Clijsters 1983). However, it is not well understood to what extent Cr-induced inhibition of photosynthesis is due to disorganization of chloroplasts' ultrastructure (Vazques et al. 1987), inhibition of electron transport, or the influence of Cr on the enzymes of the Calvin cycle. Chromate is used as a Hill reagent by isolated chloroplast. The more pronounced effect of Cr (VI) on PS I than on PS II activity in isolated chloroplasts has been reported by Bishnoi et al. (1993a, b) in peas. Chromium stress can induce three possible types of metabolic modification in plants: (i) alteration in the production of pigments, which are involved in the life sustenance of plants (e.g., chlorophyll, anthocyanin) (Boonyapookana et al. 2002); (ii) increased production of metabolites (e.g., glutathione, ascorbic acid) as a direct response to Cr stress, which may cause damage to the plants (Shanker et al. 2003a, b); and (iii) alterations in the metabolic pool to channelize the production of new biochemically related metabolites, which may confer resistance or tolerance to Cr stress (e.g., phytochelatins, histidine) (Schmfger 2001).

5.5.6 Lead Toxicity in Plants

Lead (Pb) is one of the major heavy metals of the antiquity and has gained considerable importance as a potent environmental pollutant. Among heavy metals, lead is a potential pollutant that readily accumulates in soils and sediments. Although lead is not an essential element for plants, it gets easily absorbed and accumulated in different plant parts. Leafy vegetables accumulated maximum levels of Pb, followed by root vegetables (carrot and radish), tomato, and French bean (Varalakshmi and Ganeshamurthy 2012). It exerts adverse effect on morphology, growth, and photosynthetic processes of plants. Excess lead causes a number of toxicity symptoms in plants, e.g., stunted growth, chlorosis, and blackening of root system. Lead is known to inhibit seed germination of *Spartina alterniflora* (Morzck and Funicclli 1982) and *Pinus halepensis* (Nakos 1979). Inhibition of germination may result from the interference of lead with important enzymes. Mukherji and Maitra (1976) observed 60 μM lead acetate inhibited protease and amylase by about 50 % in rice endosperm. Early seedling growth was also inhibited by lead in soybean (Huang et al. 1974), rice (Mukherji and Maitra 1976), maize (Miller et al. 1975), barley (Stiborova et al. 1987), tomato, eggplant (Khan and Khan 1983), and certain legumes (Sudhakar et al. 1992).

Lead also inhibited root and stem elongation and leaf expansion in *Allium* species (Gruenhage and Jager 1985), barley (Juwarkar and Shende 1986), and *Raphanus sativus*. The degree to which root elongation is inhibited depends upon the concentration of lead and ionic composition and pH of the medium (Goldbold and Hutterman 1986). Concentration-dependent inhibition of root growth has been observed in *Sesamum indicum* (Kumar et al. 1992). A high lead level in soil induces abnormal morphology in many plant species. For example, lead causes irregular radial thickening in pea roots and cell walls of the endodermis and lignification of cortical parenchyma (Paivoke 1983). Lead also induces proliferation effects on the repair process of vascular plants (Kaji et al. 1995). Lead administrated to potted sugar beet plants at rates of 100–200 ppm caused chlorosis and growth reduction (Hewilt 1953). In contrast, there was no visual symptom of lead toxicity in alfalfa plants exposed to 100 mg/mL (Porter and Cheridan 1981). Low amounts of lead (0.005 ppm) caused significant reduction in growth of lettuce and carrot roots (Baker 1972). Inhibitory effects of Pb^{2+} on growth and biomass production may possibly derive from effects on metabolic plant processes (Van Assche and Clijsters 1990). The primary cause of cell growth inhibition arises from a lead-induced simulation of indol-3 acetic acid (IAA) oxidation. Lead is also known to affect photosynthesis by the inhibiting activity of carboxylating enzymes (Stiborova et al. 1987). High level of Pb also causes inhibition of enzyme activities (Sinha et al. 1988a, b), water imbalance, changes in hormonal status, alterations in membrane structure and permeability and disturbs mineral nutrition (Sharma and Dubey 2005) and also induces oxidative stress by increasing the production of ROS in plants (Reddy et al. 2005). Pb inhibits the activity of enzymes at cellular level by reacting with their sulfhydryl groups (Nagajyoti et al. 2010).

5.5.7 Arsenic Toxicity in Plants

The element arsenic (As) is an environmental toxin that is found naturally in all soils. The metalloid enters into farming systems through a variety of means that include natural geochemical processes, the past and present use of As-based pesticides, mining operations, irrigation with As-contaminated groundwater, and fertilization with municipal solid wastes. Arsenic is a nonessential element and generally toxic to plants. Roots are usually the first tissue to be exposed to arsenic, where the metalloid inhibits root extension and proliferation. Upon translocation to the shoot, arsenic can severely inhibit plant growth by slowing or arresting expansion and biomass accumulation, as well as compromising plant reproductive capacity through losses in fertility, yield, and fruit production (Garg and Singla 2011). At sufficiently high concentrations, arsenic interferes with critical metabolic processes, which can lead to death. Most plants possess mechanisms to retain much of their As burden in the root. However, a genotype-dependent proportion of the As is translocated to the shoot and other tissues of the plant. Arsenate (As) is an analog of phosphate (P) and competes for the same uptake carriers in the root plasmalemma of plants (Meharg and Macnair 1992). The As tolerance in grasses results from suppression of a high-affinity P/As uptake system (Meharg and Macnair 1992). This suppression reduces As influx to a level at which plant can easily detoxify it, presumably by constitutive mechanisms (Meharg 1994). The As tolerance is achieved by a single gene encoding for the suppressed P/As transport (Meharg and Macnair 1992). Despite this clear understanding of the process controlling decrease in As uptake, tolerant grasses still assimilate As, albeit at much lower rate compared with non-tolerant. Nevertheless, assimilation over the life history of plants growing on contaminated soil can result in a very high As concentration, e.g., 3470 mg/g As in *Agrostis tenuis* and 560 mg/g As in *Holcus lanatus* (Porter and Peterson 1975).

5.5.8 Cobalt Effects on Plants

Cobalt, a transition element, is an essential component of several enzymes and coenzymes. It has been shown to affect growth and metabolism of plants, in different degrees, depending on the concentration and status of cobalt in rhizosphere and soil. Cobalt interacts with other elements to form complexes. The cytotoxic and phytotoxic activities of cobalt and its compounds depend on the physicochemical properties of these complexes, including their electronic structure, ion parameters (charge–size relations), and coordination. Thus, the competitive absorption and mutual activation of associated metals influence the action of cobalt on various phytochemical reactions. Cobalt (Co) naturally occurs in the earth's crust as cobaltite [CoAsS], erythrite [$Co_3(AsO_4)_2$], and smaltite [$CoAs_2$]. Plants can accumulate small amount of Co from the soil. The uptake and distribution of cobalt in plants are species dependent and controlled by different mechanisms (Li et al. 2004; Bakkaus et al. 2005). Very little information is available regarding the phytotoxic effect of excess cobalt. Phytotoxicity study of cobalt in barley (*Hordeum vulgare* L.), oilseed rape (*Brassica napus* L.), and tomato (*Lycopersicon esculentum* L.) has recently shown the adverse effect on shoot growth and biomass (Li et al. 2009). In addition to biomass, excess of cobalt restricted the concentration of Fe, chlorophyll, protein, and catalase activity in leaves of cauliflower. Further, high level of Co also affected the translocation of P, S, Mn, Zn, and Cu from roots to tops in cauliflower. In contrast to excess Cu or Cr, Co significantly decreased water potential and transpiration rate (Nagajyoti et al. 2010), while diffusive resistance and relative water content increased in leaves of cauliflower upon exposure to excess Co (Chatterjee and Chatterjee 2000).

5.5.9 Nickel Toxicity in Plants

Among different environmental heavy metal pollutants, Ni has gained considerable attention in recent years, because of its rapidly increasing concentrations in soil, air, and water in different parts of the world. Nickel is a transition metal and found in natural soils at trace concentrations except in ultramafic or serpentine soils. However, Ni^{2+} concentration is increasing in certain areas by human activities such as mining works, emission of smelters, burning of coal and oil, sewage, phosphate fertilizers, and pesticides (Gimeno-Garcia et al. 1996). Ni^{2+} concentration in polluted soil may range from 20- to 30-fold (200–26,000 mg/kg) higher than the overall range (10–1000 mg/kg) found in natural soil (Izosimova 2005). Although Ni is metabolically important in plants, it is toxic to most plant species when present at excessive amounts in soil and in nutrient solution. High Ni concentrations in growth media severely retards seed germinability of many crops. This effect of Ni is a direct one on the activities of amylases, proteases, and ribonucleases, thereby affecting the digestion and mobilization of food reserves in germinating seeds. At vegetative stages, high Ni concentrations retard shoot and root growth, affect branching development, deform various plant parts, produce abnormal flower shape, decrease biomass production, induce leaf spotting, disturb mitotic root tips, and produce Fe deficiency that leads to chlorosis and foliar necrosis. Additionally, excess Ni also affects nutrient absorption by roots, impairs plant metabolism, inhibits photosynthesis and transpiration, and causes ultrastructural modifications. Ultimately, all of these altered processes produce reduced yields of agricultural crops when such crops encounter excessive Ni exposures. Excess of Ni^{2+} in soil causes various physiological alterations and diverse toxicity symptoms such as chlorosis and necrosis in different plant species (Pandey and Sharma 2002; Rahman et al. 2005; Yadav 2010), including rice (Das et al. 1997). Plants grown in high Ni^{2+}-containing soil showed impairment of nutrient balance and resulted in disorder of cell membrane functions. Thus, Ni^{2+} affected the lipid composition and H-ATPase activity of the plasma membrane as reported in *Oryza sativa* shoots (Ros et al. 1992). Exposure of wheat to high level of Ni^{2+} enhanced MDA concentration (Pandolfini et al. 1992). Moreover, Gonnelli et al. (2001) reported an increase in MDA concentration of Ni^{2+}-sensitive plants compared to a Ni^{2+}-tolerant saline. Such changes might disturb membrane functionality and ion balance in the cytoplasm, particularly of K^+, the most mobile ion across plant cell membrane. High uptake of Ni^{2+} induced a decline in water content of dicot and monocot plant species. The decrease in water uptake is used as an indicator of the progression of Ni^{2+} toxicity in plants (Pandey and Sharma 2002; Gajewska et al. 2006).

5.5.10 Manganese Toxicity in Plants

Manganese (Mn) is an essential element for plant growth. It can, however, be detrimental when available in excess in the growth medium. Excess Mn in the growth medium can interfere with the absorption, translocation, and utilization of other mineral elements such as Ca, Mg, Fe, and P (El-Jaoual and Cox 1998). High Mn concentration in plant tissues can alter activities of enzymes and hormones, so that essential Mn-requiring processes become less active or nonfunctional (Epstein 1961; Horst 1988a, b). Accumulation of excessive manganese (Mn) in leaves causes a reduction of photosynthetic rate (Kitao et al. 1997a, b). Symptoms of Mn toxicity as well as the concentration of Mn that causes toxicity vary widely among plant species and varieties within species. Necrotic brown spotting on leaves, petioles, and stems is a common symptom of Mn toxicity (Wu 1994). This spotting starts on the lower leaves and progresses with time toward the upper leaves (Horiguchi 1988). With time, the speckles can increase in both number and size resulting in necrotic lesions, leaf browning, and death (Elamin and Wilcox 1986a, b). General leaf bronzing and shortening of internodes have been documented in *Cucumis sativus* (Crawford et al. 1989). Another common symptom is known as "crinkle leaf," and it occurs in the youngest leaf, stem, and petiole tissue. It is also associated with chlorosis and browning of these tissues (Wu 1994). Roots exhibiting Mn toxicity are commonly brown in color and sometimes crack (Foy et al. 1995). Chlorosis in younger leaves by Mn toxicity is thought to be caused through Mn-induced Fe deficiency (Horst 1988a, b). Excess Mn is reported to inhibit synthesis of chlorophyll by blocking an Fe-concerning process (Clarimont et al. 1986). Manganese toxicity in some species starts with chlorosis of older leaves moving toward the younger leaves with time (Bachman and Miller 1995). This symptom starts at the leaf margins progressing to the interveinal areas, and if the toxicity is acute, the symptom progresses to marginal and interveinal necrosis of leaves (Bachman and Miller 1995). In the only research on Mn toxicity of Australian native trees, *Corymbia gummifera* (red bloodwood) displayed small, chlorotic leaves that were often distorted in shape and death of terminal buds (Winterhalder 1963).

5.5.11 Iron Toxicity in Plants

Iron as an essential nutrient for all plants has many important biological roles in the processes as diverse as photosynthesis, chloroplast development, and chlorophyll biosynthesis. Iron is a major constituent of the cell redox systems such as heme proteins including cytochromes, catalase, peroxidase, and leghemoglobin and iron sulfur proteins including ferredoxin, aconitase, and superoxide dismutase (SOD). It functions to accept and donate electrons and plays important roles in the electron-transport chains of photosynthesis and respiration. But iron is toxic when it accumulates to high levels. Although most mineral soils are rich in iron, the expression of iron toxicity symptoms in leaf tissues occurs only under flooded conditions, which involves the microbial reduction of insoluble Fe^{3+} to soluble Fe^{2+} (Becker and Asch 2005). Iron toxicity is believed to be involved in several extremely complicated physiological diseases of rice under flooded conditions. These include "bronzing" in Sri Lanka and elsewhere, "Alkagare type" disorder in Japan, and "Akiochi" in Korea (Aizer et al. 1975). The Fe^{2+} excess causes free radical production that impairs cellular structure irreversibly and damages membranes, DNA, and proteins (Arora et al. 2002; de Dorlodot et al. 2005). Iron toxicity in tobacco, canola, soybean, and *Hydrilla verticillata* is accompanied with reduction of plant photosynthesis and yield and the increase in oxidative stress and ascorbate peroxidase activity (Sinha et al. 1997). Excess Fe also apparently contributes to "freckle leaf" of sugarcane in Hawaii, along with excess AI, Mn, and Zn (Clements et al. 1974). In tobacco, excess Fe produces brittle, tender, dark-brown to purple leaves which have poor burning qualities and flavor (Foy 1978). In navy bean, Fe toxicity associated with Zn deficiency produced black spots on the foliage (Foy 1978).

5.6 Conclusions

Numerous harmful health effects upon exposure to toxic heavy metals in the environment is a matter of serious concern and a universal issue. Thus, it is evident from the several research reports that the occurrence of heavy metals has toxic effects on plants, animals, and other living organisms and affects the same after certain limits. Therefore, it is well required to intensify the research programs for better understanding of heavy metal toxicity on plants and allied areas to maintain the ecological harmony of the globe. Metal toxicities in plants are often not clearly identifiable entities; instead, they may be the results of complex interactions of the major toxic ions with other essential or nonessential ions and with other environmental factors. Although excesses of different metals may produce some common effects on plants in general, there are many cases of specific differential effects of individual metals on different plant genotypes. Such occurrences must be recognized in approaching any problem of metal toxicity. It now seems likely that the phytotoxic mechanisms of a given metal ion involve different biochemical pathways in different plant species and varieties.

Plant heavy metal tolerance is a multigenic trait and controlled at multiple levels. Once the genetic control mechanisms of metal tolerance are identified, it may be possible to combine metal tolerance with other desirable traits to produce plants that are better adapted to soils polluted with heavy metals. Such problems are often not economically correctable with conventional fertilization and liming practices. Additionally, the selection of near-isogenic lines may provide valuable tools for studying specific biochemical mechanisms of tolerance. Once these mechanisms are better understood, it may be possible to select or breed, even more precisely, those plants having greatest tolerance to a particular metal toxicity. Such biochemical knowledge concerning metal tolerance mechanisms could also lead to the development of improved chemical (fertilization, liming, organic matter, etc.) and physical (tillage, drainage, etc.) management practices for problem soils.

5.7 Future Strategies

A wealth of recent studies demonstrated that plants protect themselves from heavy metal toxicity, besides other mechanisms, through an elevated level of nonenzymatic and enzymatic components of antioxidant and glyoxalase defense systems. Significant progress has been achieved in regard to heavy metal toxicity, and tolerance and different key components ensuring heavy metal tolerance in plants have been identified; however, many key questions remain unanswered. Additionally, different heavy metals appear to have different mechanisms to stimulate toxicity symptoms, and plants employ different mechanisms as resistant reactions to particular heavy metals. Likewise, the reaction response and tolerance mechanisms are also different when a plant is grown under excess heavy metals in hydroponic or actual field conditions. Therefore, it is very difficult to hypothesize a common resistance mechanism against all heavy metals and metalloids. Importantly, among the various nonenzymatic antioxidant molecules that plants usually utilize against heavy metal stress, glutathione (GSH) is the most vital one. Besides detoxification of reactive oxygen species (ROS) directly or detoxification of ROS and methylglyoxal (MG) through antioxidative and glyoxalase systems, GSH also plays additional roles, including heavy metal (HM) uptake, transport, and sequestration and formation of specific metal-binding ligands such as PCs. The reviewed literature confirms the central role of GSH metabolism in heavy metal tolerance in plants as evidenced by proteomic, genomic, and transgenic studies over a range of plant species. Although significant progress has been made in different aspects of GSH and its related enzymes in heavy metal-induced ROS and MG metabolism and heavy metal chelation in a large number of plant species, most of the studies were conducted under laboratory conditions. Therefore, to get more realistic information, we need to move from the laboratory to actual field conditions by using commercial plant species at various phases of plant growth. The increasing identification and study of the remarkable natural variation in the capacity of plants to accumulate and tolerate heavy metals are and will continue to provide a wealth of information. Therefore, concerted efforts by various research domains will further increase our understanding of the fundamental mechanisms involved in hyperaccumulation processes that naturally occur in metal hyperaccumulating plants. This should allow us to develop plants that are more ideally suited for phytoremediation of heavy metal-contaminated soils. Therefore, molecular and cellular adaptation of plant cells in response to heavy metal stress appears to be necessary to improve plant heavy metal tolerance that ultimately reduces the chance of entering heavy metal into the food chain.

Differential cultivar tolerances to metal toxicity almost certainly involve differences in the structure and function of membranes. But such differences are extremely difficult to measure in living plants. Any fractionation, extraction, or tissue fixation procedure used will introduce the possibility of measuring only artifacts. Electron microscope procedures involving the use of X-ray emission can determine the physical location of metal ions in root sections, provided that membranes have not been disturbed during tissue preparation. Comparative physiological studies need to be undertaken between closely related genotypes (near-isogenic lines, if possible) of the same species which differ widely in tolerance to a given metal toxicity. Investigations are needed on plant anatomical and rhizospheric changes responsible for the variability in absorption, translocation, and uptake of trace elements by seed and fodder of food crops. Mass awareness has to be created about pollutant elements in soil–plant–animal continuum and remedial measures to establish optimum level of elements for good health of animal and humans. A joint multidisciplinary team consisting of soil scientist, nutrition scientist, physiologists, and veterinary and medicine doctors has be constituted to establish definite quantitative association of animal and human health.

There is need for developing systematic database using GPS to monitor health hazards from heavy metal contamination and minor element toxicities in soil, plant, human, and animal chain. Maps of trace element deficiency and toxicity need to be produced to create awareness of such areas for taking remedial measures by the people, planners, and policy makers. These research efforts require close collaboration between soil scientists, plant breeders, plant physiologists, plant biochemists, and perhaps pathologists. Until recently, such cooperation has been rare in both state and central institutions concerned with agricultural research. A multidisciplinary approach is now essential if world food needs are to be met.

References

Aizer RS, Rajagopel CK, Money NS (1975) Available zinc, copper, iron, and manganese status of the acid rice soils of Kuttanad, Kerala State. Agric Res J Kerala 13:15–19

Alcantara E, Romera FJ, Canete M, De La Guardia MD (1994) Effects of heavy metals on both induction and function of root Fe (III) reductase in Fe-deficient cucumber (*Cucumis sativus* L.) plants. J Exp Bot 45:1893–1898

Alia Prasad KVSK, Pardha Saradhi P (1995) Effect of zinc on free radical and proline in *Brasica juncea* and *Cajanus cajan*. Phytochem 39:45–47

Ames BA, Shingenaga MK, Park EM (1991) In: Elmsford (ed) Oxidative damage and repair: chemical, biological and medical aspects. Pergamon Press, New York, pp 181–187

Arduini I, Godbold DL, Onnis A (1995) Influence of copper on root growth and morphology of *Pinus pinea* L. and *Pinus pinaster* Ait. seedlings. Tree Physiol 15:411–415

Arora A, Sairam RK, Srivastava GC (2002) Oxidative stress and antioxidative system in plants. Curr Sci 82:1227–1338

Aust SD, Marehouse CE, Thomas CE (1985) Role of metals in oxygen radical reactions. J Free Radi Biol Med 1:3–25

Bachman GR, Miller WB (1995) Iron chelate inducible iron/ manganese toxicity in zonal geranium. J Plant Nutr 18:1917–1929

Baker WG (1972) Toxicity levels of mercury lead, copper and zinc in tissue culture systems of cauliflowers lettuce potato and carrot. Can J Bot 50:973–976

Bakkaus E, Gouget B, Gallien JP, Khodja H, Carrot H, Morel JL, Collins R (2005) Concentration and distribution of cobalt in higher plants: the use of micro-PIXE spectroscopy. Nucl Inst Methods B 231:350–356

Balestrasse KB, Benavides MP, Gallego SM, Tomaro ML (2003) Effect on cadmium stress on nitrogen metabolism in nodules and roots of soybean plants. Func Plant Biol 30:57–64

Becker M, Asch F (2005) Iron toxicity in rice-conditions and management concepts. J Plant Nutr Soil Sci 168:558–573

Bishnoi NR, Chugh LK, Sawhney SK (1993a) Effect of chromium on photosynthesis, respiration and nitrogen fixation in pea (*Pisum sativum* L) seedlings. J Plant Physiol 142:25–30

Bishnoi NR, Dua A, Gupta VK, Sawhney SK (1993b) Effect of chromium on seed germination, seedling growth and yield of peas. Agric Ecosyst Environ 47:47–57

Blaylock MJ, Huang JW (2000) Phytoextraction of metals. In: Raskin I, Ensley BD (eds) Phytoremediation of toxic metals-using plants to clean up the environment. Wiley, New York, pp 53–70

Boonyapookana B, Upatham ES, Kruatrachue M, Pokethitiyook P, Singhakaew S (2002) Phytoaccumulation and phytotoxicity of cadmium and chromium in duckweed *Wolffia globosa*. Int J Phytoremed 4:87–100

Cakmak I, Marshner H (1993) Effect of zinc nutritional status on superoxide radical and hydrogen peroxide scavenging enzymes in bean leaves. In: Barrow NJ (ed) Plant nutrition-from genetic engineering field practice. Kluwer, Dordrecht, pp 133–137

Cargnelutti D, Tabaldi LA, Spanevello RM, Jucoski GO, Battisti V, Redin M, Linares CEB, Dressler VL, Flores MM, Nicoloso FT, Morsch VM, Schetinger MRC (2006) Mercury toxicity induces oxidative stress in growing cucumber seedlings. Chemosphere 65:999–1106

Chakravarty B, Srivastava S (1992) Toxicity of some heavy metals *in vivo* and *in vitro* in *Helianthus annuus*. Mutat Res 283:287–294

Chatterjee J, Chatterjee C (2000) Phytotoxicity of cobalt, chromium and copper in cauliflower. Environ Pollut 109:69–74

Clarimont KB, Hagar WG, Davis EA (1986) Manganese toxicity to chlorophyll synthesis in *tobacco callus*. Plant Physiol 80:291–293

Clements HF, Putnam EW, Suehisa RH, Yee GLN, Wehling ML (1974) Soil toxicities as causes of sugarcane leaf freckle, macadamia leaf chlorosis (Keaau) and Maui sugarcane growth failure. Hawaii Agric Exp Station Tech Bull 88, pp 52

Costa G, Morel JL (1994) Water relations, gas exchange and amino acid content in Cd-treated lettuce. Plant Physiol Biochem 32:561–570

Crawford TW, Stroehlein JL, Kuehl RO (1989) Manganese and rates of growth and mineral accumulation in cucumber. J Am Soc Hortic Sci 114:300–306

Cunningham RP (1997) DNA repair: caretakers of the genome? Curr Biol 7:576–579

Das P, Samantaray S, Rout GR (1997) Studies on cadmium toxicity in plants: a review. Environ Pollut 98:29–36

Davies FT, Puryear JD, Newton RJ, Egilla JN, Grossi JAS (2002) Mycorrhizal fungi increase chromium uptake by sunflower plants: influence on tissue mineral concentration, growth, and gas exchange. J Plant Nutr 25:2389–2407

de Dorlodot S, Lutts S, Bertin P (2005) Effects of ferrous iron toxicity on the growth and mineral composition of an inter specific rice. J Plant Nutr 28:1–20

De Filippis LF, Ziegler H (1993) Effect of sublethal concentrations of zinc, cadmium and mercury on the photosynthetic carbon reduction cycle of Euglena. J Plant Physiol 142:167–172

Demirevska-Kepova K, Simova-Stoilova L, Stoyanova Z, Holzer R, Feller U (2004) Biochemical changes in barley plants after excessive supply of copper and manganese. Environ Exp Bot 52:253–266

Ebbs SD, Kochian LV (1997) Toxicity of zinc and copper to *Brassica* species: implications for phytoremediation. J Environ Qual 26:776–781

Elamin OM, Wilcox GE (1986a) Effect of magnesium and manganese nutrition on muskmelon growth and manganese toxicity. J Am Soc Hortic Sci 111:582–587

Elamin OM, Wilcox GE (1986b) Effect of magnesium and manganese nutrition on watermelon growth and manganese toxicity. J Am Soc Hortic Sci 111:588–593

El-Jaoual T, Cox DA (1998) Manganese toxicity in plants. J Plant Nutr 21:353–386. doi:10.1080/01904169809365409

Epstein E (1961) Mineral metabolism of halophytes. In: Rorison IH (ed) Ecological aspects of the mineral nutrition of plants. Blackwell Publishers, Oxford, pp 345–353

Fodor A, Szabo-Nagy A, Erdei L (1995) The effects of cadmium on the fluidity and H-ATPase activity of plasma membrane from sunflower and wheat roots. J Plant Physiol 14:787–792

Fontes RLS, Cox FR (1998) Zinc toxicity in soybean grown at high iron concentration in nutrient solution. J Plant Nutr 21:1723–1730

Foy CD (1978) The physiology of metal toxicity in plants. Ann Rev Plant Physiol 29:511–566

Foy CD, Weil RR, Coradetti CA (1995) Differential manganese tolerances of cotton genotypes in nutrient solution. J Plant Nutr 18:685–706

Gajewska E, Sklodowska M, Slaba M, Mazur J (2006) Effect of nickel on antioxidative enzymes activities, proline and chlorophyll contents in wheat shoots. Biol Planta 50:653–659

Garg N, Singla P (2011) Arsenic toxicity in crop plants: physiological effects and tolerance mechanisms. Environ Chem Lett 9:303–321

Gimeno-Garcia E, Andreu V, Boluda R (1996) Heavy metals incidence in the application of inorganic fertilizers and pesticides to rice farming soils. Environ Pollut 92:19–25

Goldbold DJ, Hutterman A (1986) The uptake and toxicity of mercury and lead to spruce (*Picea abies*) seedlings. Water Air Soil Pollut 31:509–515

Goldstein S, Czapski C (1986) The role and mechanism of metal ions and their complexes in enhancing damage in biological systems or in protecting these systems from the toxicity of O_2^-. J Free Radic Biol Med 2 (1):3–11

Gonnelli C, Galardi F, Gabbrielli R (2001) Nickel and copper tolerance in three Tuscan populations of *Silene paradoxa*. Physiol Planta 113:507–514

Gopal R, Dube BK, Sinha P, Chatterjee C (2003) Cobalt toxicity effects on growth and metabolism of tomato. Commun Soil Sci Plant Anal 34:619–628. doi:10.1081/CSS-120018963

Gruenhage L, Jager IIJ (1985) Effect of heavy metals on growth and heavy metals content of *Allium Porrum* and *Pisum sativum*. Angew Bot 59:11–28

Guo J, Dai X, Xu W, Ma M (2008) Over expressing GSHI and AsPCSI simultaneously increases the tolerance and accumulation of cadmium and arsenic in Arabidopsis thaliana. Chemosphere 72:1020–1026

Halliwell B, Cutteridge JMC (1990) Role of free radicals and catalytic metal ions in human disease: an overview. Methods Enzymol 186:1–85

Han FX, Su Y, Monts DL, Waggoner AC, Plodinec JM (2006) Binding distribution, and plant uptake of mercury in a soil from Oak Ridge, Tennessee, USA. Sci Total Environ 368:753–768

Hawkes JS (1997) Heavy metals. J Chem Educ 74:1369–1374

Hegedus A, Erdei S, Horvath G (2001) Comparative studies of H_2O_2 detoxifying enzymes in green and greening barley seedings under cadmium stress. Plant Sci 160:1085–1093

Hernandez LE, Carpena-Ruiz R, Garate A (1996) Alterations in the mineral nutrition of pea seedlings exposed to cadmium. J Plant Nutr 19:1581–1598

Hewilt EJ (1953) Metal inter-relationships in plant nutrition. J Exp Bot 4:59–64

Horiguchi T (1988) Mechanism of manganese toxicity and tolerance of plants. IV. Effects of silicon on alleviation of manganese toxicity of rice plants. Soil Sci Plant Nutr 34:65–73

Horst J (1988a) Beschreibung der Gleichgewichtslage des ionenaustauschs an schwach saoren harzen mit hilfe eines models der oberflachenkomplexbildung, doctoral thesis, University of Karlsruhe, Kfk report, 4464

Horst WJ (1988b) The physiology of Mn toxicity. In: Graham RD, Hannam RJ, Uren NC (eds) Manganese in soils and plants. Kluwer Academic Publishers, Dordrecht, pp 175–188

Hossain MA, Piyatida P, Jaime A, da Silva T, Fujita M (2012) Molecular mechanism of heavy metal toxicity and tolerance in plants: central role of glutathione in detoxification of reactive oxygen species and methylglyoxal and in heavy metal chelation. J Bot vol 2012, Article ID 872875, 37 pp. doi: 10.1155/2012/872875

Huang CV, Bazzaz FA, Venderhoef LN (1974) The inhibition of soya bean metabolism by cadmium and lead. Plant Physiol 34:122–124

Huffman EWD Jr, Allaway HW (1973a) Chromium in plants: distribution in tissues, organelles, and extracts and availability of bean leaf Cr to animals. J Agric Food Chem 21:982–986

Huffman EWD Jr, Allaway WH (1973b) Growth of plants in solution culture containing low levels of chromium. Plant Physiol 52:72–75

Illan YA, Crapski C, Meisel D (1976) The one-electron transfer redox potentials of free radicals. 1. The oxygen/superoxide system. Biochim Biophys Acta 430:209–224

Israr M, Sahi S, Datta R, Sarkar D (2006) Bioaccumulation and physiological effects of mercury in *Sesbania drummondii*. Chemosphere 65:591–598

Izosimova A (2005) Modelling the interaction between calcium and nickel in the soil-plant system. FAL Agric Res Spec Issue 288:99

Jain R, Srivastava S, Madan VK, Jain R (2000) Influence of chromium on growth and cell division of sugarcane. Indian J Plant Physiol 5:228–231

Juwarkar AS, Shende GB (1986) Interaction of Cd-Pb effect on growth yield and content of Cd, Pb in barley. Indian J Environ Health 28:235–243

Kaji T, Suzuki M, Yamamoto C, Mishima A, Sakamoto M, Kozuka H (1995) Severe damage of cultured vascular endothelial cell monolayer after simultaneous exposure to cadmium and lead. Arch Environ Contam Toxicol 28:168–172

Kamal M, Ghalya AE, Mahmouda N, Cote R (2004) Phytoaccumulation of heavy metals by aquatic plants. Environ Int 29:1029–1039

Kasprzak KS (1995) Possible role of oxidative damage in metal induced carcinogenesis. Cancer Invest 13:411–430

Khan S, Khan NN (1983) Influence of lead and cadmium on growth and nutrient concentration of tomato (*Lycopersicum esculentum*) and egg plant (*Solanum melongena*). Plant Soil 74:387–394

Kitao M, Lei TT, Koike T (1997a) Effects of manganese toxicity on photosynthesis of white birch (*Betula platyphylla* var. japonica) seedlings. Physiol Plant 101:249–256

Kitao M, Lei TT, Koike T (1997b) Effects of manganese in solution culture on the growth of five deciduous broad-leaved tree species with different successional characters from northern Japan. Photosynthesis 36:31–40

Kumar G, Singh RP, Sushila (1992) Nitrate assimilation and biomass production in *Seasamum indicum* (L.) seedlings in lead enriched environment. Water Soil Pollut 215:124–215

L'Huillier L, d'Auzac J, Durand M, Michaud-Ferriere N (1996) Nickel effects on two maize (*Zea mays*) cultivars: growth, structure, Ni concentration, and localization. Can J Bot 74:1547–1554

Lee YW, Klein CB, Kargacin B, Salnikow K, Kitahara J, Dowjat K, Zhitkovich A, Christie NT, Costa M (1995) Carcinogenic nickel silences gene expression by chromatin condensation and DNA methylation: a new model for epigenetic carcinogens. Mol Cell Biol 15:2547–2557

Lee CW, Choi JM, Pak CH (1996) Micronutrient toxicity in seed geranium (*Pelargonium* x *hortorum* Bailey). J Am Soc Hortic Sci 121:77–82

Lenntech Water Treatment and Air Purification (2004) Water treatment. Lenntech, Rotterdamseweg, http://www.excelwater.com/thp/filters/WaterPurification.htm

Lewis S, Donkin ME, Depledge MH (2001) Hsp 70 expression in *Enteromorpha intestinalis* (Chlorophyta) exposed to environmental stressors. Aquat Toxicol 51:277–291

Li Z, McLaren RG, Metherell AK (2004) The availability of native and applied soil cobalt to ryegrass in relation to soil cobalt and manganese status and other soil properties. N Z J Agric Res 47:33–43

Li HF, Gray C, Mico C, Zhao FJ, McGrath SP (2009) Phytotoxicity and bioavailability of cobalt to plants in a range of soils. Chemosphere 75:979–986

Luna CM, Gonzalez CA, Trippi VS (1994) Oxidative damage caused by an excess of copper in oat leaves. Plant Cell Physiol 35:11–15

Luo Y, Han Z, Chin SM, Linn S (1994) Three chemically distinct types of oxidants formed by iron mediated Fenton reactions in the presence of DNA. Proc Natl Acad Sci U S A 91:12438–12442

Manara A (2012) Plant responses to heavy metal toxicity. In: Furini A (ed) Plants and heavy metals. Springer Briefs in Biometals. doi: 10.1007/978-94-007-4441

Marschner H (1986) Mineral nutrition of higher plants. Academic, London, p 674

Mathys W (1975) Enzymes of heavy metal-resistant and non-resistant populations of *Silene cucubalus* and their interactions with some heavy metals *in vitro* and *in vivo*. Physiol Plant 33:161–165

Meharg AA (1994) Integrated tolerance mechanisms-constitutive and adaptive plant-response to elevated metal concentrations in the environment. Plant Cell Environ 17:989–993

Meharg AA, Macnair MR (1992) Suppression of the high affinity phosphate uptake system; a mechanism of arsenate tolerance in *Holcus lanatus* L. J Exp Bot 43:519–524

Mildvan AS (1970) Metal in enzymes catalysis. In: Boyer DD (ed) The enzymes, vol 11. Academic, London, pp 445–536

Miller JE, Hassete JJ, Koppe DE (1975) Interaction of lead and cadmium of electron energy transfer reaction in corn mitochondria. Physiol Plant 28:166–171

Mohanpuria P, Rana NK, Yadav SK (2007) Cadmium induced oxidative stress influence on glutathione metabolic genes of *Camellia sinensis* (L.). O Kuntze. Environ Toxicol 22:368–374

Monni S, Salemma M, Millar N (2000) The tolerance of *Empetrum nigrum* to copper and nickel. Environ Pollut 109:221–229

Moreno-Caselles J, Moral R, Pera-Espinosa A, Marcia MD (2000) Cadmium accumulation and distribution in cucumber plants. J Plant Nutr 23:243–250

Morzck E Jr, Funicclli NA (1982) Effect of lead and on germination of *Spartina alterniflora* Loisel seeds at various salinities. Environ Exp Bot 22:23–32

Mukherji S, Maitra P (1976) Toxic effects of lead growth and metabolism of germinating rice (*Oryza sativa* L.)

seeds mitosis of onion (*Allium cepa*) root tip cells. Ind J Exp Biol 14:519–521
Nagajyoti PC, Lee KD, Sreekanth TVM (2010) Heavy metals, occurrence and toxicity for plants: a review. Environ Chem Lett 8:199–216
Nakos G (1979) Lead pollution: fate of lead in soil and its effects on *Pinus haplenis*. Plant Soil 50:159–161
Neelima P, Reddy KJ (2002) Interaction of copper and cadmium with seedlings growth and biochemical responses in *Solanum melongena*. Envin Pollut Technol 1:285–290
Nieboer E, Richardson DHS (1980) The replacement of the nondescript term heavy metals by a biologically and chemistry significant classification of metal ions. Environ Pollut B 1:3–26
Ouzounidou G (1994) Change in chlorophyll fluorescence as a result of copper treatment: dose response relations in Silene and Thlaspi. Photosynthetica 29:455–462
Paivoke H (1983) The short term effect of zinc on growth anatomy and acid phosphate activity of pea seedlings. Ann Bot 20:307–309
Panda SK, Patra HK (2000) Does chromium (III) produce oxidative stress in excised wheat leaves? J Plant Biol 27:105–110
Pandey N, Sharma CP (2002) Effect of heavy metals Co^{2+}, Ni^{2+}, and Cd^{2+} on growth and metabolism of cabbage. Plant Sci 163:753–758
Pandolfini T, Gabbrielli R, Comparini C (1992) Nickel toxicity and peroxidase activity in seedlings of *Triticum aestivum* L. Plant Cell Environ 15:719–725
Parr PD, Taylor FG Jr (1982) Germination and growth effects of hexavalent chromium in Orocol TL (a corrosion inhibitor) on *Phaseolus vulgaris*. Environ Int 7:197–202
Patra M, Sharma A (2000) Mercury toxicity in plants. Bot Rev 66:379–422
Patterson W, Olson JJ (1983) Effects of heavy metals on radicle growth of selected woody species germinated on filter paper, mineral and organic soil substrates. Can J Forest Res 13:233–238
Peralta JR, Gardea Torresdey JL, Tiemann KJ, Gomez E, Arteaga S, Rascon E (2001) Uptake and effects of five heavy metals on seed germination and plant growth in alfalfa (*Medicago sativa*) L. Bull Environ Contam Toxicol 66(6):727–734
Porter JR, Cheridan RP (1981) Inhibition of nitrogen fixation in alfalfa by arsenate, heavy metals, fluoride and simulated acid rain. Plant Physiol 68:143–148
Porter EK, Peterson PJ (1975) Arsenic accumulation by plants on mine waste (United Kingdom). Environ Pollut 4:365–371
Prasad MNV, Hagmeyer J (1999) Heavy metal stress in plants. Springer, Berlin, pp 16–20
Prasad MNV, Greger M, Landberg T (2001) Acacia nilotica L. bark removes toxic elements from solution: corroboration from toxicity bioassay using Salix viminalis L. in hydroponic system. Int J Phytoremed 3:289–300
Pryor WA (1988) Why is the hydroxyl radical the only radical that commonly adds to DNA? Hypothesis: it is a rare combination of high electrophilicity, high thermo chemical reactivity, and a mode of production that occurs near DNA. Free Radic Biol Med 4:219–223
Rahman H, Sabreen S, Alam S, Kawai S (2005) Effects of nickel on growth and composition of metal micronutrients in barley plants grown in nutrient solution. J Plant Nutr 28:393–404
Reddy AM, Kumar SG, Jyotsnakumari G, Thimmanayak S, Sudhakar C (2005) Lead induced changes in antioxidant metabolism of horsegram (*Macrotyloma uniflorum* (Lam.) Verdc.) and bengalgram (*Cicer arietinum* L.). Chemosphere 60:97–104
Reeves RD, Baker AJM (2000) Metal-accumulating plants. In: Raskin I, Ensley BD (eds) Phytoremediation of toxic metals: using plants to clean up the environment. Wiley, New York, pp 193–229
Romero-Puertas MC, Rodriquez-Serrano M, Corpas FJ, Gomez M, Del Rio LA, Sandalio LM (2004) Cadmium-induced subcellular accumulation of O_2^- and H_2O_2 in pea leaves. Plant Cell Environ 27:1122–1134
Ros R, Cook DT, Martinez-Cortina C, Picazo I (1992) Nickel and cadmium-related changes in growth, plasma membrane lipid composition, ATPase hydrolytic activity and proton pumping of rice (*Oryza sativa* L. cv. Bahia) shoots. J Exp Bot 43:1475–1481
Roseman IE, Levine RL (1987) Purification of a protease from Escherichia coli with specificity for oxidized glutamine synthetase. J Biol Chem 262(5):2101–2110
Rout GR, Sanghamitra S, Das P (2000) Effects of chromium and nickel on germination and growth in tolerant and non-tolerant populations of *Echinochloa colona* (L). Chemosphere 40:855–859
Salt DE, Blaylock M, Kumar NPBA, Dushenkov V, Ensley D, Chet I, Raskin I (1995) Phytoremediation: a novel strategy for the removal of toxic metals from the environment using plants. Biotechnology 13:468–474
Schmfger MEV (2001) Phytochelatins: complexation of metals and metalloids, studies on the phytochelatin synthase. PhD Thesis, Munich University of Technology (TUM), Munich
Scholz RW, Graham KS, Wynn MK (1990) Interaction of glutathione and a-tocopherol in the inhibition of lipid peroxidation of rat liver microsomes. In: Eddy CC, Hamilton GA, Madyastha KM (eds) Biological oxidation systems. Academic, San Diego, pp 841–867
Shah K, Dubey RS (1998) Effect of cadmium on proline accumulation and ribonuclease activity in rice seedlings: role of proline as a possible enzyme protectant. Biol Plant 40:121–130
Shanker AK, Sudhagar R, Pathmanabhan G (2003a) Growth phytochelatin SH and antioxidative response of sunflower as affected by chromium speciation. In: 2nd international congress of plant physiology on sustainable plant productivity under changing environment, New Delhi
Shanker AK, Djanaguiraman M, Pathmanabhan G, Sudhagar R, Avudainayagam S (2003b) Uptake and phytoaccumulation of chromium by selected tree

species. In: Proceedings of the international conference on water and environment held in Bhopal, India

Sharma P, Dubey RS (2005) Lead toxicity in plants. Braz J Plant Physiol 17:35–52

Singh PK, Tewari SK (2003) Cadmium toxicity induced changes in plant water relations and oxidative metabolism of *Brassica juncea* L. plants. J Environ Biol 24:107–117

Sinha SK, Srinivastava HS, Mishra SN (1988a) Nitrate assimilation in intact and excised maize leaves in the presence of lead. Bull Environ Contam Toxicol 41:419–422

Sinha SK, Srinivastava HS, Mishra SN (1988b) Effect of lead on nitrate reductase activity and nitrate assimilation in pea leaves. Bot Pollut 57:457–463

Sinha S, Guptha M, Chandra P (1997) Oxidative stress induced by iron in *Hydrilla verticillata* (i.f) royle: response of antioxidants. Ecotoxicol Environ Safe 38:286–291

Somasekharaiah BV, Padmaja K, Prasad ARK (1992) Phytotoxicity of cadmium ions on germinating seedlings of mung bean (*Phaseolus vulgaris*): involvement of lipid peroxidase in chlorophyll degradation. Physiol Plant 85:85–89

Stadtman ER (1993) Oxidation of free amino acids and amino acid residues in proteins by radiolysis and by metal-catalysed reactions. Annu Rev Biochem 62:797–821

Stadtman ER, Oliver CN (1991) Metal-catalyzed oxidation of proteins. Physiological consequences. J Biol Chem 266:2005–2008

Stiborova M, Pitrichova M, Brezinova A (1987) Effect of heavy metal ions in growth and biochemical characteristic of photosynthesis of barley and maize seedlings. Biol Plant 29:453–467

Sudhakar C, Symalabai L, Veeranjaveyuler K (1992) Lead tolerance of certain legume species grown on lead or tailing. Agric Ecosyst Environ 41:253–261

Tang SR, Wilke BM, Brooks RR, Tang SR (2001) Heavy-metal uptake by metal tolerant *Elsholtzia haichowensis* and *Commelina communis* from China. Commun Soil Sci Plant Anal 32:895–905

Thomas F, Malick C, Endreszl EC, Davies KS (1998) Distinct responses to copper stress in the halophyte, *Mesembryanthemum crystallinum*. Physiol Plant 102:360–368

Van Assche F, Clijsters H (1983) Multiple effects of heavy metals on photosynthesis. In: Marcelle R (ed) Effects of stress on photosynthesis, vol 7. Nijhoff/Junk, The Hague, pp 371–382

Van Assche F, Clijsters H (1987) Enzymes analysis in plants as a tool for assessing phytotoxicity on heavy metal polluted soils. Med Fac Landouw Rijiksuniv Gent 52:1819–1824

Van Assche F, Clijsters H (1990) Effects of metals on enzyme activity in plants. Plant Cell Environ 13:195–206

Varalakshmi LR, Ganeshamurthy AN (2009) Effect of cadmium on plant biomass and cadmium accumulation in amaranthus (*Amaranthus tricolor*) cultivars. Indian J Agric Sci 79(1):861–864

Varalakshmi LR, Ganeshamurthy AN (2012) Heavy metal contamination of water bodies, soils and vegetables in peri-urban areas: a case study in Bengaluru. J Hortic Sci 7(1):62–67

Varalakshmi LR, Ganeshamurthy AN (2013) Phytotoxicity of cadmium in radish and its effects on growth, yield and cadmium uptake. Commun Soil Sci Plant Anal 44:1444–1456

Vazques MD, Poschenrieder C, Barcelo J (1987) Chromium (VI) induced structural changes in bush bean plants. Ann Bot 59:427–438

Verma S, Dubey RS (2003) Lead toxicity induces lipid peroxidation and alters the activities of antioxidant enzymes in growing rice plants. Plant Sci 164:645–655

Warne MS, Heemsbergen D, Stevens D, McLaughlin M, Cozens G, Whatmuff M, Broos K, Barry G, Bell M, Nash D, Pritchard D, Penney N (2008) Modeling the toxicity of copper and zinc salts to wheat in 14 soils. Environ Toxicol Chem 27:786–792

Weckex JEJ, Clijsters HMM (1997) Zn phytotoxicity induces oxidative stress in primary leaves of *Phaseolus vulgaris*. Plant Physiol Biochem 35:405–410

Winterhalder EK (1963) Differential resistance of two species of eucalyptus to toxic soil manganese levels. Aust J Sci 25:363–364

Wintz H, Fox T, Vulpe C (2002) Responses of plants to iron, zinc and copper deficiencies. Biochem Soc Trans 30:766–768

Wu S (1994) Effect of manganese excess on the soybean plant cultivated under various growth conditions. J Plant Nutr 17:993–1003

Yadav SK (2010) Heavy metals toxicity in plants: an overview on the role of glutathione and phytochelatins in heavy metal stress tolerance of plants. S Afr J Bot 76:167–179

Zeid IM (2001) Responses of Phaseolus vulgaris to chromium and cobalt treatments. Biol Plant 44:111–115

Zhang WH, Tyerman SD (1999) Inhibition of water channels by $HgCl_2$ in intact wheat root cells. Plant Physiol 120:849–857

Zhou ZS, Huang SQ, Guo K, Mehta SK, Zhang PC, Yang ZM (2007) Metabolic adaptations to mercury-induced oxidative stress in roots of *Medicago sativa* L. J Inorg Biochem 101:1–9

Zhu B, Alva AK (1993) Effect of pH on growth and uptake of copper by *Swingle citrumelo* seedlings. J Plant Nutr 16:1837–1845

Zingg JM, Jones PA (1997) Genetic and epigenetic aspects of DNA methylation on genome expression, evolution, mutation and carcinogenesis. Carcinogenesis 18:869–882

Seed Priming for Abiotic Stress Tolerance: An Overview

6

K. Bhanuprakash and H.S. Yogeesha

Abstract

Heat, drought, cold, and salt stress are some of the major kinds of stresses that crops usually face under adverse weather or soil conditions. Abiotic stresses are often interrelated, either individually or in combination; they cause morphological, physiological, biochemical, and molecular changes that affect plant growth and development and ultimately yield. Various methodologies are in vogue to evolve stress-tolerant varieties either through conventional breeding or through transgenics. However, alternatively, more simple and economical practices are also in race to address this problem. Seed priming is one such farmer's friendly techniques recommended by many researchers for better crop stand establishment and growth even under adverse conditions. The present chapter deals with different seed priming methods and their scope in mitigating stress effects. Besides hydro-, osmo-, and halopriming, the relevance of nutrient priming and redox priming techniques for stress tolerance was also discussed. Further, how seed "priming-induced" biochemical and molecular changes regulate stress tolerance was amply explained in the light of the latest research work carried in this direction. Although the phenomenon of "priming-induced" stress tolerance appears complex, the present-day advanced techniques like proteomics, genomics, metabolomics, and transcriptomics made the task much simpler to understand these events clearly at subcellular level. Since priming mimics similar events happening under stress, the same can be exploited as a model system to decipher pathways that contribute stress tolerance.

6.1 Introduction

In the present era of global climate change, abiotic stresses are becoming more prevalent. Besides chilling and freezing temperatures, drought is one of the most important limiting factors of crop production. The increasing threat

K. Bhanuprakash (✉) • H.S. Yogeesha
Section of Seed Science and Technology, ICAR-Indian Institute of Horticultural Research, Hessaraghatta Lake Post, Bangaluru 560089, Karnataka, India
e-mail: kbp@iihr.ernet.in

N.K.S. Rao et al. (eds.), *Abiotic Stress Physiology of Horticultural Crops*,
DOI 10.1007/978-81-322-2725-0_6

of climate change is already having a substantial impact on agricultural production worldwide causing significant unpredictable loss in agriculture (Jakab et al. 2005) and threat to global food security (Christensen and Christensen 2007). In most plants, stress causes a variety of biochemical, physiological, and metabolic changes (Xiong and Zhu 2002), which may result in oxidative stress and affect plant metabolism, performance, and thereby the yield (Shafi et al. 2009). There is an ample evidence that poor crop stand establishment is a widespread constraint of crop production in developing countries, particularly under adverse conditions. Patchy plant stands are common, and yields are often reduced simply because there are not enough plants in the field. In addition, plants that do eventually emerge often grow slowly and are highly susceptible pests and diseases (Harris 2006). Various methodologies were adapted from time to time to achieve tolerance to stresses. These include conventional breeding methods such as selection and hybridization and modern methods such as mutation breeding, polyploidy breeding, genetic engineering, etc. Attempts were also made to produce transgenic plants which can withstand various kinds of stresses (Jisha and Puthur 2015). But they are time-consuming and demand skills and involve legal and ethical issues. The alternative solution would be more acceptable if it is simple, cost-effective, and can be adopted by the farmers without any complication, and at the same time, it should be effective in manifesting the tolerance. Clearly, anything that can be done to increase the proportion of seeds that emerge, and the rate at which they do so, will have a large impact on farmers' livelihoods (Harris 2006). In the last decade, interest has grown up in the use of pregermination treatments involving partial or total imbibitions of seeds for various crops aiming at acceleration of germination even under adverse conditions. Various terminologies like invigoration, priming, and hardening are in practice but with almost similar principle of controlling the imbibition process so that all events essential for germination will take place except germination. Drought, salinity, extreme temperatures, and oxidative stress are often interconnected and may induce similar damage. As a consequence, these diverse environmental stresses often activate similar cell signaling pathways and cellular responses. It is known that seed priming can activate these signaling pathways in the early stages of growth and result in faster plant defense responses. The exact molecular mechanism behind priming is not completely known; it is speculated that sensitization was associated with accumulation of inactive signaling proteins in primed cells. Under most abiotic stress conditions, plant metabolism is perturbed either because of inhibition of metabolic enzymes, shortage of substrate, or excess demand for specific compounds or a combination of these factors and many other reasons. Therefore, the metabolic network must be reconfigured to maintain essential metabolism and to acclimate by adopting a new steady state in light of the prevailing stress conditions. This metabolic reprogramming is also necessary to meet the demand for antistress agents including compatible solutes, antioxidants, and stress-responsive proteins (Obata and Fernie 2012), and while most of these technologies are cost-intensive, recent studies indicate microorganisms can also be used to help crops cope with climate change in a cost-effective manner (Venkateswarlu et al. 2008). Rhizosphere and endorhizosphere microorganisms are reported to help plants tolerate these abiotic stresses by a variety of mechanisms including modification of plant response at the gene level. Innovative systems biology approaches such as proteomics and metabolomics (Obata and Fernie 2012) are currently recognized as essential tools to understand the molecular mechanisms underlying plant responses to environmental stimuli and priming phenomena. Thus, priming is not a simple emulation of early imbibition stage of germination. Rather, it has been proposed that priming may impose moderate stress on seeds, which activates the stress-responsive systems that confer a "cross tolerance" on seeds to future exposure to stresses (Ligterink et al. 2007).

6.2 Abiotic Stress and Its Effect on Crop Growth and Development

Abiotic stresses are often interrelated, either individually or in combination; they cause morphological, physiological, biochemical, and molecular changes that adversely affect plant growth and productivity and ultimately yield. Heat, drought, cold, and salinity are the major abiotic stresses that induce severe cellular damage in plant species, including crop plants. Water stress or drought is a great challenge to agricultural production worldwide. Severe wilting causes an alteration in the ratio of membrane lipids, and this, in turn, may result in increased electrolyte leakage (Zwiazek and Blake 1990). Drought slows growth, induces stomatal closure, and therefore reduces photosynthesis (Nemeth et al. 2002). Organelles such as the chloroplasts and mitochondria are the main sites of ROS production in the plant cells (Breusegem et al. 2001). These effects lead to reduced photosynthesis, growth, and yield in a number of plant species (Razmjoo et al. 2008). Salinity is also considered as a major abiotic stress and significant factor affecting crop production all over the world and especially in arid and semiarid regions (Khajeh-Hosseini et al. 2003). Soil salinity reduces water availability of plant roots via negative (low) osmosis potential, as well as decrease of germination dynamics of plant seeds by ionic toxicity of Na^+ and Cl^- (Munns et al. 1988). In general, salinity affects almost every aspect of the physiology and biochemistry of plants (Cuartero et al. 2006). Of the major forms of abiotic stress plants are exposed to in nature, heat stress has an independent mode of action on the physiology and metabolism of plant cells. Various physiological injuries have been observed under elevated temperatures, such as scorching of leaves and stems, leaf abscission and senescence, and shoot and root growth inhibition or fruit damage, which consequently lead to decreased plant productivity (Vollenweider and Günthardt-Goerg 2005). High temperatures reduce plant growth by affecting the shoot net assimilation rates and thus the total dry weight of the plant (Wahid et al. 2007). In many crop species, the effects of high temperature stress are more prominent on reproductive development than on vegetative growth, and the sudden decline in yield with temperature is mainly associated with pollen infertility (Zinn et al. 2010). Heat stress which causes multifarious, and often adverse, alterations in plant growth, development, physiological processes, and yield is one of the major consequences of high temperature stress (Hasanuzzaman et al. 2013). The effects of cold stress at the reproductive stage of plants delay heading and result in pollen sterility, which is thought to be one of the key factors responsible for the reduction in grain yield of crops (Suzuki et al. 2008). The major adverse effect of cold stress in plants has been seen in terms of plasma membrane damage. This has been documented due to cold stress-induced dehydration (Steponkus et al. 1993). In addition to the well-established harmful effect of cold stress alterations in lipid composition of the biomembranes, changes in the carbohydrate metabolism (Frankow-Lindberg 2001) and the boosting of the radical scavenging potential of the cells (Baek and Skinner 2003) were reported. Taken together, cold stress results in loss of membrane integrity, leading to solute leakage. Further, cold stress disrupts the integrity of intracellular organelles, leading to the loss of compartmentalization. Exposure of plants to cold stress also causes reduction and impairing of photosynthesis, protein assembly, and general metabolic processes (Yadav 2010).

6.3 Seed Priming and Its Importance

Priming is a pre-sowing treatment that involves exposure of seeds to a low external water potential that limits hydration (controlled hydration of seed) to a level that permits pre-germinative metabolic activity to proceed but prevents actual emergence of the radical. This will ensure better field emergence and disease resistance under various adverse conditions. The purpose of priming is to reduce the germination time and improve stand and germination percentage under adverse environmental conditions. Primed seeds are used immediately but may be dried and stored for a short time for later use. According to McDonald (2000), primed seeds acquire the potential to rapidly imbibe and revive the seed metabolism thus enhancing the germination rate. In many crops, seed germination and early seedling growth are the most sensitive stages of water limitation and the water deficit may delay the onset and reduce the rate and uniformity of germination, leading to poor crop performance and yield. Therefore, the beneficial effects of priming may be more evident under unfavorable rather than favorable conditions (Parera and Cantliffe 1994). Primed seeds usually exhibit an increased germination rate, greater germination uniformity, and, at times, greater total germination percentage. These attributes have practical agronomic implications notably under adverse germination conditions (McDonald 2000).

6.3.1 Methods of Priming

Various kinds of priming are in practice depending on the need. Some well-known priming practices are as follows.

6.3.1.1 Hydropriming

This technique implies soaking the seeds in water for about a specific duration. This terminology is currently used both in the sense of steeping (imbibition in water for a short period) and in the sense of "continuous or staged addition of a limited amount of water." Hydropriming methods have practical advantages of minimal waste material produced when compared to osmo- and matrix priming.

6.3.1.2 Halo- and Osmopriming

In halopriming, seeds will be soaked in various solution of inorganic salts such as KCl, KNO_3, $CaCl_2$, $Ca(NO_3)_2$, KH_2PO_4, etc. This method is practiced for higher germination and plant emergence in salt-affected soils. In the case of osmopriming, substances like polyethylene glycol (PEG), sugar, glycerol, sorbitol, mannitol, etc. are used as osmotic solutes to develop lower water potential. As this process, unlike hydropriming, regulates water movement in much controlled fashion for longer period, this method is preferred in those crops where soaking in treatment solutions, even for shorter period, leads to germination (e.g., onion, beans, etc.).

6.3.1.3 Matrix Priming

Solid matrix priming is done using solid carriers with low matrix potentials, e.g., vermiculite, peat moss, sand, celite, etc. for slow imbibition process. In this case, seeds slowly imbibe and reach an equilibrium hydration level. After priming, the moist matrix material is removed by sieving or screening or may be partially incorporated into a coating. This process mimics the natural uptake of water by the seed from soil. Seeds are generally mixed into carrier at matric potentials from −0.4 to −1.5 MPa at 15–20 °C for 1–14 days.

6.3.1.4 Thermopriming

It is a kind of presoaking seed treatment with high and low temperature to improve germination and emergence under different environmental (low and high temperatures) conditions. This process enables seeds to germinate at temperatures lower or higher than those at which they would have not been able to germinate if untreated.

6.3.1.5 Biopriming

Treating the seed with some of the microbial agents like *Rhizobium*, *Azospirillum*, *Pseudomonas*, *Bacillus*, *Trichoderma*, *Gliocladium*, etc. is practiced in this method, for improving seed viability or vigor. Beneficial microbes are included in the priming process, either as a technique to colonize seeds or to control pathogen proliferation, during priming. Compatibility with existing crop protection seed treatments and other biologicals needs to be looked into while practicing this method. Costs of registration and other factors currently limit the commercial use of biopriming.

6.3.1.6 Drum Priming

Seeds are hydrated in a tumbling drum using precise volume of water. The amount of water is limited so that it is less than the amount needed for natural imbibition and for seed germination to occur. In this method, the seeds are evenly and slowly hydrated to a predetermined moisture content (typically <25–30 % fresh weight basis) by misting, condensation, or dribbling. Drum priming enhances seed performance without the loss of additional materials associated with the conventional osmotic priming technique.

6.3.1.7 Priming Using Growth Regulators

In this method, seeds are primed using solutions containing minute quantities of plant growth regulators like gibberellic acid, indoleacetic acid, benzyladenine, methyl jasmonate, 1-aminocyclopropane-1-carboxylic acid, etc. This method is usually followed to address seed dormancy problems or to enhance seed germination under adverse soil conditions or to reactivate impaired metabolism of aged and deteriorated seeds.

6.3.1.8 Nutrient Priming

Nutrient priming has been proposed as a novel technique that combines the positive effects of seed priming with an improved nutrient supply (Al-Mudaris and Jutzi 1999). In nutrient priming, seeds are pretreated (primed) in solutions containing the limiting nutrients instead of being soaked just in water.

6.3.1.9 Redox Priming

Cellular redox state is an important factor which regulates the key process in growth and development as well as stress tolerance. In response to any external stimuli, plants modify their redox state, and the extent of change is dependent on the nature of the stimulus itself, the dose, and the time to which the tissue is exposed (Miller et al. 2009). It is believed that if the reduced redox state is maintained, the extent of stress-induced damage can be minimized (Mittler 2002). Thiourea, hydrogen peroxide, cysteine, ascorbic acid, glutathione, and tocopherol are some of the compounds being used for this purpose.

6.3.2 Seed Priming and Biochemical and Molecular Basis of Stress Tolerance

Different seed priming methods employed to mitigate stress tolerance as reported by many researchers are shown in Table 6.1.

The various physiological, biochemical, and molecular events that took place during seed priming process are considered to have a major role in coping up with the stress effects which otherwise not seen in unprimed seeds. Non-availability of water under drought stress and inability to imbibe water under salt stress stand as major causes for the failure of germination in adverse conditions. Absorption of water is essential to initiate the process of germination and respiration, protein synthesis, mitochondria repair, DNA repair, etc. These are various crucial events that took place in the first two phases out of the three-phased model proposed by Bewley (1997) who demonstrated the time course pattern showing major events associated with germination and subsequent postgermination growth process. As discussed, the first two phases of germination will be completed in the process of priming, and hence primed seeds could finish germination even at low moisture condition unlike unprimed seeds. Many workers established the fact of higher and quick germination in primed seeds in various agricultural crops. In addition, seed priming causes the hydrolysis of abscisic acid (ABA) and the leaching of different biochemicals, e.g., coumarin, and various phenolic compounds from the seed to the priming solution (Hopkins 1995). These compounds are reported as germination inhibitors when present in the seeds. Priming, on the other hand, also improves membrane integrity of the germinating seed. Short-chain fatty acids (C6–C12) are known for their function to inhibit seed

Table 6.1 Different seed priming techniques and their effectiveness in improving growth of various crops under adverse conditions

Sl. no.	Method	Priming agent	Crop	Attributes improved	References
(A)	Osmopriming				
1.		PEG	Tomato and asparagus	Germination under saline condition	Pill et al. (1991)
2.		Mannitol	Cucumber	Germination under saline condition	Passam and Kakouriotis (1994)
(B)	Hydropriming	Water	Wheat	Germination rate under saline condition	Roy and Srivastava (1999)
1.		Water	*Triticale* spp.	Drought tolerance	Mehmet and Digdem (2008)
2.		Water	Maize	Drought tolerance	Janmohammadi et al. (2008)
(C)	Thermopriming	Chilling treatment	*Brassica* spp.	Germination under saline condition	Sharma and Kumar (1999)
1.		PEG at 20 °C	Carrot	Low and high temperature stress	Marcia et al. (2009)
(D)	Chemical priming	Gibberellins	Okra and pearl millet	Growth and yield under saline condition	Vijayaraghavan (1999)
1.		Solution of inorganic salts	Broad bean	Growth under saline condition	Sallam (1999)
2.		Salicylic acid	Muskmelon	Growth under drought stress	Ahment et al. (2007)
3.		NaCl	Muskmelon	Growth under saline stress	Yeaoung et al. (1996)
4.		Salts	Amaranthus	Drought tolerance	Moosavi et al. (2009)
5.		Brassinosteriods	Tomato	Drought tolerance	Behnamnia et al. (2009)
6.		NaCl	Sugarcane	Drought tolerance	Vikas Yadav et al. (2009)
7.		GA_3	Rapeseed	Drought tolerance	Li et al. (2010)
8.		Polyamines	Rice	Drought tolerance	Muhammad et al. (2009)
9.		Salts	Onion	Drought tolerance	Arrvin and Kazemi Poor (2003)
10.		Paclobutrazol	Tomato	Drought tolerance	Souza Machado et al. (1999)
11.		Chitosan	Maize	Low temperature tolerance	Gnan et al. (2009)
12.		Putrescine	Tobacco	Chilling tolerance	Xu et al. (2011)
(E)	Biopriming				
1.		*Pseudomonas fluorescens* strain (Pf_1)	Green gram	Enhance the water stress resistance	Saravankumar et al. (2011)
		Bacillus subtilis strain (EPB5)			
3.		Inoculation with *Rhizobium* and *Pseudomonas*	Maize	Salt tolerance	Bano and Fatima (2009)

(continued)

Table 6.1 (continued)

Sl. no.	Method	Priming agent	Crop	Attributes improved	References
4.		*Pseudomonas* spp. strain (GAP-P45)	Sunflower	Increased the survival, plant biomass, and root-adhering soil/ root tissue ratio of sunflower seedlings subjected to drought stress	Sandhya et al. (2009)
5.		*Pseudomonas* sp. strain (AKM-P6)	Sorghum	Enhanced the tolerance of seedlings to elevated temperatures	Ali et al. (2009)
6.		*Pseudomonas fluorescens* isolates	Pearl millet	Improved growth of the plants and resistance against downy mildew	Niranjan raj et al. (2004)
7.		*Trichoderma* spp.	Soybean	Control of damping of soybean caused by *C. truncatum* of soybean	Begum et al. (2010)
8.		*Pseudomonas fluorescens*	Sunflower	Control of *Alternaria* blight	Rao et al. (2009)
9.		*Pseudomonas fluorescens*	Chickpea	Control *Fusarium* wilt in tomato	Vidhyasekaran and Mutamilan (1995)

germination. During imbibition, the position of certain fatty acids associated with seed dormancy is changed and does not revert upon drying (Bewley and Black 1982). Maize seed priming with micronutrients, i.e., Zn, Mn, and Fe, significantly increased germination rate and concentrations of primed nutrients in young seedlings grown under low root zone temperature conditions (Imran et al. 2013).

Besides germination, early and healthy seedling development is essential for optimal growth and maximum yield in agricultural crops, and only strong and well-developed seedlings can tolerate several biotic and abiotic stresses during various growth stages. At this juncture seed reserves and their mobilization play a major role in germination and early seedling development. During early establishment stages, seedlings obtain mineral nutrients partly from seed reserves and partly from soil (Rengel and Graham 1995), and nutrients stored in the seeds are degraded by enzymes and transported to the growing embryo (Fincher 1989). A young seedling passes through various stages in nutrient uptake during growth, based on the internal and external nutrient supplies and on crop nutrient requirements (Scaife and Smith 1973). In the beginning, the plant utilizes nutrients from the seed reserves; it is the stage when external sources have a partial effect, and afterwards nutrient uptake by roots is responsible for plant growth with no further contribution from seed reserves (Whalley et al. 1966). Therefore, sufficient mineral nutrient reserves in the seed are necessary to maintain seedling growth until the root system starts supplying the nutrients. Seed priming increases seed reserve utilization, seedling dry weight, and seed reserve depletion percentage in mountain rye (Ansari et al. 2012) and wheat (Soltani et al. 2006). In monocotyledon plants like wheat (Soltani et al. 2006) and mountain rye (Ansari et al. 2012), gibberellic acid after synthesis in the scutellum migrates in to the aleurone layer. Mobilization of seed reserves happens when suitable enzyme system works. Enzymes like amylases, lipases, proteases, etc. are essential for breakdown of seed reserves. Seed priming with polyamines was effective in improving the emergence and seedling growth in hybrid sunflower (Farooq et al. 2007), and reports show that polyamines can improve stress tolerance in moderately drought-tolerant plants, and these roles are mainly related to alleviation of oxidative damage (Davies 2004). PAs applied to seeds by priming appreciably improved the drought tolerance, as evident from seedling height and their fresh and dry weight. Seed priming with PAs was effective in producing the vigorous seedlings (Farooq et al. 2007), which were obviously better able to withstand adversaries of drought. As evident from close correlations of growth and net photosynthesis with water relation attributes, it is believed that PAs have the ability to improve the cell water status thereby allowing rice to grow better under limited moisture supply conditions. However, having potential to act as free radical scavengers, PAs can protect the membranes and other macromolecules from oxidative damages (Besford et al. 1993) and thus can stabilize biological membranes under stressful conditions, and the membrane permeability and H_2O_2 and MDA contents of leaves under drought were significantly increased (Roberts et al.1986).

Priming led to an increased solubilization of seed storage proteins like the beta-subunit of the 11-S globulin in *Beta vulgaris* L. (Bourgne et al. 2000), reduction in lipid peroxidation, and enhanced antioxidative activity in seeds of *Momordica charantia* L. (Yeh et al. 2005) and *Zea mays* L. (Randhir and Shetty 2005). Recently Afzal et al. (2008) observed that the priming-induced salt tolerance was associated with improved seedling vigor, metabolism of reserves as well as enhanced K^+ and Ca^{2+} and decreased Na^+ accumulation in wheat plants. The higher salt tolerance of plants obtained from primed seeds is attributed to the higher capacity for osmotic adjustment and maintenance of ionic homeostasis by promoting K^+ and Ca^{2+} accumulation (Farhoudi and Sharifzadeh 2006). Salt priming also led to acclimation of *Bruguiera cylindrica*, a halophyte, to harsh conditions of PEG-induced desiccation stress (Atreya et al. 2009). Higher activities of catalase and superoxide dismutase were observed in salt-primed plants suggesting their protective role against oxidative stress.

A proteome analysis of seed germination during priming in the model plant *Arabidopsis thaliana* by MALDI-TOF spectrometry identified those proteins which appear specifically during seed hydropriming and osmopriming (Gallardo et al. 2004). Among these are the degradation products of the storage protein 12S-cruciferin β-subunits. Other reserve mobilization enzymes such as those for carbohydrates (α and β-amylases) and lipids mobilization (isocitrate lyase) are also activated during priming (Sung and Chang 1993). These results indicate that priming induces the synthesis and initiates activation of enzymes catalyzing the breakdown and mobilization of storage reserves, though most of the nutrient breakdown and utilization occur postgerminative after the radical emergence. Corbineau et al. 2000 observed that imbibition of tomato seeds in PEG results in sharp increases in adenosine triphosphate (ATP), energy charge (EC), and ATP/ADP (adenosine diphosphate) ratio. Significant increase in the number of mitochondria in response to priming was also reported in osmoprimed leek cells, although these were not correlated to respiration levels. The proteomic analysis also reveals that α- and β-tubulin subunits, which are involved in the maintenance of the cellular cytoskeleton and are constituents of microtubules involved in cell division, are abundant during priming (Varier et al. 2010).

Another protein detected by the proteomic analysis, whose abundance specifically increases during hydropriming is catalase which is synthesized in response to this stress to minimize cell damage. In addition to catalase, levels of superoxide dismutase, another key enzyme quenching free radicals, also increase during priming (Vari et al. 2003). Increased levels of these free radical scavenging enzymes due to the oxidative stress during priming could also protect the cell against membrane damage owing to lipid peroxidation occurring naturally. The abundance of low molecular weight heat shock proteins (LMW HSPs) of 17.4 and 17.7 kD specifically increased in osmoprimed seeds in the MALDI-TOF spectrometry analysis (Gallardo et al. 2004). LMW HSPs are reported to have molecular chaperone activity (Lee et al. 1995). In osmopriming, seeds are soaked in osmotica, viz., polyethylene glycol (PEG) and mannitol, which result in incomplete hydration, and an osmotic stress situation is created. This explains the abundance of heat shock proteins, which are known to accumulate in high amounts during any kind of stress. Similarly, the enzyme L-isoaspartyl protein methyltransferase, which repairs damage to cellular proteins, is reported to increase in response to priming (Kester et al.1997). Thus, it appears that one of the ways in which priming is effective at the subcellular level is by conferring protection to the cellular proteins damaged by any kind of stress (Varier et al. 2010).

Besides good germination and seedling establishment, proper growth and development of plant determines the yield status under abiotic stress condition. Factors like soil water, temperature, nutrition, air temperature, and disease incidence are known to effect growth and development of plants. Abiotic factors severely affect soil temperature. Soil temperature has a vital role in determining the soil moisture contents, rate constants of different chemical reactions, as well as nutrient transport and availability in the soil. Besides that, it also affects plant physiological processes involved in nutrient uptake, root growth and the soil microbial activities (Pregitzer and King 2005). Root growth is severely affected by soil temperatures (Kasper and Bland 1992). Therefore, low root zone temperature (RZT) stress can be associated with decrease in soil temperature responsible for inhibited root growth, reduced availability, and transport of mineral nutrient in the soil for optimal plant growth.

Due to inhibition of root growth under abiotic stress, plants fail to absorb many micro- and macronutrients that are essential for plant metabolic activity to exhibit proper growth and development. Abiotic stress factors such as low soil water contents and low soil temperature have been reported to increase in Zn deficiency symptoms (Lindsay and Norvell 1978). Zn is involved in a vast range of physiological and biochemical processes, and zinc is the only metal, with functions in all six enzyme classes including oxidoreductases, transferases, hydrolases, lyases, isomerases, and ligases in biological systems (Sousa et al. 2009). Hence, Zn deficiency causes membrane lipid peroxidation, increases membrane permeability (Cakmak and Marschner 1988a) and oxidative stress (Cakmak 2000), decreases carbohydrate metabolism (Marschner 1995), and impairs the photosynthetic rate (Burnell 1990). RNase increase under Zn deficiency results in stunted plant growth (Dwivedi and Takkar 1974). Abiotic stress tolerance in transgenic plants was increased by the overexpression of CuZn-SOD (Kim et al. 2010).

In case of salt stress and under alkaline conditions, the presence of Fe ions in the soil solution is severely reduced due to the formation of Fe hydroxides, oxyhydroxides, and oxides (Lemanceau et al. 2009). Cytochromes are well-known heme proteins and are essential components of the redox systems in chloroplasts (Hil 1954) and for the electron transport (Cava et al. 2008). Reduced photosynthetic activity is a typical physiological response of plants to Fe deficiency (Larbi et al. 2004), which might be due to reduction in size of the chloroplast and protein contents per chloroplast (Terry and Abadia 1986), reduction in chloroplast production (Perur et al. 1961), and low concentration of rubisco protein (Timperio et al. 2007). The role of catalase and peroxidase as scavenging enzymes in counteracting the ill effects of ROS under abiotic stress is well documented; the activity of these enzymes declines under the low Fe supply (Chouliaras et al. 2004). Catalases play an important role in the detoxification of photorespiratory hydrogen peroxide (H_2O_2) to water and signal transduction pathways of various stress responses (Verslues et al. 2007).

Manganese is essential, and production of MnSOD in the chloroplast of numerous transgenic plants has increased abiotic stress tolerance, e.g., salinity (Tanaka et al. 1999) and drought (Wang et al. 2005a). Under severe Mn deficiency, concentration and contents of chlorophyll are significantly low in the leaves, and ultrastructure of thylakoids is considerably changed and thus photosynthesis is severely depressed in higher plants (Yu et al. 1999a). Under Mn deficiency, high seed phosphorus reserves have been shown to improve early seedling development and increased dry matter production in various crops, such as spring wheat (Derrick and Ryan 1998), barley (Zhang et al. 1990), rice (Ros et al. 1997), and clover (Thomson and Bolger 1993).

Beneficial effects of microorganisms such as *Pseudomonas*, *Bacillus*, *Arthrobacter*, *Pantoea*, *Burkholderia*, *Rhizobium*, etc. in enhancing the tolerance of crops such as sunflower, maize, wheat, chickpea, groundnut, spices, and grapes to drought, salinity, heat stress, and chilling injury under controlled conditions were well documented (Ait Barka et al. 2006). The introduced microorganisms in the rhizosphere enhance soil aggregation by production of EPS thereby improving the water availability to plants during dry periods (Sandhya et al. 2009), induce the synthesis of heat shock proteins and osmoregulants such as proline and glycine betaine, and help in maintenance of cell membrane integrity (Bano and Fatima 2009).

Conrath et al. (2006) proposed that priming could involve accumulation of signaling proteins or transcription factors. Epigenetic changes would enable stress imprints to be left in the plant for long periods (Bruce et al. 2007) which involves modification of DNA activity by methylation, histone modification, or chromatin remodeling without alterations of nucleotide sequence (Madlung and Comai 2004). A few genes are expressed to the same extent in osmoprimed seeds which resembles stress induction. These include genes for serine carboxypeptidase (involved in reserve protein mobilization and transacylation) and cytochrome b (involved in the mitochondrial electron transport). Timmusk and Wanger (1999) reported that inoculation of *Paenibacillus polymyxa* confers drought tolerance in *Arabidopsis thaliana* through induction of a drought-responsive gene ERD15.

Specific stress-responsive system appears to be enhanced in primed seeds. For example, osmopriming induced accumulation of stress proteins, such as late embryogenesis abundant (LEA) proteins and heat shock proteins (HSP) (Catusse et al. 2011). A gene-encoding peroxiredoxin, an antioxidant protein, has also been upregulated during ABA priming in canola (*Brassica napus*) (Li et al. 2005). In tomato, *LeGOLS-1* was upregulated specifically during ABA and PEG priming (Downie et al. 2003); *GOLS* encodes for *galactinol synthase* (GOLS, EC 2.4.1.123), a putative enzyme in raffinose family oligosaccharides (RFO) biosynthesis pathway, and is known to express well under cold, drought, salinity, and osmotic stress in several species (Shinozaki and Shinozaki 2007). Microarray analysis of primed tomato seeds further suggests that genes involved in cell defense cover a major percentage of all genes that are upregulated during priming (Ligterink et al. 2007).

Some specific compounds are also known to play a vital role in providing abiotic stress tolerance. H_2S (hydrogen sulfide) treatment was known to overcome the deleterious effects of salt and nonionic osmotic stress by controlling oxidative and nitrosative cellular damage through increased performance of antioxidant mechanisms and the coordinated regulation of the SOS pathway. Quantitative real-time RT-PCR gene expression analysis of key antioxidant (*cAPX*, *CAT*, *MnSOD*, *GR*), ascorbate and glutathione biosynthesis (*GCS*, *GDH*, *GS*), transcription factor (*DREB*), and salt overly sensitive (SOS) pathway (*SOS2-like*, *SOS3-like*, *SOS4*) genes suggests that H_2S plays a pivotal role in the coordinated regulation of multiple transcriptional pathways (Christou et al. 2013). The reduction in the growth of green gram seedlings subjected to NaCl/polyethylene glycol (PEG) stress is alleviated by seed priming with chemical BABA, which enhanced photosynthetic pigment content and photosynthetic and mitochondrial activities and also modified the chlorophyll a fluorescence-related parameters. Moreover, priming with B-aminobutyric acid (BABA) reduced malondialdehyde content in the seedlings and enhanced the accumulation of proline, total protein, total carbohydrate, nitrate reductase activity, and activities of antioxidant enzymes like guaiacol peroxidase and superoxide dismutase (Jisha and Puthur 2015). 50 mM seed priming with glycine betaine-5 enhanced chilling temperature and drought tolerance in various turf grasses during the germination and seedling growth stage. Seed priming with optimum concentration of salicylic acid (150 ppm) and ascorbic acid (50 ppm) significantly enhanced drought tolerance in *Matricaria aurea* due to accumulation of higher amounts of chlorophyll, carotenoid, and anthocyanin content compared to unprimed (Zhang et al. 2014).

The phenomenon of priming-induced stress tolerance was well understood as discussed above. However and of late, to identify the key molecules in relation to cell signaling, ROS protection, ion transport and regulation, researchers are embarked on advanced techniques like genomics, proteomics, metabolomics, and transcriptomics to unravel the mysteries and understand the process of stress tolerance further. Since the process of priming mimics stress, primed seeds could act as a model system to decipher the events of stress-responsive system.

References

Afzal I, Rauf S, Basra SMA, Murtaza G (2008) Halopriming improves vigor metabolism of reserves and ionic contents in wheat seedlings under salt stress. Plant Soil Environ 54(9):382–388

Ahment K, Murat U, Ali Riza D (2007) Treatment with acetyl salicylic acid protects muskmelon seedlings against drought stress. Acta Physiol Plant 29:503–508

Ait Barka E, Nowak J, Clement C (2006) Enhancement of chilling resistance of inoculated grapevine plantlets with a plant growth promoting rhizobacterium *Burkholderia phytofirmans* strain PsJN. Appl Environ Microbiol 72(11):7246–7252

Ali SKZ, Sandhya V, Minakshi Grover Kishore N, Venkateswara Rao L, Venkateswarlu B (2009) *Pseudomonas* sp strain AKM–P6 enhances tolerance of sorghum seedlings to elevated temperatures. Biol Fertil Soils 46:45–55

Al-Mudaris MA, Jutzi SC (1999) The influence of fertilizer based seed priming treatment on emergence and seedling growth of Sorghum bicolor and Pennisetum glaucum in pot trials under greenhouse conditions. J Agron Crop Sci 182:135–141

Ansari O, Choghazardi HR, Sharif-Zadeh F, Nazarli H (2012) Seed reserve utilization and seedling growth of treated seeds of mountain ray (*Secale montanum*) as affected by drought stress. Cercet Agronomice Moldova 2(150):43–48

Arrvin MJ, Kazemi Poor N (2003) Response of onion cultivation to drought and salinity stress at germination stage and seed priming by chemicals to improve germination. Iran J Hortic Sci Technol 4:95–104

Atreya A, Vartak V, Bhargava S (2009) Salt priming improves tolerance to desiccation stress and to extreme salt stress in *Bruguiera cylindrica*. Int J Integr Biol 6(2):68–73

Baek KH, Skinner DZ (2003) Alteration of antioxidant enzyme gene expression during cold acclimation of near-isogenic wheat lines. Plant Sci 165:1221–1227

Bano A, Fatima M (2009) Salt tolerance in *Zea mays* (L) following inoculation with *Rhizobium* and *Pseudomonas*. Biol Fertil Soils 45:405–413

Begum MM, Sariah M, Puteh AB, Zainal Abidin MA, Rahman MA, Siddiqui Y (2010) Field performance of bio primed seeds to suppress *Colletotrichum truncatum* causing damping off and seedling strand of soybean. Biol Control 53:18–23

Behnamnia M, Kalantari M, Rezanejad F (2009) Exogenous application of brassinosteriods alleviates drought induced oxidative stress in *Lycopersicon esculentum*. Gen Appl Plant Physiol 35:22–24

Besford RT, Richardson CM, Campos JL, Tiburcio AF (1993) Effects of polyamines on stabilization of molecular complexes in thylakoid membranes of osmotically stressed oat leaves. Planta 189:201–206

Bewley JD (1997) Seed germination and dormancy. Plant Cell 9:1055–1066

Bewley JD, Black M (1982) Physiology and biochemistry of seeds, vol 2, Viability Dormancy and Environmental Control. Springer, Berlin

Bourgne S, Job C, Job D (2000) Sugarbeet seed priming: solubilization of the basic subunit of 11-S globulin in individual seeds. Seed Sci Res 10:153–156

Breusegem FV, Vranova E, Dat JF, Inze D (2001) The role of active oxygen species in plant signal transduction. Plant Sci 161:405–414

Bruce TJA, Matthes MC, Napier JA, Pickett JA (2007) Stressful "memories" of plants: evidence and possible mechanisms. Plant Sci 173:603–608

Burnell JN (1990) Immunological study of carbonic anhydrase in C and C plants using antibodies to maize cytosolic and spinach chloroplastic carbonic anhydrase. Plant Cell Physiol 31:423–427

Cakmak I (2000) Possible role of zinc in protecting plant cells from damage by reactive oxygen species. New Phytol 146:185–205

Cakmak I, Marschner H (1988) Increase in membrane permeability and exudation in roots of zinc deficient plants. J Plant Physiol 132:356–361

Catusse J, Meinhard J, Job C, Strub J, Fischer U, Pestsova E, West- hoff P, Van Dorsselaer A, Job D (2011) Proteomics reveals potential biomarkers of seed vigor in sugarbeet. Proteomics 11:1569–1580

Cava F, Zafra O, da Costa MS, Berenguer J (2008) The role of the nitrate respiration element of *Thermus thermophilus* in the control and activity of the denitrification apparatus. Environ Microbiol 10:522–533

Chouliaras V, Therios I, Molassiotis A, Patakas A, Diamantidis G (2004) Effect of iron deficiency on gas exchange and catalase and peroxidase activity in citrus. J Plant Nutr 27:2085–2099

Christensen JH, Christensen OB (2007) A summary of the PRUDENCE model projections of changes in European climate by the end of this century. Clim Change 81:7–30

Christou A, Manganaris GA, Papadopoulos I, Fotopoulos V (2013) Hydrogen sulfide induces systemic tolerance to salinity and non-ionic osmotic stress in strawberry plants through modification of reactive species biosynthesis and transcriptional regulation of multiple defence pathways. J Exp Bot 64(7):1953–1966

Conrath U, Beckers GJM, Flors V, Garcia-Agustin P, Jakab G, Mauch F, Newman MA, Pieterse CMJ, Poinssot B, Pozo MJ, Pugin A, Schaffrath U, Ton J, Wendehenne D, Zimmerli L, Mauch-Mani B (2006) Priming: getting ready for battle. Mol Plant Microbe Interact 19:1062–1071

Corbineau F, Ozbingol N, Vineland D, Come D (2000) Improvement of tomato seed germination by osmopriming as related to energy metabolism. In: Black M, Bradford KJ, Vasquez-Ramos J (eds) Seed biology advances and applications: proceedings of the sixth international workshop on seeds. CABI Cambridge, Merida, pp 467–474

Cuartero J, Bolarin MC, Asins MJ, Moreno V (2006) Increasing salt tolerance in tomato. J Exp Bot 57:1045–1058

Davies PJ (2004) The plant hormones: their nature occurrence and function. In: Davies PJ (ed) Plant hormones biosynthesis signal transduction action. Kluwer, Dordrecht

Derrick JW, Ryan MH (1998) Influence of seed phosphorus content on seedling growth in wheat: implications for organic and conventional farm management in South East Australia. Biol Agric Hortic 16:223–237

Downie B, Gurusinghe S, Dahal P, Thacker RR, Snyder JC, Nonogaki H, Yim K, Fukanaga K, Alvarado V, Bradford KJ (2003) Expression of a *GALACTINOL SYNTHASE* gene in tomato seeds is up-regulated before maturation desiccation and again after imbibition whenever radicle protrusion is prevented. Plant Physiol 131:1347–1359

Dwivedi R, Snehi Takkar PN (1974) Ribonuclease activity as an index of hidden hunger of zinc in crops. Plant and Soil 40:173–181

Farhoudi R, Sharifzadeh F (2006) The effects of NaCl priming on salt tolerance in canola (*Brassica napus* L) seedlings grown under saline conditions. Indian J Crop Sci 1(1–2):74–78

Farooq M, Basra SMA, Hussain M, Rehman H, Saleem BA (2007) Incorporation of polyamines in the priming media enhances the germination and early seedling growth in hybrid sunflower (*Helianthus annuus* L). Int J Agric Biol 9:868–872

Fincher GB (1989) Molecular and cellular biology associated with endosperm mobilization in germinating cereal grains. Annu Rev Plant Physiol Plant Mol Biol 40:305–346

Frankow-Lindberg BE (2001) Adaptation to winter stress in nine white clover populations: changes in non-structural carbohydrates during exposure to simulated winter conditions and 'spring' regrowth potential. Ann Bot 88:745–751

Gallardo K, Job C, Groot SPC, Puype M, Demol H, Vandekerekhove J, Job D (2004) Proteomics of *Arabidopsis* seed germination and priming. In: Nicholas G (ed) The biology of seeds: recent advances. CABI, Cambridge, pp 199–209

Gnan Y, Hu J, Wang X, Shao C (2009) Seed priming with chitosan improves maize germination and seedling growth in relation to physiological enhances under low temperature stress. J Zhejiang Univ Sci B 10 (6):427–433

Harris D (2006) Development and testing of onfarm seed priming. Adv Agron 90:129–178

Hasanuzzaman M, Nahar K, Fujita M (2013) Extreme temperatures oxidative stress and antioxidant defense in plants. In: Vahdati K, Leslie C (eds) Abiotic stress – plant responses and applications in agriculture. In Tech, Rijeka, pp 169–205

Hil R (1954) The cytochrome b component of chloroplasts. Nature 174:501

Hopkins WG (1995) Introduction to plant physiology, 4th edn. Wiley, New York

Imran MA, Mehmood V, Römheld G, Neumann (2013) Nutrient seed priming improves seedling development and increases grain yield of maize exposed to low root zone temperatures during early growth. Eur J Agron 49:141–148

Jakab G, Ton J, Flors V, Zimmerli L, Me'traux JP, Mauch-Mani B (2005) Enhancing *Arabidopsis* salt and drought stress tolerance by chemical priming for its abscisic acid responses. Plant Physiol 139:267–274

Janmohammadi M, Moradi Dezfuli P, Sharifzadeh F (2008) Seed invigoration techniques to improve germination and early growth of inbred line of maize under salinity and drought stress. Gen Appl Plant Physiol 34:215–226

Jisha KC, Puthur JT (2015) Seed priming with BABA (β-amino butyric acid): a cost-effective method of abiotic stress tolerance in *Vigna radiata* (L) Wilczek. Protoplasma. doi:10.1007/s00709-015-0804-7

Kasper TC, Bland WI (1992) Soil temperature and root growth. Soil Sci 154:290–299

Kester ST, Geneve RL, Houtz RL (1997) Priming and accelerated ageing effect L-isoaspartylmethyltransferase activity in tomato (*Lycopersicon esculentum* Mill) seed. J Exp Bot 48:943–949

Khajeh-Hosseini M, Powell AA, Bimgham IJ (2003) The interaction between salinity stress and seed vigor during germination of soybean seeds. Seed Sci Technol 31:715–725

Kim TH, Bohmer M, Hu H, Nishimura N, Schroeder JI (2010) Guard cell signal transduction network:

advances in understanding abscisic acid CO_2 and Ca^{2+} signaling. Annu Rev Plant Biol 61:561–591

Larbi A, Abadía A, Morales F, Abadía J (2004) Fe resupply to Fe-deficient sugar beet plants leads to rapid changes in the violaxanthin cycle and other photosynthetic characteristics without significant de novo chlorophyll synthesis. Photosynth Res 79:59–69

Lee GJ, Pokala N, Vierling E (1995) Structure and in vitro molecular chaperone activity of cytosolic small heat shock proteins from pea. J Biol Chem 270:10432–10438

Lemanceau P, Bauer P, Kraemer S, Briat JF (2009) Iron dynamics in the rhizosphere as a case study for analyzing interactions between soils plants and microbes. Plant Soil 321:513–535

Li F, Wu X, Tsang E, Cutler AJ (2005) Transcriptional profiling of imbibed *Brassica napus* seed. Genomics 86:718–730

Li Z, Lu GY, Zhang YK, Zou CS, Cheng Y, Zheng PY (2010) Improving drought tolerance of germinating seeds by exogenous application of GA_3 in rapeseed. Seed Sci Technol 38:432–440

Ligterink WJ, Kodde M, Lammers H, Dassen AHM, van der Geest RA, De Maagd RA, Hilhorst HWM (2007) Stress-inducible gene expression and its impact on seed and plant performance: a microarray approach. In: Adkins S, Ashmore S, Navie SC (eds) Seeds: biology development and ecology. CAB International, Wallingford, pp 139–148

Lindsay WL, Norvell WA (1978) Development of a DTPA soil test for zinc iron manganese and copper. Soil Sci Soc Am J 42:421–428

Madlung A, Comai L (2004) The effect of stress on genome regulation and structure. Ann Bot 94:481–495

Marcia DP, Denise C, Luiz A, Eusdarardo FA (2009) Primed carrot seeds performance under water and temperature stress. Sci Agric 662:174–179

Marschner H (1995) Mineral nutrition of higher plants, 2nd edn. Academic, San Diego

McDonald MB (2000) Seed priming. In: Black M, Bewley JD (eds) Seed technology and its biological basis. Sheffield Academic Press, Sheffield, pp 287–325

Mehmet Y, Digdem K (2008) Alleviation of osmotic stress of water and salt in germination and seedling growth of triticale with seed priming treatments. Afr J Biotechnol 7(13):2156–2162

Miller G, Suzuki N, Ciftci-Yilmaz S, Mittler R (2009) Reactive oxygen species homeostasis and signaling during drought and salinity stresses. Plant Cell Environ 33:453–467

Mittler R (2002) Oxidative stress antioxidants and stress tolerance. Trends Plant Sci 9:405–410

Moosavi A, Tavakkol Afshari R, Sharif-Zadeh F, Aynehband A (2009) Seed priming to increase salt and drought stress tolerance during germination in cultivated species of *Amaranth*. Seed Sci Technol 37:781–785

Muhammad F, Abdul W, Dong JL (2009) Exogenous applied polyamines increases drought tolerance of rice by improving leaf water status photosynthesis and membrane properties. Acta Physiol Plant 31:937–945

Munns RA, Gardrer ML, Rawson HM (1988) Growth and development in NaCl treated plants II Do Na^+ or Cl^- concentrations in dividing or expanding tissue determine growth in barley. Aust J Plant Physiol 15:529–540

Nemeth M, Janda T, Hovarth E, Paldi E, Szali G (2002) Exogenous salicylic acid increases polyamine content but may decrease drought tolerance in maize. Plant Sci 162:569–574

Niranjan RS, Shetty NP, Shetty HS (2004) Seed biopriming with *Pseudomonas fluorescens* isolates enhances growth of pearl millet plants and induces resistance against downy mildew. Int J Pest Manag 50(1):41–48

Obata T, Fernie AR (2012) The use of metabolomics to dissect plant responses to abiotic stresses. Cell Mol Life Sci 69:3225–3243

Parera CA, Cantliffe DJ (1994) Presowing seed priming. Hortic Rev 16:109–114

Passam HC, Kakouriotis D (1994) The effect of osmo conditioning on germination and emergence and early plant growth of cucumber under saline conditions. Sci Hortic 57:233–240

Perur NG, Smith RL, Wiebe HH (1961) Effect of iron chlorosis on protein fraction on corn leaf tissue. Plant Physiol 36:736–739

Pill WG, Freett JJ, Morneau DC (1991) Germination and seedling emergence of primed tomato and asparagus seeds under adverse conditions. Hortic Sci 26:160–1162

Pregitzer KS, King JS (2005) Effects of soil temperature on nutrient uptake. In: Bassiri Rad H (ed) Nutrient acquisition by plants: an ecological perspective. Springer, Berlin

Randhir R, Shetty K (2005) Developmental stimulation of total phenolics and related antioxidant activity in light- and dark-germinated corn by natural elicitors. Process Biochem 40:1721–1732

Rao MSL, Kulkarni S, Lingaraaju S, Nadaf HL (2009) Bio priming of seeds: a potential tool in the integrated management of *Alternaria* blight of sunflower. Helia 32:107–114

Razmjoo K, Heydarizadeh P, Sabzalian MR (2008) Effect of salinity and drought stresses on growth parameters and essential oil content of *Matricaria chamomile*. Int J Agric Biol 10:451–454

Rengel Z, Graham RD (1995) Importance of seed Zn content for wheat growth on Zn deficient soil. II. Grain yield. Plant Soil 173:267–274

Roberts DR, Dumbroff EB, Thompson JE (1986) Exogenous polyamines alter membrane fluidity in bean leaves- a basis for their potential misinterpretation of their true physiological role. Planta 167(3):395–401

Ros C, Bell RW, White PF (1997) Effect of seed phosphorus and soil phosphorus applications on early growth of rice (*Oryza sativa* L) cv IR66. Soil Sci Plant Nutr 43:499–509

Roy NK, Srivastava AK (1999) Effect of presoaking seed treatment on germination and amylase activity of wheat under salt stress condition. Rachis 18:46–51

Sallam HA (1999) Effect of some seed soaking treatments on growth and chemical components on faba bean plants under saline conditions. Ann Agric Sci 44:159–171

Sandhya V, Ali SKZ, Grover M, Reddy G, Venkateswaralu B (2009) Alleviation of drought stress effects in sunflower seedlings by the exopolysaccharides producing *Pseudomonas putida* strain GAP – P45. Biol Fertil Soils 46:17–26

Saravanakumar D, Kavino M, Raguchander T, Subbiah P, Samiyappan R (2011) Plant growth promoting bacteria enhance water stress resistance in green gram plants. Acta Physiol Plant 33:203–209

Scaife MA, Smith R (1973) The phosphorus requirement of lettuce. II. A dynamic model of phosphorus uptake and growth. J Agric Sci 80:353–361

Shafi M, Bakht J, Hassan MJ, Raziuddin M, Zhang G (2009) Effect of cadmium and salinity stresses on growth and antioxidant enzyme activities of wheat (*Triticum aestivum* L). Bull Environ Contam Toxicol 82:772–776

Sharma PC, Kumar P (1999) Alleviation of salinity stress during germination in *Brassica juncea* by pre sowing chilling treatments to seeds. Biol Plant 42:451–455

Shinozaki K, Shinozaki YK (2007) Gene networks involved in drought stress response and tolerance. J Exp Bot 58:221–227

Soltani A, Gholipoor M, Zeinali E (2006) Seed reserve utilization and seedling growth of wheat as affected by drought and salinity. Environ Exp Bot 55:195–200

Sousa SF, Lopes AB, Fernandes PA, Ramos MJ (2009) The zinc proteome: a tale of stability and functionality. Dalton Trans 38:7946–7956

Souza-Machado V, Pitblado R, Ali A, May P (1999) Paclobutrazol in tomato for improved tolerance to early transplanting and early harvest maturity. Acta Hortic 487:139–144

Steponkus PL, Uemura M, Webb MS (1993) A contrast of the cryostability of thc plasma membrane of winter rye and spring oat-two species that widely differ in their freezing tolerance and plasma membrane lipid composition. In: Steponkus PL (ed) Advances in low-temperature biology, vol 2. JAI Press, London, pp 211–312

Sung FJM, Chang YH (1993) Biochemical activities associated with priming of sweet corn seeds to improve vigor. Seed Sci Technol 21:97–105

Suzuki K, Nagasuga K, Okada M (2008) The chilling injury induced by high root temperature in the leaves of rice seedlings. Plant Cell Physiol 49:433–442

Tanaka Y, Hibino T, Hayashi Y, Tanaka A, Kishitani S, Takabe T, Yokota S, Takabe T (1999) Salt tolerance of transgenic rice overexpressing yeast mitochondrial Mn-SOD in chloroplasts. Plant Sci 148:131–138

Terry N, Abadia J (1986) Function of iron in chloroplasts. J Plant Nutr 9:609–646

Thomson CJ, Bolger TP (1993) Effects of seed phosphorus concentration on the emergence and growth of subterranean clover (*Trifolium subterraneum* L). Plant and Soil 156:285–288

Timmusk S, Wagner EGH (1999) The plant growth promoting rhizobacterium *Paenibacillus polymyxa* induces changes in *Arabidopsis thaliana* gene expression: a possible connection between biotic and abiotic stress responses. Mol Plant-Microbes Interact 12:951–959

Timperio AM, D'Amici GM, Barta C, Loreto F, Zolla L (2007) Proteomics pigment composition and organization of thylakoid membranes in iron-deficiency spinach leaves. J Exp Bot 13:3695–3710

Vari A, Mitrabinda S, Dadlani M, Sharma SP (2003) Physiological and biochemical changes associated with osmopriming in maize seeds Paper presented in II International Congress of Plant Physiology, 8—12 January 2003, New Delhi p 113

Varier A, Vari AK, Dadlani M (2010) The subcellular basis of seed priming. Curr Sci 99(4):45–456

Venkateswarlu B, Desai S, Prasad YG (2008) Agriculturally important microorganisms for stressed ecosystems: challenges in technology development and application. In: Kachatou-rians GG, Arora DK, Rajendran TP, Srivastava AK (eds) Agriculturally important microorganisms, vol 1. Academic World, Bhopal, pp 225–246

Verslues PE, Batelli G, Grillo S, Agius F, Kim YS, Zhu J, Agarwal M, Katiyar-Agarwal S, Zhu JK (2007) Interaction of SOS2 with nucleoside diphosphate kinase 2 and catalases reveals a point of connection between salt stress and H_2O_2 signaling in *Arabidopsis thaliana*. Mol Cell Biol 27:7771–7780

Vidhyasekharan P, Muthamilan M (1995) Development of formulations of *Pseudomonas fluorescens* for control of chickpea wilt. Plant Dis 79:782–786

Vijayaraghavan H (1999) Effect of seed treatment with PGR on bhendi grown under sodic soil condition. Madras Agric J 86:247–249

Vikas Yadav P, Sujata B, Surasanna (2009) Halopriming imparts to tolerance to salt and PEG induced drought stress in sugarcane. Agric Ecosyst Environ 134:24–28

Vollenweider P, Günthardt-Goerg MS (2005) Diagnosis of abiotic and biotic stress factors using the visible symptoms in foliage. Environ Pollut 137:455–465

Wahid A, Gelani S, Ashraf M, Foolad M (2007) Heat tolerance in plants: an overview. Environ Exp Bot 61:199–223

Wang FZ, Wang QB, Kwon SY, Kwak SS, Su WA (2005) Enhanced drought tolerance of transgenic rice plants expressing a pea manganese superoxide dismutase. J Plant Physiol 162:465–472

Whalley RDB, Mckell CM, Green LR (1966) Seedling vigor and the early non-photo synthetic stage of seedling growth in grasses. Crop Sci 6:147–150

Xiong L, Zhu JK (2002) Molecular and genetic aspects of plant responses to osmotic stress. Plant Cell Environ 25:131–139

Xu SC, Hu J, Li YP, Ma WG, Zheng YY, Zhu SJ (2011) Chilling tolerance in *Nicotiana tabacum* induced by seed priming with putrescine. Plant Growth Regul 63 (3):279–290

Yadav SK (2010) Cold stress tolerance mechanisms in plants: a review. Agron Sustain Dev 30:515–527

Yeaoung YR, Wilson JR, Murray GA (1996) Germination performance and loss of LEA proteins during musk-melon seed priming. Seed Sci Technol 24:429–439

Yeh YM, Chiu KY, Chen CL, Sung JM (2005) Partial vacuum extends the longevity of primed bitter gourd seeds by enhancing their anti-oxidative activities during storage. Sci Hortic 104:101–112

Yu Q, Osborne LD, Rengel Z (1999) Increased tolerance to Mn deficiency in transgenic tobacco over producing superoxide dismutase. Ann Bot 84:543–547

Zhang M, Nyborg M, Mcgill WB (1990) Phosphorus concentration in barley (*Hordeum vulgare* L) seed: influence on seedling growth and dry matter production. Plant Soil 122:79–83

Zhang Q, Rue K, Mueller J (2014) The effect of glycine betaine priming on seed germination of six turfgrass species under drought salinity or temperature stress. Hortic Sci 49:1454–1460

Zinn KE, Tunc-Ozdemir M, Harper JF (2010) Temperature stress and plants sexual reproduction: uncovering the weakest links. J Exp Bot 61:1959–1968

Zwiazek JJ, Blake TJ (1990) Effects of preconditioning on electrolyte leakage and lipid-composition in black spruce (*Picea mariana*) stressed with polyethylene-glycol. Physiol Plant 79:71–77

Part II

Abiotic Stress Tolerance in Horticultural Crops: Vegetables

7 Tomato

A.T. Sadashiva, Aradhana Singh, R. Punith Kumar, V. Sowmya, and Dominic P. D'mello

Abstract

Tomato (*Solanum lycopersicum* L.) is an important vegetable crop widely grown in both temperate and tropical regions. Abiotic stresses like heat, cold, salinity and flood have become major constraints for the tomato growth and yield. This chapter deals with the effects of the various stresses on tomato crop and the technologies developed to cope with these stresses to get optimum yield and quality.

7.1 Introduction

Abiotic stress is any change in surrounding environmental factors that may have adverse effect on plant growth and crop productivity. Some of the environmental factors like extreme temperature, drought, flooding, heavy metal toxicity, salinity and oxidative stress may affect the natural phenomenon of plant, due to overall changes in molecular, biochemical, physiological, and morphological characters. It is estimated that nearly two thirds of the yield potential is significantly reduced due to these constraints. To overcome these problems, strategies to be adopted by the plant breeders include exploration, collection and evaluation of the germplasm, selection and breeding of new varieties/hybrids, production of genetically modified crops and exogenous use of osmoprotectants.

Tomato (*Solanum lycopersicum* L.) belongs to Solanaceae family, mainly cultivated for its edible fruits and commercially grown vegetable crop all over the world. Though the available germplasm have got tolerance for various biotic and abiotic stresses, but under extreme climatic conditions, crop losses will be severe. Domestication, selection and different breeding methods have narrowed down the genetic diversity in cultivated tomato, due to which modern cultivars are vulnerable to many biotic and abiotic stresses because of narrow genetic base. Therefore, developing new high-yielding cultivars which are tolerant to various abiotic stresses will have great impact on the production of tomato.

A.T. Sadashiva (✉) • A. Singh • R.P. Kumar
V. Sowmya • D.P. D'mello
Division of Vegetable Crops, ICAR-Indian Institute of Horticultural Research, Hessaraghatta Lake Post, Bengaluru 560089, Karnataka, India
e-mail: ats@iihr.ernet.in

N.K.S. Rao et al. (eds.), *Abiotic Stress Physiology of Horticultural Crops*,
DOI 10.1007/978-81-322-2725-0_7

7.2 Impact of Abiotic Stresses on Tomato

Plant growth and development and yield potential of crop are affected/limited by several factors like drought, salinity and extreme temperature. The plant induces a complex cellular and molecular response for different stresses to prevent harm and safeguard, but often at the detriment of growth and yield (Herms and Mattson 1992).

7.2.1 Heat Stress

Heat stress is the major constraint in tomato cultivation. The vegetative and reproductive phases are adversely affected by heat stress, with consequent reduction in yield and fruit quality (Abdul-Baki 1991). Optimal temperature for tomato is between 21 and 24 °C, depending on the crop stage. Few degrees above optimal will lead to decrease in fruit set. Temperature more than 35 °C will affect the germination, flowering, fruit set and fruit ripening in tomato.

7.2.1.1 Effect of Heat Stress on Vegetative Growth

Photosynthesis is one of the physiological processes that is very sensitive to high temperature and it can be avoided before other incidence of the stress is identified (Berry and Bjorkman 1980). The plants which are grown under elevated temperature will be shorter in height because of short internodes (Heuvelink 1989; Langton et al. 1997). The effect of high temperature stress on tomato is more during reproductive growth compared to vegetative growth. Wide variability exists amongst the tomato genotypes for heat stress tolerance (Abdelmageed and Gruda 2007).

7.2.1.2 Effect of Heat Stress on Fruit Set

Tomato productivity is considerably affected by the heat stress causing reduced fruit set and lower fruit quality. At higher temperatures, tomato performance will be very poor due to abnormal development of male and female reproductive tissues, imbalance in hormone, a decrease in level of carbohydrates and pollination deficit (Abdalla and Verkerk 1968; Peet et al. 1997). Hazra et al. (2007) summarized the adverse effects of heat stress on tomato fruit set failure which include flower drop, lack of flower development, less pollen production, dehiscence and viability, ovule abortion and poor viability, decreased carbohydrate availability and other reproductive abnormalities. Hanna and Hernandez (1982) observed significant increment in flower drop during reproductive stage of tomato due to high temperature. Developmental changes in the anther such as lack of opening of the stromium and poor pollen formation were observed specially in the epidermis and endothecium (Sato et al. 2002). In tomato, various reproductive processes like meiosis in both male and female organs, pollen germination and pollen tube growth, ovule viability, stigmatic and style positions, number of pollen grains retained by the stigma, fertilization and postfertilization processes, growth of the endosperm, proembryo and fertilized embryo were adversely affected by high temperature (Foolad 2005). Also, occurrences of an exerted style which may prevent self-pollination are the common effects of heat stress on reproductive process in tomato. Even moderately high temperatures at crucial stages of plant development lead to exhaustion of available carbohydrates due to depletion of sink and source strength resulting in lower fruit set and other parameters related to yield (Sato et al. 2006).

7.2.1.3 Effect of Heat Stress on Fruit Growth and Ripening

Elevation in temperature often increases the fruit growth rate, but it has a greater effect in hastening maturity, and, as a result, the final mean weight of tomato fruits is reduced (Hurd and Grave 1985). Tissue softening, red colour development and climacteric production of ethylene were delayed by high temperature in tomato fruits (Cheng et al. 1988). Significant reduction in fruit growth period and hastened ripening resulting in lower fruit quality has been observed in tomato (Adams et al. 2001). High temperature during ripening reduces the development of lycopene resulting in dull-coloured fruits (Shivashankar et al. 2014).

7.2.2 Chilling/Cold Stress

Plant growth and development is seriously affected by chilling. Low temperature affects plants by causing dehydration of the cells and tissues through the crystallization of the cellular water (Pearce 2001). Under low temperature, a decrease in the conductivity of membrane, an increase in viscosity of water, and the stomata regulation would result in water stress (Sarad et al. 2004). Different phenotypic symptoms in reaction to cold stress include poor germination, stunted seedlings, chlorosis, reduced leaf expansion and wilting and may lead to death of tissues. Plant growth and development of tomato is also severely affected by cold. For example, delayed onset, decreased rate of seed germination and poor crop establishment and performance are observed. At later stages, cold stress results in reduced plant growth and development, poor flower development and fruit set and substantial reduction in fruit yield.

7.2.3 Salt Stress

The tomato growth and yield is enormously affected by salt stress. The tomato germplasm, lines and cultivars show varied response to salt stress. Bolarin et al. (1991, 1993) observed remarkable decrease in fresh and dry weight of tomato shoots in response to salinity stress. Shoot length is an important indicator for a vast range of tomato genotype tolerance under salt stress (Cruz et al. 1990). The lower calcium uptake has been linked with decreased transpiration rate rather than competition effects with Na^+ (Cuartero and Fernandez-Munoz 1999). Gumi et al. (2013) reported that the tomato plants exhibited a decline in number of leaves, length of leaf and dry matter accumulation along with increased NaCl concentration and free proline content. Salinity stress also caused increase in cytotoxic ions (Na^+ and Cl^-) and Ca^{2+} with corresponding reduction in K^+ concentrations.

7.2.4 Water Stress

The plant water status determines several physiological processes of the quantity produce and quality of crop growth. The severity and duration of water stress affects plant growth, yield and fruit quality of tomato. It is the most important abiotic factor that damages the chloroplast, thereby causing photosynthesis and inactivation of electron transfer reaction in turn leading to the formation of active oxygen species (AOS) that cause photooxidative damage (Asada 1999). Drought stress will damage the protoplast which leads to loss of turgor, alteration of membrane potentials and osmotic stress. Huge loss of water from the cells will make the membrane disintegrate and will have adverse effects on metabolic processes. The highest demand for water by tomato plant is during flowering stage and thus regular irrigation during this stage is vital for optimum yield. Wudiri and Henderson (1985) observed that under greenhouse conditions, severe water stress (25 % ET) decreased the fruit set by more than 90 % in cultivar 'VF145B-7879' compared to a reduction of 40 % in cv. 'Saladette' showed better tolerance due to its ability to roll leaves under high evaporative demand, thus maintaining high leaf water status. Extreme water stress (40 % of PC) by 24 % will affect the height of the plant, 18 % of stem diameter and 32 % of chlorophyll concentration compared to the control, and 69 % decrease in yield was noted in the most stressed plants of tomato (Sibomana et al. 2013). In tomato plants earlier growth stage (20-day stage) has more inhibition of water stress as compared to later stage (30-day stage), and the exposed plants exhibit a significant reduction in leaf water potential, activity of nitrate reductase, photosynthesis rate, chlorophyll and relative water content, carbonic anhydrase and membrane stability index (Hayata et al. 2008).

7.2.5 Flooding Stress

Flooding is an environmental stress that significantly affects the crop growth and production. Waterlogging and flooding are common in rainfed ecosystems, especially in soils with poor drainage. The main problem associated with flooding is lack of oxygen. Tomato is identified as a sensitive crop for flooding stress. Lower leaves of the stem are the main indicators of flooding injury, where symptom includes yellowing and senescence of the leaf which often appear during 4–6 days after flooding. Flooding in tomato plants results in reduced leaf elongation, leaf epinasty and formation of adventitious roots, and an increase in diffusive resistance consequently results in wilting of leaves. Severity of flood stress increases with gradual rise in temperature; wilting and death of tomato plant is due to short period of flooding at high temperatures (Kuo et al. 1982). Kramer (1951) reported that the middle leaves of tomato showed epinastic curvature 24–48 h after flooding, and tomato plants will recover from flood stress through the formation of new adventitious roots. In flooded tomato plants, stomatal closure is a protective mechanism where it is controlled by root-to-shoot signalling regulation. Stomatal closure is due to increased leaf water potential, less photosynthetic activity and shortage of oxygen in the roots. So this protective mechnism decreases the damage to the plant system by increasing the root resistance to water entry from the soil (Else et al. 2009). Flooding tolerance in tomato scions could be increased by grafting them on eggplant rootstocks (Bhatt et al. 2015). Accumulation of endogenous ethylene causes injury to the plants (Drew 1979). Under low oxygen levels, an increased production of an ethylene precursor, 1-aminocyclopropane-1-carboxylic acid (ACC), is stimulated in the roots.

7.3 Biochemical Parameters

The plants which are subjected to stresses for a longer duration experience altered growth which changes the quality of the products.

7.3.1 Heat Stress

An increase in ABA level was observed when tomato plants were subjected to constant stressful temperatures (Daie 1980). Tomato plants develop a strong defence mechanism through the accumulation of phenolic compounds against stress at 35 °C, which is above optimal temperature of 25 °C (Rivero et al. 2001). Heat stress decreases the sugar concentration of tomato (Rosales et al. 2007). Under heat stress, heat shock proteins in the cytoplasm protect the protein biosynthesis (Miroshnichenko et al. 2005). HSPs LMW HSP21 has been reported to have a dual role in tomato as it protects photosystem II from oxidative damage and is involved in the colour change of fruit stored at low temperatures (Neta-Sharir et al. 2005).

7.3.2 Cold Stress

Low temperatures induce numerous changes in cellular components, including the extent of unsaturated fatty acids (Cossins 1994), composition of glycerolipids (Lynch and Thompson 1982), changes in protein and carbohydrate composition and the activation of ion channels (Knight et al. 1996). Sucrose and other simple sugars that occur with cold acclimation help in stabilizing the membrane as these molecules protect membranes against freezing damage.

7.3.3 Salt Stress

Higher salt tolerance was observed in wild tomato (*Lycopersicon pennellii*) when compared to cultivated tomato (*L. esculentum*) and this is mainly due to increased activities of antioxidant enzymes, viz. APX (ascorbate peroxidase), SOD (superoxide dismutase) and POD (guaiacol peroxidase) (Mittova et al. 2002). Accumulation of proline was more in salt-sensitive species of tomato than in tolerant wild relatives. Incubation of tomato plants in low concentration of salicylic acid for long time enabled plants to tolerate salt stress caused by 100 mM sodium chloride. In the treated plants, sodium ions accumulated in the leaf tissues will function as osmolytes without the well-known detrimental effects of the excess sodium (Tari et al. 2002).

7.3.4 Water Stress

The contents of the sugar and acids, viz. malic acid, citric acid and ascorbic acid, increased in tomato fruits with water stress, thus improving the fruit quality (Nahar and Gretzmacher 2002). Significant increase in fructose, glucose, sucrose in fruits and proline content in leaves was also reported and it showed the tendency of tomato to adjust osmotically to water stress. Increase in ascorbate concentration has been reported in tomato under drought stress (Favati et al. 2009).

7.4 Improvement in Abiotic Stress Tolerance

Numerous attempts have been made to enhance abiotic stress tolerance in tomato through biotechnological techniques, breeding approaches and cultivation practices. The timely perception of the stress in order to respond in a rapid and efficient manner is a critical step in plant defence which leads to stimulating the complex signalling cascades of defence against various stresses (Abou Qamar et al. 2009).

7.4.1 Cultural Practices

In tomato Whitaker (1994) confirmed that the plants which undergo high temperature stress before exposure to chilling stress resulted in less incidence and extent of chilling damage to fruit and other plant parts. To manage the effect of salt stress, Cuartero et al. (2006) suggested three cultural practices which are found to be significant in tomato: pre-exposure of seedlings to NaCl or drought helps in the adaptation of the adult plants to salinity, application of mist to tomato plants grown in Mediterranean conditions enhances the yield and growth in saline conditions, and grafting the cultivars on suitable rootstocks could decrease the impact of salinity. Wargent et al. (2013) pretreated tomato seedlings with 'Alethea' (plant health activator technology) and subsequently exposed them to 100 mM salinity stress, resulting in maximum reduction of 47 % in net photosynthetic rate at 8 days following sodium chloride treatment. In Alethea pretreated tomato seedlings, reaction to salt stress was distinctly lower during experimental period. The expression of genes associated to biotic stress, cell wall synthesis, redox signalling, ethylene signalling and photosynthetic processes was affected in Alethea-treated plants.

7.4.2 Grafting Technology

The adverse effects of flooding along with soil-borne diseases cause severe crop losses in tomato. Identification of suitable rootstocks through breeding and evaluation is in infancy because the traits associated with abiotic stress are complex and multigenic. The use of physiological and genetic markers during the selection process is still lagging behind though they are necessary to accelerate the identification of rootstocks with beneficial traits. In 2003 AVRDC suggested to use eggplant rootstocks VI046103 (EG195) and VI045276 (EG203) for grafting tomato scions to provide enhanced intensity of tolerance and resistance against bacterial wilt, flooding, root-knot nematodes and different soil-borne diseases (Keatinge et al. 2014). The advantages of grafting tomato plants (cv. Arka Rakshak) with few eggplant (*S. melongena* L.) rootstocks (Mattu Gulla, Arka Keshav, BPLH-1 and Arka Neelkanth) against flooding were reported by Bhatt et al. (2015), and they noticed that grafted plants showed better survival, better fruit yield and increased physiological adaptation over the ungrafted and self-grafted plants.

7.4.3 Breeding for Improved Abiotic Stress Tolerance

Breeding strategies include utilization of genetic variability in the existing germplasm, variability created by interspecies hybridization, somaclonal variation of cell and tissue culture and induced mutations. The presence of genetic variability for resistance is the important criterion for crop improvement. Introgression has been done for abiotic stress from tolerant wild relatives to cultivated tomato. Sources of genetic tolerance to various abiotic stresses are associated within the relative wild species, including *L. pennellii, L. chmielewski, L. peruvianum, L. chilense, L. hirsutum, L. parviflorum, L. cheesmanii and L. pimpinellifolium* (Foolad 2005). The breeding techniques, such as backcross method, pedigree method, single seed descent, and single plant selection, will exploit only additive gene action; in these methods F_2 will not be productive to enhance heat-tolerant character (Hazra et al. 2009) and selection of fruit set under high temperature should be purely on replication trial in F_3 and subsequent generations.

7.4.3.1 Heat Stress Tolerance

One of the best approaches for breeding to high temperature stress tolerance is the selection of varieties. Heat-tolerant tomato varieties have the capability to set fruit at higher temperatures than any other varieties (AVRDC 2001; Warner and Erwin 2005). Evaluating the germplasm for flowering and fruit set helps in identification of heat tolerance in tomatoes since these two characteristics are sensitive to heat and directly related to yield (Berry and Uddin 1988; Hanna and Hernandez 1982). Chandra and Gupta (1994) screened many tomato accessions and identified 46 moderately heat-tolerant/heat-tolerant accessions including the wild species. AVRDC has reported two QTLs in chromosomes 6 and 12 for heat resistance in tomato (Kadirvel 2010).

7.4.3.2 Cold Stress Tolerance

The tomato cultivar which has tolerance to cold stress should have the efficiency to germinate vegetative growth and maximum fruit set at low temperatures. 'LA1777' (*Solanum habrochaites*) has the ability to sustain chilling stress due to its inherent characteristics (Cao et al. 2014).

7.4.3.3 Salt Stress Tolerance

Several wild species within *Lycopersicon* have many useful genes, whereas cultivated species have very less variability for salt stress (Rick 1979). Sources of salt stress tolerance have been identified amongst and within the wild species of tomato: *Solanum pimpinellifolium, Solanum cheesmaniae, Solanum habrochaites, Solanum peruvianum, Solanum pennellii* and *Solanum chmielewski* (Rao et al. 2013). General growth and physiological and biochemical traits may be used as effective approaches to screen the sensitive and tolerant varieties of tomato exposed to salt stress (Fariduddin et al. 2012), and amongst the selected cultivars of tomato (K-21, PKM-1, Pusa Gaurav, Pusa Ruby, selection N5, Hera research and S-22), K-21 was identified as the most tolerant and S-22 was the most sensitive cultivar to salt stress. An interspecific cross between *Lycopersicon pimpinellifolium* and *Lycopersicon esculentum* that the fruit yield of the hybrid was more sensitive to increasing salt than either parents (Lyon 1941) and the rest of the crosses between cultivated and wild tomato also showed complex genetics. Heterosis was evident under saline conditions in the stem elongation in hybrids of *Solanum lycopersicum* and *Solanum esculentum* developed with three wild species (*Solanum peruvianum, Solanum cheesmanii* and *Solanum pennellii*) (Tal and Shannon 1983). The stem elongation characteristic of parent *S. pennellii* showed dominance in the hybrids but not with *S. cheesmanii* as a parent. F1 hybrid between *S. esculentum* and *S. pennellii* exhibited hybrid vigour for total dry matter production under salinity conditions (Saranga et al. 1991). Intergeneric crosses of tomato exhibited various QTLs linked with fruit yield under salinity (Breto et al. 1994); however, some of the QTLs related with tolerance were later found to be dependent on the parentage of the cross (Monforte et al. 1997). The QTLs linked to tolerance differ with stage of plant growth and also with the level of tolerance to germination (Foolad et al. 1997, 1998, 2001; Foolad and Chen 1999). Therefore, identification of suitable QTLs based on the parentage and stage of plant development is important in the selection of genotypes for crop improvement programmes for salinity tolerance.

7.4.3.4 Drought Stress Tolerance

Drought tolerance potential in a crop can be enhanced through either selection or breeding. Several accessions of wild taxa have showed resistance for drought, viz. LA1401, LA0429, LA3661, LA1334, LA2680, LA2133, LA1421, LA3657, LA1416, LA1335 and RF4A. Chandra and Gupta (1994) reported that *S. pimpinellifolium* need more days to express wilting, thus exhibiting tolerance to drought situation; wild species have also been shown to possess abiotic stress tolerance (*Lycopersicon esculentum* var. cerasiforme, tolerant to water stress; *S. hirsutum*, tolerant to drought).The Indian Institute of Horticultural Research has developed a drought-tolerant line (RF4A) through interspecific hybridization with *S. pennellii*. Shamim et al. (2014) reported five tolerant lines: 'Lyallpur-1', '17889','10584/G' and '*S. pennellii*' and '*S. chilense*' as wild; three moderately tolerant lines, 'Punjab Chuhara', 'Pusa Ruby' and 'Ailsa Craig'; and three sensitive lines, 'Roma', 'Avinash-2' and 'Ratan', and concluded that the tolerant genotypes accumulated more biomass under stress conditions.

7.4.4 Biotechnological Techniques for Improved Abiotic Stress Tolerance

Realizing the importance of protein stability under stress, improvement has been done using genetic engineering tools.

7.4.4.1 Genetic Engineering

Compared to marker-assisted selection and traditional breeding programmes, genetic engineering techniques (directly introducing less amount of genes) are more interesting and rapid advance towards improving stress tolerance. Introgression of genes using genetic engineering principles, which are known to be required in stress response, will help to develop strong tools for better improvement of abiotic stress tolerance in tomato combined with stress physiology. Introducing and improving abiotic stress tolerance by transferring one or more stress-responsive genes from different species could be a successful approach to improve the ability of crop plants in less-productive agricultural areas.

Transgenic tomato plants expressing yeast SAMDC (key regulatory enzymes in the biosynthesis of polyamines) have the capability to build 1.7–2.4-fold higher levels of spermine and spermidine compared to wild-type plants at high temperature stress and increase antioxidant enzyme activity and the safeguarding of membrane lipid peroxidation. This later enhances the effectiveness of CO_2 assimilation at high temperature stress (38 °C) when compared with wild-type plants. The nuclear-encoding gene of HSP, *Hsa32*, has been cloned in tomato (Liu et al. 2006). Lin et al. (2006) detected 14 RAPD markers linked to heat-tolerant characters in tomatoes under heat stress using bulk segregant analysis.

Transgenic tomato (Gisbert et al. 2000) plants expressing the HAL1 gene have limited salt tolerance as a result of preserving more K^+ than the control plants under salinity condition. Transgenic tomato plants that overexpress AtNHX1 to confer salt tolerance by compartmentalizing Na^+ ions in the vacuoles accumulated abundant quantities of the transporter in the tonoplast and showed substantially improved salt tolerance (Zhang and Blumwald 2001). The transgenic tomato crop can grow, flower and set fruit in the presence of 200 mM NaCl.

Patade et al. (2013) reported that *Nicotiana tabacum* osmotin (Nt Osm) transgenic tomato imparted cold tolerance to plants.

The CBF1 cDNA when introduced in tomato under the control of a CaMV 35S promoter showed better tolerance to chilling, drought and salt stress but had dwarf phenotype and reduction in the fruit set and seed number (Hsieh et al. 2002). Klay et al. (2014) showed that *Sl-ERF.B.3* (*Solanum lycopersicum* ethylene response factor *B.3*) gene that encodes for a tomato transcription factor of the ERF family is an abiotic stress-responsive gene, which is induced by cold, heat and flooding, but downregulated by salinity and drought.

7.4.4.2 miRNA Technology

miRNAs play a major role in the abiotic stress response. Through degradome sequencing, most of the target genes for miRNAs were identified in tomato (Cao et al. 2014), and many targeted genes that expressed miRNAs strongly respond to chilling stress by inhibiting the anti-stress protein expression, antioxidant enzyme and genes present in cell wall formation.

7.5 Conclusion and Future Thrusts

Tomato is considered as commercially grown vegetable crop in the world and is moderately tolerant to various abiotic stresses. For successful growth of tomato crop under environmental condition, tolerance may be needed at all major stages of plant development, including seed germination, vegetative stage, and flowering and fruit production. Each stage of the plant may need a different screening protocol and simultaneous or sequential screening may be impractical or impossible. In addition to the breeding approaches for developing tolerant genotypes, tolerant rootstocks are also used for flooding and water stress tolerances. Screening of wild relatives of tomato and also other related species as rootstocks for abiotic stress tolerance may be intensified.

References

Abdalla AA, Verkerk K (1968) Growth, flowering and fruit set of the tomato at high temperature. Neth J Agric Sci 16:71–76

Abdelmageed AHA, Gruda N (2007) Influence of heat shock pretreatment on growth and development of tomatoes under controlled heat stress conditions. J Appl Bot Food Qual 81:26–28

Abdul-Baki AA (1991) Tolerance of tomato cultivars and selected germplasm to heat stress. J Am Soc Hortic Sci 116(6):1113–1116

Abou Qamar S, Luo H, Laluk K, Mickelbart VM, Mengiste T (2009) Crosstalk between biotic and abiotic stress responses in tomato is mediated by AIM1 transcription factor. Plant J 58:1–13

Adams SR, Cockshull KE, Cave CRJ (2001) Effect of temperature on the growth and development of tomato fruits. Ann Bot 88:869–877

Asada K (1999) The water-water cycle in chloroplasts: scavenging of active oxygens and dissipation of excess photons. Annu Rev Plant Physiol Plant Mol Biol 50:601–639

AVRDC (2001) Asian vegetable research and development center. In: AVRDC report 2000, Shanhua, Tainan, vii+ pp 152

Berry J, Bjorkman O (1980) Photosynthetic response and adaptation to temperature in higher plants. Ann Rev Plant Physiol 31:491–543

Berry SZ, Uddin MR (1988) Effect of high temperature on fruit-set in tomato cultivars and selected germplasm. HortSci 23:606–608

Bhatt RM, Upreti KK, Divya MH, Srilakshmi Bhat, Pavithra CB, Sadashiva AT (2015) Interspecific grafting to enhance physiological resilience to flooding stress in tomato (*Solanum lycopersicum* L.). Sci Hortic 182:8–17

Bolarin MC, Cuartero EG, Cruz V, Cuartero J (1991) Salinity tolerance in four wild tomato species using vegetative yield-salinity response curves. J Am Soc Hortic Sci 116:286–290

Bolarin MC, Perez-Alfocea F, Cano EA, Estan MT, Carol M (1993) Growth, fruit yield and ion concentration in tomato genotypes after pre- and post-emergence salt treatments. J Am Soc Hortic Sci 118:655–660

Breto MP, Asins MJ, Carbonell EA (1994) Salt tolerance in Lycopersicon species. III. Detection of QTLs by means of molecular markers. Theor Appl Genet 88:395–401

Cao X, Wu Z, Jiang F, Zhou R, Yang Z (2014) Identification of chilling stress-responsive tomato microRNAs and their target genes by high-throughput sequencing and degradome analysis. BMC Genomics 15:1130

Chandra U, Gupta PN (1994) Evaluation of tomato germplasm adaptable to abiotic stress conditions of Northern India. Indian J Plant Genet Res 7(2):165–172

Cheng TS, Floros JD, Shewfelt RL, Chang CJ (1988) The effect of high-temperature stress on ripening of

tomatoes (*Lycopersicon esculentum*). J Plant Physiol 132(4):459–464
Cossins AR (1994) Homeoviscous adaptation of biological membranes and its functional significance. In: Cossins AR (ed) Temperature adaptation of biological membranes. Portland Press, London, pp 63–76
Cruz V, Cuartero J, Bolarin MC, Romero M (1990) Evaluation of characters for ascertaining salt stress responses in *Lycopersicon* species. J Am Soc Hortic Sci 115:1000–1003
Cuartero J, Fernandez-Munoz R (1999) Tomato and salinity. Sci Hortic 78:83–125
Cuartero J, Bolarιin MC, Asιins MJ, Moreno V (2006) Increasing salt tolerance in the tomato. J Exp Bot 57 (5):1045–1058
Daie J (1980) Determination of the temperature response curves for abscisic acid and its derivatives in economically important horticultural crops, PhD thesis, Utah State University, Logan
Drew MC (1979) Plant responses to anaerobic conditions in soil and solution culture. Curr Adv Plant Sci 36:1–14
Else MA, Janowiak F, Atkinson CJ, Jackson MB (2009) Root signals and stomatal closure in relation to photosynthesis, chlorophyll a fluorescence and adventitious rooting of flooded tomato plants. Ann Bot 103:313–323
Fariduddin Q, Mir BA, Ahmad A (2012) Physiological and biochemical traits as tools to screen sensitive and resistant varieties of tomatoes exposed to salt stress. Braz J Plant Physiol 24(4):281–292
Favati F, Lovelli S, Galgano F, Miccolis V, Di Tommaso T, Candido V (2009) Processing tomato quality as affected by irrigation scheduling. Sci Hortic 122:562–571
Foolad MR (2005) Breeding for abiotic stress tolerances in tomato. In: Ashraf M, Harris PJC (eds) Abiotic stresses: plant resistance through breeding and molecular approaches. The Haworth Press, New York, pp 613–684
Foolad MR, Chen FQ (1999) RFLP mapping of QTLs conferring salt tolerance during vegetative stage in tomato. Theor Appl Genet 99:235–243
Foolad MR, Stoltz T, Dervinis C, Rodriguez RL, Jones RA (1997) Mapping QTLs conferring salt tolerance during germination in tomato by selective genotyping. Mol Breed 3:269–277
Foolad MR, Chen FQ, Lin GY (1998) RFLP mapping of QTLs conferring salt tolerance during germination in an interspecific cross of tomato. Theor Appl Genet 97:1133–1144
Foolad MR, Zhang LP, Lin GY (2001) Identification and validation of QTLs for salt tolerance during vegetative growth in tomato by selective genotyping. Genome 44:444–454
Gisbert C, Rus AM, Bolarin MC, Lopez-Coronado M, Arrillaga I, Montesinos C, Caro M, Serrano R, Moreno V (2000) The yeast HAL1 gene improves salt tolerance of transgenic tomato. Plant Physiol 123:393–402
Gumi AM, Aliero AA, Shehu K, Danbaba A (2013) Effects on growth, biochemical parameters and ion homeostasis in *Solanum lycospersicum* L. (Cv. Dan eka). Cent Eur J Exp Biol 2(3):20–25
Hanna HY, Hernandez TP (1982) Response of six tomato genotypes under summer and spring weather conditions in Louisiana. Hortic Sci 17:758–9
Hayata S, Hasana SA, Fariduddina Q, Ahmad A (2008) Growth of tomato (*Lycopersicon esculentum*) in response to salicylic acid under water stress. J Plant Interact 3(4):297–304
Hazra P, Samsul HA, Sikder D, Peter KV (2007) Breeding tomato (*Lycopersicon esculentum* Mill) resistant to high temperature stress. Int J Plant Breed 1(1):31–40
Hazra P, Ansary SH, Dutta AK, Balacheva E, Atanassova B (2009) Breeding tomato tolerant to high temperature stress. Acta Hortic 830:241–248
Herms DA, Mattson WJ (1992) The dilemma of plants to grow or defend. Q Rev Biol 67:283–335
Heuvelink E (1989) Influence of day and night temperature on the growth of young tomato plants. Sci Hortic 38:11–22
Hsieh TH, Lee JT, Charng YY, Chan MT (2002) Tomato plants ectopically expressing Arabidopsis CBF1 show enhanced resistance to water deficit stress. Plant Physiol 130:618–626
Hurd RG, Grave CJ (1985) Some effects of air and root temperatures on the yield and quality of glasshouse tomatoes. J Hort Sci 60:359–371
Kadirvel P (2010) Molecular breeding in vegetable crops: opportunities and challenges presented. In: 2nd national workshop on marker-assisted selection for crop improvement. ICRISAT, Patancheru
Keatinge JDH, Lin LJ, Ebert AW, Chen WY, Hughes J'A, Luther GC, Wang JF, Ravishankar M (2014) Overcoming biotic and abiotic stresses in the *Solanaceae* through grafting: current status and future perspectives. Biol Agric Hortic Int J Sustain Prod Syst 30(4):272–287
Klay I, Pirrello J, Riahi L, Bernadac A, Cherif A, Bouzayen M, Bouzid S (2014) Ethylene response factor *Sl-ERF.B.3* is responsive to abiotic stresses and mediates salt and cold stress response regulation in tomato. Sci World J Article ID 167681, 12 pages. http://dx.doi.org/10.1155/2014/16768
Knight H, Trewavas AJ, Knight MR (1996) Cold calcium signaling in Arabidopsis involves two cellular pools and a change in calcium signature after acclimation. Plant Cell 8:489–503
Kramer PJ (1951) Causes injury to plants resulting from flooding of the soil. Plant Physiol 26:722–736
Kuo DG, Tsay JS, Chen BW, Lin PY (1982) Screening for flooding tolerance in the genus *Lycopersicon*. HortSci 17(1):6–78
Langton FA, Lumsden PJ, Horridge J (1997) Are gibberellins involved in temperature-mediated stem

extension responses in tomato? Acta Hortic 435:105–112

Lin KH, Lo HF, Leo SP, Kuo CG, Chen JT, Yeh WL (2006) RAPD markers for the identification of yield traits in tomatoes under heat stress via bulked segregant analysis. Hereditas 143:142–154

Liu N, Ko S, Yeh KC, Charng Y (2006) Isolation and characterization of tomato Hsa32 encoding a novel heat-shock protein. Plant Sci 170:976–985

Lynch DV, Thompson GA Jr (1982) Low temperature-induced alterations in the chloroplast and microsomal membrane of Dunaliella salina. Plant Physiol 69:1369–1375

Lyon C (1941) Responses of two species of tomatoes and the F1 generation to sodium sulphate in the nutrient medium. Bot Gazette 103:107–122

Miroshnichenko S, Tripp J, Nieden U, Neumann D, Conrad U, Manteuffel R (2005) Immunomodulation of function of small heat shock proteins prevents their assembly into heat stress granules and results in cell death at sublethal temperatures. Plant J 41:269–281

Mittova V, Guy M, Tal M, Volokita M (2002) Response of the cultivated tomato and its wild salt-tolerant relative *Lycopersicon pennellii* to salt-dependent oxidative stress: increased activities of antioxidant enzymes in root plastids. Free Radic Res 36:195–202

Monforte AJ, Asins MJ, Carbonell EA (1997) Salt tolerance in *Lycopersicon* species. V. Does genetic variability at quantitative trait loci affect their analysis? Theor Appl Genet 95:284–293

Nahar K, Gretzmacher R (2002) Effect of water stress on nutrient uptake, yield and quality of tomato (*Lycopersicon esculentum* Mill.) under subtropical conditions. Die Bodenkultur 53(1):45–51

Neta-Sharir I, Isaacson T, Lurie S, Weiss D (2005) Dual role for tomato heat shock protein 21: protecting photosystem ii from oxidative stress and promoting color changes during fruit maturation. Plant Cell 17:1829–1838

Patade VY, Khatri D, Kumari M, Grover A, Gupta SM, Ahmed Z (2013) Cold tolerance in Osmotin transgenic tomato (*Solanum lycopersicum* L.) is associated with modulation in transcript abundance of stress responsive genes. Springer Plus 2:117

Pearce RS (2001) Plant freezing and damage. Ann Bot 87:417–424

Peet MM, Willits DH, Gardner R (1997) Response of ovule development and postpollen production processes in male-sterile tomatoes to chronic, sub-acute high temperature stress. J Exp Bot 48(306):101–111

Rao ES, Kadirvel P, Symonds RC, Ebert AW (2013) Relationship between survival and yield related traits in *Solanum pimpinellifolium* under salt stress. Euphytica 190:215–228

Rick CM (1979) Potential improvement of tomatoes by controlled introgression of genes from wild species. In: Proceedings, conference on broadening genetic base crops. Pudoc, pp 167–173

Rivero RM, Ruiz JM, Garcia PC, Lopez-Lefebre LR, Sanchez E, Romero L (2001) Resistance to cold and heat stress: accumulation of phenolic compounds in tomato and watermelon plants. Plant Sci 160:315–321

Rosales MA, Rubio-Wilhelmi MM, Castellano R, Castilla N, Ruiz JM, Romero L (2007) Sucrolytic activities in cherry tomato fruits in relation to temperature and solar radiation. Sci Hortic 113:244–249

Sarad N, Rathore M, Singh NK, Kumar N (2004) Genetically engineered tomatoes: new vista for sustainable agriculture in High altitude regions. In: 4th international crop Science Congress, Brisbane

Saranga Y, Zamir D, Marani A, Rudich J (1991) Breeding tomatoes for salt tolerance field-evaluation of *Lycopersicon* germplasm for yield and dry-matter production. J Am Soc Hortic Sci 116:1067–1071

Sato S, Peet MM, Thomas JF (2002) Determining critical pre- and post-anthesis periods and physiological processes in Lycopersicon esculentum Mill. exposed to moderately elevated temperatures. J Exp Bot 53:1187–1195 [PubMed]

Sato S, Kamiyama M, Iwata T, Makita N, Furukawa H, Ikeda H (2006) Moderate increase of mean daily temperature adversely affects fruit set of *Lycopersicon esculentum* by disrupting specific physiological processes in male reproductive development. Ann Bot 97:731–738

Shamim F, Saqlan SM, Habib-Ur-Rehman A, Waheed A (2014) Screening and selection of tomato genotypes/cultivars for drought tolerance using multivariate analysis. Pak J Bot 46(4):1165–1178

Shivashankara KS, Pavithra KC, Laxman RH, Sadashiva AT, George Christopher M (2014) Genotypic variability in tomato for total carotenoids and lycopene content during summer and response to post harvest temperature. J Hortic Sci 9:98–102

Sibomana IC, Aguyoh JN, Opiyo AM (2013) Water stress affects growth and yield of container grown tomato (*Lycopersicon esculentum* Mill) plants. GJBB 2 (4):461–466

Tal M, Shannon MC (1983) Salt tolerance in two wild relatives of the cultivated tomato: responses of *Lycopersican esculentum*, *L. cheesmani*, *L. peruvianum*, *Solanum pennelli* and F1 hybrids of high salinity. Aust J Plant Physiol 10:109–117

Tari I, Jolan Csiszar J, Szalai G, Horváth F, Pécsváradi A, Kiss G, Szepesi A, Szabo M, Erdei L (2002) Acclimation of tomato plants to salinity stress after a salicylic acid pre-treatment. Acta Biol Szeged 46(3–4):55–56

Wargent JJ, Pickup DA, Paul ND, Roberts MR (2013) Reduction of photosynthetic sensitivity in response to abiotic stress in tomato is mediated by a new generation plant activator. BMC Plant Biol 13:108. doi:10.1186/1471-2229-13-108

Warner RM, Erwin JE (2005) Naturally occurring variation in high temperature induced floral bud abortion across *Arabidopsis thaliana* accessions. Plant Cell Environ 28:1255–1266

Whitaker BD (1994) A reassessment of heat treatment as a means of reducing chilling injury in tomato fruit. Postharvest Biol Technol 4:75–83

Wudiri BB, Henderson DW (1985) Effects of water stress on flowering and fruit set in processing-tomatoes. Sci Hortic 27(3–4):189–198

Zhang HX, Blumwald E (2001) Transgenic salt-tolerant tomato plants accumulate salt in foliage but not in fruit. Nat Biotechnol 19:765–768

8 Onion

N.K. Srinivasa Rao

Abstract

Onion (*Allium cepa* L.) is among the most important horticultural crops grown worldwide for its culinary preparations. The onion is the second most widely cultivated horticultural crop in the world. Limited water availability, increasing temperature, salinity and flooding are the major environmental factors in sustaining and increasing the productivity of onion worldwide. In onion, water is the main limiting factor for low productivity. Onions are considered a shallow-rooted crop and are more sensitive to water stress during bulb formation and enlargement than during the vegetative stage. In rainy season, the production of vegetables is often limited due to excessive moisture brought about by heavy rains. Temperature limits the range and production. Day length and temperature influence bulb formation in onions. Onions are sensitive to salt, are relative excluders of both Na and Cl and are sensitive to sulphate. Physiologically, excessive soil salinity imposes initial water deficit that results from the relatively high solute concentrations in the soil, causes ion-specific pores resulting from altered K^+/Na^+ ratio and leads to a build-up in Na^+ and Cl^- concentrations that are detrimental to plants. Understanding the physiological, molecular and biochemical processes in relation to these changing climatic factors including increasing CO_2 constitutes the first step toward the development of strategies for designing stress-resistant genotypes of onion. Effects of environmental factors-water, temperature salinity and elevated CO_2 on whole plant physiology of onion crops as expressed by growth, yield, quality and photosynthetic features are discussed.

8.1 Introduction

Out of 15 vegetables listed by FAO, onion (*Allium cepa* L.) falls second only to tomato in terms of total annual world production (Pathak

N.K.S. Rao (✉)
Division of Plant Physiology and Biochemistry, ICAR-Indian Institute of Horticultural Research, Hessaraghatta Lake Post, Bengaluru 560089, Karnataka, India
e-mail: nksraoiihr@gmail.com

N.K.S. Rao et al. (eds.), *Abiotic Stress Physiology of Horticultural Crops*,
DOI 10.1007/978-81-322-2725-0_8

2000). Onion crop requires cool weather during the early development of bulbs. Environmental factors influence development, growth and biological yield of plants primarily by affecting their physiology. Bulb onion is a shallow-rooted, biennial plant that is grown as an annual. It has long, hollow leaves with widening, overlapping bases. The tubular leaf blades are flattened on the upper surface, and the stem of the plant also is flattened. Roots arise from the bottom of the growing bulb. Leaf initiation stops when the plant begins to bulb. The base of each leaf becomes one of the "scales" of the onion bulb, so the final bulb size depends in part on the number of leaves present at bulb initiation. The leaf base begins to function as a storage organ at bulb initiation, so the size of the leafy part of the plant also influences bulb size. Thus, the more leaves present and the larger the size of the plant at the onset of bulb initiation, the larger will be the bulbs and the greater will be the crop yield. Plants grown from large onion sets (small mature bulbs) bulb earlier than plants grown from smaller sets of the same age or from seed. The chronological sequence of crop maturity for the different types of onion propagules is (first) dry onion sets (second), transplants from seed and (last) direct-sown seed. Night temperatures below 10 °C for a 2–3-week period will induce bolting (seed stalk formation) in onions after the 7–10-leaf stage. However, little bolting occurs if temperatures are around 21 °C. High temperatures during early growth also induce bolting.

8.2 Climate and Day Length Requirements for Bulbing

Day length and temperature influence bulb formation in onions. Also, before bulbing can occur, a certain amount of vegetative growth is required before the plant can respond to day length. At a specific threshold, past the "juvenile" stage of leaf growth, the plant becomes sensitive to the bulbing stimulus that is triggered if the days are long enough. When day length is at or greater than the threshold for bulb initiation of the particular cultivar (cultivated variety) of onion being grown, bulbing occurs if the average daily temperature is 16 °C or above and the average night temperature is 16–28 °C. If the requirement for day length is not met (i.e. the days are not long enough when the onion plant is physiologically mature), leaf production continues without bulb formation. The bulbing response is stronger when night-time temperatures are low, and also with larger plants. Light intensity, light quality and other factors interact with temperature and day length to influence the bulbing response of onion cultivars. For example, with warm weather and bright days, onions bulb at shorter day lengths than when the days are cool and overcast. Excessive nitrogen applications near this time may delay the bulbing response, even if the critical day length period occurs at the right stage of crop growth. Optimal temperatures for onions are in the range of 15–20 °C during early growth and 20–27 °C during bulb development. Once bulbing has been initiated, the growth of the bulb is influenced by temperature. Research indicates that onions maturing under hot conditions will have a lower dry matter content than those that mature under cooler conditions, and in general the lower the dry matter content and the fleshier the onion, the less pungent it is.

8.3 Impact of Abiotic Stresses

Excessive rains, humidity, temperature and salinity are critical factors of risk to onion cultivation. Onion is highly sensitive to water and heat stress. Both bulb and seed production are adversely affected due to these stresses.

8.3.1 Water Stress

Frequent irrigations to maintain high soil moisture are required to produce high yields of onion (Al-Jammal et al. 2000; Hegde 1986; Jones and Johnson 1958; Kadayifci et al. 2005; Rana and Sharma 1994; Shock et al. 1998, 2000a). The crop requires 350–500 mm of water over the growing season (FAO 2013); hence, adequate moisture possibly through irrigation is important in the production of onions. Onion root system is shallow so it extracts very little water from depths beyond 60 cm. Most of the water is from the top 30 cm of soil. Thus, upper soil areas must be kept moist to stimulate root growth and provide adequate water for the plant. Onion yield and grade are very responsive to careful irrigation scheduling and maintenance of optimum soil moisture (Shock et al. 1998, 2000).

Onion is cultivated under both irrigated and nonirrigated conditions. The reasons for the lower yield of onion are many including inadequate management practices, insufficient amount and unfavourable arrangement of precipitation in the growing season, as well as poor water management applied to onions grown from seed. The actual evaluation of stress related to the yield due to soil water deficit during the onion-growing season can be obtained by the estimation of the yield response factor (Ky) that represents the relationship between a relative yield decrease (1–Ya/Ym) and a relative evaporation deficit (1–ETa/ETm). Doorenbos and Kassam (1986) estimated that the average value of Ky is 1.5 during the onion-growing season. Values of Ky in the growing season (Ky 1.78) indicate that onion is highly sensitive to water stress under the climate conditions.

Both onion yield and size have been found to be reduced by temporary water stress. Jones and Johnson (1958) found that withholding irrigations for 2 weeks in any of three periods during the season resulted in yield reductions. Van Eeden and Myburgh (1971) irrigated onion at 60 % of the available soil water and found that yield was reduced when the soil moisture depletion was allowed to reach 90 % of the available water during the second half of the season. Using rain shelters, Dragland (1974) showed that onion yield reductions were more pronounced with 3 weeks of water stress at the beginning of the season than in the middle or end of the season. Irrigating onion at a higher soil water tension of 30, 50 or 70 kPa during the last 5 weeks of the season compared with 20 kPa reduced yield and size (Shock et al. 2000b). Sorensen and Grevsen (2001) found that drought stress during the last 3 weeks of the season reduced onion yields. Furthermore, when water stress was imposed 30 days after transplanting for a period of 15 days, leaf area and bulb growth were considerably decreased with a reduction of 17–26 % in onion yield (Bhatt et al. 2006).

The soil moisture gradually reduced when water stress was imposed for a period of 1, 2, 3 and 4 weeks in two cultivars of onion, viz. Arka Kalyan and Agrifound Dark Red (ADR), under field conditions (Srinivasa Rao et al. 2010). It was 22.0 % at 0 day, 13.0 % at 1 week stress, 9.5 % at 2 weeks stress, 6.5 % at 3 weeks stress and 5.5 % at 4 weeks stress (Table 8.2). There was substantial decrease in leaf area, root length, bulb fresh and dry mass and bulb yield under stress (Table 8.1). The reduction in leaf area was more than 40.0 % after 2 weeks stress, and it decreased to 53.0 % after 4 weeks stress. Root growth was affected in both the cultivars, but the affect was more in ADR as indicated by 44–63.0 % reduction compared to Arka Kalyan which had a reduction of 29.0–53.0 % under stress. The decrease in root growth was more after 3 weeks stress in both cultivars. There was a greater reduction in photosynthetic rate (84.0–92.0 %), total plant dry matter accumulation (32.0–45.2 %) and bulb dry matter (44.4–54.0 %) after 3 weeks stress (Table 8.1 and Fig. 8.1) in both cultivars, but the percent decrease was more in ADR than Arka Kalyan. The results indicated that the reduction in these physiological parameters under different duration resulted in a reduction in bulb size (Plate 8.1). The bulb yield reduction was higher in ADR (49.0 %) compared to Arka Kalyan (34.0 %) at 4 weeks stress (Table 8.2). The results showed that the soil moisture of 5.5 % was found to be critical as it affected the physiological parameters and the bulb yield. The cultivar Arka Kalyan performed better than ADR under stress.

In a trial with several vegetable crops, Singh and Alderfer (1966) observed that soil water stress at any growth stage leads to reduction in marketable yield. They further observed that with regard to yield reduction, onions are more sensitive to water stress during bulb formation

Table 8.1 Leaf area, root length, bulb fresh weight, bulb dry weight and total plant dry weight of onion as affected by different water stress duration

Stress duration (week)	Treatment	RWC	Leaf area	Root length	Bulb fresh wt.	Bulb dry wt.	Total dry wt.
Arka Kalyan							
One	Control	78.7	948.5	14.5	42.8	3.97	11.46
	Stress	68.3	804.5	10.3	27.4	3.12	9.91
Two	Control	78.0	1093.1	12.2	69.5	5.84	15.36
	Stress	65.0	938.4	7.8	39.0	4.44	12.02
Three	Control	79.6	1072.4	14.3	85.8	9.49	19.28
	Stress	64.1	853.0	8.3	66.9	7.32	12.89
Four	Control	78.5	1609.7	14.0	127.6	14.06	27.95
	Stress	61.7	772.7	7.3	90.1	9.58	15.47
ADR							
One	Control	76.5	992.6	15.3	33.5	3.33	10.20
	Stress	72.1	913.5	8.5	26.4	2.77	9.93
Two	Control	76.5	1438.1	14.8	49.1	4.80	15.10
	Stress	68.6	838.6	6.5	38.1	3.21	10.16
Three	Control	78.2	1516.5	12.0	89.7	12.37	18.14
	Stress	66.5	838.2	5.0	71.0	7.88	11.50
Four	Control	78.3	1967.3	11.0	156.1	16.39	31.82
	Stress	64.2	922.6	4.2	84.3	8.78	14.68

Table 8.2 Soil moisture (%) and plant yield (g plant^{-1}) of onion as affected by different water stress durations

Cultivar	Control	One week stress	Two weeks stress	Three weeks stress	Four weeks stress
Soil moisture	22.0	13.0	9.5	6.2	5.5
Arka Kalyan	111	108.4	97.3	96.1	73.0
ADR	132	130.0	104.0	90.0	67.6

and enlargement than during the vegetative stage. Dragland (1974) reported that, when compared to an unstressed control treatment, an imposed 3-week-long drought early in the season reduced onion yield more than when the 3-week drought was imposed near the end of the growing season. Van Eeden and Myburgh (1971) found that water stress imposed late in the season (at 84–103 days after transplanting) reduced onion total yield by 15 % when compared to the yield with no water stress. Managing onion crop with less than adequate irrigation water availability is a challenge. Options of (1) deficit irrigation (DI), i.e. 20 % or 40 % less water application, at one of the growth stages of the crop and (2) controlled deficit of 20 % or 40 % on all growth stages, i.e. regulated deficit irrigation (RDI), were explored for maximising irrigation water use efficiency (IWUE) of onion under deficit water application through subsurface drip irrigation (Patel and Rajput 2013). Deficit (or regulated deficit) irrigation is one of the methods of maximising IWUE per unit of applied water. In DI treatments, the crop was provided the irrigation with 60 % and 80 % of crop evapotranspiration (ETc) creating water stress of 40 % and 20 %, respectively, at developmental, bulb formation and bulb maturity. In case of RDI treatments, 20 % and 40 % water stress was created throughout the crop season by applying the irrigation water at 80 % and 60 % ETc. The maximum yield (44.7 t ha^{-1}) was obtained in the full irrigation treatment (T1). In RDI, 20 % and 40 % deficit water application saved 19.2 % and 41.7 % water and resulted in 20 % and 32 % reduction in yield, respectively. In DI, 20 % water deficit in the growth stages of second, third and fourth saved 2.1, 13.2 and

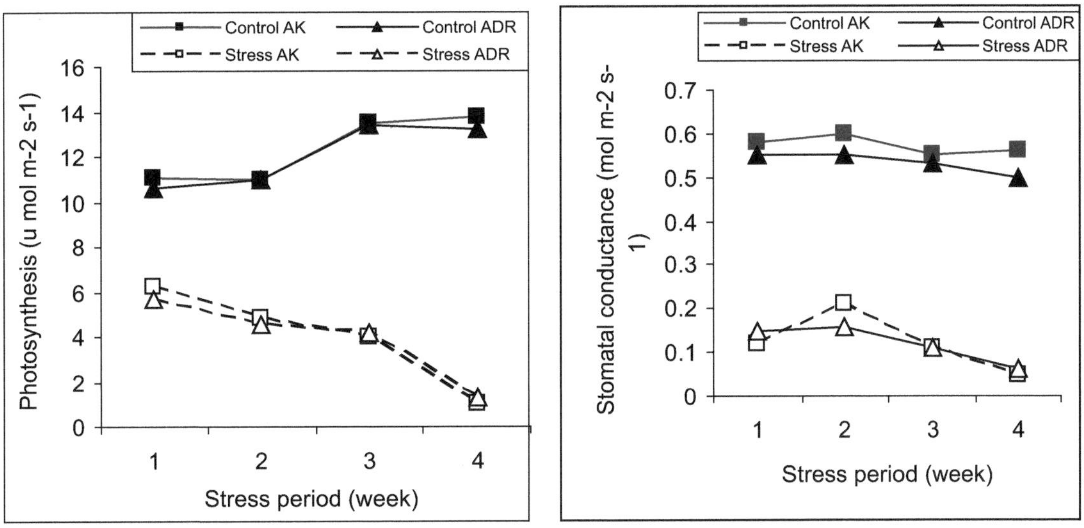

Fig. 8.1 Response of photosynthesis and stomatal conductance in onion under water stress

4.6 % of water with 19.8, 18.3 and 11.2 % reduction in yield, respectively, in comparison to full irrigation water application. This suggests that RDI is the better option of water saving than DI. Saving of water through RDI may be used to irrigate additional cropped area. Strategy suggested for productions of onion crop can be adopted in large scale to offset high cost of onion, which is the cause of concern for all stakeholders.

Studies on the effect of scheduling of drip irrigation on the growth, yield and water use efficiency of onion (cv. Telagi Red) revealed that scheduling of drip irrigation at shorter intervals, i.e. at 1 day and 2 days, significantly increased the growth parameters like plant height (61.58 and 60.25 cm, respectively), number of leaves (8.81 and 8.71, respectively), leaf area (466.47 and 462.02 cm^2, respectively), leaf area index (4.15 and 4.11, respectively) and neck girth (1.38 and 1.37 cm, respectively) as compared to 3 days interval (Bagali et al. 2012). The crop like onion performs better when irrigation is given on depletion of 15–20 % soil moisture of the field capacity. The shorter interval of irrigation at 1 day and 2 days ensures the moisture in the crop root zone nearly at 15–20 % depletion from field capacity. Drip irrigation system maintains soil physical conditions in congenial form for plant's growth by maintaining optimum soil water balance around plant bases. The bulb yield of the onion increased significantly with shorter interval of surface drip irrigation. Irrigation scheduled at 1 day interval recorded significantly higher bulb yield (46.93 t ha^{-1}) over 3 days interval (42.80 t ha^{-1}), and it was at par with 2 days interval (46.47 t ha^{-1}). Increase in the bulb yield is mainly attributed to positive association between yield and yield-contributing parameters like bulb weight and size in terms of equatorial and polar diameter of the bulb. Irrigation with 1 day and 2 days interval significantly influenced the equatorial diameter of the bulb which determines the bulb weight and that in turn decides the bulb yield. The shorter interval of irrigation ensures optimum growth of the crop by assuring balanced water and nutrient supply throughout the crop growth period (Bagali et al. 2012). Similar result for bulb yield was reported by Quadir et al. (2005). Drought stress imposed at the vegetative stage significantly decreased chlorophyll a content, chlorophyll b content and total chlorophyll content. The lack of effects on the chlorophyll a/b ratio indicates that chlorophyll a is more sensitive to drought than chlorophyll b (Mafakheri et al. 2010). They also play a critical role in the assembly of the light-harvesting complex and in the radiationless

Plate 8.1 Bulb size of onion as affected by different duration of water stress under field conditions (*C* control, *1* = 1 week stress, *2* = 2 weeks stress, *3* = 3 weeks stress and *4* = 4 weeks stress)

dissipation of excess energy (Streb et al. 1998). Carotenoid content decreased under severe water stress, significantly indicating that drought caused considerable oxidative stress by accumulation of ROS.

Onion single centredness is an important characteristic for the food industry. Onion ring manufacturing efficiency is reduced when onions are multiple centred. Axillary branching in onion, which occurs when a new lateral growing point forms adjacent to the existing one, leads to multiple growing points resulting in multiple-centred bulbs (Brewster 1994; Shock et al. 2007). Onions are sensitive to the formation of multiple centres with water stress at the four-leaf to late six-leaf stages. Environmental and management factors may influence onion single centredness by inducing axillary branching. Short interruptions of irrigation, resulting from the need to perform tractor operations in furrow-irrigated onion or from system failures with drip irrigation, might reduce onion single centredness under commercial production conditions. In a 3-year study with drip-irrigated onions in the Columbia Basin of Washington, translucent scale is an internal bulb defect in which one or more rings or scales acquire a translucent or watery appearance, rendering the onion unmarketable. Werner and Harris (1965) found that high air and soil temperatures during the growing season induced translucent scale. Higher soil and air temperatures might be induced by an interruption of irrigation and the resulting lack of evaporative cooling by the soil and plant leaves.

8.3.2 Flooding

Flooding is a major environmental stress that severely limits crop productivity, and it has become a major problem worldwide. Vegetable production occurs in both dry and wet seasons in the tropics. However, production is often limited during the rainy season due to excessive moisture brought about by heavy rain. Most vegetables are highly sensitive to flooding, and genetic variation with respect to this character is limited. In general, damage to vegetables by flooding is due to the reduction of oxygen in the root zone which inhibits aerobic processes. In tropical and subtropical regions, excessive rainfall is the major constraint for onion production. The yield penalty resulting from water logging may vary between 15 % and 80 %, depending on the species, soil type and duration of the stress. In a study on the effect of flooding for 7 days on onion cv. Arka Kalyan, there was 75.0–86.0 % reduction in leaf photosynthesis, 28.0–46.0 % in total plant fresh and 26.0–47.0 % in dry mass production under flooding in onion. The leaf area reduction of 25 % and 51 % was observed in once and continuously flooded plants, respectively. The leaf senescence was also more in the flooded plants. The bulb initiation stage was found to be sensitive (Plate 8.2) as the continuous flooding at this stage resulted in maximum reduction in bulb size (27.2 %) and bulb yield (48.3 %) (Srinivasa Rao et al. 2009).

The flooding imposed at bulb initiation stage in onion showed that plants grown in raised beds recorded photosynthetic rate 6.0–16.0 % higher than plants grown in channel after 6 days flooding. The bulb dry matter and bulb yield were higher in raised beds as compared to plants grown in the channel. The 6 days flooding was found to be critical at bulb initiation stage, and cultivar ADR was affected more compared to Arka Kalyan. There was a decrease in anthocyanin and ferric reducing antioxidant potential (FRAP) and increase in total flavonoids due to flooding. The raised bed cultivation may be adopted for onion in flood-prone areas (Plate 8.3).

During excess moisture availability, the bulbs are being harvested early and spread on the bunds to save the bulbs from rotting (Plate 8.4).

8.4 Temperature

Temperature is another important environmental parameter which affects the bulb initiation and formation in onions. The cardinal temperatures for optimum seedling growth, vegetative growth, before bulbing and bulb development are 20–25, 13–24, 15–21 and 20–25 °C, respectively. Very low temperatures at bulb development stage favour bolting. Sudden rise in temperature results in early maturity of the crop during winter season, thereby reducing bulb size.

Temperature variations have been shown to influence the rate of vegetative growth (Butt 1968; Brewster 1979; Seabrook 2005), leaf initiation and emergence (de Ruiter 1986). Temperature also influences the bulbing response, and varieties differ in how strongly they are influenced. High temperatures reduce the minimum day length required for bulbing, while if temperatures are too low, a variety may not bulb despite an otherwise suitable day length. Given a suitable day length, bulbing will start earlier and the plants mature sooner under warmer conditions. Given similar sowing dates, temperature is likely to be the major reason for differences in maturity date between seasons and between sites at the same latitude. In the tropics, it is temperature rather than day length which controls the timing of bulbing. Temperature is also important because of its overall

Plate 8.2 Plant response to flooding in onion

Plate 8.3 Raised bed cultivation of onion during rainy season

effects on the growth rate of the plant. Trials in the United Kingdom showed that the growth rate of onion seedlings increases by more than four-fold between 10 and 19 °C. There is a smaller decrease to the maximum growth rate at about 23–27 °C and a marked decrease at 31 °C. Hence, cold winter and hot summer temperatures are not good for growth.

Plate 8.4 Early harvesting and spreading of onion on the field

8.4.1 Photoperiod and Temperature

Onion growth and development could be described under five growth stages, namely, germination, leaf growth, bulbing or bulb initiation, bulb growth and maturation. The duration of these five stages varies for short-day, intermediate and long-day types. It is probable that the response of tropical onions to short-day variations in day length may be further influenced by temperature. Tropical onion varieties can be classified as "short-day onions" because these plants will initiate and form bulbs in <12 h photoperiods (day length) and are suitable for warm climates (Rabinowitch and Currah 2002).

Responses of the Eritrean onion cultivars to photoperiod and temperature on bulbing were compared (Tesfay et al. 2011) Plants were grown in growth rooms under combinations of photoperiod (11.5, 12 and 12.5 h) and day/night temperatures (25/12, 30/15 and 35/18 °C). Growth responses were determined by growing degree day (GDD) base and the broken-stick regression model. There were relationships between bulb initiation and rate of leaf area growth under 12 and 12.5 h. Under the 12 h photoperiod, cultivars needed 343, 482 and 597 GDD before bulb initiation and 405, 432 and 431 GDD to increase rate of leaf area development at 25/12, 30/15 and 35/18 °C, respectively. Under a 12.5 h photoperiod, cultivars needed 344, 423 and 432.2 GDD to initiate bulbing and 140, 411 and 579 GDD to increase leaf growth rates at 25/12, 30/15 and 35/18 °C, respectively. The 25/12 °C temperature induced bulb initiation at lower GDD compared to the 30/15 °C and 35/18 °C temperatures. Early bulb initiation under the 11.5 h photoperiod suggests that variations may exist with respect to bulb initiation and formation under natural short-day conditions. The plants can start bulbing under short day lengths (11.5 h) but would not form a bulb unless a critical day length is attained. Instead, they remain vegetative and produce thick-neck bulbs. Similar observations were made by Uzo and Currah (1990) who reported that under short-day conditions, onion plants produced new leaves indefinitely without bulb formation, while at longer day lengths bulbs were formed. Thus, under natural short-day conditions, day lengths shorter than a certain minimum value may cause plants to remain continually vegetative, and no bulbing will occur. The bulb ratio increased at the 25/12 °C earlier than at the 30/15 and 35/18 °C. It would be expected that growth under cooler conditions (25/12 °C) would result in production of thick-necked (poor quality) onions. Wickramasinghe et al. (2000) found that at the lowest (17–22 °C) temperatures tested, bigger bulbs with thick necks were produced. This may be due to changes in bulb structure at low temperature. The thermal time presentation of development of bulb ratio indicates that the cultivars were sensitive to temperature. However, plants should be exposed to a minimum photoperiod to produce bulbs. Bulbing of the onion cultivars was regulated more by temperature than photoperiod as determined by GDD. This agrees with Abdalla (1967), Robinson (1973) and Currah (1985) who reported that bulbing of onions in the tropics is regulated more by temperature than day length. However, in those studies, temperature measurements did not consider the effect of base temperature. In addition, the broken-stick regression model was able to pinpoint the onset of bulbing with respect to GDD. All cultivars responded in a similar manner to the photoperiod and temperature combinations. In conclusion, this study demonstrates that short-day variations in day length under natural short-day conditions may influence bulb initiation and development of "short-day onion" plants. Growing degree days and the use of broken-stick regression provide a novel approach that can be used to gain insights on the relationship between growing degree days, bulb initiation and development of "short-day onion" plants under natural short-day conditions.

8.4.2 Response of Onion to Temperature Gradient on Growth and Biomass

Response of onion cultivars to temperature gradient inside polycarbonate temperature gradient tunnels (providing about +3.0 °C from outside temperatures) was studied (Srinivasa Rao 2014). Temperature inside the tunnel was 30.7 °C (front), 28.8 °C (middle) and 27.6 °C (end). At 40 DAP, in cv. Arka Kalyan, the leaf area was reduced by 40 % with an increase of 3 °C. At 60 and 75 DAP, a reduction of 38 and 42.9 % was observed. There was a reduction of 5.9, 11.63 and 16.33 % in total biomass at 40, 60 and 75 DAP due to the increase in temperature by 3 °C. There was no reduction in total biomass by an increase of 1.2 °C in both the cultivars. There was a reduction of 31.85, 11.73 and 27.98 % in the photosynthetic rate at 40, 60 and 75 DAP due to the increase in temperature by 3 °C. In plants of Arka Kalyan, there was no reduction in the photosynthetic rate at 40 DAP. There was no reduction in the photosynthetic rate by an increase of 1.2 °C. Higher reduction of stomatal conductance was observed during the bulbing stage. Bulb fresh weight was 64.9, 75.1 and 80.6 g in A. Kalyan, and in cv. A. Niketan it was 65.10, 61.46 and 64.46 g at 27.6, 27.8 and 30.7 °C, respectively.

8.5 Salinity

The salinisation of soils dedicated to agriculture, caused by the accumulation of salts in irrigation water, causes these soils to become increasingly unproductive (Lima and Bull 2008). When irrigation waters have a high concentration of salts and there is no possibility of exporting these brackish waters to a sink, they can accumulate and cause damage (Santos et al. 2009). Such accumulation can limit the germination and development of various species (Barroso et al. 2010), leading to morphological, cellular, biochemical and molecular alterations that hinder the agricultural yield in response to the decrease in the water potential of the soil solution induced by the high osmolarity (Lima and Bull 2008). In addition, ionic toxicity promotes an imbalance in the absorption of essential nutrients, causing metabolic disorders, which inhibit growth (Maia et al. 2012). Salt stress can also lead to excess intracellular production of reactive oxygen species (ROS) such as the superoxide radical ($O_2^{\cdot -}$), the hydroxyl radical ($OH^{\cdot}$), hydrogen peroxide (H_2O_2) and singlet oxygen ($^1O_2$) (Stanisavljević et al. 2011). Tolerance to salinity is specific for each species or cultivar. Vegetables have a high sensitivity to the effects of NaCl (Zhu 2002), which hinders growth because of its toxic and osmotic effects, respectively, causing accumulation of ions in the protoplasm and physiological drought (Kader and Lindberg 2010; Deuner et al. 2011).

Onions are sensitive to salt, are relative excluders of both Na and Cl and are sensitive to sulphate. Little genetic variation has been detected even though many cultivars have been tested. Tolerance is high at germination, is very low during seedling growth and increases again at about the three- to five-leaf stage. Leaves change from rich green to dull blue-green with salt stress and leaf tips express burn symptoms typically associated with salinity stress. Onions are sensitive to salt, are relative excluders of both Na and Cl and are sensitive to sulphate. Little genetic variation has been detected even though many cultivars have been tested. Tolerance is high at germination, is very low during seedling growth and increases again at about the three- to five-leaf stage. Leaves change from rich green to dull blue-green with salt stress, and leaf tips express burn symptoms typically associated with salinity stress.

8.5.1 Germination

The germination and development of several plant species can be limited when those plants are grown in soils with high salinity, which reduces seedling viability and vigour, as well as activates the antioxidant defense system. Wannamaker and Pike (1987) studied the germination and growth response to salinity of five onion cultivars commonly grown in Texas. Using $NaCl/CaCl_2$ solutions (1:1 by weight), they found that germination was unaffected at ECi up to 20 day m^{-1} but was drastically reduced thereafter with no discernible cultivar difference. After 8 days, solutions of 30 ± 35 days m^{-1} reduced germination by 50 % in all the cultivars. Seedlings of three onion cultivars exposed to different concentrations of NaCl (0, 40, 80, 120 and 160 mM) were evaluated for viability, vigour and antioxidant enzyme activity. For all cultivars, viability and vigour decreased in parallel with increasing NaCl concentrations, whereas antioxidant enzyme activity increased. High NaCl concentrations have a negative effect in the physiological quality of onion seeds, resulting in lower seedling growth rates and increased antioxidant enzyme activity (Ryang et al. 2009). The smaller size of the seedlings in solutions with higher salt concentrations can be explained by the fact that, as the seeds absorb water with highly soluble salts, these salts become toxic and, consequently, cause physiological disturbances, decreasing the water potential and the germinability of the seeds (Carvalho and Kazama 2011).

8.5.2 Growth and Biomass

Bernstein and Ayers (1953b) tested the salt tolerance of five onion cultivars ('Yellow Sweet Spanish', 'Texas Early Grano', 'San Joaquin', 'Crystal Wax' and 'Excel') in field plots at the US Salinity Laboratory. Initial yield decline started at a threshold EC of 1.4 days m^{-1}, and 50 % yield reduction (C50) was at 4.1 days m^{-1}. Bernstein and Ayers (1953b) noted that the osmotic potential of the expressed sap increased with salinity without a significant concomitant increase in sucrose or reducing sugar. Bulb ion content increased as a function of applied salts (Na^{+}, Ca^{2}, Cl^{-}) and percent dry weight increased. Salinity decreased bulb diameter, bulb weight, root growth, plant height and the number of leaves per plant. Onions may mature a week earlier when grown under saline conditions. Pasternak et al. (1984) hypothesised that sensitivity during early growth stages may be due to the small and shallow rooting system of young plants. No research has been conducted to determine whether rooting systems can be genetically modified to improve tolerance or even if variability exists for this character. If variability can be introduced, it must be done without affecting the commercial quality of the bulb.

In addition to the toxicity, the reduction in growth, characterised by the decrease in the length of the plant and less accumulation of dry mass, also is promoted by the decrease of osmotic potential, which causes a water deficit and alterations in the balance of K^{+}/Na^{+} and other nutrients (Willadino and Camara 2010). With the increase in the salinisation of the irrigation water, there was a decrease in the accumulation of dry mass of the seedlings of the onion cultivars. The problems caused by salt stress, as seen in the dry mass analysis, indicate that salinity decreases biomass production, in addition to altering the partition of photoassimilates between the different plant parts (Silva et al. 2007). This differential partition can contribute to the adaptation of plants to salt stress. This process compensates the reduction of the leaf area dedicated to carbon assimilation as the acceleration of the metabolic processes that are necessary for the adjustment of the plant, depending on the stress level (Munns and Tester 2008). The seedlings that were subjected to the higher concentrations of NaCl had less vigour. Of the enzymes involved in the removal of ROS, SOD is usually the first line of defense against oxidative stress (Pompeu et al. 2008). The relative performance of onion, viz. Pusa Red, Pusa Madhvi and Pusa White Flat, on saline soils with increasing levels (0.3, 4, 8, 12 and 16 days/m) of electrical conductivity (EC) was assessed under greenhouse conditions (Yadav et al. 1998). Growth and bulb yield of all varieties were not affected adversely up to a salinity level of 4 days/m. Highest salinity (16 days m^{-1}) reduced plant height, number of leaves/plant, leaves' fresh and dry weight and bulb yield by about 40–50 % as compared to the control (0.3 days m^{-1}). Reduction was of higher magnitude at early growth stages in all the varieties. Bulb maturity has been hastened by a week due to increasing levels of salinity. Onion Pusa Red performed relatively better as compared to others.

8.6 Impact of Elevated Carbon Dioxide

Onion was grown in the field within temperature gradient tunnels (providing about −2.5 to +2.5 °C from outside temperatures) maintained at either 374 or 532 μmol mol-1 CO_2 (Wheeler et al. 2004). Maximum rate of leaf area expansion increased with mean temperature. Leaf area duration and maximum rate of leaf area expansion were not significantly affected by CO_2. The light-saturated rates of leaf photosynthesis (Asat) were greater in plants grown at normal than at elevated CO_2 concentrations at the same measurement CO_2 concentration. Acclimation of photosynthesis decreased with an increase in growth temperature and with an increase in leaf nitrogen content at elevated CO_2. The ratio of intercellular to atmospheric CO_2 (Ci/Ca ratio) was 7.4 % less for plants grown at elevated compared with normal CO_2. Asat in plants grown at elevated CO_2 was less than in plants grown at normal CO_2 when compared at the same Ci. Hence, acclimation of photosynthesis was due both to stomatal acclimation and to limitations to biochemical CO_2 fixation. Carbohydrate content of the onion bulbs was greater at elevated than at normal CO_2. In contrast, carbohydrate content was less at elevated compared with normal CO_2 in the leaf sections in which CO_2 exchange was measured at the same developmental stage. Therefore, acclimation of photosynthesis in fully expanded onion leaves was detected despite the absence of localised carbohydrate accumulation in these field-grown crops. The higher rates of photosynthesis observed under elevated CO_2 growth conditions underlie the growth stimulation to CO_2 previously observed (Daymond et al. 1997). The response to higher CO_2 growing conditions was greater at higher temperatures and is consistent with the theory of Long (1991) that the photorespiration response to temperature will cause an interaction between CO_2 and temperature on the net rate of photosynthesis.

Response of onion to elevated CO_2 (550 ppm) was studied in open-top chambers (Srinivasa Rao et al. 2009). Shoot dry weight increased throughout the crop growth period under elevated CO_2. The shoot dry weight increase at 550 ppm was highest at 70 days after planting (80 %) over chamber control. Leaf area increased from 20 to 90 days after planting in all the treatments. Under elevated CO_2, maximum leaf area of 11.27 dm^2 was observed compared to Ch control (5.97 dm^2). Elevated CO_2 levels enhanced the total biomass at all growth stages. Total biomass was maximum at 100 days after planting (18.6 g/plant) compared to control (13.40 g/plant). The

Table 8.3 Photosynthesis (μmol $m^{-2}s^{-1}$), stomatal conductance (mol $m^{-2}s^{-1}$) and transpiration rate (mol $m^{-2}s^{-1}$) of onion at elevated and ambient CO_2

Days after planting	Photosynthesis	Stomatal conductance	Transpiration
550 ppm			
20	8.654	0.124	2.822
40	11.818	0.198	6.936
60	21.108	0.205	15.30
80	17.740	0.710	11.354
100	16.000	0.332	9.160
365 ppm			
20	8.216	0.220	3.318
40	9.896	0.294	9.314
60	12.890	0.396	19.045
80	9.480	1.022	12.568
100	9.720	0.698	10.840

percent increment in total biomass at elevated CO_2 varied from 52 to 81 % at different growth stages. Increase in photosynthetic rate varied from 19.4 to 64.60 % during bulb development stage (Table 8.3). There was a reduction in stomatal conductance, indicating better water use efficiency at elevated CO_2. A yield increase of 25.9 % was observed over control at elevated CO_2. Yield increase was mainly due to increase in the bulb size. The data on quality parameters showed that there is a decrease in the antioxidant content at elevated CO_2. Total phenols were higher and a decrease in flavonoids content was observed (Table 8.4).

8.6.1 Interaction Effect of Elevated CO_2 and Water Stress on Onion

The influence of CO_2 on the water use of crops may well prove to be the most important benefit of increased CO_2 concentrations for agriculture. Response of onion cvs. Arka Kalyan and Agrifound Dark Red (ADR) to elevated CO_2 (550 ppm) and water stress was studied in open-top chambers (Srinivasa Rao et al. 2012). Water stress was imposed at 45 and 75 days after planting. All the chambers were maintained at 550 ppm. Two chambers were used as irrigated control, and two chambers were used for imposing water stress. The elevated CO_2 concentrations were maintained throughout the growth period and monitored continuously during experimental period. There was a decrease in leaf area and total dry matter content in both the varieties due to water stress. When water stress was imposed at 45 days after planting, cv. Arka Kalyan had the maximum leaf area of 809.67 cm^2 after 4 weeks of water stress. LAI varied from 2.97 to 5.40 in Arka Kalyan and 3.18–4.80 under water stress in ADR. Total dry weight was highest in Arka Kalyan (5.30 g) under water stress conditions (Table 8.5). When water stress was imposed at 75 days after planting, a decrease of 48.1 and 5.6 % in leaf area was observed in Arka Kalyan and ADR, respectively, compared to irrigated control. Dry weight was highest in Arka Kalyan (11.26 g). Harvest index which is a reflection of translocation efficiency to bulbs was higher under elevated CO_2 and water stress conditions (82 %). In irrigated, it was 719 and 75.6 % in Arka Kalyan and ADR, respectively. There was a yield reduction of 41.0 and 38 % in the two cultivars, respectively, when water stress was imposed at 45 days after planting. When water stress was imposed at 75 days after planting, yield reduction varied from 6 to 10 %. Increased CO_2 concentration resulted in decreased photosynthesis, stomatal conductance and transpiration under both irrigated and water stress conditions. A decrease was higher under water stress conditions (26–46 % at different growth stages).

Table 8.4 Quality parameters in onion and tomato at elevated and ambient CO_2

Parameter	OTC (550 ppm)	OTC (365 ppm)
Anthocyanin	21.29	20.68
(mg/100 g dry weight)	20.50	20.79
Mean	20.90	20.73
Total phenols	495.44	471.96
	498.33	470.80
Mean	496.89	471.38
Flavonoids	23.53	25.96
	21.06	32.88
Mean	22.29	29.42
Antioxidant	135.37	162.01
	136.69	162.46
Mean	136.03	162.24

Table 8.5 Effect of elevated CO_2 and water stress at 45 DAP on leaf area, LAI, shoot, leaf, bulb dry weight and total dry weight

			Dry weight (g)				
Stress – Arka Kalyan	Leaf area (cm^2)	LAI	Shoot	Leaf	Bulb	Total dry weight	Bulb FW
46 DAT	463.09	3.09	0.34	1.13	0.14	1.61	2.08
54 DAT	404.85	2.7	0.72	1.83	0.87	3.52	6.92
61 DAT	445.17	2.96	0.54	1.93	2.27	4.76	17.46
68 DAT	809.97	5.4	0.81	2.48	1.93	5.30	18.77
ADR							
46 DAT	385.89	3.02	0.48	1.47	0.18	2.13	2.57
54 DAT	477.48	3.18	0.68	2.15	0.38	3.32	3.92
61 DAT	466.99	3.09	0.54	1.98	1.01	3.53	12.03
68 DAT	655.69	4.37	0.62	1.88	1.60	4.20	18.24
SEm	107.62	0.72	0.21	0.074	0.325	0.624	3.82
			Dry weight (g)				
Irrigated – Arka Kalyan	Leaf area (cm^2)	LAI	Shoot	Leaf	Bulb	Total dry weight	Bulb FW
46 DAT	463.09	3.09	0.34	1.13	0.15	1.62	2.16
54 DAT	638.33	4.26	0.99	1.83	1.41	6.05	14.67
61 DAT	612.21	4.08	0.90	1.93	2.20	5.98	21.43
68 DAT	765.79	5.11	1.11	2.48	4.15	8.72	35.28
ADR							
46 DAT	486.23	4.32	0.48	1.15	0.19	1.88	2.54
54 DAT	578.24	3.85	0.80	2.58	0.68	4.16	7.85
61 DAT	483.88	6.08	0.72	2.24	1.29	4.24	15.49
68 DAT	442.40	5.96	1.65	1.96	4.22	10.98	38.56
SEm	76.41	0.30	0.213	0.62	0.79	1.70	7.42

A decrease in transpiration rate varied from 14.9 to 25.53 % under irrigated conditions at elevated CO_2. Under water stress conditions, it varied from 16 to 29 %. The reduction in stomatal conductance had a consequent beneficial effect on leaf water potential (−1.9 M.Pa). SLW increased under high CO_2 and water stress (54.4 g/m^2) compared to plants grown under irrigated and elevated CO_2 (43.0 g/m^2). WUE was increased by 19 % in CO_2 enrichment.

8.7 Conclusions

Excessive rains, humidity, temperature and salinity are critical factors of risk to onion cultivation. Onion is highly sensitive to water and heat stress. Both bulb and seed production are adversely affected due to these stresses. The soil moisture is critical for onion production as it affects the physiological parameters and the bulb yield. Managing onion crop with less than adequate irrigation water availability, maximising irrigation water use efficiency and alleviating the effects of excess moisture are a challenge. Temperature induces variations in leaf number, plant height and leaf area and affects bulbing, and these responses may be used in the development of superior cultivars for tropical conditions. Growing degree days and the use of broken-stick regression provide a novel approach that can be used to gain insights on the relationship between growing degree days, bulb initiation and development of "short-day onion" plants under natural short-day conditions. Mechanisms of tolerance to salinity specific for cultivars and modifying rooting systems genetically to improve salt tolerance should be addressed. The strategies must identify the gene tolerant to high temperature, flooding and drought and salinity.

References

Abdalla AA (1967) Effect of temperature and photoperiod on bulbing of common onion (*Allium cepa* L.) under arid tropical conditions of the Sudan. Exp Agric 3:137–142

Al-Jammal MS, Sammis TW, Ball S, Smeal D (2000) Computing the water production function for onion. Agric Water Manag 46:29–41

Bagali AN, Agali HB, Patil MB, Guled, Patil RV (2012) Effect of scheduling of drip irrigation on growth, yield and water use efficiency of onion (*Allium cepa* L.). Karnataka J Agric Sci 25(1):116–119

Barroso CM, Franke LB, Barroso IB (2010) Substrato e luz na germinação das sementes de rainha-do-abismo. Hortic Bras 28:236–240

Bernstein L, Ayers AD (1953) Salt tolerance of five varieties of onions. Proc Am Soc Hortic Sci 62:367–370

Bhatt RM, Srinivasa Rao NK, Veere Gowda R (2006) Response of bulb onion (*Allium cepa* L.) to water stress: photosynthesis, stomatal conductance and osmotic adjustment. Indian J Hortic 63(3):20–25

Brewster JL (1979) The response of growth rate to temperature in seedlings of several *Allium* crop species. Ann Appl Biol 93:351–357

Brewster JL (1994) Onions and other vegetable alliums. CAB International, Wallingford

Butt AM (1968) Vegetative growth, morphogenesis and carbohydrate content of the onion plant as a function of light and temperature under field and controlled conditions. Meded Landbouwhogeschool Wagening Ned 68:29

Carvalho LC, Kazama EH (2011) Efeito da salinidade de cloreto de potássio (KCL) na germinação de sementes e crescimento de plântulas de pepino (Cucumis sativus L.). Enciclopédia Biosfera Cent Científico Conhecer Goiânia 7(13):429–435

Currah L (1985) Review of three onion improvement schemes in the tropics. Trop Agric 62:131–136

Daymond AJ, Wheeler TR, Hadley P, Ellis RH, Morison JI (1997) Effects of temperature, CO_2 and their interaction on the growth, development and yield of two varieties of onion (*Allium cepa* L.). J Hortic Sci 72:135–145

De Ruiter JM (1986) The effect of temperature and photoperiod on onion bulb growth and development. P Agron Soc NZ 16:93–100

Deuner C, Maia MS, Deuner S, Almeida A, Meneghello GE (2011) Viabilidade e atividade antioxidante de sementes de genótipos de feijão-miúdo submetidos ao estresse salino. Rev Bras Sementes 33(4):711–720

Doorenbos J, Kassam AH (1986) Yield response to water. FAO Irrig. and Drain, Paper 33. Rome, pp 193

dos Santos PR, Ruiz HA, Neves JCL, de Almeida EF, Freire MBGS, Freire FJ (2009) Germinacao, vigor e crescimento de cultivares de feijoeiro em solucoes salinas. Rev Brasil Engen Agríc Ambient 13:882–889

Dragland S (1974) Nitrogen and water requirements in onions. Forskning Forsok Landbruket 26:93–113

Ells JE, McSay AE, Soltanpour PN, Schweissing FC, Bartolo ME, Kruse EG (1993) Onion irrigation and nitrogen leaching in the Arkansas Valley of Colorado. Hortic Tech 3(2):184–187

FAO (2013) Crop water information: onion. Available online with updates at http://www.fao.org/nr/water/crop info_onions.html

Hegde DM (1986) Effect of irrigation regimes on dry matter production, yield, nutrient uptake and water use of onion. Indian J Agron 31:343–348

Jones ST, Johnson WA (1958) Effect of irrigation at different minimum levels of soil moisture and of imposed drought on yield of onions and potatoes. J Am Soc Hortic Sci 71:440–445

Kadayifci A, Tuylu GI, Ucar Y, Cakmak B (2005) Crop water use of onion (*Allium cepa* L.) in Turkey. Agric Water Manag 72:59–68

Kader MA, Lindberg S (2010) Cytosolic calcium and pH signaling in plants under salinity stress. Plant Signal Behav 5:233–238

Koriem SO, El-Koliey MMA, Wahba MF (1994) Onion bulb production from "Shandwell 1" sets as affected by soil moisture stress. Assiut J Agric Sci 25:185–193

Lima MDB, Bull LT (2008) Produção de cebola em solo salinizado. Rev Brasil Engen Agrí Ambi 12:231–235

Long SP (1991) Modification of the response of photosynthetic productivity to rising temperature by atmospheric CO_2 concentrations: has its importance been underestimated? Plant Cell Environ 14:729–739

Mafakheri A, Siosemardeh A, Bahramnejad B, Struik PC, Sohrabi Y (2010) Effect of drought stress on yield, proline and chlorophyll contents in three chickpea cultivars. Aust J Crop Sci 4(8):580–585

Maia JM, Ferreira-Silva SL, Voigt EL, de Macedo CEC, Ponte LFA, Silveira JAG (2012) Atividade de enzimas antioxidantes e inibição do crescimento radicular de feijão caupi sob diferentes níveis de salinidade. Acta Bot Bras 26:342–349

Munns R, Tester M (2008) Mechanisms of salinity tolerance. Annu Rev Plant Biol 59(651):681

Pasternak D, De Malach Y, Borovic I (1984) Irrigation with brackish water under desert conditions I. Problems and solutions in production of onions (*Allium cepa* L.). Agric Water Manag 9:225–235

Patel N, Rajput TBS (2013) Effect of deficit irrigation on crop growth, yield and quality of onion in subsurface drip irrigation. Int J Plant Prod 7(3):1735–6814

Pathak CS (2000) Hybrid seed production in onion. J New Seeds 1:89–108

Pompeu GB, Gratão PL, Vitorello VA, Azevedo RA (2008) Antioxidant isoenzyme responses to nickel-induced stress in tobacco cell suspension culture. Sci Agríc 65:548–552

Quadir M, Boulton A, Ekman J, Hickey M, Hoogers R (2005) Influence of drip irrigation on onion yield and quality, vol 170. IREC Farmers' Newsletter, Australia, pp 29–31

Rabinowitch HD, Currah L (2002) Allium crop science: recent advances. CAB International, Wallington, p 551

Rana DS, Sharma RP (1994) Effect of irrigation regime and nitrogen fertilization on bulb yield and water use of onion (*Allium cepa* L). Indian J Agric Sci 64:223–226

Robinson JC (1973) Studies on the performance and growth of various short-day onion varieties (*Allium cepa* L.) in the Rhodesian Lowveld in relation to sowing. 1. Growth analysis. Rhod J Agric Res 11:51–68

Ryang S, Woo S, Kwon S, Kim S, Lee SH, Kim K, Lee D (2009) Changes of net photosynthesis, antioxidant enzyme activities, and antioxidant contents of Liriodendron tulipifera under elevated ozone. Photosynthetica 47:19–25

Seabrook JEA (2005) Light effects on the growth and morphogenesis of potato (*Solanum tuberosum*) in vitro: a review. Am J Potato Res 82:353–367

Shock CC, Feibert EBG, Saunders LD (1998) Onion yield and quality affected by soil water potential as irrigation threshold. Hortic Sci 33:188–191

Shock CC, Feibert EBG, Saunders LD (2000a) Irrigation criteria for drip irrigated onions. Hortic Sci 35:63–66

Shock CC, Feibert EBG, Saunders LD (2000b) Onion storage decomposition unaffected by late-season irrigation reduction. HortTechnology 10:176–178

Shock CC, Feibert EBG, Saunders LD (2007) Short-duration water stress decreases onion single centers without causing translucent scale. Hortic Sci 42 (6):1450–1455

Silva RN, Lopes NF, Moraes DM, Pereira AL, Duarte GL (2007) Physiological quality of barley seeds submitted to saline stress. Rev Brasil Sementes 29:40–44

Singh R, Alderfer RB (1966) Effects of soil moisture stress at different periods of growth of some vegetable crops. Soil Sci 101(1):69–80

Sørensen JN, Grevsen K (2001) Sprouting in bulb onions (*Allium cepa* L.) as influenced by nitrogen and water stress. J Hortic Sci Biotechnol 76:501–506

Srinivasa Rao NK (2014) Response of onion to temperature gradient on growth and biomass. In: final report of ICAR Emeritus Scientist Project Impact of Climate change and Climate Variability on Onion Productivity, Production and Quality, pp 18–23

Srinivasa Rao NK, Laxman RH, Bhatt RM (2009) Impact of elevated carbon dioxide on growth and yield of onion and tomato. In: Aggarwal PK (ed) Global climate change and Indian Agriculture. ICAR, New Delhi, pp 35–37

Srinivasa Rao NK, Laxman RH, Bhatt RM (2010) Extent of impact of flooding and water stress on growth and yield of onion and tomato. In: Aggarwal PK (ed) Annual progress report of ICAR network project impact, adaptation and vulnerability of Indian Agriculture to climate change 2009–10. pp 111–112

Srinivasa Rao NK, Laxman RH, Bhatt RM (2012) Interaction effect of elevated CO_2 and water stress on onion. National Symposium on Climate Change and Indian Agriculture-Slicing down the uncertainties held from 22–23 January 2013 at CRIDA, Hyderabad

Stanisavljević NS, Nikolić DB, Jovanović ŽS, Samardžić JT, Radović SR, Maksimović VR (2011) Antioxidative enzymes in the response of buckwheat (*Fagopyrum esculentum* Moench) to complete submergence. Arch Biol Sci 63:399–405

Streb P, Shang W, Feierabend J, Bligny R (1998) Divergent strategies of photoprotection in high-mountain plants. Planta 207:313–324

Tesfay SZ, Bertling I, Odindo AO, Greenfield PL, Workneh TS (2011) Growth responses of tropical onion cultivars to photoperiod and temperature based on growing degree days. Afr J f Biotechnol 10 (71):15875–15882

Uzo JO, Currah L (1990) Cultural systems and agronomic practices in tropical climates. In: Rabinowitch HD, Brewster JL (eds) Onions and allied crops.

II. Agronomy, biotic interactions, pathology, and crop protection. CRC Press, Boca Raton, pp 49–62
Van Eeden FJ, Myburgh J (1971) Irrigation trials with onions. Agroplantae 3:57–62
Wannamaker MJ, Pike LM (1987) Onion responses to various salinity levels. J Am Soc Hortic Sci 112:49–52
Werner JL, Harris CM (1965) Factors influencing the incidence of translucent scale of stored onion bulbs. J Am Soc Hortic Sci 87:341–354
Wheeler TR, Daymond AJ, Morison JIL, Ellis RH, Hadley PP (2004) Acclimation of photosynthesis to elevated CO2 in onion (*Allium cepa*) grown at a range of temperatures. Ann Appl Biol 144:103–111
Wickramasinghe UL, Wright CJ, Currah L (2000) Bulbing responses of two cultivars of red tropical onions to photoperiod, light integral and temperature under controlled growth conditions. J Hortic Sci Biotech 75:304–311
Willadino L, Camara TR (2010) Tolerância das plantas à salinidade: aspectos fisiológicos e bioquímicos. Enciclopédia Biosfera, Centro Científico Conhecer. Goiânia 6(11):2
Yadav SS, Singh N, Yadav BR (1998) Effect of different levels of soil salinity on growth and yield of onion (Allium cepa L). Ind J Hortic 55:243–247

Capsicum (Hot Pepper and Bell Pepper) 9

K. Madhavi Reddy, K.S. Shivashankara, G.A. Geetha, and K.C. Pavithra

Abstract

Capsicum is one of the most widely consumed vegetables and is also used as a spice for its pungency. Many species of *Capsicum* are being cultivated worldwide. *Capsicums* are considered as a commercial crop for their economical value. However, the yield of the crop suffers severely due to abiotic and biotic factors. Among the various abiotic stresses, low moisture and high temperature are the major constraints for productivity. Flowering and fruit-set stages were most sensitive for the abiotic stresses. Wide genotypic variability for abiotic stress tolerance is present in *Capsicum* species. Significant differences in fruit-set, yield, photosynthetic rates, stomatal conductance, total chlorophyll content, proline, and root characteristics have been reported. Pollen germination and fertilization were better in the tolerant *capsicum* varieties. Higher activities of hexokinase, phosphofructokinase and invertase in the pollen were responsible for the better pollen germination in the tolerant genotypes. Lesser oxidative stress and better antioxidative enzymes were observed as one of the tolerance mechanisms in the *capsicum* genotypes. Breeding for abiotic stress tolerance has also been attempted and a few cultivars have also been developed in hot pepper using root characteristics and yield as selection parameters under stress. Transgenic plants have also been developed using osmotin gene overexpression and also by overexpressing certain transcription factors. Enriching the soil microbia with mycorrhiza, *Arthrobacter* and *Bacillus* species has given encouraging results in giving abiotic stress tolerance. The use of chemicals like brassinolides, sitosterols, and nitric oxide has been found to improve the

K. Madhavi Reddy (✉)
Division of Vegetable Crops, ICAR-Indian Institute of Horticultural Research, Hessaraghatta Lake Post, Bengaluru 560089, Karnataka, India
e-mail: kmr14@iihr.ernet.in

K.S. Shivashankara • G.A. Geetha • K.C. Pavithra
Division of Plant Physiology and Biochemistry, ICAR-Indian Institute of Horticultural Research, Hessaraghatta Lake Post, Bengaluru 560089, Karnataka, India

N.K.S. Rao et al. (eds.), *Abiotic Stress Physiology of Horticultural Crops*,
DOI 10.1007/978-81-322-2725-0_9

tolerance in *capsicums*. In spite of these management practices, crop improvement, transgenic approaches to overcome the abiotic stresses and emphasis for rootstock breeding is lacking. Therefore, future studies should focus on root and flower traits and also on rootstocks to overcome the abiotic stress in addition to the stress management practices.

9.1 Introduction

Capsicum is a versatile plant being grown as a vegetable and spice crop, also commonly known as chilli, hot pepper, bell pepper, sweet pepper, bird's eye pepper and paprika, and belongs to the family Solanaceae. The genus *Capsicum* is comprised of more than 200 varieties grouped into more than 30 species, out of which five are domesticated: *Capsicum annuum* L., *Capsicum baccatum* L., *Capsicum chinense* Jacq., *Capsicum frutescens* L., and *Capsicum pubescens* Ruiz and Pav. Among them, *Capsicum annuum* L. is by far the best known and of greatest economic importance since it presents a largest distribution worldwide and it is usually consumed either raw or cooked and used as additives in the food industry. It has become an indispensable commodity in every cuisine due to its pungency, spicy taste (capsaicinoids), appealing color (carotenoids), and flavor. India is the world's largest producer and exporter of chilli and contributes 25 % of world's total chilli production. Chilli continued to maintain the lead in the spice export basket in India, accounting for 347,000 tonnes in quantity and Rs. 35.17 billion in value. However, the chilli productivity is highly erratic due to both abiotic and biotic constraints during the crop period. Temperature extremes and inadequate water availability are among the dominant abiotic stress factors that drastically reduce the growth and productivity of chilli crop. With the impending climate change, by 2050, the mean temperature of the earth is expected to rise by 2–5 °C or more, according to the Intergovernmental Panel on Climate Change (IPCC). Moreover, tropical and subtropical regions of the globe are expected to experience more variability in rainfall. Developing nations, including India, are increasingly suffering major losses in crop yield due to chronic and acute drought conditions in crop-growing areas. Therefore, it is necessary to identify and develop abiotic stress-tolerant varieties to cope with this upcoming problem.

High temperature influences many aspects of plant physiology and growth, which in turn may have a direct or indirect effect on fruit-set. High temperatures strongly influence reproductive development and yield. Temperature and drought stress can induce oxidative stress, which might create differential synthesis of pepper fruit carotenoids across distinct environments. Abiotic stress tolerance in chili is complex, expression of which depends on interaction of different morphological, physiological, and biochemical characters. With the inherent complexity of drought or heat stress tolerance, it is opined that a trait-based breeding approach will be helpful in developing abiotic stress-tolerant varieties, for which simplifying through targeting the important related traits contributing for abiotic stress tolerance, and understanding the genetic mechanism will be essential for improving the drought or heat tolerance.

Abiotic stress is the most important constraint to realize the potential yield of chilli crop as it is mainly cultivated under rain-fed conditions in the country. Moisture stress during the critical crop growth period accounts for about 30–70 % yield loss, hence the fluctuations in its annual productivity. Plants have evolved diverse adaptive strategies to cope with water-limited environments. These mechanisms range from cellular-level tolerance to adaptation in response to stress and/or specific growth behavior and morphological characteristics that avoid stress effects. Relevance of different traits in crop species for abiotic stress tolerance is stomatal conductance, leaf temperature, and photosynthetic efficiency (Lawlor and Cornic 2002), partitioning and stem reserve utilization (Bernier et al. 2007), osmotic adjustment (Zhang et al. 1999; Serraj and Sinclair 2002), relative water content in leaves (Madhusudan et al. 2002), canopy-air temperature difference (Kashiwagi et al. 2008), cell membrane stability (Tripathy et al. 2000), osmolyte accumulation in roots (Sharp et al. 2004), greater root length and diameter (Chang et al. 1972), deep root system (Shashidhar et al. 2000; Sinclair and Muchow 2001; Venuprasad et al. 2002), thick root system (Chang et al. 1986), root weight to shoot weight ratio (Fukai and Copper 1995; Prince and Tomos 1997), enhanced root system (Prince and Tomos 1997), root penetrating ability (Ray et al. 1996), increased root characteristics (Serraj and Sinclair 2002), and more roots in the crown region (Kinyua et al. 2003).

Pepper (*Capsicum annuum* L.) is the second most widely consumed vegetable in the world and an excellent source of many essential nutrients for humans, especially vitamin C, phenolic compounds, flavonoids, tocopherols (vitamin E), carotenoids (provitamin A), capsaicinoids, and calcium. Additionally, some pepper cultivars contain significant quantities of capsaicinoids, a group of pungent phenolic-derived compounds with strong physiological and pharmacological properties. Thus, the growing global demand of pepper fruits implies several strategies to increase crop production and fruit quality or promote the investigation to improve the plant resistance to environmental stresses (Jimenez-Garcia et al. 2014). However, fruit growth and production were affected by drought and high temperature stresses. Therefore, in developing the technologies to overcome the abiotic stress effects on productivity of *capsicum*, understanding the mechanisms of tolerance, sources of tolerance, and methods to incorporate the traits into the better horticulture background assumes importance. This chapter focuses mainly on the physiological, biochemical, and molecular mechanisms of stress tolerance in *capsicums*, traits of tolerance, and breeding methodologies in developing tolerant varieties.

9.2 Drought Stress

Water is the most limiting factor for crop production, and consequently crops have to deal with it frequently. Drought-induced loss in crop yield probably exceeds losses from all other causes, since both the severity and duration of the stress are critical. It is especially important in countries where crop agriculture is largely rain fed (Ludlow and Muchow 1990). During drought season, the inflow of water diminishes leading to soil drying, which consequently affects all the growth processes dependent on turgor pressure, such as cell elongation and division (Larcher 2006).

9.2.1 Morphological and Physiological Responses

Drought response includes changes in growth, yield, membrane integrity, pigment content, osmotic adjustment, water relations, and photosynthetic activity (Anjum et al. 2011). Drought stress refers to a situation when the amount of water in the plant is not sufficient to meet the transpiration requirements, which leads to altered visible symptoms such as leaf curl and plant wilt, which also causes the delay in flowering. The reduction in plant growth is related to the water stress-induced stomatal closure resulting in lower assimilation and reduced leaf area and therefore the growth. Water stress also reduces ion transport to the root surface and from root to shoot which further decreases the plant growth. The effect of moisture stress on different plant growth parameters such as drastic reduction in yield (Cafer et al. 2006; Showemimo and Olarewaju 2007; Kirada et al. 2007), flower abscission (Aloni et al. 1994), significant reduction in fruit weight and dry matter accumulation (Cantore et al. 2000; Della Coasta and Gianquinto 2002), significant reduction in plant height (Thakur et al. 2000), plant spread, increase in root-to-shoot ratio and root length (Kulkarni and Phalke 2009), significant reduction in leaf area, relative water content (Kulkarni and Phalke 2009; Hu et al. 2010), increased leaf water potential, stomatal conductance, photosynthetic efficiency, transpiration efficiency and chlorophyll content (Stefana et al. 2003; Hu et al. 2010), increased capsaicinoids (pungency) (Bosaland and Votava 2002; Sung et al. 2005), reduction in the fruit biochemical traits such as color (red, yellow and total carotenoids) and pungency (total capsaicinoids and capsaicin), and significant increase in nordihydrocapsaicin content (Naresh 2015) has been reported. Drought-stressed plants expressed a decrease in leaf water potential implying a difference in plant water status (González-Dugo et al. 2007). The leaf area which is an indicator of canopy photosynthetic rates, shoot-to-root ratio, and dry fruit yield was significantly decreased as a result of drought stress (Ismail et al. 2002).

Significant impact on growth and yield was observed when chili was subjected to water stress (50 % of field capacity level) (Gunawardena and De Silva 2014). Reduction in total yield by drought was reported by many workers in *capsicums* (Della Costa and Gianquinto 2002; Yildirim 2010), and the yield reduction was mainly due to reduced leaf area, plant growth, stomatal conductance, assimilation rate, water use efficiency, fruit yield and quality, leaf relative water content, and macro-nutrition (Cantore et al. 2000; Kirnak et al. 2003; R'Him and Radhouane 2015). Significant reduction in plant height, number of branches, number of leaves, leaf area, as well as dry weight of roots, stems, and leaves was also observed under water stress conditions (El-Ghinbihi and Hassan 2007). In addition to the reduction in growth parameters, water stress also negatively affected the yield parameters like the number of fruits and fruit size (Sezen et al. 2011). Adverse effects of water stress can be reduced by planting the seedlings which were exposed to mild water stress. The adaptation was due to better root-to-shoot ratio of the seedlings which were exposed to mild water stress (Nieto-Garibay et al. 2010). Water stress was also found to enhance the development and anthesis of primary flowers but not the secondary flowers which showed increased abscission due to water stress (Hawa 2003; Abdulmalik et al. 2012). Water stress tolerance in pepper (*Capsicum annuum* L.) cultivars was exhibited by the better photosynthetic rate and reduced oxidative stress which was mainly due to higher photorespiration, non-photochemical quenching, cytochrome, and alternate oxidase respiration (Hu et al. 2010).

9.2.2 Biochemical Responses

Environmental factors such as temperature, light, water stress, and soil nutrient have been reported to affect the capsaicinoid content in peppers (Murakami et al. 2006). Capsaicinoid content is increased under drought stress (Phimchan et al. 2012) with some of the increase resulting from varietal differences (Estrada et al. 1999). In addition, the stress level and biochemical responses related to yield and pungency level were good criteria for evaluating hot pepper cultivars with high capsaicinoid levels under drought stress (Sung et al. 2005). Hot peppers vary widely in pungency level. Some of the hottest cultivars are found in *C. chinense* that include 'Habanero', 'Red Savina', and 'Bhut Jolokia' with pungency levels up to more than one million Scoville heat units (SHUs) (Bosland and Baral 2007). Habanero pepper plants under water stress had increased capsaicin and dihydrocapsaicin concentrations and reduced capsaicin synthase activity in fruits of stressed plants compared with control plants (Ruiz-Lau et al. 2011). Drought stress induced a sudden increase in antioxidant activity and reduction of soluble compound accumulation along with activation of antioxidant enzymes (Asadi et al. 2012).

Lower levels of superoxides and hydrogen peroxide and increased activity of superoxide dismutase and ascorbate peroxidase were noticed in the tolerant cultivars of pepper (*Capsicum annuum* L.) when compared to the susceptible (Hu et al. 2010). Photosynthetic pigments (chlorophyll and carotenoids), total soluble sugars, and total carbohydrate concentration were significantly decreased in response to water stress treatments compared with untreated plants. The concentration of total free amino acids, total protein, phenoloxidase, and peroxidase activities as well as mineral concentrations (N, P and K^+) was negatively affected by water stress treatments (El-Ghinbihi and Hassan 2007). Lobato et al. (2009) suggest that there is a significant and negative relationship between chlorophyll and total soluble carbohydrates, and the hypothesis was concluded. Klar et al. (2006) worked on biochemical changes during water stress on plants and revealed that the peroxidase enzyme activity was one of the efficient indicators of water stress in pepper plants. AlHarbi et al. (2014) reported that the vegetative and fruiting growth stages seem to be more tolerant to deficit irrigation than the reproductive stage. Deficit irrigation in vegetative and fruiting stages increased total yield and fruit quality characteristics, particularly total soluble solid, total sugars, and vitamin C. Capsaicin content was also increased by the deficit irrigation methods in Tunisian chillies (Mani 2014).

Water stress induces the activation of the phenol biosynthetic pathway under deficit irrigation conditions. Mild water stress increased the phenolic content in the leaves of pepper plants (Estiarte et al. 1994). Increased activity of nitrate reductase and accumulation of proline were observed in the leaves of two *C. annuum* varieties during water stress (Gaikwad et al. 2013).

Biochemical changes associated with salinity and metal toxicity have been reported in *capsicums*. Most of these changes are similar to the water stress-induced response of the plants. Low salinity stress increased the chlorophyll content, soluble sugars, and proline. However, the highest level of salt stress decreased the activities of superoxide dismutase (SOD), catalase (CAT), peroxidase (POD), ascorbate peroxidase (APX), and glutathione reductase (GR) enzymes in *capsicums* (Abu-Muriefah 2015).

Application of sitosterol alleviated the stress effects in terms of increased enzyme activities and reduced lipid peroxidation of the membranes (Abu-Muriefah 2015). Exposure of seedlings to nitric oxide (NO) treatment increased the activities of the antioxidant enzymes and reduced the lipid peroxidation of the membranes upon imposition of salinity stress in *capsicums* (Nalousi et al. 2012). Exogenous application of 24-epibrassinolide (0.5 mg/L) was also found to be better for increasing the shoot growth and leaf water status under salinity conditions (Houimli et al. 2008). Mycorrhizal application to the soil has been reported to improve the RWC, protein, total chlorophyll, and carotenoid content of pepper plants during salt application (Cekic et al. 2012). Application of brassinolides, sitosterols, and NO can alleviate the effects of water stress also in *capsicums*. Pepper plants inoculated with *Arthrobacter* sp. and *Bacillus* sp. tolerated osmotic stress exposure up to 45 % PEG indicating that soil microbes can be explored for giving further tolerance to the pepper plants (Sziderics et al. 2007).

9.2.3 Molecular Responses

Functional genomics has played a key role in making plant breeding more efficient in the development of crop varieties tolerant to abiotic stresses. A number of genetic and cellular events that occur under such stresses have been widely documented (Zhu 2002). In addition to various molecular markers, expressed sequence tags (ESTs) are currently used as a fast and efficient method of profiling genes expressed in various tissues, cell types, or developmental stages. The genes, thus, discovered will be utilized in their transfer to commercially important crops through marker-assisted selections or transgenic breeding (Medini et al. 2009). Signal transduction pathways and cellular events that occur under such unfavorable growth conditions have been widely documented.

Drought stress is one of the major limitations to crop productivity worldwide due to its multigenic nature, making the production of transgenic crops a challenging prospect. The hot pepper genome will enable the advancement of new breeding technologies through the exploration on genome-wide associations and genome selection studies on horticulturally important traits such as fruit size, yield, pungency, tolerance to abiotic stresses, and nutritional content (Kim et al. 2014). Comparative transcriptome analysis also found the differentially expressed genes in pepper between the control plants watered once per day and the drought stress-treated plants. The upregulated genes involved in drought stress response were helpful in accelerating the process of vegetable crop breeding with tolerance to environmental stresses.

Park et al. (2003) studied expression analysis which showed that Ca-LEAL1 was strongly activated by drought and plays an important role in the regulation of the stress gene in hot pepper plants. Isbat et al. (2009) reported that CaBI-1 gene was upregulated during water stress in chili pepper and was reported to be associated with the water stress tolerance (Mohammad et al. 2009). Overexpression of tobacco osmotin gene in chilli pepper plants increased the total chlorophyll, proline, glycine betaine, APX, SOD, DHAR, MDHAR, GR, and relative water content when compared to the wild plants under salinity stress. Transgenic plants also showed better yield under stress conditions (Subramanyam et al. 2011), indicating that the osmotin gene overexpression can be used to induce tolerance for water stress conditions also. CaDREB-LP1 has been identified as the transcription factor associated with the drought/salinity response in *Capsicum annuum* (Hong and Kim 2005). Constitutive expression of abiotic stress-inducible hot pepper CaXTH3, which encodes a xyloglucan endotransglucosylase/hydrolase homolog, improved drought and salt tolerance in transgenic *Arabidopsis* plants (Cho et al. 2006). Some of the genes which were differentially expressed in *Capsicum* under salinity stress are given below (Maurya et al. 2015):

CADREBLP1 – *Capsicum annuum* DREB-LIKE PROTEIN 1 (upregulated)
CAOSM1 – *Capsicum annuum* OSMOTIN-LIKE PROTEIN (upregulated)
CAPROX1 – *Capsicum annuum* PROLINE OXIDASE/DEHYDROGENASE (downregulated)
CACHI2 – *Capsicum annuum* CHITINASE CLASSII (upregulated)
CABI-1 – *Capsicum annuum* BAX INHIBITOR1 (upregulated)
CAPIP2 – *Capsicum annuum* AQUAPORIN (downregulated)
CARMA1H1 – *Capsicum annuum* RING DOMAIN CONTAINING PROTEIN (upregulated)
CAKR1 – *Capsicum annuum* CYS-3-HIS ZINC FINGER PROTEIN1 (upregulated)

9.2.4 Breeding for Drought Tolerance

Conventional breeding has played a considerable role during the last century for drought tolerance crop improvement in many crops (Sreenivasulu et al. 2007). In order to address the problem of climate change-triggered environment conditions, particularly drought situation (low moisture stress) and heat stress (high temperature stress), attempts have been made to breed vegetable varieties suitable for rain-fed cultivation and summer cultivation. As a result, several vegetable varieties have been successfully released for commercial cultivation. Although plant breeders have made ample progress through conventional breeding in developing drought-tolerant chilli (G4, Arka Lohit, LCA334) cultivars, the approach is complicated. Due to its complexity, empirical breeding for drought tolerance is a highly time-consuming, cost-, labor-intensive approach, and to achieve desired gain, a large number of selections and breeding cycles may be required. To avoid the above limitations, prior knowledge of gene effects on drought tolerance-related traits is needed to formulate the best breeding strategy to achieve the required goals in limited time. Heterosis for drought tolerance-related traits such as root length, root-to-shoot ratio, root dry weight and volume, total dry matter, and chlorophyll content has been studied in field crops such as rice, wheat, and maize (Abd Allah 2009; Hassan et al. 2011; Zhai et al. 2013). However, no work on these lines has been reported in *Capsicums*.

The high correlation of RWC and osmotic adjustment with the yield under stress is reported in peppers (Stefania et al. 2003; Kulkarni and Phalke 2009; Hu et al. 2010). In chilli, the effect of drought stress on morphophysiological traits was drastic, and it significantly reduced the yield. Correlation studies among the drought tolerance-related traits indicated that yield under drought stress is significantly correlated with drought susceptibility index (DSI), shoot dry weight, relative water content (RWC), and osmotic adjustment (Naresh 2015).

Drought resistance as such is not a simple trait governed by one or more genes and is apparently determined by a number of morphophysiological traits, which are being independently controlled by more than one or two genes (Fukai and Cooper 1995). Under drought stress, it is more realistic to carry out selection process on the basis of those putative characters, which serve as effective tools for screening for drought tolerance. However, confirmatory evidences are still lacking the traits of drought tolerance either at vegetative or reproductive phase that are governed by the same set of genes (Morgan 1984).

Drought tolerance-related traits such as root volume, shoot dry weight, root-to-shoot ratio, RWC, chlorophyll content, and all gas exchange parameters such as net photosynthetic rate, transpiration rate, stomatal conductance, and instantaneous water use efficiency (IWUE) were high in the crosses in which the two inbred lines, *viz*., IHR 4506 and IHR 4507, were used (Naresh 2015). This indicates that lines with different selection history will lead to good hybrid combination for the contrasting environments, which is due to favorable allele combinations between conventionally selected inbred lines and inbred lines selected for drought tolerance (Betran et al. 2003).

The ANOVA revealed highly significant variation for the different drought tolerance-related traits studied. The yield under drought stress is the ultimate trait which has to be maintained under drought conditions, and among the 45 F_1 hybrids evaluated, significant heterosis was observed for yield under drought stress in 33 (73.3 %) and 16 (35.5 %) hybrids, which recorded significant midparent and better parent heterosis, respectively, indicating the dominance nature and possibility of exploitation through heterosis breeding. The F_1 hybrid, IHR 3849 × IHR 3476, recorded highest midparent (555.23 %) and better parent (494.65 %) heterosis under drought stress (Naresh 2015), and such a high midparent heterosis (MPH) under stress is due to the poor performance of inbred lines under severe drought stress (Betran et al. 2003). The significant heterosis was reported for secondary traits of drought tolerance such as root length, root volume, root dry weight, shoot dry weight, root-to-shoot ratio, yield, harvest index, and drought susceptibility index in the study. The physiological traits such as RWC, chlorophyll content, and osmotic adjustment and gas exchange parameters such as stomatal conductance, photosynthetic rate, transpiration rate, and IWUE also showed significant desirable heterosis (Naresh 2015). There are no reports for heterosis of physiological traits in *capsicums*.

Under drought stress, both additive and non-additive genetic components were significant for all the traits studied except for root-to-shoot ratio, RWC, chlorophyll a, total chlorophyll, and other gas exchange parameters. Drought susceptibility index is controlled by dominance gene effect. Yield per plant is under the control of dominance gene effect under stress in the *Capsicum* sp. (Naresh 2015). The gene effects under stress for the all the traits except for plant height, root length, root volume, root dry weight, root-to-shoot ratio, chlorophyll a, chlorophyll b, and total chlorophyll content were significant for dominance components, and for other traits it is controlled by both additive and dominance components. The gas exchange parameters were found to be controlled by dominance gene effects in *Capsicum* species (Naresh 2015).

9.2.4.1 Breeding Strategy for Improving Root Traits

A more robust root system (deeper root length, large root volume, high root dry weight, and higher root-to-shoot ratio) was observed in chili germplasm IIHR 4502 (*C. chinense*) coupled with thick lateral roots which was not present in other genotypes. Overall, the root traits found to be complex in inheritance as indicated by the significance of the different genetic components. Both additive and nonadditive (dominance and epistasis) are important in controlling the root traits, so the reciprocal recurrent selection is the best breeding method to improve these traits. The modified methods such as bulk-pedigree method, which was effectively used by barley breeding for drought tolerance at the International Center for Agricultural Research in the Dry Areas (ICARDA), Syria, can also be effectively used in improving and isolating the best pure lines for root traits imparting drought tolerance in *Capsicum* species. And for introgressing the robust root system from wild species, *C. chinense* maintenance of large populations is necessary in the breeding program because in the adapted and exotic crosses, segregating loci are expected to be more as the number of homozygous genotypes in segregating population is a geometric function of number of segregating loci. Once the best lines are developed using the above said breeding strategies, then highly vigorous F_1 can be produced as these traits are mainly controlled by nonadditive components predominantly dominance gene effects in the *Capsicum* species. Another best strategy nowadays that is gaining lots of importance is rootstock breeding which can also be utilized for the protected cultivation of high-value bell peppers; the high value seedlings can be grafted onto the robust rooted rootstock to increase the efficient utilization of inputs. But before advising this strategy, it has to be evaluated using different rootstocks on the growth and yield as some are incompatible and negative effects would also occur.

9.3 Temperature Stress

Heat stress is a major factor influencing the productivity and adaptation of wild and cultivated plants. Pepper plants are originally from tropical regions and require high temperature conditions for their development. Consequently, the optimum growth temperature is between 25 and 30 °C. Change in temperatures affects a variety of physiological functions and morphological development. When temperature falls below 15 °C or exceeds 32 °C, growth is usually retarded and yield decreases (Mercado et al. 1997). Low temperature affects vegetative development and reproductive growth by disturbing the function of the flower female organs and the number of viable pollen grains per flower. Thus, fruits from plants that have been set under low night temperatures (14°C or less) usually are deformed and seedless, causing significant economical losses (Polowick and Sawhney 1985; Pressman et al. 1998). Plants possess major tolerance mechanisms such as activation of ion transporters, proteins, osmoprotectants, antioxidants, and other factors involved in signaling cascades and transcriptional control to offset stress-induced biochemical and morphological alterations (Hasanuzzaman et al. 2013). Plant survival under high temperature stress depends on the ability to perceive the temperature stimulus, generate and transmit the signal, and initiate appropriate morphological and biochemical changes.

9.3.1 Morphological and Physiological Changes

The visible morphological changes in bell peppers upon exposure to environmental stresses during the flowering and fruiting period are the abscission of flowers and flower buds (Saha et al. 2010). Abscission of flowers in hot peppers under high temperature and humidity conditions proceeds independently of the processes of pollination and fertilization (Usmana et al. 1999).

This loss of reproductive structures can result in serious yield reduction and constitutes a major risk factor in pepper production. High temperature has an adverse effect on fruit-set in bell pepper (Dorland and Went 1947). Poor fruit-set was believed to be one of the major barriers to the tropical adaptation of bell pepper. Song et al. (1976) observed decreased fruit-set in pepper as temperatures were raised from 18/13°C to 23/18°C and 33/28°C. Heat-induced yield reduction was documented in *Capsicum*.

Successful fruit-set depends on several reproductive processes including pollen germination and tube growth. Tolerant genotypes can be selected based on better fruit-set and yield. Selection of tolerant genotypes can also be made based on pollen germination characteristics at different temperatures in *Capsicum* using the cumulative temperature response index (CTRI) concept. Mean minimum, optimum, and maximum temperatures were 15.2, 30.7, and 41.8°C for pollen germination and 12.2, 31.2, and 40.4°C for pollen tube growth, respectively, across the species. *Capsicum annuum* cv. Mex Serrano from Mexico was identified as tolerant; *C. chacoense* cv. 1312 and *Capsicum* sp. cv. Cobanero from Argentina and Guatemala, respectively, as intermediate; and *C. frutescens* cv. Early Spring Giant from China, *C. annuum* cv. Long Green from South Korea, *Capsicum* sp. cv. NM89C130, and *C. pubescens* cv. 90002 from Guatemala as sensitive to high temperatures (Raja Reddy and Kakani 2007). The sensitivity of developmental stages to high temperature in chilli pepper (*Capsicum annuum* L. var. Shishito) at different periods of development after anthesis adversely affected fruit growth, seed yield, and seed quality (Pagamas and Nawata 2008).

Much attention has been paid to the impact of high temperatures on fruit growth and seed production. High temperatures during fruit development resulted in reduced fruit and seed yield in bell pepper (Aloni et al. 2001). The number of species adversely affected by heat stress suggests that flower development and seed filling stages after anthesis are most sensitive to heat stress, leading to a reduction in fruit and seed yield. High temperatures frequently occur after anthesis and strongly influence reproductive development and yield. Erickson and Markhart (2001) reported that under high temperature conditions (33.8°C), bell pepper plants continued to produce flowers and these flowers did not abscise upon opening; however, high temperature during post-pollination (33.8°C for 2 days) inhibited fruit-set (Erickson and Markhart 2002), suggesting that fertilization is sensitive to high temperature stress. In bell pepper (*Capsicum annuum* L. var. annuum), flowers that reached anthesis during the high temperature exposure had reduced fruit set when exposed to 33°C for 48 or 120 h. Flower bud length and pollen viability also were reduced when exposed to 33°C for 120 h. High post-pollination temperatures inhibited fruit-set, suggesting that fertilization is sensitive to high temperature stress (Erickson and Markhart 2002). High temperatures (36/27.8°C) during seed development in chili pepper do not induce conspicuous abortion of flowers but result in a significant reduction in seed production and increase the numbers of abnormal seeds (Pagamas and Nawata 2007). However, the causes of the increase in the proportion of abnormal seeds induced by heat stress and the sensitivity of seed developmental stages to high temperature conditions have not been elucidated. It may be due to the improper fertilization due to poor germination of pollens. Pollen sugar metabolism was reported to be affected by the high temperatures due to reduced activity of invertases (Aloni et al. 2001). Reduction in pollen hexokinase and phosphofructokinase activities was also reported to be responsible for the lower germination rate at higher temperatures (Karni and Aloni 2002). Higher temperatures were also reported to alter the time taken for flowering and fruiting (Garruña-Hernández et al. 2012).

A field evaluation of heat-tolerant chilli genotypes from AVRDC was carried out in Nepal during summer. A wide variability was observed for fruit set, yield, and other yield parameters, indicating that the major parameter for heat tolerance is the fruit set (Dahal et al. 2006). Variability in fruit set could be due to pollen germination and fertilization differences between the genotypes. In another experiment using the heat tolerant genotypes from AVRDC, the evaluation was done at two temperatures in phytotrons. In this trial, also the major difference between the genotypes was in the fruit set (Saha et al. 2010). Tolerant genotypes showed more proline accumulation than the susceptible genotypes. Temperatures of 32–34°C were found to affect vegetative as well as reproductive growth in chilli crop (Gunawardena and De Silva 2014). Temperature effect depends on the phenophases of the crop. High temperature at vegetative and at flowering stages did not affect the leaf area and dry mass of the plants. But high temperature (40°C) at fruiting stage reduced leaf area and dry mass by 51 % and 58 %, respectively, when compared to 30°C in chili plants (Garruña-Hernández et al. 2014). Lower temperatures of less than 20°C also reported to reduce the growth-related traits in *capsicum* (De Swart et al. 2006). Slower growth rate of fruits at lower temperatures during autumn was found to produce bigger sized fruits. The effects of growth season indicated that the pepper fruit morphology and physiology were closely related to the number of seeds per fruit in autumn, but to a less extent in summer (Saha et al. 2010).

9.3.2 Biochemical Changes

Peppers are cultivated in most of the temperate and tropical areas in the world. They are considered to be good sources of various nutritional compounds such as carotenoids, flavonoids, and mineral elements. Usman et al. (2014) reported that there was a highly significant variation among the genotypes in response to high temperature in terms of photosynthesis rate, plant height, fruit length, fruit weight, number of fruits, and yield per plant. Yield per plant showed strong to moderately positive correlations ($r = 0.23$–0.56) at phenotypic level while at genotypic level correlation coefficient ranged from 0.16 to 0.72 for CMT, plant height, fruit length, and number of fruits. Thanopoulos et al. (2013) worked on bell peppers grown in controlled conditions in the Mediterranean region which were frequently subjected to high temperatures in summer and low temperatures in winter and revealed that bell pepper fruits produced in autumn were larger in size but lower in nutritional value (less vitamin C). Slower growth rate of fruits at lower temperatures during autumn was found to produce bigger sized fruits, however, with lower vitamin C content. The heat-tolerant lines produced higher amount of proline indicating the role of proline in expressing the heat-tolerant capability of sweet pepper genotypes concerned (Saha et al. 2010). Low temperature exposure of pepper plants increased the accumulation of total soluble proteins, proline, phenolic compounds, and chlorophyll content compared to normal temperature (Koc et al. 2010). Leaves from pepper (*Capsicum annuum* L.) plants exposed to low temperature (8°C) for different time periods (1–3 days) exhibited increased lipid peroxidation and the soluble nonenzymatic antioxidant ascorbate and glutathione and the activity of the main NADPH-generating dehydrogenases (Airaki et al. 2012).

Two pepper (*Capsicum annuum* L.) cultivars tolerant to heat stress did not show an increase in the reactive oxygen species production due to increased photorespiration rate, non-photochemical quenching, cytochrome respiratory pathway activity, alternative respiratory pathways, and activities of SOD and APX (Hu et al. 2010). High temperature was found to increase the content of total capsaicinoids for the varieties Serrano, Puya, Ancho, Guajillo, and Bell pepper, while the content of total capsaicinoids and Scoville heat units (SHU) for the varieties De árbol and Jalapeño decreased (González-Zamora et al. 2013). Bouvier et al. (1998) demonstrated that reactive oxygen species induce the expression of carotenoid synthesis genes in pepper fruit. Higher levels of the xanthophyll compounds, such as lutein and zeaxanthin, provide the benefit of increased heat tolerance (Demmig-Adams and Adams 2002).

9.3.3 Molecular and Biotechnological Strategies for the Development of Temperature Stress Tolerance in *Capsicum*

Chilli pepper is an important horticultural crop that can surely benefit from plant biotechnology. However, although it is a Solanaceae member, developments in plant cell, tissue, and organ culture as well as on plant genetic transformation have lagged far behind those achieved for other members of the same family such as tobacco (*Nicotiana tabacum*), tomato (*Lycopersicon esculentum*), and potato (*Solanum tuberosum*). Cell and tissue cultures have been applied successfully to the selection of variant cells exhibiting increased resistance to abiotic stresses, but no plants exhibiting the selected traits have been regenerated (Ochoa-Alejo and Ramirez-Malagon 2001).

Along with different physiological and biochemical mechanisms, molecular approaches are helping to understand the concept of heat stress tolerance in plants. Plants tolerate such stresses by modulating multiple genes and by coordinating the expression of genes in different pathways. *CaAbsil* gene was found to express in hot pepper in response to many abiotic stresses like mannitol or NaCl, low temperature, wounding, and also to ABA and oxidative stress. Results suggest that *CaAbsil* plays a role in multiple responses to abiotic stresses (Seong and Wang 2008). In an attempt to determine a cold defense mechanism in plants, we have attempted to characterize changes occurring in the expression of cold-regulated transcript levels in the hot pepper (*Capsicum annuum*). Forty-two genes were upregulated by cold stress and were identified as variety of transcription factors, including a family of four ethylene-responsive element binding protein (EREBP, designated CaEREBP-C1 to CaEREBP-C4) genes, a bZIP protein (CaBZ1), RVA1, Ring domain protein, HSF1, and the WRKY (CaWRKY1) protein (Hwang 2005). Reduced tolerance to cold stress in transgenic *Arabidopsis* overexpressing a *Capsicum annuum* multiprotein bridging factor 1 (CaMBF1) indicates that the gene needs to be suppressed for stress tolerance (Guo et al. 2014). Isbata et al. (2009) studied that *CaBI-1* is a *BI-1* gene in *Capsicum annuum* of which expression in tobacco plants induced under heat stress and endows tolerance to heat stress. Antioxidant enzymes like ascorbate peroxidase (APOX), glutathione reductase (GR), and superoxide dismutase (SOD) activity level were higher in transgenic Virginia pine plants overexpressing the CaPF1 gene, which may protect cells from the oxidative damage caused by stresses, compared to the controls (Tang et al. 2005). High temperature exposure of hot pepper plants induced *CaRZFP1*, a zinc finger-type putative transcription factor (Zeba et al. 2006). Strong induction of the CASAR82A gene was observed under various biotic and abiotic stresses in *Capsicum*. This is also induced by ethylene, salicylic acid, hydrogen peroxide, methyl jasmonate, indole-3-acetic acid, abscisic acid, benzothiadiazole, and DL-β-n-aminobutyric acid (Lee and Hwang 2003).

Small heat shock proteins are a kind of main synthetically heat shock proteins in plants under heat stress, and they play an important role in heat resistance in plants. The small heat shock proteins function as molecular chaperones and can bind partially denatured proteins, thereby preventing irreversible protein aggregation during stress conditions. They play a crucial role in protecting plants against abiotic stresses (Sun et al. 2002; Wang et al. 2004) and prevent thermal denaturation of glucose oxidase and citrate synthase, which are the two critical enzymes in oxidative phosphorylation (Jakob et al. 1993). Zhu et al. (2011) indicated that the expression of *CaHSP24* small heat shock protein gene in pepper was induced by heat stress and the expression level was markedly higher in the heat-resistant cultivar than in the heat-sensitive cultivar. The gene could also enhance the viability of *E. coli* under heat stress. It is concluded that *CaHSP24* might be a stress-related gene in *Capsicum* under abiotic stress conditions. Differential expression of heat shock transcription factor (CaHsfA2) between the susceptible and tolerant types was reported under high temperature conditions. Expression of *Hsf* gene was also varying between the organs within the plant types. Higher expression in tolerant types was in flowers; however, in susceptible types higher expression was in leaves and lowest was in flowers. This might be one of the reasons for higher flower abscission and lower fruit set in susceptible types (Guo et al. 2014).

conventional plant breeding. More emphasis should be given to reduce the flower drop and for increasing the pollen germination using biotechnological approaches. Genotypic variability for root characters and for better expression of heat shock transcription factors in anthers needs to be established.

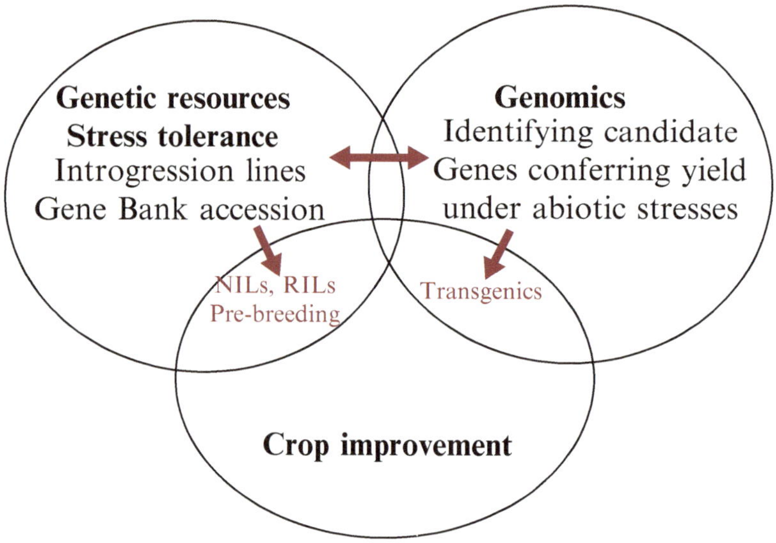

Sreenivasulu et al. 2007

9.4 Future Strategies

Lots of work has been done in understanding the abiotic stress responses in plants and identifying the traits related to the tolerance. Morphological, physiological, biochemical, and molecular analyses have played a major role in understanding various abiotic stress responses in plants. In recent years, molecular biology and biotechnology have appeared as a promising tool for overcoming stresses in plants; however, to date the progress has been limited in pepper crops. Omics approaches have proved to be the most straightforward and potential biotechnological applications for improving abiotic stress tolerance in pepper which lack proper resources for

9.5 Conclusion

Pepper crop faces various abiotic stresses of seasonal occurrence which influence plant growth and productivity. Abiotic stress even at moderate levels significantly decreased growth and biomass production of pepper plants. Drought and high temperature stress majorly affect the photosynthetic pigments, some of the metabolites, biomass production, and finally yield in pepper. The adverse effects of stress on the growth and productivity of pepper can be mitigated by genetic engineering advances for improving tolerance to abiotic stresses. However, more efforts are needed in this direction especially in horticultural crops to survive with existing environmental challenges.

References

Abd Allah AA (2009) Genetic studies on leaf rolling and some root traits under drought conditions in rice (*Oryza sativa* L.). Afr J Biotech 8(22):6241–6248

Abdulmalik MM, Olarewaju JD, Usman IS (2012) Effects of moisture stress on flowering and fruit set in sweet pepper (*Capsicum annuum* L.) cultivars. Prod Agric Technol 8(1):191–198

Abu-Muriefah SS (2015) Effect of sitosterol on growth, metabolism and protein pattern of pepper (*Capsicum annuum* L.) plants grown under salt stress conditions. Intl J Agric Crop Sci 8(2):94–106

Airaki M, Leterrier M, Mateos RM, Valderrama R, Chaki M, Barroso JB, Del Río LA, Palma JM, Corpas FJ (2012) Metabolism of reactive oxygen species and reactive nitrogen species in pepper (*Capsicum annuum* L.) plants under low temperature stress. Plant Cell Environ 35:281–295

AlHarbi AR, Saleh AM, Al-Omran AM, Wahb-Allah MA (2014) Response of bell-pepper (*Capsicum annuum* L.) to salt stress and deficit irrigation strategy under greenhouse conditions. In: Blok C et al (ed) Proceedings of IS on growing media & soilless cultivation. Acta Hortic ISHS 1034, Leiden, The Netherlands, pp 443–450

Aloni B, Karni L, Zaidman Z, Riov Y, Huberman M, Goren R (1994) The susceptibility of pepper (*Capsicum annuum* L.) to heat-induced flower abscission. Possible involvement of ethylene. J Hortic Sci 69:923–928

Aloni B, Peet M, Phart M, Karni L (2001) The effect of high temperature and high atmospheric CO_2 on carbohydrate change in bell pepper (*Capsicum annuum*) pollen in relation to its germination. Physiol Plant 112:505–512

Anjum SA, Xie X, Wang L, Saleem MF, Man C, Lei W (2011) Morphological, physiological and biochemical responses of plants to drought stress. Afr J Agric Res 9:2026–2032

Asadi S, Lebaschy MH, Khourgami A, Rad AHS (2012) Effect of drought stress on the morphology of three *Salvia sclarea* populations. Ann Biol Res 3 (9):4503–4507

Bernier J, Kumar A, Ramaiah V, Spaner D, Atlin G (2007) A large effect QTL for grain yield under reproductive drought stress in upland rice. Crop Sci 47:507–516

Betran FJ, Ribaut JM, Beck D, Leon GD (2003) Genetic diversity, specific combining ability and heterosis in tropical maize hybrids under stress and non-stress environments. Crop Sci 43:797–806

Bosland PW, Baral B (2007) 'Bhut Jolokia'-the world's hottest known chile pepper is a putative naturally occurring interspecific hybrid. HortSci 42:222–224

Bosland PW, Votava EJ (2002) Peppers: vegetable and spice *Capsicum*. CABI, New York

Bouvier F, Backhaus RA, Camara B (1998) Induction and control of chromoplast-specific carotenoid genes by oxidative stress. J Biol Chem 273:30651–30659

Cafer G, Irfan E, Akincik K, Ucan S, Akinci SG (2006) Response of red hot pepper plant (*Capsicum annuum* L.) to the deficit irrigation. Akdeniz Uuniversiteisi Zirrat Fakultesi Dergisi 19:131–138

Cantore F, Boari, Caliandro A (2000) Effect of split-root-system water stress on physiological and morphological aspects of pepper (*Capsicum annuum* L.). Acta Horticult 537:321–328

Çekiç FO, Ünyayar S, Ortaş I (2012) Effects of arbuscular mycorrhizal inoculation on biochemical parameters in *Capsicum annuum* grown under long term salt stress. Turk J Bot 36:63–72

Chang TT, Loresto GC, Tagumpay O (1972) Agronomic and growth characteristics of upland and lowland rice varieties. In: Rice breeding. IRRI, Los Banos, pp 645–661

Chang TT, Armento-S-Tp JL, Mao CX, Peiri R, Loresto GC (1986) genetic studies on the components of drought resistance in rice. In: Rice genetics, IRRI, Los Banos, pp 389–398

Cho SK, Kim JE, Park JA, Eom TJ, Kim WT (2006) Constitutive expression of abiotic stress-inducible hot pepper *CaXTH3*, which encodes a xyloglucan endotransglucosylase/ hydrolase homolog, improves drought and salt tolerance in transgenic *Arabidopsis* plants. FEBS Lett 580:3136–3144

Dahal KC, Sharma MD, Dhakal DD, Shakya SM (2006) Evaluation of heat tolerant chilli (*Capsicum annuum* L.) genotypes in Western Terai of Nepal. J Inst Agric Anim Sci 27:59–64

De Swart EAM, Marcelis LFM, Voorrips RE (2006) Variation in relative growth rate and growth traits in wild and cultivated *Capsicum* accessions grown under different temperatures. Research gate, pp 1–28

Della Costa L, Gianquinto G (2002) Water stress and water-table depth influence yield, water use efficiency, and nitrogen recovery in bell pepper: lysimeter studies. Aust J Agric Res 53:201–210

Demmig-Adams BD, Adams W (2002) Antioxidants in photosynthesis and human nutrition. Science 298:2149–2153

Dorland RE, Went FW (1947) Plant growth and controlled condition viii. Growth and fruiting of chili peppers (*C. annuum*). Am J Bot 34:393–401

El-Ghinbihi FH, Hassan MI (2007) Effect of some natural extracts and ascorbic acid as foliar spray on growth, leaf water contents, chemical composition and yield of pepper plants grown under water stress conditions. Minufiya J Agric Res 32(3):683–710

Erickson AN, Markhart AH (2001) Flower production, fruit set, and physiology of bell pepper during elevated temperature and vapor pressure deficit. J Am Soc Hortic Sci 126(6):697–702

Erickson AN, Markhart AH (2002) Flower developmental stage and organ sensitivity of bell pepper (*Capsicum annuum* L.) to elevated temperature. Plant Cell Environ 25:123–130

Estiarte M, Filella I, Serra J, Penuelas J (1994) Effects of nutrient and water stress on leaf phenolic content of peppers and susceptibility to generalist herbivore *Helicoverpa armigera* (Hubner). Oecologia 99:387–391

Estrada B, Pomar F, Diaz J, Merino F, Bernal MA (1999) Pungency levels in fruit of the Padro'n pepper with different water supply. Sci Hortic 81:385–39

Fukai S, Cooper M (1995) Development of drought-resistant cultivars using physio-morphological traits in rice. Field Crop Res 40:67–86

Gaikwad SE, Shinde RA, Thoke RB, Aparadh VT (2013) Potential of drought stress in two varieties of capsicum annum grown in Maharashtra. Intl Res Pharm Chem 3(2):303–307

Garruña-Hernández R, Canto A, Mijangos-Cortés JO, Islas I, Pinzón L, Orellana R (2012) Changes in flowering and fruiting of Habanero pepper in response to higher temperature and CO_2. J Food Agric Environ 10:802–808

Garruna-Hernandez R, Orellana R, Larque-Saavedra A, Canto A (2014) Understanding the physiological responses of a tropical crop (*Capsicum chinense Jacq.*) at high temperature. PLoS One 9(11):1–9

Gonzalez VD, Orgaz F, Federes E (2007) Response of pepper to deficit irrigation for paprika production. Sci Hortic 114:77–82

González-Zamora A, Sierra-Campos E, Luna-Ortega JG, Pérez-Morales R, Ortiz JCR, García-Hernández JL (2013) Characterization of different *Capsicum* varieties by evaluation of their capsaicinoids content by high performance liquid chromatography, determination of pungency and effect of high temperature. Molecules 18:13471–13486

Gunawardena MDM, De Silva CS (2014) Identifying the impact of temperature and water stress on growth and yield parameters of Chilli (Capsicum annuum L). OUSL J 7:25–42

Guo W, Chen R, Du X, Zhang Z, Yin Y, Gong Z, Wang G (2014) Reduced tolerance to abiotic stress in transgenic *Arabidopsis* overexpressing a *Capsicum annuum* L. multiprotein bridging factor 1. BMC Plant Biol 14:138

Hasanuzzaman M, Nahar K, Alam MM, Roychowdhury R, Fujita M (2013) Physiological, biochemical, and molecular mechanisms of heat stress tolerance in plants. Int J Mol Sci 14:9643–9684

Hassan HM, El-Abd AB, El-Baghday NM (2011) Combining ability for some root physiological and grain quality traits in rice under water deficit conditions. J Agric Res 37(2):239–256

Hawa ZJ (2003) Effects of water stress on anthesis and flower abscission in the glasshouse sweet pepper (*Capsicum annuum* L.). In: Challenges in plant productivity and food security in changing environment. Proceedings of the 14th Malaysian Society of Plant Physiology Conference, Pahang (Malaysia), 23–25 Sept 2003, pp 108–116

Hong JP, Kim WT (2005) Isolation and functional characterization of the Ca-D REBLP1 gene encoding a dehydration-responsive element binding-factor-like protein 1 in hot pepper (*Capsicum annuum* L. cv. Pukang). Planta 220:875–888

Houimli SIM, Denden M, Hadj SBL (2008) Induction of salt tolerance in pepper (*Capsicum annuum*) by 24-epibrassinolide. Eur Asia J Bio Sci 2:83–90

Hu WH, Xiao Y, Azeng JJ, Hu XH (2010) Photosynthesis, respiration and antioxidant enzymes in pepper leaves under drought and heat stresses. Biol Plant 54 (4):761–765

Isbat M, Zeba N, Kim SR, Hong CB (2009) A BAX inhibitor-1 gene in *Capsicum annuum* is induced under various abiotic stresses and endows multi-tolerance in transgenic tobacco. J Plant Physiol 166:1685–1693

Ismail MR, Davies WJ, Awad MH (2002) Leaf growth and stomatal sensitivity to ABA in droughted pepper plants. Sci Hortic 96:313–327

Jakob U, Gaestal M, Engel K, Buchner J (1993) Small heat shock proteins are molecular chaperones. J Biol Chem 268(3):1517–1520

Jimenez-Garcia SN, Vazquez-Cruz MA, Garcia-Mier L, Guevara-Gonzalez RG, Torres-Pacheco I, Ocampo-Velazquez RV, Cruz-Hernandez A, Feregrino-Perez AA (2014) Changes in the quantity of phenolic compounds in peppers (*Capsicum annuum* L.) sprinkled with elicitors under cold stress. J Chem Biol Phys Sci 4(5):11–17

Karni L, Aloni B (2002) Fructokinase and hexokinase from pollen grains of bell pepper (*Capsicum annum* L.): possible role of pollen germination under conditions of high temperature and CO_2 enrichment. Ann Bot 90:607–611

Kashiwagi J, Krishnamurthy L, Gaur PM, Chandra S, Upadhyaya HD (2008) Estimation of gene effects of the drought avoidance root characteristics in chickpea (*Cicer arietinum* L.). Field Crop Res 105:64–69

Kim et al (2014) Genome sequence of the hot pepper provides insights into the evolution of pungency in *Capsicum* species. Nat Genet 46(3):270–279

Kinyua RJ, Njoka, Gesimba, Birech (2003) Selection of drought tolerant bread. Wheat genotypes using root characteristics at seedling stage. Int J Agric Rural Dev 4:9–15

Kirada C, Topcu S, Cetin M, Dasgan HY, Kaman H, Topaloglu F, Derici MR, Ekici B (2007) Prospects of partial root zone irrigation for increasing irrigation water use efficiency of major crops in the Mediterranean region. Ann Appl Biol 150:281–291

Kirnak H, Kaya C, Higgs D, Tas I (2003) Responses of drip irrigated bell pepper to water stress and different nitrogen levels with or without mulch cover. J Plant Nutr 26(2):263–277

Klar AE, Jadoski SO, Lima PP (2006) Peroxidase activity as an indicator of water stress in sweet pepper plants. Irrig Botucatu 11(4):441–447

Koc E, Islek C, Üstün AS (2010) Effect of cold on protein, proline, phenolic compounds and chlorophyll content of two pepper (*Capsicum annuum* L.). GUJ Sci 23(1):1–6

Kulkarni M, Phalke S (2009) Evaluating variability of root size system and its constitutive traits in hot pepper (*Capsicum annum* L.) under water stress. Sci Hortic 120(2):159–166

Larcher W (2006) Ecofisiologia vegetal (trans: Prado CHBA). Rima, São Carlos

Lawlor DW, Cornic G (2002) Photosynthetic carbon assimilation and associated metabolism in relation to water deficits in higher plants. Plant Cell Environ 25 (2):275–294

Lee SC, Hwang BK (2003) Identification of the pepper SAR8.2 gene as a molecular marker for pathogen infection, abiotic elicitors and environmental stresses in *Capsicum annuum*. Planta 216:387–396

Lobato AKS, Luz LM, Costa RCL, Tan DKY, Bonato CM, Silva MHL, Oliveira Neto CF, Silva LI (2009) Relationship between chlorophyll a and total soluble carbohydrates in pepper submitted to water deficiency. J Anim Plant Sci 5(2):515–526

Ludlow MM, Muchow RC (1990) A critical evaluation of the traits for improving crop yield in water limited environments. Adv Agron 43:107–153

Madhusudan KV, Giridarakumar S, Ranganayakul GS, Reddy PC, Sudhakar C (2002) Effect of water stress on some physiological responses in two groundnut (*Arachis hypogea* L.) cultivars with contrasting drought tolerance. J Plant Biol 29:199–202

Mani F (2014) Effect of water stress on some biochemical and physiological characters of chili pepper (*Capsicum frutescens* L.). JECET 4(1):75–83

Maurya VK, Srinivasan R, Nalini E, Ramesh N, Gothandam KM (2015) Analysis of stress responsive genes in Capsicum for salinity responses. Ann Res Rev Biol 6(1):66–78

Medini H, Elaissi A, Farhat F, Khouja ML, Chemli R, Harzallah-Skhiri F (2009) Seasonal and geographical influences on the chemical composition of *Juniperus phoenicea* L. essential oil leaves from the Northern Tunisia. Chem Biodivers 6:1378–1387

Mercado JA, Reid MS, Valpuesta V, Quesada MA (1997) Metabolic changes and susceptibility to chilling stress in *Capsicum annuum* plants grown at suboptimal temperature. Aust J Plant Physiol 24:759–767

Mohammad IL, Naheed ZI, Seong RK, Choo BH (2009) A BAX inhibitor-1 gene in *Capsicum annuum* is induced under various abiotic stresses and endows multi-tolerance in transgenic tobacco. J Plant Physiol 166:1685–1693

Morgan JM (1984) Osmoregulation and water stress in higher plants. Ann Rev Plant Physiol 35:299–319

Murakami K, Ido M, Masuda M (2006) Fruit pungency of 'Shishito' pepper as affected by a dark interval in continuous fluorescent illumination with temperature alteration. J Shita 18:284–289

Nalousi AM, Ahmadiyan S, Hatamzadeh A, Ghasemnezhad M (2012) Protective role of exogenous nitric oxide against oxidative stress induced by salt stress in bell-pepper (Capsicum annum L.). Am-Eurasian J Agric Environ Sci 12(8):1085–1090

Naresh P (2015) Genetic and molecular analyses for resistance to viruses (cucumber mosaic virus & chilli veinal mottle virus), drought tolerance and fruit quality traits in chilli (*Capsicum annuum* L.). Ph. D. thesis, COH, Bengaluru, UHS, Bagalkot

Nieto-Garibay A, Murillo-Amador B, Troyo-Diéguez E, García-Hernández JL, Ruíz-Espinoza FH (2010) Water stress in two capsicum species with different domestication grade. Trop Subtrop Agroecosyst 12:353–360

Ochoa-Alejo N, Ramirez-Malagon R (2001) In vitro chili pepper biotechnology. In vitro cell. Dev Biol Plant 37:701–729

Pagamas P, Nawata E (2007) Effect of high temperature during the seed development on quality and chemical composition of chili pepper seed. Jpn J Trop Agric 51:22–29

Pagamas P, Nawata E (2008) Sensitive stages of fruit and seed development of chili pepper (*Capsicum annuum* L. var. Shishito) exposed to high-temperature stress. Sci Hortic 117:21–25

Park J, Cho SK, Kim JE, Chung HS, Hong J, Hwang B, Hong CB, Kim WT (2003) Isolation of cDNAs differentially expressed in response to drought stress and characterization of the Ca-LEAL1 gene encoding a new family of atypical LEA-like protein homologue in hot pepper (*Capsicum annuum* L. cv. Pukang). Plant Sci 165:471–481

Phimchan P, Techawongstien S, Chanthai S, Bosland PW (2012) Impact of drought stress on the accumulation of capsaicinoids in capsicum cultivars with different initial capsaicinoid levels. HortSci 47(9):1204–1209

Polowick PL, Sawhney VK (1985) Temperature effects on male fertility and flower and fruit development in *Capsicum annuum* L. Sci Hortic 25:117–127

Pressman E, Moshkovitch H, Rosenfeld K, Shaked R, Gamliel B, Aloni B (1998) Influence of low night temperatures on sweet pepper flower quality and the effect of repeated pollinations, with viable pollen, on fruit setting. J Hortic Sci Biotechnol 73:131–136

Prince AH, Tomos AD (1997) Genetic dissection of root growth in rice (*Oryza sativa* L.). II: mapping quantitative trait loci using molecular markers. Theor Appl Genet 95:143–152

R'Him T, Radhouane L (2015) Growth and yield responses of two Tunisian pepper (*Capsicum annuum* L.) varieties to salinity and drought stress. Int J Innov Sci Res 14(2):159–167

Raja Reddy K, Kakani VG (2007) Screening *Capsicum* species of different origins for high temperature tolerance by in vitro pollen germination and pollen tube length. Sci Hortic 112:130–135

Ray JD, Yu L, Mccouch SR, Chapoux MC, Wang G, Nguyen H (1996) Mapping quantitative trait loci associated with root penetration ability in rice. Theor Appl Genet 92:627–633

Ruiz-Lau N, Medina-Lara F, Minero-Garcia Y, Zamudio-Moreno E, Guzman-Antonio A, Echevarria-Machado-I, Martinez-Estevez M (2011) Water deficit affects the accumulation of capsaicinoids in fruits of *Capsicum Chinense Jacq*. HortSci 46(3):487–492

Saha SR, Hossain MM, Rahman MM, Kuo CG, Abdullah S (2010) Effect of high temperature stress on the performance of twelve sweet pepper genotypes. Bangladesh J Agric Res 35(3):525–534

Sashidhar VR, Nagalakshmi V, Mahesh JK, Prasad TG (2000) Should plants keep their canopy cool or allow themselves to grow warm under stress. Curr Sci 78:786–789

Seong ES, Wang M (2008) A novel CaAbsil gene induced by early-abiotic stresses in Pepper. BMB Rep 41(1):86–91

Serraj R, Sinclair TR (2002) Osmolyte accumulation: can it really help increase crop yield under drought conditions. Plant Cell Environ 25:333–341

Sezen SM, Yazar A, Tekin S, Eker S, Kapur B (2011) Yield and quality response of drip-irrigated pepper under Mediterranean climatic conditions to various water regimes. Afr J Biotechnol 10(8):1329–1339

Sharp RE, Poroyko V, Hejlek LG, Spollen WG, Springer GK, Bohnert HJ, Bguyen HT (2004) Root growth maintenance during water deficits: physiology to functional genomics. J Exp Bot 55:2343–2351

Showemimo FA, Olarewaju JD (2007) Drought tolerance indices in sweet pepper (*Capsicum annum* L.). Int J Plant Breed Genet 1(1):29–33

Sinclair, Muchow (2001) System analysis of plant traits to increase grain yield on limited water supplies. Agron J 93:263–270

Song KW, Park SK, Kim CK (1976) Studies on the flower abscission of hot pepper. Res Rept Off Rural Dev 18:9–32

Sreenivasulu N, Sopory SK, Kavi Kishor PB (2007) Deciphering the regulatory mechanisms of abiotic stress tolerance in plants by genomic approaches. Gene 388:1–13

Stefania DP, Ruggiero C, Barbieri G (2003) Physiological responses of pepper to salinity and drought. J Am Soc Hortic Sci 128(1):48–54

Subramanyam K, Sailaja KV, Subramanyam K, Muralidhara Rao D, Lakshmidevi K (2011) Ectopic expression of an osmotin gene leads to enhanced salt tolerance in transgenic chilli pepper (*Capsicum annuum* L.). Plant Cell Tissue Organ Cult 105:181–192

Sun WN, Van Montagu N, Verbruggen N (2002) Small heat shock proteins and stress tolerance in plants. Biochem Biophys Acta 1577:1–9

Sung U, Chang YY, Ting NL (2005) Capsaicin biosynthesis in water stress hot pepper fruits. Bot Bull Acad Sin 46:35–42

Sziderics AH, Rasche F, Trognitz F, Sessitsch A, Wilhelm E (2007) Bacterial endophytes contribute to abiotic stress adaptation in pepper plants (*Capsicum annuum* L.). Can J Microbiol 53:1195–1202

Tang W, Charles TM, Newton RJ (2005) Overexpression of the pepper transcription factor CaPF1 in transgenic Virginia pine (*Pinus virginiana* Mill.) confers multiple stress tolerance and enhances organ growth. Plant Mol Biol 59:603–617

Thakur PS, Anju T, Kanaujia SP (2000) Reversal of water stress effects. I. Mulching impact on the performance of Capsicum annuum under water deficit. Ind J Hortic 57(3):250–254

Thanopoulos C, Akoumianakis KA, Passam HC (2013) The effect of season on the growth and maturation of bell peppers. Int J Plant Prod 7(2):1735–8043

Tripathy JN, Zhang J, Robin S, Nguyen HT (2000) QTLs for cell-membrane stability mapped in rice (*Oryza sativa* L.) under drought stress. Theor Appl Genet 100:1197–1202

Usman MG, Rafii MY, Ismail MR, Malek MA, Latif MA (2014) Heritability and genetic advance among chili pepper genotypes for heat tolerance and morphophysiological characteristics. The Sci World J 2014:1–14

Usmana IS, Mamatb AS, Mohd HSZS, Aishahc HS, Anuarc AR (1999) The non-impairment of pollination and fertilization in the abscission of chilli (*Capsicum annuum* L. Var. Kulai) flowers under high temperature and humid conditions. Sci Hortic 79:1–11

Venuprasad R, Shashidhar HE, Hittalmani S, Hemamalini GS (2002) Tagging quantitative trait loci associated with grain yield and root morphological traits in rice (*Oryza sativa* L.) under contrasting moisture regimes. Euphytica 128:293–300

Wang WX, Vinocur B, Shoseyov O, Altman A (2004) Role of plant heat-shock proteins and molecular chaperones in the abiotic stress response. Trends Plant Sci 9(5):244–252

Yildirim M (2010) Water management in coastal areas with low quality irrigation water for pepper growth. J Coast Res 26(5):869–878

Zeba N, Ashrafuzzaman M, Hong CB (2006) Molecular characterization of the *Capsicum annuum* RING zinc finger protein 1 (*CaRZFP1*) gene induced by abiotic stresses. J Plant Biol 49:484–490

Zhai R, Feng Y, Wang H, Zhan X, Shen X, Weiming W, Zhang Y, Chen D, Gaoxing, Yang Z, Cao L, Cheng S (2013) Transcriptome analysis of rice root heterosis by RNA-Seq. BMC Genom 14:19

Zhang J, Nguyen HT, Blum A (1999) Genetic analysis of osmotic adjustment in crop plants. J Exp Bot 50 (332):291–302

Zhu JK (2002) Salt and drought stress signal transduction in plants. Annu Rev Plant Biol 53:247–273

Zhu W, Lu M, Gong Z, Chen R (2011) Cloning and expression of a small heat shock protein gene *CaHSP24* from pepper under abiotic stress. Afr J Biotechnol 10(25):4968–4976

Part III

Abiotic Stress Tolerance in Horticultural Crops: Fruit Crops

10 Mango

R.H. Laxman, C.J. Annapoornamma, and Geeta Biradar

Abstract

Mango is well adapted to both tropical and subtropical conditions, and its cultivation extends from Australia to Spain. The crop has adapted to wide growing conditions wherein different rainfall patterns, temperature and soil conditions prevail. However, the best areas for commercial production are with a cool or dry period prior to flowering, abundant soil moisture and moderately hot temperatures (30–33 °C) during fruit development. Although mango has adapted to various environmental conditions, it experiences unfavourable conditions during various phenological stages. Due to its perennial nature, mango tree experiences abiotic stresses, during various developmental stages, individually and also in combination. The frequency and duration of the abiotic stresses are likely to vary under climate change conditions. Temperature has a dominant influence on mango. The frequency, extent of flowering, fruit growth and quality of mango are influenced by temperature. Mango is sustained by rainfall and evenly distributed rainfall is essential for economic production of mango. Occurrence of water stress at flowering stage is advantageous to mango. Under tropical conditions, water stress not only induces flowering but also hastens it. Water stress prevents vegetative flushing and provides more time for accumulation of floral stimulus. Mango growth and gas exchange are adversely affected by flooding, but the trees develop hypertrophic lenticels on stems immediately above the waterline to survive prolonged flooding stress. Saline conditions are not favourable for profitable mango cultivation as the toxic ions cause nutrient imbalance and reduce growth, fruit bearing and yield. However, several rootstock cultivars have exhibited differential tolerance to salinity stress. Tolerant

R.H. Laxman (✉) • C.J. Annapoornamma • G. Biradar
Division of Plant Physiology and Biochemistry,
ICAR-Indian Institute of Horticultural Research,
Hessaraghatta Lake Post, Bengaluru 560089,
Karnataka, India
e-mail: laxmaniihr@gmail.com

N.K.S. Rao et al. (eds.), *Abiotic Stress Physiology of Horticultural Crops*,
DOI 10.1007/978-81-322-2725-0_10

rootstocks could be made use in salt-affected soils to overcome the adverse effects on growth and yield. Since the mango growth, development and final yields are determined by various abiotic factors, thorough understanding of the effects of many abiotic factors, influencing growth, development, phenology and physiology of mango, becomes very essential. Hence, the various abiotic factors affecting mango are discussed hereunder.

10.1 Introduction

Mango is an important fruit crop being grown in more than 110 countries located in tropical and subtropical regions. Indo-Burma region is considered to be the centre of origin (Vavilov 1926; Singh 1976). In India, mango cultivation is believed to have taken place since 4,000 years, and it is designated as national fruit. The occurrence of diverse flavour, taste and shades of colour has brought 'king of fruits' status to this fruit. The distribution of mango from the centre of origin to other parts took place mainly by the travellers and invaders. However, spread in many parts of the world happened during the fifteenth and sixteenth centuries.

Mango grows very well in tropical regions with seasonal rains. The cultivation of mango extends from 39°S, Australia (Crane et al. 1997), to 36°N in Spain (Galen Sauco 1997). It is being cultivated and produced in large quantities in countries like India, China, Thailand, Indonesia, Mexico, Philippines, Pakistan, Nigeria, Guinea and Brazil. India is the largest producer of mango followed by China. Though mango is a tropical tree, it endures a wide range of temperatures. It grows well both in low (25 cm) and high (250 cm) rainfall areas. Though mango has adapted to both tropical and subtropical climatic conditions, it performs well in regions from sea level to 600 m altitude. In India, mango has adapted to the varied climatic conditions from the tropical south to the sub-mountainous regions of north India (30°N) and altitudes up to 1400 m. It grows at an altitude of about 1400 m, provided that high humidity, rain or frosts are not prevailing during flowering (Majumder and Sharma 1985). In far-western Nepal, orchards at the altitude of 1400–1490 m are also reported (Subedi et al. 2008). But at the altitudes above 1500 m, the prevailing lower temperatures and frost limit the cultivation of mango crop.

Though mangos are cultivated throughout the tropics and subtropics, the best areas for commercial mango fruit production are those possessing a cool or dry period prior to flowering, abundant soil moisture and moderately hot temperatures (30–33 °C) during fruit development (Chacko 1986). Although mango is grown at locations from sea level to 1,250 m elevation, commercial production is difficult in areas above 700 m. In sub-Himalayan valleys, mango is found to flourish well in protected sites and at lower elevations of 300–700 m. It gives profitable yield in semiarid conditions, especially with irrigation. The congenial temperature range favourable for mango cultivation during growing season is 24–30 °C (Mukherjee 1953). In India, the regions with annual rainfall of 75 cm and above support cultivation of mango with little or no irrigation. However, it is found to grow in diverse areas receiving rainfall in the range from 2500 mm to even 250 mm. The mango cultivars are classified as monoembryonic and polyembryonic ecotypes. Monoembryonic types evolved in dry subtropical monsoonal regions and the polyembryonic types in the consistently hot, humid tropics of Southeast Asia (Mukherjee 1972). Since the growth of shoot, root, flower and fruit are dependent on cultivars and climatic conditions, the relative performance of various cultivars would be primarily determined by the climate of growing locations.

Although mango has adapted to various environmental situations, it undergoes certain unfavourable conditions during the growth and development. The potential yield is determined by the interaction of endogenous and various environmental factors during the tree growth and development (Schaffer et al. 1994). Because of its perennial nature, the tree experiences one or the other abiotic stresses during various phenological stages. The occurrence of abiotic stresses like limited and excess moisture, salinity and low- and high-temperature episodes would influence the growth and development. These factors influence the mango crop growth not only individually but in interaction with each other. Under climate change conditions, the periodicity and extent of these abiotic stresses are likely to increase. The adverse effect of abiotic stresses on the morphology, phenology and physiology of mango would ultimately hinder the realisation of its genetic yield potential. As the growth and development is determined by the combined effects of different abiotic factors, their role in successful growth and also the adverse effects they cause on the proper growth and development need to be thoroughly understood. Such information would help in efficient management of the trees and realisation of the potential yields under most favourable conditions and relatively better yields under adverse abiotic stress situations. Many studies have been attempted to decipher the adverse effects of abiotic stresses like temperature, salinity, excess and limited water. The researchers have attempted to understand the effect of abiotic factors' influence on growth, development, phenology and physiology of mango. The present chapter discusses the various abiotic factors and their effects on physiology, growth and development of mango.

10.2 Temperature

Temperature has a dominant influence on growth and yield, through its effect on physiological processes and phenological events during the plant growth cycle. Frequency and extent of flowering, fruit growth and quality of mango are influenced by temperature. Since it is adapted to a wide range of temperatures, mango is cultivated in tropics as well as subtropics. Since the annual average temperatures suitable for the optimal growth and development of mango are between 22 and 27 °C, the occurrence of sub- and supra-optimal temperatures during various stages of mango growth and development leads to varied responses.

10.2.1 Growth

Vegetative and flowering flushes in mango are highly influenced by variations in temperature. Mango in general has a vegetative bias, and it becomes stronger with increases in temperature, provided that nutrients and water are non-limiting. Mango vegetative shoots have episodic growth pattern, and both plant and environmental factors exert a strong control on the synchronisation of growth phases in the tree. Studies have shown that both soil and air temperatures play an important role in the tree growth. Soil temperatures have a strong effect on the growth patterns of 'Irwin' trees grafted on Turpentine rootstocks (Yusof et al. 1969). Periodic shoot growth, alterations between phases of activity and dormancy, occurred when soil temperatures were held at 27 or 32 °C for 120 days, but at 21 °C an extended dormant period was observed. In monoembryonic cultivar Ruby, germination was rapid between 25 and 40 °C and seedling growth was optimal at 30 °C (Corbineau et al. 1986), and a temperature of 40 °C was lethal to young seedlings.

The cessation of mango shoot growth was observed at approximately 15 °C. Whiley et al. (1989) quantified the daily temperature for zero shoot growth for ten mango cultivars to be

15 °C, and above this temperature shoot extension occurred. The day and night temperatures beyond this base temperature have been observed to increase the vegetative flushes, number and size of leaves. As temperatures increased during a 20-week exposure period, the number of vegetative flushes increased from 1.02 at 20 °C day/15 °C night to 3.20 at 30 °C day/25 °C night. Further, the temperature was also found to have effect on number and size of leaves. Among the ten cultivars tested, at 20/15 °C, an average of 7.1 leaves were observed in a shoot but at 30/25 °C increased to 13.6. Mean leaf size was 300 % larger at 30/25 °C compared with leaves grown at 20/15 °C. Whiley et al. (1989) reported that temperature influences dry matter distribution also. In their study, dry matter partitioning to roots was greatest when trees were grown at 15/10 °C, where shoot growth was suppressed. And as temperatures increased, the partitioning to roots declined but correspondingly it increased for stem and leaves. There was a greater allocation of dry matter to the roots of 'Irwin' at higher temperatures of 25/20 and 30/25 °C, which is a reflection of its reduced flushing and semidwarf stature. Though higher yields are obtained in frost-free subtropical latitudes, having marked dry season, high heat accumulation and the extreme heat, drought and high evaporative demand reduces the production.

Mango due to its perennial nature responds to changes in temperatures differently than annual crops (Litz 2009). Being a perennial tree, even under extreme conditions, mango tree would survive and bounce back in the succeeding growth season. Mature trees can withstand −4 °C temperature for a few hours with a limited damage. However, juvenile trees were killed after 13 h at −4 to −6 °C (Campbell et al. 1977). Hence, in mango, various phenological events are strongly under the influence of environmental conditions.

10.2.2 Floral Initiation and Development

Temperature is an important environmental stimulation for mango flowering, and the effect of cool temperature on flowering is independent of photoperiod (Nunez-Elisea and Davenport 1995). Low temperatures have been shown to promote reproductive morphogenesis in mango. The floral induction which is driven through the floral promoter is temperature dependent (Davenport 2007), and this floral induction is triggered by temperatures below 16 °C (Schaffer et al. 1994). Cool night temperatures (15 °C) followed by <20 °C day temperatures are observed to strongly influence the developmental fate of mango buds (Ou 1980, 1982). Whiley et al. (1989, 1991), in their studies conducted with both mono- and polyembryonic cultivars, observed vegetative induction at 30 °C day and 25 °C night temperatures and floral induction at 15 °C day and 10 °C night temperatures. The extent of flowering is influenced by the prevailing low temperatures. The flowering in axillary buds of 'Haden' increased from 18 % to 100 % when trees were transferred to temperatures of 31/25 °C following 1–3 weeks at 19/13 °C (Shu and Sheen 1987). 'Tommy Atkins' trees flowered within 10 weeks when held at day/night temperatures of 18/10 °C, whereas trees held at 30/25 °C produced vegetative growth and did not flower (Nunez-Elisea et al. 1993).

The temperatures lower than 15 °C cause the inhibition of shoot growth but are favourable for panicle growth. Whiley et al. (1989) and Pongsomboon (1991) reported that panicle growth and development is sustainable at temperatures lower than 15 °C. They separately observed the growth of panicles occurring at base temperatures of 12.5 °C, though no vegetative shoots were produced. Though the optimum temperature for growth is in the range of 24–30 °C, the base temperatures for panicle growth and shoot growth are 12.5 and 15 °C, respectively. The ratio between male and hermaphrodite flowers present on the panicles is temperature dependent. Singh et al. (1966) observed that the panicles that emerged late on Dashehari trees had up to seven times the percentage of perfect flowers than the early emerging panicles on the same trees, when the mean maximum and minimum temperatures at the time of emerge were

20.7/3.7 and 27.3/13.1 °C for early and late emerging panicles, respectively. The percent hermaphrodite flowers in some south Indian cultivars was much less when they were grown in northern India due to the lower maximum and minimum temperatures during the period of panicle development (Singh et al. 1965). The percentage of hermaphrodite flowers in south Indian cultivars was greatest in late emerging panicles which coincided with higher temperatures (Ramaswamy and Vijayakumar 1992). Panicles which developed during December to January, when the prevailing temperatures were comparatively low, usually had higher proportion of male flowers (Singh et al. 1974), thus indicating that the proportion of hermaphrodite flowers increase at higher temperatures. It is also evident that any climatic variability would influence the variations in vegetative and reproductive cycles resulting in adverse effects on the productivity.

The low temperatures have the adverse effects on the pollen development and viability. In cv. 'Kensington', the most sensitive phase to the degree and duration of temperature stress was that from meiosis to the pre-vacuolate microspore (about 3-day duration at 25/20 °C) though vacuolated microspores were also sensitive to low temperature. Pollen viability of <50 % was observed at night temperatures below 10 °C. The temperature between 15 and 33 °C during the same phase was found to be optimum for pollen development with 70–85 % pollen viability. A linear negative correlation between pollen viability percentage and days with mean night temperatures lower than 10 °C during the above sensitive phase under field conditions was observed (Issarakraisila and Considine 1994). Low temperatures (diurnal maximum temperature <20 °C) caused reduction in pollen viability and pollen tube growth consequently low fertilisation rate in cv. Tainong 1 (Huang et al. 2010). A 60 % reduction in style length occurred at daily temperatures <15 °C (Issarakraisila et al. 1992). Young (1955) observed that viability of pollen was poor at the beginning of anthesis when field temperatures were low (<4.4 °C) but improved substantially mid-bloom. The extreme temperature events cause poor pollination and fertilisation, and studies have shown the development of parthenocarpic fruits. Field studies in Mexico in cv. 'Haden' demonstrated that the development of parthenocarpic fruit was related to low temperatures during fruit set (Lakshminarayana and Aguilar 1975). The occurrence of parthenocarpic fruit in mango cultivars has also been recorded in Israel when temperatures reached 44 °C at a critical stage of fruit development.

10.2.3 Photosynthesis

In mango, temperatures not only inhibit the vegetative growth but cause considerable reductions in photosynthesis rates. In 'Kensington', temperature is known to have considerable influence on photosynthesis (Isarakraisila et al. 1991). Exposure of container-grown 'Nam Doc Mai' trees to temperatures of 15/10 °C, day and night temperatures, affected the photosynthesis rates (Pongsomboon 1991), thus suggesting that the temperatures below 15 °C not only affect shoot growth but also the photosynthesis rates. Whiley et al. (1999) observed that the low temperatures caused the reduction in light saturation from 1,284 μmol quanta m^{-2} s^{-1} to 1,180 μmol quanta m^{-2} s^{-1} and also the reduction in the quantum efficiency from 0.042 μmol CO_2 μmol quanta m^{-2} s^{-1} to 0.025 μmol CO_2 μmol quanta m^{-2} s^{-1}. The rate of photosynthesis at saturating photosynthetic photon flux also decreased from 15.2 μmol CO_2 m^{-2} s^{-1} to 8.8 μmol CO_2 m^{-2} s^{-1}.

The low-temperature effects on photoinhibition have been studied by chlorophyll fluorescence measurements. In 'Kensington' though the leaves recorded average Fv/Fm of 0.80 in summer, it was considerably reduced during winter. The photochemical conversion efficiency of PS II of mango is dependent on the embryony of cultivars. The polyembryonic cultivars, at temperatures below 10 °C during winter, had lower Fv/Fm ratios (0.54–0.60) compared to monoembryonic cultivars (0.65–0.71), probably due to different centres of evolution

(Whiley and Schaffer 1997). The previous night temperature plays an important role in causing photoinhibition in mango. Nir et al. (1997) observed that the minimum night air temperature of 2–15 °C followed by the midday temperature of 24–30 °C reduced maximum quantum efficiency of PSII (F_v/F_m). It was also observed that the chilling night-induced photoinhibition of PSII was accompanied by reduction in both photosynthesis and stomatal conductance. Thus, the low temperatures cause the considerable reductions in photosynthesis and photoinhibition.

10.3 Water Stress

Most of the moisture requirement of mango is met from the rainfall in many growing regions. As it is known to survive without rain or irrigation for more than 8 months (Gandhi 1955), it is considered as relatively drought tolerant (Singh 1977; Chacko 1986). The evenly distributed and average annual rainfall of 635 mm is essential for economic production of mango (Chacko 1986). However, water stress is known to cause current season's reductions in vegetative growth and photosynthesis (Yan and Chen 1980). Several studies have emphasised the importance of irrigation for better growth, fruit retention, fruit size and higher yields. Azzouz et al. (1977) reported that in Egypt fruit yield and size in field-grown 'Hindi' mangos increased with more frequent irrigation. In South Florida field, 'Tommy Atkins' trees grown with 7 or 14 irrigation showed that irrigation at 7-day intervals resulted in the greatest yield with the largest fruit. And fruit size was increased during the early period of harvest (Larson et al. 1989).

10.3.1 Flowering, Fruit Set and Development

In mango, mature leaves produce floral stimulus, whereas immature leaves are rich source of floral inhibitors. Water stress may promote flowering in mango by increasing the total amount of floral stimulus produced by the canopy. In tropical climates, the floral induction occurs in the presence of minimum daily temperatures greater than 20 °C, which are otherwise considered conducive to the expression of vegetative growth. Under tropical conditions, flowering occurs after a brief period of water stress (Scholefield et al. 1986). Flowering is also observed to be early in water-stressed trees than unstressed ones. The primary impact of water stress is to prevent vegetative flushing during stress period, which may provide more time for accumulation of floral stimulus (Schaffer et al. 1994). Several studies have also shown that the water stress advanced floral bud break by nearly 2 weeks, promoted growth of floral buds and delayed the development of vegetative buds (Whiley 1986; Nunez-Elisea and Davenport 1994; Schaffer et al. 1994; Chaikiattiyos et al. 1994).

Water stress has considerable influence on the fruit set, its retention and subsequent growth. Singh and Arora (1965) studied fruit drop in field-grown 'Dashehari' under weekly and three weekly intervals. During the first 6 weeks, weekly irrigation reduced fruit drop compared with three weekly irrigation treatments. However, there was greater fruit drop in the weekly irrigation treatment during the latter stages of fruit development. Thus, providing irrigation during the early 4–6 weeks following the fruit set becomes important for better fruit size and yield, as at this stage cell division and cell wall development take place (Whiley and Schaffer 1997). The slight reductions in plant water status at this stage would have a bearing on the growth and retention of fruits (Pongsomboon 1991). Thus, in areas where the prolonged dry seasons occur during flowering and fruiting period, irrigation is very essential to ameliorate the adverse effects of drought stress (Whiley and Schaffer 1997). Studies have also been attempted to quantify the effects of different levels of irrigation through improved methods under water-limiting conditions. Spreer et al. (2009) conducted a study to assess the response of mango trees to varying amounts of available water with four irrigation treatments, CO (100 % of ETc), regulated deficit irrigation (RDI) (50 % of ETc), partial root-zone drying (PRD) (50 % of ETc) and no irrigation

(NI). They observed that over 4 years, the average yield in different irrigation treatments was 83.35 kg/tree (CO), 80.16 kg/tree (RDI), 80.85 kg/tree (PRD) and 66.1 kg/tree (NI). Water-use efficiency (WUE) calculated as yield per volume of irrigation water was always significantly higher in the deficit irrigation treatments as compared to the control. The study indicated that in normal years, the yields of the two deficit irrigation treatments (RDI and PRD) did not differ significantly, while in a dry year, yields under PRD were higher than under RDI. And in a year with early rainfall, RDI yields were more than PRD. In PRD-irrigated trees, mangoes had a bigger average fruit size and a more favourable fruit size distribution during all the years. Thus, the study concluded that deficit irrigation strategies can save considerable amounts of water without affecting the yield to a large extent, possibly increasing the average fruit weight, apparently without negative long-term effects. The drought stress caused a great reduction in the emergence of vegetative flushes during the stress period. Number of leaves per flush, flush length and weight, leaf water contents and root growth were also reduced due to drought stress. A number of malformed panicles and percentage of malformation were minimised in the plants in which drought stress was imposed. Complete drought stress proved better to reduce the intensity of malformation than partial drought stress (Tahir et al. 2003).

10.3.2 Water Relations

Water deficit is one of the factors that cause detrimental effects on mango production. Studies on the onset of water deficit in containerised 'Kensington' showed that the loss of turgor (wilting) occurred in young expanding leaves at water potential (Ψ_l) −1.2 MPa, while mature leaves did not lose turgor until Ψ_l of −1.75 MPa. Permanent damage to oldest leaves (necrotic areas around the midrib) occurred at Ψ_l −3.2 Mpa. At −3.45 Mpa Ψ_l, the damage was nonreversible and it corresponded to 77 % relative water content (Pongsomboon and Whiley 1992). The critical value of leaf water potential above which permanent wilting occurs is −3.50 MPa (Pongsomboon 1991). During the drying cycle, osmotic potential declined by about −1.2 Mpa, indicating osmotic adjustment occurred as water deficit developed. And the osmotic adjustment was attributed to active rather than passive changes, which were probably mediated through the latex. A linear correlation between gs and Ψ_l during the development of water stress in 'Kensington'-containerised mango trees was observed. Reich and Borchert (1988) in the studies with five tropical tree species observed that stomatal response in mango significantly reduces the rate of development of internal water deficit compared with the other species. The radial expansion of mango trunks continued when most of the other tree species were shrinking. Yan and Chen (1980) observed that in container-grown mango trees, stomatal conductance was sensitive to water deficit and declined as water potential of the leaf decreased. In cv. 'Kensington', even under normal water application, stomatal conductance was sensitive to increases in VPD (Pongsomboon 1991). Water stress imposed by withholding irrigation for 15 days in seedlings of cvs. Choke Anand and Khieo Sawoei resulted in progressive decreases in leaf relative water content, net photosynthesis and stomatal conductance. The water stress effect was more pronounced in full sunlight than 50 % sunlight. It caused less pronounced inhibition of photosynthesis in cv. Choke Anand than in cv. Khieo Sawoei under both light regimes. After rehydration, the recovery was also relatively quicker in cv. Choke Anand than in cv. Khieo Sawoei. Complete recovery in photochemistry was observed after 5 days of rewatering, but photosynthesis did not show complete recovery (Elsheery and Cao 2008). The increasing water stress caused reductions in stomatal conductance, transpiration rate, photosynthesis rate and intercellular CO_2 concentration in 6-month-old mango rootstock seedlings (Luvaha et al. 2007). In container-grown mango rootstocks, imposition of water stress by withholding irrigation for 7–18 days, it was observed that the rootstocks Starch, Peach and

Kensington reached photosynthesis rates to negligible levels by seventh, eighth and ninth day, respectively. On rewatering, Starch recovered in 1 day, Peach in 3 days and Kensington in 4 days. Though the rootstock, Mylepelian, took 16 days to reach the negligible photosynthesis rates, it recovers in just 2 days after rewatering, thus showing differential responses to water stress (Laxman 2015). Hence, information on the response of different rootstocks to moisture stress is available, and it should be used for developing adaptation options.

10.4 Flooding

Mango is considered to be moderately flood-tolerant species (Schaffer et al. 1994), and many mango-growing areas experience periodic waterlogging and flooding situations (Whiley and Schaffer 1997). The events of excess rainfall would cause stagnation of water in mango orchards and trees experience flooding stress. Flooding affects growth and gas exchange and mango trees have developed mechanisms to overcome the anoxic conditions.

10.4.1 Growth

Anaerobic soil conditions, due to flooding, caused decline in vegetative growth of mango trees (Larson et al. 1989). A small percentage of containerised mango trees died after roots were submerged in water for 10 days, but no additional death occurred after roots were submerged for up to 110 days. Vegetative growth of mango trees generally declines if trees are flooded for more than 2–3 days. Containerised trees growing in limestone soil, when flooded for more than 110 days, showed 94 % reduction in shoot extension growth, while flooding for approximately 10 days resulted in 57 % reduction in shoot extension growth (Larson et al. 1991a). When 2-year-old mango cv. Langra trees were subjected to waterlogging stress for 55 days flowering occurred (Kohli and Reddy 1985). The effect of flood irrigation during fruit bud differentiation on flowering of 19-year-old trees of mango cv. Alphonso showed that the percentage of flowering and vegetative shoots was 24 and 16.3 % in irrigated trees while it was 16 and 23 % in control. The percentage of dormant shoots was more or less the same, in irrigated and control trees. Withholding water or flood irrigation during fruit bud differentiation has no effect on flowering in mango cv. Alphonso (Anon 2000). Flooding for more than 14 days significantly reduced root dry weight, resulting in an increased shoot to root ratio (Larson et al. 1991a).

10.4.2 Gas Exchange

Physiological responses of mango within 2–3 days after flooding include decrease in net CO_2 assimilation, stomatal conductance and transpiration rate (Larson et al. 1991a; Schaffer et al. 1992). Net CO_2 assimilation, stomatal conductance and transpiration rate declined, and internal CO_2 concentration increased after 2 days of flooding in container-grown mango cv. Tommy Atkins trees (Larson et al. 1991a). It was also observed that the effect of short-term flooding on leaf gas exchange is reversible. Gradual increase in net CO_2 assimilation, stomatal conductance and transpiration rate was observed in trees which were removed from flooding conditions after 28 days. Although prolonged flooding inhibits gas exchange of mango, trees subjected to short durations of 1–7-day flooding can exhibit increased stomatal conductance and net photosynthesis (Larson et al. 1992). This may in part be due to increased availability of micronutrients in flooded soils. In calcareous soils of Florida, Fe and Mn have been observed to be 10–50 times more extractable as a result of flooding (Larson et al. 1991b). Leaf chlorophyll content also significantly increased when Fe-deficient trees were flooded. And the flooding increased Mn content in mango leaves regardless of the Fe status of the plant prior to flooding. Thus, the increased leaf gas exchange rate may be due to, at least in part, increased Mn and chlorophyll concentrations in flooded trees. Whiley and Schaffer (1997) concluded that the reduced net CO_2 assimilation rates and presumably higher root respiration rates under flooded conditions limit the availability of carbon-based assimilates required for growth.

10.4.3 Anatomical and Morphological Changes

Mango trees develop hypertrophic (swollen) lenticels on stems immediately above the waterline to survive prolonged flooding stress (Whiley and Schaffer 1997). Mango trees which did not exhibit lenticel hypertrophy died within a few days of flooding, and the trees that survived stem lenticels hypertrophied within 10 days (Larson et al. 1991c, 1993). When these lenticels were sealed with petroleum jelly or silicone grease, within 3 days trees died. The initial stages of lenticel hypertrophy in mango are characterised by the development of intercellular spaces in the phellem tissue and production of additional phellem tissue by increased phellogen activity (Larson et al. 1991a). Hypertrophic lenticels may serve as excretory sites for the elimination of potentially toxic compounds such as ethanol, acetaldehyde and ethylene produced in roots under flooded condition due to anaerobic metabolism (Larson et al. 1993; Whiley and Schaffer 1997). The studies in several wood species indicate that the increased intercellular spaces may function as organs for increased oxygen absorption and transport to the roots (Kozlowski 1984).

Hypertrophic lenticels may confer flood tolerance through enhanced O_2 diffusion to roots (Whiley and Schaffer 1997). Schaffer et al. (2006) opined that the observations on the hypertrophied stem lenticels' development could be a useful technique for screening and selection of flood-tolerant mango cultivars. Floodwater temperature affects the development of hypertrophic lenticels. When temperature was 15 °C, hypertrophic lenticels did not develop, whereas the trees flooded in water at 22.5 and 30 °C developed. However, the number increased more rapidly for trees submerged in 30 °C than trees submerged in 22.5 °C floodwater. The inhibition of lenticel hypertrophy as 15 °C appears to be due to temperature-mediated reductions in respiration and metabolic processes (Larson et al. 1991c). Mango trees maintained in highly oxygenated water and given exogenous ethylene developed many hypertrophic lenticels, whereas plants in the highly oxygenated water without exogenous ethylene developed fewer or no hypertrophic lenticels. Thus, it appears that ethylene plays a role in promotion of stem lenticel hypertrophy in flooded mango trees, and floodwater oxygen content can regulate stem lenticel hypertrophy and ethylene evolution in mango (Larson et al. 1993). Schaffer et al. (1994) observed development of adventitious roots above the waterline in container-grown mango trees when flooded for long periods. However, it is not observed in field-grown mango trees where flooding periods to the extent of longer than 1 week do not occur under normal production conditions (Whiley and Schaffer 1997).

10.5 Salinity

The higher amount of chlorides and sulphates of the calcium, magnesium and sodium in salt-affected soils is not tolerated by most of the crop plants. The large amounts of dissolved salts in the soil solutions cause both osmotic stress to the roots and also toxicity when taken up by the plants. Mango is classified as sensitive species for salinity as salt effects have been observed at EC lower than 1 mS cm^{-1} (Eber 2000). For profitable mango cultivation, the saline conditions are not favourable (Majumder and Sharma 1985). The toxic ions which accumulate in mango leaves have been observed to cause nutrient imbalance, reduce uptake of major nutrients, cause injury to leaf, inhibit growth, fruit bearing and ultimately yield.

10.5.1 Growth

The adverse effects of salinity on physiology of plants would ultimately lead to growth suppression. Typically growth decreases more or less linearly as salinity increases beyond a threshold level and the effect is similar for a range of different salts. The salt ions accumulate in toxic amounts in leaves of mango which in turn inhibit growth, cause leaf injury and nutrient imbalance and reduce uptake of major nutrients. Salt toxicity is known to limit mango productivity in arid environments. The increasing levels of salinity caused reductions in growth characters like height, number of leaves, leaf area and stem thickness (Ahmed and Ahmed 1997). The mango ecotypes, mono- and polyembryonic cultivars, show considerable variations in their response to salinity stress. Studies show that the polyembryonic genotypes appear to have greater tolerance to salinity compared to monoembryonic types (Jindal et al. 1975; Kadman et al. 1976). Polyembryonic mango genotype, Kurakkan, is reported to have salt tolerance, and the salt tolerance level of genotypes is also variable. Leaf scorching after 6 months and subsequent death occurred when mono- and polyembryonic cultivars were irrigated for 2 years with 10 m mol l^{-1} of chloride (Kadman et al. 1976). Some seedlings which did not show damage or showed only slight toxicity symptoms were mostly from 13-1 polyembryonic cultivar or related types (Schmutz and Ludders 1993).

10.6 Gas Exchange

Salt stress affects physiological parameters, viz. leaf water potential, stomatal conductance and photosynthesis. Schmutz (2000) in a greenhouse experiment observed the decline in maximum photosynthesis after 6 days from the beginning of the application of 15 mM NaCl for 3 days followed by 30 mM NaCl for 3 days. The CO_2 assimilation and transpiration decreased in grafted mango plants due to 30 mM NaCl salinity. The reduction was highest in the Turpentine grated on 13-1 rootstock combination (Schmutz and Ludders 1999).

10.7 Mineral Uptake and Nutrition Imbalance

The salinity level more than 0.2 % adversely influenced the uptake of nutrients (Ahmed and Ahmed 1997). Mango generally accumulated 2.5–3.0 times more sodium than other species in both old and young leaves (Samra 1985). Higher chloride ion concentration in soil caused increased reduction in nitrogen content of mango leaves than by SO_4 ions (Jindal et al. 1979a, b), which may be due to their specific effect on inhibition of NO_3 absorption and higher absorption of ammonium. The uptake of N, P, K, Ca, Mg, Zn and Fe was adversely affected with increasing levels of salinity; after 29 days of NaCl treatment, leaf sodium and chloride concentrations were higher in 13-1 than Turpentine (Schmutz and Ludders 1994). Lower Na and Cl contents but higher K content in roots and higher Ca and Mg contents in leaves were observed on *Mangifera zeylanica* compared to *Mangifera indica* cv. 13-1. The higher K/Na ratio was found in all plant parts especially in roots of *Mangifera zeylanica* (Schmutz and Ludders 1999). The increase salinity of irrigation water from 0.7 to 5.7 dS m^{-1} caused the reductions in N, K, Ca and Mg contents in leaves but did not affect contents of P and S. It is observed that some rootstocks perform better under saline conditions. The rootstock, Espada, had higher contents of Na and P and lower Ca compared to Manguito. However, the contents of Na and Na/K ratio in both cultivars increased with increase in water salinity, and it was found to be higher in rootstock Espada (Silva et al. 2004).

Several rootstock cultivars have exhibited differential tolerance to salinity stress. Differences in accumulation of Na and Cl were observed in two mango rootstocks, Gomera-1 and Gomera-3 (Duran Zuazo et al. 2003). Gomera-3 accumulated higher amounts of Cl and Na and was more sensitive than Gomera-1. Gomera-1 was tolerant, due to its capacity to restrict the uptake and transport of Cl and Na ions from root system to the above-ground parts. Dubey et al. (2006, 2007) and Srivastav et al. (2009) studied salt tolerance limit of different polyembryonic mango rootstocks and reported that Olour and Kurukkan were tolerant to salt stress. The polyembryonic rootstock cvs. Bappakai and Olour (Anonymous 1989) have been identified as moderately tolerant to salt stress. The chloride concentration in tolerant seedlings was higher compared to susceptible seedlings, indicating the physiological tolerance to chloride concentrations in leaf tissues. But the sodium tolerance shown by 13-1 is due to the exclusion of sodium from shoots and accumulation in vacuoles of root cells (Schmutz and Ludders 1993). The highest reduction in shoot growth and the highest leaf Cl contents were also observed in the same rootstock combination. The study showed that 13-1 was not able to protect the salt-sensitive scion Turpentine from the adverse effects of salinity (Schmutz and Ludders 1999). Hence, making use of the tolerant rootstocks in salt-affected soils could be one of the strategies to overcome the adverse effects on growth and yield of the preferred mango cultivars.

10.8 Conclusions

Mango being a tropical fruit tree is being cultivated in both tropical and subtropical conditions, and the diverse soil, rainfall and temperatures prevailing in these regions play an important role in realising the potential yields. Due to the perennial nature and manifestation of vegetative and reproductive phases in flushes, mango tree is subjected to many abiotic stresses. The dominant influence of low temperature and its requirement for flowering induction determine the extent of flowering and yield. Though the role of low-temperature requirement is very well understood through many basic studies, under climate change conditions, the mango crop is likely to experience different temperature regimes. The variability in precipitation patterns is also likely to subject the mango orchards to experience limited and/or excess water stress situations at different phenological stages. The soil salinisation caused due to indiscriminate use of water is likely to be aggravated under high evaporative demands posed by increasing temperatures under changing climatic conditions. Thus, under climate change conditions, abiotic stresses like extreme temperatures, excess and limited moisture and salinity stresses are likely to increase in periodicity and duration. Hence, the efforts should be focused on monitoring the incidence of extreme events and emerging climate scenarios to identify the abiotic stresses likely to occur individually or in combination in mango-growing regions. Though the existing knowledge identifies the extent of damage caused by various abiotic stresses, further studies are needed to quantify the likely impacts and also develop adaptation strategies for sustaining yields.

References

Ahmed AM, Ahmed FF (1997) Effect of saline water irrigation and cycocel on growth and uptake of some elements of Taimour and Alphonso mango seedlings. Ann Agric Sci Moshtohor 35(2):901–908

Anonymous (1989) News, Indian Institute of Horticultural Research 10(4): 1–3

Anonymous (2000) Annual report, Indian Institute of Horticultural Research, pp 19–20

Azzouz S, El-Nokrashy MA, Dahshan IM (1977) Effect of frequency of irrigation on tree production and fruit quality of mango. Agric Res Rev 55:59

Campbell CW, Knight RJ, Zareski NL (1977) Freeze damage to tropical fruits in southern Florida in 1977. Fla State Hortic Soc 90:254–257

Chacko EK (1986) Physiology of vegetative and reproductive growth in mango (*Mangifera indica* L.) trees. In: Proceedings of 1st Australian Mango Research Workshop, Cairns, QLD, Australia. CSIRO, Melbourne, pp 54–70

Chaikiattiyos S, Menzel CM, Rasmusssen TS (1994) Florida induction in tropical fruit trees, effect of temperature and water supply. J Hortic Sci 69:397–415

Corbineau F, Kante M, Come D (1986) Seed germination and seedling development in the mango (*Mangifera indica* L.). Tree Physiol 1:151

Crane JH, Bally I, Mosqueda-Vaquez RV, Tomer E (1997) Crop production. In: Litz RE (ed) The mango. CAB International, Wallingford, pp 203–256

da Silva JM, Gheyi HR, Fernandes PD, de Oliveira FHT, Soares FAL (2004) Macro and micro nutrient and sodium contents in leaves of mango rootstock with

different levels of salinity. Proc Interam Soc Trop 47:213–217
Davenport TL (2007) Reproductive physiology of mango. Braz J Plant Physiol 19(4):363–376
Dubey AK, Srivastav M, Singh R, Pandey RN, Deshmukh PS (2006) Response of mango (*Mangifera indica*) genotypes to graded levels of salt stress. Indian J Agric Sci 76:670–672
Dubey AK, Srivastav M, Sharma YK, Pandey RN, Deshmukh PS (2007) Dry mass production and distribution of nutrients in two mango rootstocks as affected by salinity. Indian J Hortic 64:385–390
Duran-Zuazo VH, Martinez-Raya A, Aguilar Ruiz J (2003) Salt tolerance of mango rootstock. Span J Agric Res 1:67–78
Eber G (2000) Salinity problems in sub-tropical fruit production. Acta Hortic 531(99):105
Elsheery NI, Cao KF (2008) Gas exchange, chlorophyll fluorescence, and osmotic adjustment in two mango cultivars under drought stress. Acta Physiol Plant 30:769–777
Galen Sauco V (1997) Mango world production (outside Israel, Egypt and India). Acta Hortic 455:15–22
Gandhi SR (1955) The mango in India. Farm Bull 6 Indian Council Agric Res
Huang JH, Ma WH, Liang LY, Zhang WX, Wang ZJ, Cai ZJ, Wen SX (2010) Effects of low temperature on sexual reproduction of Tainong 1 mango. Sci Hortic 126(1):109–119
Isarakraisila M, Considine JA, Turner DW (1991) Pattern of vegetative and reproductive growth of mango trees in a warm temperature region of Western Australia. Acta Hortic 291:188
Issarakraisila M, Considine JA, Turner DW (1992) Seasonal effects on floral biology and fruit set of mangoes in a warm temperature region of Western Australia. Acta Hortic 321:626
Issarakraisila M, Considine JA (1994) Effects of temperature on pollen viability in mango cv. 'Kensington'. Ann Bot 73:231–240
Jindal PC, Singh JP, Gupta OP (1975) Screening of mango seedlings for salt tolerance. Haryana J Hortic Sci 4:112–115
Jindal PC, Singh JP, Gupta OP (1979a) Effect of salinity on the mineral nutrients in mango seedlings. Indian J Agric Sci 49:105–109
Jindal PC, Singh JP, Gupta OP (1979b) Mineral nutrient status of different mango cultivars under varying salt status in soil. Haryana J Hortic Sci 8:39–41
Kadman A, Gazit S, Ziv G (1976) Selection of mango rootstocks for adverse water and soil conditions in arid areas. Acta Hortic 57:81–88
Kohli RR, Reddy YTN (1985) Effect of stress in induction of flowering in grafted mango plants. Indian J Hortic 42(1–2):41–44
Kozlowski TT (1984) Responses of woody plants to flooding. In: Kozlowski TT (ed) Flooding and plant growth. Academic, Orlando, pp 129–163
Lakshminarayana S, Aguilar PH (1975) Effect of orchard heating in reducing parthenocarpic fruits in 'Haden' mango. Fla State Hortic Soc 88:502
Larson KD, Schaffer B, Davies FS (1989) Effect of irrigation on leaf water potential, growth and yield of mango trees. Proc Fla State Hortic Soc 102:226–228
Larson KD, Schaffer B, Davies FS (1991a) Flooding leaf gas exchange and growth of mango trees in containers. J Am Soc Hortic Sci 116:156
Larson KD, Graetz DA, Schaffer B (1991b) Flood-induced chemical transformation in calcareous agricultural soils of South Florida. Soil Sci 152:33
Larson KD, Davies FS, Schaffer B (1991c) Floodwater temperature and stem lenticel hypertrophy of *Mangifera indica* L. Am J Bot 78:1397–1403
Larson KD, Schaffer B, Davies FS, Sanchez CA (1992) Flooding, mineral nutrition and net gas exchange of mango trees. Sci Hortic 52:113–124
Larson KD, Schaffer B, Davies FS (1993) Floodwater oxygen content, lenticel hypertrophy, and ethylene evolution from mango trees. J Exp Bot 44:665–671
Laxman RH (2015) Unpublished data. ICAR-Indian Institute of Horticultural Research, Bengaluru
Litz RE (2009) The mango, 2nd edn, Botany, Production and Uses. CABI International, UK
Luvaha E, Netondo GW, Ouma G (2007) Physiological responses of mango (*Mangifera indica*) rootstock seedlings to water stress. J Agric Biol Sci 2(4–5):6–12
Majumder PK, Sharma DK (1985) Mango. In: Bose TK (ed) Fruits of India: tropical and sub-tropical. NayaPrakash Publishers, Calcutta, pp 69–123
Mukherjee SK (1953) The Mango-its botany, cultivation, uses, and future improvement, especially as observed in India. Econ Bot 7:130
Mukherjee SK (1972) Origin of mango (*Mangifera indica*). Econ Bot 26:260
Nir G, Ratner K, Gussakovsky E, Shahak Y (1997) Photoinhibition of photosynthesis in mango leaves: effect of chilly nights. Acta Hortic 455:288–235
Nunez-Elisea R, Davenport TL (1994) Flowering of mango trees in containers as influenced by seasonal temperature and water stress. Sci Hortic 58:57–66
Nunez-Elisea R, Davenport TL (1995) Effect of leaf age, duration of cool temperature treatment, and photoperiod on bud dormancy release and floral initiation in mango. Sci Hortic 62:63–73
Nunez-Elisea R, Davenport TL, Caldeira ML (1993) Bud initiation and morphogenesis in 'Tommy Atkins' mango as affected by temperature and triazole growth retardants. Acta Hortic 341:192
Ou SK (1980) Temperature effect on flowering and fruiting in the 'Irwin' mango (Mangifera indica L.). J Agric Res China 29:301–308
Ou SK (1982) Temperature effect on differential shoot development of mango during flowering period. J Agric Res China 31:209–212
Pongsomboon W (1991) Effect of temperature and water stress on tree growth. PhD thesis Kasetsart University, Bangkok

Pongsomboon, Whiley (1992) Development of water stress and stomatal closure in juvenile mango (*Mangifera indica* L.) stress. Acta Hortic 321:496

Ramaswamy N, Vijayakumar M (1992) Studies of the effects of flowering and fruiting behaviour of South Indian mango cultivars in Abstract IV. International Mango Symposium, Miami Beach, p 47

Reich PB, Borchert R (1988) Changes with leaf age in stomatal function and water status of several tropical tree species. Biotropica 20:60

Samra JS (1985) Comparative sodium accumulation and its toxicity in mango, guava and ber. Indian J Hortic 42 (3/4):178–183

Schaffer B, Anderson PC, Ploetz RC (1992) Response of fruit crops to flooding. Hortic Rev 13:257–313

Schaffer B, Whiley AW, Crane JH (1994) Mango. In: Schaffer B, Andersen PC (eds) Handbook of environmental physiology of fruit crops, vol 2, Sub tropical and tropical crops. CRC Press, Boca Raton, pp 165–197

Schaffer B, Davies FS, Crane JH (2006) Responses of subtropical and tropical fruit trees to flooding in calcareous soil. Hortic Sci 41(3):549–555

Schmutz U (2000) Effect of salt stress (NaCl) on whole plant CO_2 gas exchange in mango. Acta Hortic 609:269–276

Schmutz U, Ludders P (1993) Physiology of saline stress in one mango rootstock. Acta Hortic 341:160–167

Schmutz U, Ludders P (1994) Salt tolerance in mango (*Mangifera indica* L.) rootstocks. Angew Bot Ber 5:189–193

Schmutz U, Ludders P (1999) Effect of NaCl salinity on growth, leaf gas exchange and mineral composition of grafted mango rootstocks (var. '13-1' and 'Turpentine'). Gartenbauwissenschaft 64:60–64

Scholefield PB, Oag DR, Sedgley M (1986) The relationship between vegetative and reproductive development in mango in Northern Australia. Aust J Agric Res 37:425–433

Shu ZH, Sheen TF (1987) Floral induction in axillary buds of mango (*Mangifera indica* L.) as affected by temperature. Sci Hortic 31:81

Singh LB (1976) Mango. In: Simmonds NW (ed) Evolution of crop plants. Longman, London, pp 7–9

Singh LB (1977) Mango. In: de T Alvim P, Kozlowski TT (eds) Ecophysiology of tropical crops. Academic, New York, pp 479–485

Singh RN, Arora KS (1965) Some factors affecting fruit drop in mango (*Mangifera indica* L.). Indian J Agric Sci 35:196

Singh RN, Mujumder PK, Sharma DK (1965) Studies on the bearing behaviour of some South Indian varieties of mango (*Mangifera indica* L.) under North Indian conditions. Trop Agric 42:171

Singh RN, Mujumder PK, Sharma DK (1966) Sex-expression in mango (*Mangifera indica* L.) with reference to prevailing temperature. Proc Am Soc Hortic Sci 89:228

Singh RN, Majumder PK, Sharma DK, Sinha GC, Bose PC (1974) Effect of de-blossoming on the productivity of mango. Sci Hortic 2:399–403

Spreer W, Ongprasert S, Hegele M (2009) Yield and fruit development in mango (*Mangifera indica* L. cv. Chok Anan) under different irrigation regimes. Agric Water Manag 96:574–584

Srivastav M, Dubey AK, Singh AK, Singh R, Pandey RN, Deshmukh PS (2009) Effect of salt stress on mortality, reduction in root growth and distribution of mineral nutrients in Kurukkan mango at nursery stage. Indian J Hortic 66:28–34

Subedi A, Bajracharya J, Joshi BK, Gupta SR, Regmi HN, Sthapit BR (2008) Locating and managing the mango (*Mangifera indica* L.) genetic resources in Nepal. PGR Newsl 155:52–61

Tahir FM, Ibrahim M, Hamid K (2003) Effect of drought stress on vegetative and reproductive growth behaviour of mango (*Mangifera indica* L.). Asian J Plant Sci 2:116–118

Vavilov NI (1926) Studies on the origin of cultivated plants. Bull Appl Bot XVI(2):139–248

Whiley AW (1986) Crop management review. Proc of the first Australian mango research workshop, Australia, Melbourne, pp 186–195

Whiley AW, Schaffer B (1997) Stress physiology. In: Litz RE (ed) The mango: botany, production and uses. CAB International, New York, pp 147–174, 198 Madison Avenue

Whiley AW, Rasmussen TS, Saranah JB, Wolstenholme BN (1989) Effect of temperature on growth, dry matter production and starch accumulation in ten mango (Mangifera indica L.) cultivars. J Hortic Sci Biotechnol 64:753–765

Whiley AW, Rasmussen TS, Wolstenholme BN, Saranah JB, Cull BW (1991) Interpretation of growth responses of some mango cultivars grown under controlled temperature. Acta Hortic 291:22–31

Whiley AW, Searl C, Schaffer B, Wolstenholme BN (1999) Cool orchard temperatures or growing trees in containers can inhibit leaf gas exchange of avocado and mango. J Am Soc Hortic Sci 124:46–51

Yan SS, Chen CY (1980) The effect of soil moisture on the photosynthesis, stomatal aperture and growth of mango (*Mangifera indica* L.). J Sci Eng 17:275

Young TW (1955) Influence of temperature on growth of mango pollen. Fla State Hortic Soc 68:308

Yusof IM, Buchanan DW, Gerber JF (1969) The response of avocado and mango to soil temperature. J Am Soc Hortic Sci 94:619

Grapes

11

Satisha Jogaiah

Abstract

In many of the semiarid and arid climatic regions where grape is largely grown on commercial scale, abiotic stresses such as soil and water salinity and water scarcity are the major constraints. Drought and salinity stress can cause a variety of symptoms common to other major stresses such as light, heat and nutrient deficiency, and the symptoms are very specific to time and geographical location. In grapevines there are several combinations of mechanisms which can help to tolerate most of these stresses. Since abiotic stress tolerance in grapes is controlled by multigenes, it is very difficult to understand the stress tolerance at molecular level. Poor vine growth and severe foliar damage due to excess salt accumulation coupled with drastic reduction in productive life span of own-rooted grapevines necessitated the use of rootstocks to combat these abiotic stresses also. Many of the grape rootstocks are known to possess drought- and salt-tolerant traits which can be seen on grafted scions through several mechanisms at both cellular and whole-plant levels. This chapter focuses on such mechanism of grapevines (directly by vines or indirectly by rootstocks) to overcome adverse situations of these stresses at morphological, physio-biochemical, nutritional and molecular level. In recent years, with scenario of climate change, some of the mechanisms adapted by grapevines to tolerate flooding stress are also reviewed.

11.1 Introduction

Semiarid tropical regions, where most of the grape cultivation is concentrated in the world, are subjected to frequent moisture stress which usually accompanies with soil salinity. In addition, frost and chilling injury also occurs in early

S. Jogaiah (✉)
Division of Fruit crops, ICAR-Indian Institute of Horticultural Research, Hessaraghatta Lake Post, Bengaluru 560089, Karnataka, India
e-mail: satilata@gmail.com; jsatisha@iihr.ernet.in

N.K.S. Rao et al. (eds.), *Abiotic Stress Physiology of Horticultural Crops*,
DOI 10.1007/978-81-322-2725-0_11

spring in most of temperature grape-growing countries. These stresses often limit grape production and productivity. Heat and light stresses are associated with moisture stress which imposes additional strain on vines. Thus, limited water availability, soil and water salinity, high temperature, high solar radiation and water logging are the major threats to grape cultivation in this order of priority which significantly limits grape cultivation in world scenario. Several field and pot experiments have been conducted for the last many decades all over the world by simulating the stress conditions to observe the morphological, physio-biochemical and molecular responses of grapevines to such stresses. Some of the common morphological responses are variation in development of leaf area and root development which balances canopy water requirement with respect to demand and supply. Thus, when vines are exposed to series of abiotic stresses, there will be series of physiological, morphological, biochemical and molecular changes happening in the system, which severely affects vine growth and productivity (Wang et al. 2005). Moisture, salinity, high-/low-temperature and oxidative stresses are often associated with each other and can express similar cellular damage.

Oxidative stress usually accompanies high temperature and drought and may cause denaturation of proteins (both structural and functional). Drought stress can cause a variety of symptoms common to other major stresses such as light, heat, high salt concentration and nutrient deficiency, and the symptoms are very specific to time and geographical location (Roy et al. 2011). In plants there are several combinations of mechanisms which can help to tolerate most of these stresses. At molecular level, there may be expression of some genes. Since abiotic stress tolerance in grapes is controlled by multigenes, it is very difficult to understand the stress tolerance at molecular level. Most of the area under grape cultivation in the major grape-growing regions of the world is grafted on rootstocks to combat several biotic and abiotic stresses. It has been also taken care to discuss the effect of rootstocks on grafted scion varieties (Sharma and Upadhyay 2004), which can tolerate to major abiotic stresses like drought and salinity which affect vine physiology at morphological, physio-biochemical and molecular level under such abiotic stresses.

11.2 Morphological Mechanisms During Moisture and Salt Stress

11.2.1 Moisture Stress

In grapes, water stress doesn't always imply negative effects. Regulation of water stress through several agronomic practices like the use of rootstocks, tillage and specialized irrigation techniques such as partial root zone drying (PRD) and regulated deficit irrigation (RDI) has been largely used to balance vine vegetative and reproductive growth with the aim to control berry quality especially in wine grapes. It is a well-known fact that the plants experiencing water stress undergo several morphological and physiological changes. In the case of grapes, its anatomy is well adapted to water deficit with xylem characteristics which can influence their capacity to transport water in the plant system. Grafting commercial varieties on rootstocks helps to overcome water-deficit effects. An efficient root system would be an advantage for grapevines as it can improve water and nutrient uptake and is one of the ways for exploring water resource from deeper soil strata.

Many studies have shown that grapevine's response to water stress involves decreased leaf expansion and inter-nodal elongation (Schultz and Matthews 1988; Cramer et al. 2007) and reduction in yield (Chaves and Oliveira 2004). In terms of drought tolerance, rootstocks are known to enable the growth of scion varieties and function normally when there is scarcity of water. Grapevines are considered as relatively tolerant to water stress, as they are known to possess larger xylem vessels in comparison to other crops (Comas et al. 2010; Serra et al. 2013). The larger xylem vessels may result in more xylem cavitation which can affect whole-plant hydraulic conductance which may be related to plant's mechanism to adapt to water stress conditions involving regulation of stomatal conductance (Ruggiero et al. 2012). The pattern of root development in a rootstock is also known to influence shoot growth pattern under moisture stress conditions. When vines sense water stress signals, rootstocks which are more vigorous and drought tolerant may exhibit rapid growth of root system in later period in wetter soil conditions, while rootstocks which are less vigorous and less drought tolerant would produce more roots in deeper layer of soils in the early growing season (Bauerle et al. 2008). The root spread is also influenced by soil water availability, type of irrigation, canopy management, trellis system and rootstocks. Serra et al. (2013) also showed that climatic conditions can also affect the pattern of new root growth. They also concluded that drought-tolerant grapevine rootstocks produced more new roots in different layers of soil during dry and hot season, allowing the grapevines to increase the uptake of water than drought-sensitive rootstocks. The influence of rootstocks on uptake of different nutrients is one of the attributes (Bavaresco and Lovesolo 2000; Ruhl 2000; Keller et al. 2001; Kodur et al. 2010) to influence shoot growth and vine productivity. In a 2-year-old potted grapevine cultivar Chardonnay/SO4 maintained at 35 and 90 % field capacity, it revealed that continuous water stress reduced vine growth, and more dry matter accumulated in roots. Soil water deficiency induced adaptive features in leaves like high epicuticular wax, increased prostrate hair density, small stomata and low average leaf area, preferential allocation of dry matter to the roots and few leaves on the lateral shoots. These changes were attributed to optimizing the use of assimilates and increased root/leaf ratio and water-use efficiency (Palliotti et al. 2001).

Soil moisture stress is known to influence grape leaves' stomatal number per unit area and stomatal pore size. Same scions grafted on different rootstocks can have different stomatal density and pore sizes (Serra et al. 2013). Significant differences were observed for stomatal density and size when Pinotage wine grapes are grafted onto different rootstocks, whereas vines grafted onto 140 Ru recorded lower stomatal density but bigger pore diameter than those grafted onto 110R and 1103P rootstocks. Some reports say that rootstocks which are found to be drought tolerant would have lower stomatal density in their leaves compared to drought-sensitive rootstocks (Scienza and Boselli 1981). The difference in hydraulic conductance of different rootstocks will affect whole vine water status, thereby affecting leaf growth, which in turn induces differences in stomatal density and pore size as suggested by Xu and Zhou (2008).

With the sensing of stress by roots, the cells respond through cell division, growth and differentiation and transmit some signals to shoots in general and leave in particular which results in reduction of shoot growth through stomatal closure by chemical signals. Although reduction in leaf and shoot growth in response to moisture stress is the earlier sign, the decrease in root growth is considerably lesser than the reduction in shoot growth (Dry et al. 2000). This results in increased root to shoot length ratio which provides sufficient water and nutrient to the shoots. Grapevine tolerance to drought is always limited to the ability to produce new roots where soil moisture is available (Morlat and Jacquet 1993). Moisture-stressed roots continue to grow into deeper strata of the soil in search of moisture, while the roots in the irrigated vines mostly proliferate in topper soil layer (Bauerle et al. 2008). Though increase in root length resulting in the root to shoot length ratio in one of the mechanisms of drought tolerance, inhibition of shoot growth is another mechanism to cope up with the moisture stress. This includes inhibition of internode elongation, leaf expansion, etc. (Hardie and Martin 2000; Schultz and Mathews 1988). Reduction in the vessel size and xylem hydraulic conductivity (Lovisolo et al. 2002) prevents excessive water loss and helps in preventing xylem embolism (Salleo and Lo Gullo 1993).

11.2.2 Salinity Stress

The use of salt-tolerant rootstocks is a proven practice to alleviate the possible adverse effects of salinity stress in major grape-growing regions of the world (Walker et al. 2002). But according to Mullins et al. (1996), salt tolerance has been mainly attributed with the ability of different cultivars and rootstocks and their compatibility in addition to different stock-scion combination to restrict Na or Cl entry into the shoot (Tregeagle et al. 2006). The ability of the rootstocks to tolerate salinity lies in the restricted uptake of sodium into the plant system or sequestration of sodium in vacuole or older leaves (Paranychianakis and Angelakis 2007). A lot of genetic variability has been reported for salt tolerance among grape cultivars as reported by many workers in the past (Pandey and Divate 1976; Upadhyay et al. 2013). Screening of several old-world grape varieties for salinity tolerance revealed (Nauriyal and Gupta 1967) Anab-e-Shahi as the more susceptible to salt injury than Cheema Sahebi and Bangalore Purple when grown in soil at 0.1, 0.3 and 0.5 % NaCl and Na_2SO_4 in soil solution. Alsaidi et al. (1987) reported that increasing salt concentration in the soil decreased the average shoot length, shoot diameter, number of roots per cuttings, length of roots and dry weight of leaves, stems and roots of cuttings of cvs. Abbasi and Kambli. They concluded that the cuttings of cv. Abbasi tolerated the EC of 9.75 mmhos cm^{-1} in the first season and 5.20 mmhos cm^{-1} in the second season, while the cuttings of cv. Kambli did not tolerate the saline conditions.

11.3 Physiological and Biochemical Mechanisms of Moisture and Salinity Stress

11.3.1 Moisture Stress

With the onset of moisture and salinity stress in vines, major physiological and biochemical processes such as photosynthesis, protein synthesis, respiration, utilization of energy, synthesis of secondary metabolites, etc., get affected. Physiological studies with respect to gas exchange parameters also revealed increased water-use efficiency at single-leaf level in rootstocks 110R, 99R and 1103P, belonging to *Vitis berlandieri* × *Vitis rupestris* crosses (Satisha and Prakash 2006; Satisha et al. 2006, 2007). In grapevines, drought tolerance during severe moisture stress is by several mechanisms such as stomatal closures, decrease of cell growth and photosynthesis and accumulation of osmolytes and proteins. Drought tolerance with maintaining high water potential in vine system includes reduction of water loss through transpiration and increase in uptake of water as suggested by Chaves and Oliveira (2004). Stomatal conductance is the degree of stomatal opening and the rate per unit area at which gases diffuses through stomata. The flux of CO_2 and water vapour are proportional to conductance of the stomata. Although Davies and Lakso (1978) have suggested that stomatal adjustment results from decrease in the threshold water potential for initiation of ABA synthesis, there is no exact support for their hypothesis. As in most crop plants, the relationship between photosynthesis and leaf water potential and/or relative water content in field-grown vines depends on conditions during plant growth and measurement. The reduction in photosynthesis due to reduced water availability is mainly due to stomatal closure which is associated with stomatal limitation during mild water stress and non-stomatal water limitation during severe water stress (Satisha and Prakash 2006; Lovisolo et al. 2010).

Maintenance of turgor is the key element for cell enlargement and growth of vines. Turgor is also an important process for opening of stomata (Kramer and Boyer 1995). The maintenance of turgor in water-stressed plants is driven by a process called as osmotic adjustment, the phenomenon by which some of the solutes, usually referred to as compatible solutes, start accumulating in growing cells as their water potential reduces drastically (Morgan 1984; Neumann et al. 1988). The osmotic potential arising from the net accumulation of solutes thus helps in maintaining turgor of tissues which allows growth of cells and tissues continuously even under low water potential. Some of the compatible solutes reported in grapes are proline, trehalose, sucrose, inositol, potassium, etc. The active net accumulation of solutes indicated that grapevines generally are able to adjust osmotically when subjected to moisture stress conditions. Glucose and fructose were involved in such adjustment (During 1984) in addition to some of the other osmotically active compounds such as amino acids or ions, which normally participate in osmotic adjustment of grapevines (Turner and Jones 1980). Zamboni et al. (1986) reported the closure of stomata in *Vitis riparia* immediately with the onset of moisture stress, while *Vitis berlandieri* adjusted the leaf osmotic potential in order to maintain cellular turgor to overcome adverse effects of moisture stress. Differences among many grape cultivars in response to water stress have been documented by several workers in the past. When cultivar 'Trollinger' vines were subjected to moisture stress, they responded to it by reducing assimilation rate (During 1988), but cultivar 'Riesling' was osmotically adjusted, resulting in high turgor potential than 'Sylvaner' vines (During and Loveys 1982). Within the varieties there was varying degree of osmotic adjustment and changes in water-use efficiency when subjected to moisture stress (During 1984).

Continuous period of water deficits results in eventual establishment of a stable elevated concentration of ABA within the vine system (Satisha et al. 2006). But further acceleration of its metabolism either maintains or slowly increases the ABA concentration. Conjugates and metabolites of ABA accumulate with prolonged stress, but little is known about this. Free ABA is the other major compound of significance, and it is apparent that the plant possesses a mechanism to limit the accumulation in prolonged period of stress (Jorrey 1976). Abscisic acid concentration is known to increase in stressed roots, and it is transported to the leaves where it reduces the photosynthesis and partially transpiration (During et al. 1997). A tenfold increase in ABA concentration was obtained in drying roots, while the ABA concentration in partially dried roots was increased by 60 % compared to fully irrigated vines. The reduction in stomatal conductance of partially irrigated vines was attributed to increased xylem sap ABA concentration and increased xylem sap pH. The ability to produce more ABA during water stress is a varietal behaviour. Four varieties grown on Fercal rootstock and subjected to 50 % moisture stress behaved significantly differently in terms of stomatal conductance and rate of photosynthetic assimilation, and the growth reduction in these parameters was associated with highest increase in ABA (Regina and Carbonneau 1997). Satisha et al. (2007) could show the reduction in cytokinin level and increased ABA accumulation in 50 % moisture-stressed plants of most of the rootstocks such as Dogridge, Salt Creek and *Vitis champini* clone as compared to fully irrigated vines.

11.3.2 Salinity Stress

Some of the morphological mechanisms at whole-plant level can overcome the adverse effects of salt stress; the salt tolerance can also occur at cellular level (Yokoi et al. 2002). Osmotic regulation is one of the mechanisms to avoid osmotic stress. The maintenance of cellular turgor under stress conditions is due to accumulation of metabolites such as glycine betaine, proline, mannitol and soluble sugars (Heidari Sharif Abad 2002; Houimli 2010). Proline is a strong source to store carbon and nitrogen and a purifier of free radicals. Proline also maintains the structure of cell membrane and proteins (Jalili Marandi et al. 2009). Another role of proline is maintaining the buffering capacity of cells in terms of salinity (Heidari Sharif Abad 2002). Fozouni et al. (2012) demonstrated that in response to hydroponically grown own-rooted *Vitis vinifera* table grape cultivars for different salt concentrations, proline accumulation increased significantly with increased salinity. Salisbury and Ross (1992) reported that not only *Vitis* spp. but also varieties within a species showed different degrees of salt tolerance.

Cavagnaro et al. (2006) in comparison of Argentinean and European grape cultivars showed that Argentinean cultivars have better performance under saline conditions even at increased salt concentrations. According to Levitt (1980) salt-induced growth inhibition was accompanied by one or more metabolic disturbances. The most dramatic disturbance was on photosynthesis. Most of the research on the Sultani Çekirdeksiz grapevine cultivar had shown that NaCl treatment decreased photosynthesis (Downton et al. 1990; Prior et al. 1992). Behboudian et al. (1986) had also concluded that lower photosynthesis due to salinity was correlated with chlorophyll content.

Salt-induced osmotic stress led to a rapid decline in water and the osmotic potential of cells and a decrease in both cell volume and cell expansion (Neumann et al. 1988). The most dramatic and readily measurable whole-plant response to salinity is a decrease in stomatal aperture. Closure of stomata is the primary cause for reduction in photosynthetic activity in conditions of weaker saline conditions, while under stronger salinity conditions, the reduction in photosynthesis is mainly due to biochemical and destructive reactions at cellular levels (Munns 2002). Reduction in rate of photosynthesis in Sultanina grapes was reported by Fisarakis et al. (2004). Under saline conditions, photosynthetic rate decreased because salinity causes the ion accumulation in different parts of the plant which exerts toxic effects on physiological processes in plant. Ashraf and Ahmad (2000) reported that leaf area and photosynthesis rate reduced while leaf diffusive resistance increased in plants growing under salt stress. Jogaiah et al. (2014) in their studies on screening of rootstocks for moisture and salinity stress could see the increased osmotic adjustment in 110R, Dogridge, Salt Creek and B2-56 rootstocks which they attributed to increased sugar and other compatible solutes (Rodrigues et al. 1993). Among the rootstock varieties tested for salinity tolerance, maximum plant mortality was recorded in St. George and 1613C rootstocks at salinity level of 8 ds m^{-1}, while least mortality was recorded on Dogridge followed by that of Salt Creek (Yohannes 2006).

11.3.3 Differential Uptake of Nutrients to Reduce Salt Injury

Grapevine rootstocks differ in their root cation-exchange capacity (CEC) and thereby vary in their ability to absorb plant nutrients and exclude toxic ions. Sauer (1968) reported that the level of chloride in the petioles of vines of Sultana vines grafted on Dogridge and 1613 resulted in about 50 % reduction in the chloride content than that in own-rooted vines of Sultana. Two to three times more accumulation of chloride was reported by Palaniappan (1986) in the petioles of own-rooted Anab-e-Shahi and Thompson Seedless grapevines as compared to those grafted on rootstocks such as Salt Creek, 1613 and Dogridge as these rootstocks excluded sodium and chloride under saline conditions. Chloride content in the petioles of Thompson Seedless on Dogridge rootstock was restricted to 0.44 % as against 1.44 % in the petioles of own-rooted vines, when they were irrigated with saline water of $5dSm^{-1}$ strength. Ramsey had been found to be one of the most effective chloride excluders (Sykes and Newman 1987). In Thompson Seedless vineyards raised on Dogridge rootstock, Diagnostic Recommended Integrated System (DRIS) indices were instrumental in demonstrating a greater affinity for Na by Dogridge rootstock (Sharma et al. 2006). Under saline irrigation, vines grafted on Dogridge, rootstock has shown the tendency to accumulate sodium in excess, leading to K deficiency, reduced fruitfulness and death of vines (Sharma et al. 2010, 2011). In contrast, Thompson Seedless vines grafted on 110R (*Vitis berlandieri x Vitis rupestris*) rootstock sustained the yield over a period of time with less accumulation of sodium ions (Satisha et al. 2010; Sharma and Upadhyay 2008). Also, rootstock 110R imparted early and uniform sprouting in Thompson Seedless vines along with increased fruitfulness when irrigated with saline water having EC of 1.8 ds m^{-1} (Jogaiah et al. 2013).

Sharma and Upadhya (2008) demonstrated that vines grafted on B2-56 rootstock (clone of 110R) had significantly lower Na concentration at all the stages of sampling, indicating better Na exclusion compared to other two stock-scion combinations. Paranychianakis and Angelakis (2007) in their studies on the effects of saline water on performance of Sultanina grafted on three rootstocks could see greater leaf burn on 41B rootstock than on 1103P and 110R, indicating their restricted uptake of both sodium and chloride. Munns (2002) and Storey et al. (2003) have proposed that the sequestration of ions in roots, and the prevention of their transport to the shoot in the xylem, is a mechanism for salinity tolerance. Fisarakis et al. (2001) found consistently higher accumulations of Cl and Na in roots as compared to the leaves of Sultana vines and suggested that the capability to store Na in roots is a tolerance characteristic of rootstocks. Dardeniz et al. (2006) in their studies to determine the salt tolerance of four rootstocks could conclude 41B as most resistant followed by that of 140Ru and 1103P, while 5BB is least resistant based on fresh and dry root and shoot weight and also salinity-tolerant rate and salt index rate. In contrast to this, Arbazadeh and Dutt (1987) studied seven rootstocks for salt tolerance at 0.45, 2.5 and 5.0 dSm^{-1}. Based on per cent reduction in growth, leaf area and trunk diameter, Salt Creek was found to be most tolerant followed by that of Kober 5BB, SO4 and 1613C, while 41B was found to be most susceptible.

Most of the grape rootstocks appear to exclude sodium and chloride but at varying levels. Comparison of chloride concentration in xylem with concentration of soil solution clearly indicated the better chloride exclusion of 140Ru as compared to that of K51-40 (Tregeagle et al. 2006). The higher concentration of chloride in the xylem of K51-40 over the season leads to excessive accumulation of chloride in the grape juice. Using ^{36}chloride as a tracer, Tregeagle (2007) could link the better exclusion of chloride in own-rooted 140 Ru than in K51-40 to reduced transport of chloride into xylem. Salinity stress also creates nutritional imbalances in grapevines (Sharma et al. 2011). Upreti and Murti (2010), in their studies to see the response of root growth to salinity in different grape rootstocks, showed increased root Na and decreased K content in all the rootstocks, while St. George exhibited highest Na content and Na:K ratio. In in vitro screening of rootstocks for salinity tolerance, Troncoso et al. (1999) found a direct relation between salt concentration in media and the Na:K ratio in tolerant and sensitive cultivars. Though the in vitro culture is the suitable method to select salt-tolerant varieties, it needs to be verified in the field experiments to confirm their tolerance (Skene and Barlass 1988). In vitro responses of grape rootstock genotypes (Dogridge, SO4, H-144 and 3309C) to NaCl led to increased protein contents, proline, K^+ and Na^+ in tissue, while contents of chlorophyll and total soluble sugars declined (Alizadeh et al. 2010).

11.3.4 Accumulation of Compatible Solutes and Antioxidants

The accumulation of soluble sugars and other low molecular compounds which are usually called as compatible solutes is responsible for reducing osmotic potential, thus allowing the cell turgor to maintain as much as possible with higher values. This process, known as osmotic adjustment, has been clearly demonstrated in vines which were gradually subjected to water stress, either in leaves (During 1984) or in roots (During and Dry 1995). Sugars, proline, glycine betaine and other organic solutes are believed to improve salt tolerance by contributing to osmotic balance and preserving enzyme activity in the presence of toxic ions (Greenway and Munns 1980; Grumet et al. 1985). Gibberd et al. (2003) in their studies on the effect of salinity on transpiration efficiency (TE) could observe highest TE in Sultana vines grafted on Ramsey than on own-rooted Ramsey vines under salinity treatment of 80 mol m^{-3}. The variation in TE was associated with rootstock which was attributed to differences in uptake and the energy requirement associated with ion partition and the formation of compatible solutes as proline, glycine betaine, sugar alcohols, etc.

Proline is known to exhibit indirect protective role due to its antioxidant properties in addition to its direct effect to stabilize the macromolecules and their hydration layers (Bonhert et al. 2006). The two- or threefold increase in proline accumulation in *V. vinifera* leaves was consistent with an increase in transcript abundance for delta 1-pyrroline-5-carboxylate synthetase (P5CS), the enzyme that catalyzes the first two steps in the proline biosynthetic pathway. Increased transcript abundance of proline dehydrogenase also accompanied the induction of P5CS transcripts presumably to facilitate the removal of excess proline accumulation which can be toxic to the cell if allowed to over-accumulate (Deuschle et al. 2004). Glucose and fructose exhibited increased accumulation in response to water-deficit stress, as has been observed in *Arabidopsis* (Rizhsky et al. 2004). However, unlike *Arabidopsis*, sucrose amounts did not become elevated in *Vitis vinifera* in response to water deficit. Fozouni et al. (2012) demonstrated that when own-rooted vines of four table grapes were grown hydroponically at different salt concentration, there was increased accumulation of proline with increasing concentration. According to Bhardwaj et al. (2007), plants have evolved antioxidant systems to protect themselves from these stresses which includes low molecular weight antioxidants such as glutathione, ascorbate and carotenoids and free radical scavengers such as superoxide dismutase, catalase, ascorbate peroxidise, guaiacol peroxidise and glutathione reductase. Baneh et al. (2013) subjected four different table grape cultivars to different concentrations of NaCl stress and could observe significant increase of soluble sugars and proline content which acts as osmoprotectants with marked increase in salt concentration. Of the four genotypes studied, they could see the increased activity of catalase in only two genotypes, indicating that genetic variation also exists among genotypes in tolerating stress severity at biochemical levels.

Reactive oxygen species (ROS) developed during salinity stress can cause photo-oxidative damage when its production exceeds that of antioxidants (Foyer et al. 1994). Some of the antioxidants synthesized in response to salinity stress include enzymes, proteins, etc. These compounds play a major role in quenching those ROS or to limit their damaging effects by quenching them. Grape cultivar Razaki when treated with 12 mM NaCl showed significant increase in glutathione peroxidase and catalase activity even on the 7th day after treatment of salts compared to that of 1616 rootstocks (Yildirim et al. 2004).

11.4 Role of Polyamines in Drought and Salinity Stress Tolerance

Polyamines are small molecules, positively charged at physiological conditions which are essential for life in most of the eukaryotes (Takahashi and Kakehi 2010; Moschou et al. 2012; Calzadilla et al. 2014). Though they are involved in several phenomenon in plant growth and development, in recent years many of the researchers have described the protective mechanisms of polyamines during plant's response to abiotic stresses such as salt, drought, low and high temperatures, flooding, heavy metals, etc. (Yamaguchi et al. 2006; Zawoznik et al. 2011; Menendez et al. 2012; Wimalasekera et al. 2011 etc.). In many crops it has been reported that the up- and downregulations of enzymes are involved in polyamine synthesis in response to drought and salt stress. The important polyamines reported in grapes which play an important role in abiotic stress tolerance include putrescine, spermine and spermidine. Usually, under abiotic stress conditions, plants produce excess endogenous ABA to activate downstream expression and other physiological responses (Klingler et al. 2010). Toumi et al. (2010) showed that ABA enhanced accumulation of polyamine such as putrescine, spermine and spermidine in grapes, resulting in secondary protective mechanisms such as closure of stomata. Reactive oxygen species (ROS) including H_2O_2

and hydroxyl radical (OH˙) production is largely produced when several abiotic stresses were experienced by plants, and thus ROS homoeostasis is related to stress-activated oxidative stress (Shi et al. 2013). Polyamine catabolism-derived H_2O_2 production has been shown to induce plant cell death, which is also one of the defence mechanisms during abiotic stress (Tisi et al. 2011). Klingler et al. (2010) suggested that ABA-dependent amino oxidase-derived H_2O_2 affected stomatal conductance in grapevines.

Several workers have summarized the mechanisms by which polyamines could alleviate abiotic stresses like drought, salt and cold stress which includes membrane stabilization, osmotic and ion homoeostasis, antioxidant activity and their interaction with macromolecules like hormones and other transcription factors. Polyamine catabolism-derived reactive oxygen species mediated cell growth under salt stress and alteration in gene expression by stimulating phosphorylation of regulatory proteins. Enhancement of stress tolerance in grapes through modulating endogenous polyamines by genetic engineering or exogenous application of polyamines has been studied by several workers (Wi et al. 2006; Farooq et al. 2009). They also opined that tolerant grape varieties can be developed through manipulation of polyamine biosynthesis. In seedlings of Neo-Muscat grapes subjected to osmotic and salt stress, significant accumulation of putrescine was observed than in control vines. Temporary induction of the genes encoding arginine decarboxylase (pVvADC) and late induction of gene encoding spermine synthase (pVvSPMS) were observed by Liu et al. (2012). Increased level of polyamines was recorded in salt-treated grape rootstocks compared to that of control vines wherein higher concentration of spermine and spermidine was recorded in Dogridge and Salt Creek rootstocks while putrescine was more in St. George rootstock (Upreti and Murti 2010; Upreti et al. 2012). Klingler et al. (2010) suggested that ABA-dependent amine oxidase-derived H_2O_2 is involved in ABA-induced stomatal regulation under the water-deficit stresses and by PCD (Tisi et al. 2011; Moscham et al. 2012), while on the other hand, polyamine-modulated stress suggested ROS homeostasis and oxidative damage by activating some antioxidant enzyme activities including SOD, catalase and peroxidases (Tavladoroki et al. 2012).

11.5 Molecular Mechanisms of Abiotic Stress Tolerance

Moisture and salinity stress tolerance is a multigenic trait as many several processes are involved in the tolerance mechanism. The drought tolerance in grapes is also controlled by multigenes since it is a quantitative trait (Bartles and Sunkar 2005) which makes it very difficult to understand the vine response to water stress at molecular level. A lot of research work is going in the field of molecular biology to see the influence of drought and salinity stress at genetic level in several processes such as synthesis of compatible solutes, polyamines, antioxidant synthesis, ion transport and sequestration, etc. Several genes/cDNA encoding proteins involved in stress tolerance have been reported in crop species. Examination of molecular and genetic mechanisms of abiotic stress tolerance in plants is of great interest to plant biologists. Though several genes related to drought response have been identified in grapes, their physiological and biochemical relevance is not understood clearly (Chaves et al. 2003). Marguerit et al. (2012) in their studies have identified a genomic region of the grapevine rootstock which plays a major role in improving water uptake capacity and controls transpiration on grafted scion varieties. Many studies have been able to identify some of the aquaporins which regulate stomatal conductance by either hydraulic or chemical signals from roots to shoots. Dehydrins have been the most studied proteins which accumulate under moisture stress in most of the crop plants. Dehydrins are intracellular stabilizers acting at both nuclear and cytoplasmic levels (Yan et al. 2012). Though the functions of dehydrin proteins are not clear, there is evidence to suggest that they stabilized the plasma membrane as chaperons by altering solvent properties of the cell wall during the event of moisture/salt stress (Puhakainen et al. 2004). Some of the transcriptions factors, genes and proteins involved in overcoming the adverse effects of abiotic stresses in grapevines are briefly discussed below:

11.5.1 Aquaporins

Aquaporins are the membrane proteins of a larger family of intrinsic proteins which produce pores in the cell membranes. They are known to play a major role in movement of water in cells transversely by facilitating the transport of water and thus control root hydraulic conductivity (Vandeleur et al. 2009). Some of the important membrane proteins are plasma membrane intrinsic proteins (PIP), of which specifically PIP2s are known for their maximum water permeability under soil moisture stress conditions. Some of the aquaporins like VvPIP 1;1 and VvPIP 2;2 have been identified in the roots of grapevines. In response to moisture stress, VvPIP1;1 are known to regulate water transport from root system to match transpirational demand. Serra et al. (2013), in their studies on 110R rootstocks subjected to moisture stress, showed expression of some eight aquaporins in leaves and roots to limit water transpiration from leaves and to accelerate water uptake from roots.

Aquaporins play a major role in controlling movement of water through roots (Maurel et al. 2008) and also through the neighbouring living tissues of root xylem vessels (Kaldenhoff et al. 2008). Many of the grapevines' aquaporins were characterized by several workers which were known to express either in roots or shoots (Glissant et al. 2008). The role of aquaporins under moisture stress conditions were studied in many of the grapevine cultivars that were genetically different in their water-use efficiency. VvPIP1:1 transcript expression was known to increase under water stress in Chardonnay but not in Grenache variety of wine grapes (Vandeleur et al. 2009). The importance of these studies implies that root system can effectively play a major role in influencing moisture stress tolerance at molecular level. Lovisolo et al. (2008) and Lovisolo and Schubert (2006) in their different experiments have shown that in the drought-resistant grape rootstocks, aquaporins regulating decrease of hydraulic conductivity are more important than cavitation of xylem vessels. Thus, they suggested that the transpiration reduction by ABA promoted aquaporin is mediated by xylem embolism after moisture stress. In some studies, it has been shown that roots regained their growth and activity after drought stress through specific genes that adapt root development to these resistant conditions (Paltronieri et al. 2011). Picaud et al. (2003) have characterized few plasma membrane intrinsic proteins in general and identified their role in the water transport during moisture stress.

Very few of these genes were identified and characterized in grapes under salinity conditions either in own-rooted commercial varieties or in grafted vines. Upadhyay et al. (2012) could observe increased expression of Na^+/H^+ antiporter in 110R rootstock at combined moisture and salinity stress situation compared to that in Dogridge, Salt Creek and 1613C rootstocks. Thus, 110R could sequester more sodium in root vacuole which is evident from lesser accumulation of sodium in leaves of 110R rootstock. In studies on growth response and aquaporin expression in grape genotypes under salinity stress, the expression of VvPIP 2.2 gene decreased significantly in leaves and roots of cv. Gharashani, but it was higher in roots than in leaves capable of transporting water from roots to leaves. However, in roots and leaves of Shiraz cultivar, its expression was increased, indicating its more readiness to face salinity stress. Thus, they concluded that Shiraz is more sensitive genotype while Gharashani is moderately tolerant (Mohammadkhani et al. 2012).

Controlling the initial uptake of Cl^- (and Na^+) at the root-soil interface would limit the total amount of these ions entering the plant and subsequently transported to shoots. It is likely that in salt-tolerant species, several processes will be operating, and the key trait responsible for lower Cl- accumulation in shoots may change depending on the external NaCl concentration. Plants that limit uptake of toxic ions and maintain normal ranges of nutrient ions could be more salt tolerant than those that do not restrict ion accumulation and lose nutrient balance. Selective ion uptake mechanisms capable of discrimination between chemically similar ions such as Na^+ and K^+ could have adaptive value. The mechanisms responsible for ion discrimination are probably through receptors which are located in the membranes of tissues and various organelles throughout the plant (Kuiper 1968; Bliss et al. 1984).

11.5.2 Dehydrins

Since drought stress tolerance is a complex trait, a long list of drought-responsive genes could be seen from several past research findings (Cramer et al. 2013). Of the several genes induced during the events associated with drought stress, only few of them are effectively involved in stress tolerance. Dehydrins are one such group of dehydration-responsive protein types (Hanin et al. 2011). These belong to LEA D11 family, which accumulate during late embryogenesis reported in several higher plants and microorganisms like yeast, algae and cyanobacteria (Rorat et al. 2006). Several types of dehydrin proteins are reported (Vaseva et al. 2014) which are known to accumulate in response to low-temperature and moisture stress or one which is induced by ABA-mediated pathway. Dehydrins are known to protect plant proteins and membranes from moisture loss during water stress (Sadder and Doss 2014). Sometimes they are known to bind with metals and act against oxidative stress (Rorat 2006). Hanana et al. (2014) in their studies could isolate dehydrin genes from Cabernet Sauvignon grapes which are usually expressed at the late embryogenesis in seeds which encodes for a protein of 124 amino acids (VVDhn). They could not identify these genes under normal conditions in any of the vegetative tissues but could be seen in their expression during seed development (late embryogenesis) under high-salt and moisture stress and also with application of ABA. The expression of VVDhn was seen only in resistant varieties like Razegui but not in sensitive varieties like Syrah under such stresses. Choi et al. (2013) in their studies on transcriptional analysis of Dehydrin1 gene in response to dehydration stress could show the earlier expression of the DHN1a and DHN1b genes in *Vitis flexuosa*, a wild grape variety, than in cultivars such as Campbell Early and Tamnara (*Vitis vinifera*). Thus, they concluded that *Vitis flexuosa* could be a good source for stress-tolerance breeding for enhancing drought stress tolerance.

11.5.3 Peridoxins

Peroxiredoxins are ubiquitous family of non-heme thiol peroxidises which catalyzes the reduction of hydroxyl peroxide, alkyl hydroperoxides and peroxynitrite to water, alcohol or nitrite. They are usually redox-sensitive proteins that can endure reversible oxidation and reduction and thus shift from on to off and vice versa depending on the cellular redox state. They are known to be involved in balancing signal cascades mediated by reactive oxygen species and in dissipating excessively absorbed energy and protecting the photosynthetic apparatus against oxidative damage (Broin et al. 2002). Vidigal et al. (2013) subjected the 4-month-old vines of Touriga grape cultivar to several abiotic stresses like high irradiance, heat and moisture stress and analysed the expression of genes involved in peroxidoxin synthesis. They could identify the overexpression of genes encoding chloroplastic (vv2cysprx01 and vvprxII-1) and mitochondrial (vvprxII i) peroxidoxin isoforms and thus could cope up with these abiotic stresses.

11.5.4 Osmotin

Osmotin is one such protein accumulated during severe water stress in many plant species. Osmotin and osmotin-like proteins have been classified as members of the family of five pathogenesis-related (PR) proteins (Bol et al. 1990; Kumar et al. 2015). Their expression has been mostly linked to various developmental processes in plant system and is also induced by several stress conditions of biotic and abiotic nature (Nelson et al. 1992; Zhu et al. 1995). Loulakakis (1997a) described nucleotide sequence of the full-length VvOSM1 cDNA clone from grapevines. Later, Loulakakis (1997b) observed the higher expression pattern of the same gene in grapevine tissues of cv. Sultanina in response to abscisic acid, ethylene and osmotic stresses, thus suggesting regulation of osmotin-like genes by the stress signals. Agaoglu et al. (2004) in molecular characterization of grape rootstocks and varieties under salt stress could identify more expression of osmotin genes in scion cultivars than in rootstocks cultivars. Among scions the highest osmotin expression was observed in Tahannebi while it was highest in rootstock 1616C and 99R. Carbonell-Bejerano et al. (2013) observed the induction of an osmotin transcript (VIT_02s0025g04340), indicating its putative function in activating osmoprotection in response to elevated temperatures. However, conversely, Reinth et al. (2014), in their studies on transcriptome responses in green and ripening grapes, could see the downregulation of three osmotin-coding genes during heat stress and questioned about the actual role of these genes in response to heat stress in grapes. In large-scale proteomic analysis of grapevine leaf apoplastic fluids, Delaunois et al. (2013) could identify major portion of the PR proteins expressed in response to heat stress as osmotin-like proteins in apoplastic fluid and attributed their role in several abiotic stress like drought, salinity and heat stress.

11.5.5 Calcium-Dependent Protein Kinase

Calcium-dependent protein kinases (CDPKs), which are the most important Ca^{2+} sensors in plants, are known to play one of the key roles in plant adaptation to abiotic stress. CDPK is a multigene family of enzymes. Analysis of CDPK gene expression under various abiotic stress conditions would help identify those CDPKs that might play important roles in plant adaptation to abiotic stress. Dubrovina (2012) focused on studying CDPK gene expression under osmotic, water-deficit and temperature stress conditions in a wild-growing grapevine *Vitis amurensis* Rupr., which is known to possess high adaptive potential and high level of resistance against adverse environmental conditions. Healthy *V. amurensis* cuttings (excised young stems with one healthy leaf) were subjected to several stresses like water deficit, osmotic stress, heat stress and cold stress. The total level of CDPK gene expression increased under salt and decreased under low-temperature stress conditions. Analysis of the cDNA sequences identified eight different CDPK genes (VaCDPK1a, 1e, 1d, 2a, 3a, 3b, 3c, 3d). The prevalent CDPK transcript was VaCDPK3a under both non-stress and abiotic stress conditions. Under high-salt conditions, VaCDPK1d, 1e, 3b and 3d transcripts were upregulated. Under high-mannitol conditions, expression of VaCPK1e and 3b was upregulated, while expression of VaCDPK1d, 3c and 3d was only slightly induced. Under water deficit, expression of only VaCDPK3b and 3c genes was induced. Cold stress induced expression of VaCDPK2a and 3d genes, while heat stress induced expression of VaCDPK1a, 1d, 1e, 2a, 3a and 3c genes. Taken together, the data showed that the VaCDPK genes are transcriptionally regulated by osmotic, water-deficit and temperature stresses. The differential expression of the VaCDPK genes during osmotic, water-deficit and temperature stresses is suggestive of their involvement in the underlying signal transduction pathways.

11.5.6 WRKY Transcription Factors

WRKY gene super family is one of the largest transcription factor gene families, which is known to play key role in regulation of many of the transcriptional reprogramming associated when plants experience abiotic stresses (Chen et al. 2012). Many of the researchers have focused on the functional analysis of WRKY factors in plants' response to abiotic stresses. Some of the functions of WRKY gene family/ transcription factors during abiotic stresses include ABA signalling, NaCl and mannitol tolerance (Shang et al. 2010), JA signalling (Skibbe et al. 2008), sugar signalling (Mangelsen et al. 2010), tolerance to osmotic stress (Song et al. 2009), etc. Expression pattern of VvWRKY genes in response to grapevines (cv. Pinot noir 40024) subjected to abiotic stress was studied using microarray data (Wang et al. 2014). They could see the upregulation of a large group of VvWRKY genes like Vv WRKY08, 14, 16 and 24 in the very first hours after imposition of water stress. However, majority of the genes induced by water and salt stress were upregulated after 24 h. In the same study they concluded that the hormones involved in the plant defence signalling pathways are mediated by the induction of WRKY transcription factors. The expression profile of VvWRKY08 was compared with that of *Arabidopsis* WRKY's involved stress response, and they could see the similar expression pattern of AtWRKYs 18, 40, 60 when subjected to ABA and other abiotic stresses. When vines of Florida hybrid bunch grapes 'Suwannee' were subjected to drought stress for 20 days, there was expression of cDNA transcripts at 10–15 days. Some of the transcripts which were differentially expressed were sucrose synthase, actin, AP2 transcription activator, cytochrome b genes, WRK-type TF, putative aquaporins, protein kinase, ABF3, isoprene synthase and water stress-inducible proteins. Of these, about three transcription factors like ABF3, isoprene synthase and water stress-inducible genes were grouped as newly synthesized transcripts in grapevines (Katam et al. 2013).

11.5.7 Plant Hormones

Continuous period of water deficits results in eventual establishment of a stable elevated concentration of ABA within the vine system (Satisha et al. 2006). But further metabolism accelerates so that the concentration of ABA remains constant or increases slowly. Conjugates and metabolites of ABA accumulate with prolonged stress, but little is known about this. Free ABA is the other major compound of significance, and it is apparent that the plant possesses a mechanism to limit the accumulation in prolonged periods of stress (Jorrey 1976). Abscisic acid concentration is known to increase in stressed roots, and it is transported to the leaves where it reduces the photosynthesis and partially transpiration (During et al. 1997). Tenfold increase in ABA concentration was obtained in drying roots, while the ABA concentration in partially dried roots was increased by 60 % compared to fully irrigated vines. The reduction in stomatal conductance of partially irrigated vines was attributed to increased xylem sap ABA concentration and increased xylem sap pH. The ability to produce more ABA during water stress is a varietal behaviour. Four varieties grown on Fercal rootstock and subjected to 50 % moisture stress behaved significantly different in terms of stomatal conductance and rate of photosynthetic assimilation, and the growth reduction in these parameters was associated with highest increase in ABA (Regina and Carbonneau 1997). Satisha et al. (2007) could show the reduction in cytokinin level and increased ABA accumulation in 50 % moisture-stressed plants of most of the rootstocks such as Dogridge, Salt Creek and *Vitis champini* clone as compared to fully irrigated vines. Drought stress stimulates the production of abscisic acid and the occurrence of both ABA-dependent and ABA-independent pathways involved in drought response (Yamaguchi-Shinozaki and Shinozaki 2006). ABA increases drought and salinity tolerance by activating several stress-responsive genes that encode enzymes for the biosynthesis of osmolytes. Thus, regulating the synthesis of several plant hormones is also one of the mechanisms to overcome the salinity hazards. The predominant hormones include abscisic acid, ethylene, jasmonates, etc. (Parida and Das 2005). Similar to gibberellins and auxins, jasmonic acid (JA)-dependent gene activation involves hormone-induced degradation and transcriptional repressor, the ZIM/tify domain protein (Vanholme et al. 2007). Jasmonic acids are involved in plant response to abiotic stress. The activity of JA response is regulated by antagonistic crosstalk with salicylic acid signalling (Balbi and Devoto 2008). Ismail et al. (2012), in their studies on salt tolerance of two grape rootstock species through jasmonate-mediated pathway, analysed response of jasmonate ZIM/tify domain protein, Na^+/H^+ exchanger, stilbene synthase and resveratrol synthase. Suppression of jasmonate signalling by phenidone and aspirin blocked the induction of JAZ/tify transcripts. In the salt-tolerant line (*Vitis rupestris*), NHX1 is induced while the formation of ROS is suppressed as compared to that of sensitive line (*Vitis riparia*). In this study *Vitis rupestris* responded more efficiently to salinity stress because it could harbour higher density of sensitive calcium signal channels than of *Vitis riparia*.

Salicylic acid, a group of phenolic growth regulators, is known to play major physiological and biochemical functions under abiotic stresses with specific role during heat stress (Raskin 1992). It is known to counteract the adverse effect of heat stress by upregulating antioxidant mechanisms. In salicylic acid treated grapevines subjected to heat stress (Wang et al. 2010a) could maintain high photosynthetic rate by maintaining higher Rubisco activity even after recovery from heat stress. The salicylic acid treatment also protected mesophyll tissues of the leaves. Wang et al. (2007) reported that SA treatment can maintain at higher photosynthesis in grape leaves under several abiotic stresses like moisture, heat and salinity.

11.5.8 Stilbene

This phytoalexin compound is a low molecular plant secondary metabolite produced in response to several biotic and abiotic stresses. They belong to a family of phenylpropanoid produced in a limited number of plant species including grapes (Parage et al. 2012). Resveratrol and viniferins are grape stilbenes present in all organs of vine system. In response to stress, they mostly accumulate in leaves and berry skin (Wang et al. 2010a). In its biosynthetic pathway, the phenylalanine ammonia lyase (PAL) catalyzes the first step of phenylpropanoid pathway by conversion of phenylalanine to cinnamic acid which subsequently yields p-coumaric acid via cinnamic acid-4-hydroxylase. The last steps of phenylpropanoid pathway are categorized by chalcone (CHS) and stilbene (STS) synthases that are related to plant-specific polyketide synthase and are also considered as key enzymes in the biosynthesis of flavonoids and in stilbene phytoalexin, respectively. Manipulation of these enzymes involved in stilbene synthesis can be a good tool to develop grape varieties which can tolerate abiotic stresses.

11.6 Mechanisms of Heat Tolerance in Grapevines

Worldwide, the increase in global temperature depends on the emission of greenhouse gases particularly atmospheric CO_2. In most of the grapes' production areas, the air temperature during noon hours can reach more than 40 °C which is very critical at important phenological stages of growth and development of vines. Extreme temperatures may endanger berry quality and economic returns (Schultz 2007). Direct injuries of high temperature include denaturation of proteins coupled with aggregation and increased fluidity of membrane lipids. The other effects include enzyme inactivation in cellular organelles, inhibition of protein synthesis, protein degradation and loss of membrane integrity (Howarth et al. 2005). The consequence of increased air temperature will accelerate the grapevine phenology, leading to a reduction in the vegetative and reproductive period (Seguin and Cortazar 2005). Heat stress in association with some other stresses like moisture and salinity stresses results in the exhibition of some defence mechanisms such as expression of genes that could not be seen under favourable conditions.

One of the major consequences of heat stress is the excess generation of reactive oxygen species (ROS), which results in oxidative stress. The major defence mechanisms in plants to overcome heat stress are to alter their metabolism in various ways in response to heat tolerance particularly by producing compatible solutes that are able to organize protein and cellular structures, maintain cell turgor by osmotic adjustment and modify the antioxidant system to re-establish the cellular redox balance and homoeostasis. In recent years it was shown that a series of HSP will be induced with elevated levels of salicylic acid under heat stress situations (Pan et al. 2006; Wang et al. 2006). Wan et al. (2007) showed that a membrane-associated enzyme, PIP_2-specific phospholipase C, was activated by heat accumulation and involved in the development of thermotolerance in grape berries subjected to thermal stress. Wan et al. (2007) showed the increased activity of phospholipase D (PLD) enzyme as well as accumulation of PLD mRNA and synthesis of new PLD proteins during the early period of heat acclimation. When vines of Aragonez (syn. Tempranillo) were subjected to heat stress of 40 °C coupled with moisture stress, Rocheta et al. (2013) could identify 1,594 genes as differentially expressed between control and treated vines which they could group into ten major functional categories. The number of genes coding for heat shock proteins and transcription factors expressed solely in response to heat stress. HSP 70/Dnak are essential for normal cell functions. They play as molecular chaperons expressed constitutively. The induction of HSPs depends on the temperature at which each species ordinarily grows. The ubiquitination pathway is closely associated with abiotic stress tolerance, and F-box proteins recognize substrate through the SKP-1-Cullin-F-box complex for degradation by ubiquitin-26s proteosome (Mazzucotelli et al. 2008). F-box proteins were known to respond sharply to heat stress in grapes. When vines are exposed to water stress, their primary responses are mediated by variation in physiological and biochemical compositions. These may result in synthesis of several proteins which play an important role in these responses (Kumar et al. 2009). Many studies have shown that enhanced expression of transcription factors such as CBF/DRE family is related to defence signal transduction in the berries in response to several environmental stimuli (Ablett et al. 2000). Some of the genes associated with functional roles in the flavonoid and anthocyanin pathway and aroma biosynthesis in grapes were expressed in the skin and seeds when vines were exposed to moisture stress and heat stress (Deluc et al. 2006; Grimplet et al. 2006). The first report on characterization of HSP90s in grapevine was published by Banilas et al. (2012) as they are known to be important gene family as molecular chaperons. The phylogenetic analysis of cytoplasmic and three organelle-specific *Vitis vinifera* HSP90 (VvHSP90) revealed the high sequence homology with that of *Arabidopsis*. The severity and duration of heat stress influenced the vine system in a complex manner through expression of VvHSP90.1a, while other VvHSP90s were constitutively regulated. Liu et al. (2012) in their studies on transcriptome analysis of grapevines cv. Cabernet Sauvignon subjected to heat stress could identify upregulation of several candidate genes encoding the components of calcium- or calmodulin-mediated signal pathways, including calnexin and calmodulin, and CDPKs were also heat- or recovery-regulated, suggesting a role of Ca^{2+}-mediated signals in the heat stress response. They also suggested that bZIP transcription factor (basic leucine zipper) is heat regulated. In the same study, ethylene-responsive transcriptional co-activator (ERTCA) gene also upregulated in grape leaves subjected to heat stress followed recovery period indicating its probable role in heat tolerance. Wang and Li (2006, 2007) have reported that SA alleviates heat damage of plants by upregulating the antioxidant system.

11.7 Mechanisms of Flooding Tolerance in Grapevines

Flooding is a natural disturbance affecting crop growth and productivity worldwide due to its detrimental effects (Colmer and Voesenek 2009). With the changing climatic scenario, vineyards in the most of regions are subjected to unnatural rains at critical vine growth period. If the soil remains waterlogged for a longer period, it affects vine health in several different ways. A lack of oxygen in the soil over an extended period will result in root and plant death. Flooding may also cause the leaching of nutrients, loss of mulch material and crusting of the soil, and these effects will require management. In addition, shallow, stagnant water can heat up quickly in hot weather and kill plants, so it is important to remove excess water as quickly as possible.

Excess soil water causes a severe decrease in the oxygen diffusion rate into the soil. When soils are subjected to flood, due to lack of oxygen, the respiration of roots and microorganisms depletes and the environment becomes first hypoxic and later anoxic (Colmer et al. 1998; Bailey-Serres and Voesenek 2008; Wegner 2010). With the increase in flooding time, there is reduction of the soil redox potential and accumulation of many of toxic compounds (Fiedler et al. 2007). Therefore, reduced oxygen and accumulation of toxic chemicals are the big constraints that vines experience during flooding stress conditions. Striegler et al. (1993) subjected grape rootstocks to flooding stress under greenhouse conditions and could record reduced rate of shoot elongation which they identified as sensitive parameter to flooding stress. Rootstocks such as St. George, 3309C and Riparia Gloire were found to be more tolerant, while others like Kober 5BB, Seyval and Cynthiana were susceptible cultivars. They also hypothesized that the increased tolerance of some cultivars to the stress might be due to aerenchyma formation in the root cortex as opined by Smirnoff and Crawford (1983) and Colmer et al. (1998).

Flooding induces morphological changes in roots and shoots. The formation of adventitious roots is a common response of flood-tolerant cultivars. The adventitious roots which are known to have high precocity may help in vines to uptake water and nutrients without any hindrance, thus replacing the role of older roots (Kozlowski and Pallardy 1984). In grapevines subjected to flooding stress, many new adventitious roots could be seen near the aerated soil surface and sometimes on the cordons also. Jackson (2004) reviewed the formation of aerial roots under flood stress and postulated three hypothesis of aerial root formation, viz. stimulation of outgrowth of pre-existing root primordial in shoot base, induction of new root system that involves initiation of root primordial and their subsequent outgrowth and placing roots at the soil surface involving the re-orientation of the root extension. Some of the biochemical mechanisms reported in grapevine which take place during flood stress are increased proline and alcohol dehydrogenase activities which were highest in SO4, 101-14 Mgt and 3309C rootstock and which were least in 1103P, 140Ru and 110R rootstocks (Yan 2013). He also reported that the flood-tolerant rootstocks like SO4 promoted anaerobic metabolism of pyruvate by increasing the anaerobic respiration by increasing the enzyme activity of lactate dehydrogenase, alcohol dehydrogenase, etc., Choi et al. (2012) studied the gene expression of native Korean *Vitis flexuosa* grapevines, where they could record upregulation of more than 9000 transcripts under flood stress and concluded that *Vitis flexuosa* is flood tolerant. In in vitro studies in the gene expression of grapes in response to hypoxia (Tesniere et al. 1993), some RNA species of unknown identity and function increased severalfold, majority of which are alcohol ADH or GADPH which induced two- to threefold in hypoxia-stressed grape cell cultures.

References

Ablett E, Seaton G, Scott K, Shelton D, Graham MW, Baverstock P, Lee LS, Henry R (2000) Analysis of grape ESTs: global gene expression patterns in leaf and berry. Plant Sci 159:87–95

Agaoglu YS, Ergul A, Aras S (2004) Molecular characterization of salt stress in grapevine cultivars (*Vitis vinifera* L) and rootstocks. Vitis 43:107–110

Alizadeh M, Singh VSK, Patel B, Bhattacharya RC, Yadav BP (2010) *In vitro* responses of grape rootstocks to NaCl. Biol Plant 54:381–385

Alsaidi IH, Shakir IA, Hüssein AJ, Saidiq J (1987) Effect of salinity on the rooting of cuttings of Abbasi and

Kambli cultivars (*Vitis vinifera* L.). Ann Agric Sci 32:1581–1600

Arbazadeh F, Dutt G (1987) Salt tolerance of grape rootstock under greenhouse conditions. Am J Enol Vitic 38:95–99

Ashraf M, Ahmad S (2000) Influence of sodium chloride on ion accumulation, yield components, and fiber characteristics in salt-tolerant and salt-sensitive lines of cotton (*Gossypium hirsutum* L.). Field Crops Res 66:115–127

Bailey–Serres J, Voesenek LACJ (2008) Flooding stress: acclimations and genetic diversity. Annu Rev Plant Biol 59:313–339

Balbi V, Devoto A (2008) Jasmonate signalling network in *Arabidopsis thaliana*: crucial regulatory nodes and new physiological scenarios. New Phytol 177:301–318

Baneh HD, Attari H, Hassani A, Abdollahi R (2013) Salinity effects on the physiological parameters and oxidative enzymatic activities of four Iranian grapevines (*Vitis vinifera* L.) cultivar. Int J Agric Crop Sci 9:1022–1027

Banilas G, Korkas E, Ebglezos V, Nisiotou AA, Hatzopoulos P (2012) Genome-wide analysis of the heat shock protein 90 gene family in grapevine (Vitis vinifera L.). Aust J Grape Wine Res 18:29–38

Bartles D, Sunkar R (2005) Drought and salt tolerance in plants. Crit Rev Plant Sci 24:23–58

Bauerle TL, Smart DR, Bauerle W, Stockert CM, Eissenstat DM (2008) Root foraging in response to heterogeneous soil moisture in two grapevines that differ in potential growth rate. New Phytol 179:857–866

Bavaresco L, Lovisolo C (2000) Effect of grafting on grapevine chlorosis and hydraulic conductivity. Vitis 39:89–92

Behboudian NM, Walker RR, Torokfalvy E (1986) Effects of water stress and salinity on photosynthesis of Pistachio. Sci Hortic 29:251–261

Bhardwaj R, Arora N, Sharma P, Arora HK (2007) Effects of 28-homobrassinolide on seedling growth, lipid peroxidation and antioxidative enzyme activities under nickel stress in seedlings of *Zea mays* L. Asian J Plant Sci 6:765–772

Bliss RD, Platt-Aloia KA, Thompson WW (1984) Changes in Plasmalemma organization in cowpea radicle during imbibition in water and NaCl solutions. Plant Cell Environ 7:606–609

Bohnert HJ, Gong Q, Li P, Ma S (2006) Unraveling abiotic stress tolerance mechanisms—getting genomics going. Curr Opin Plant Biol 9:180–188

Bol JF, Linthorst HJM, Cornelissen BJC (1990) Plant pathogenesis-related proteins induced by virus infection. Annu Rev Phytopathol 28:113–138

Broin M, Cuine S, Eymery F, Rey P (2002) The plastidic 2- cysteine peroxiredoxin is a target for a thioredoxin involved in the protection of the photosynthetic apparatus against oxidative damage. Plant Cell 14:1417–1432

Calzadilla PI, Gazquez A, Maiale SJ, Ruiz OA, Benardina MA (2014) Polyamines as indicators and modulators in the abiotic stress in plants. In: Anjum NA, Gill SS, Gill R (eds) Plant adaptation to environmental changes. CAB International, United Kingdom

Carbonell-Bejerano P, Santa Maria E, Torres-Perez R, Royo C, Lijavetzky D, Bravo G, Aguirreolea J, Sanchez-Diaz M, Antolín M, Martinez-Zapater J (2013) Thermotolerance responses in ripening berries of *Vitis vinifera* L. Cv Muscat Hamburg. Plant Cell Physiol 54(7):200–216

Cavagnaro JB, Ponce MT, Guzman J, Cirrincione MA (2006) Argentinean cultivars of *Vitis vinifera* grow better than European ones when cultured *in vitro* under salinity. Biol Cell 30:1–7

Chaves MM, Oliveira MM (2004) Mechanisms underlying plant resilience to water deficits: prospects for water-saving agriculture. J Exp Bot 55:2365–2384

Chaves MM, Maroco JP, Pereira JS (2003) Understanding plant responses to drought – from genes to the whole plant. Funct Plant Biol 30:239–264

Chen L, Song Y, Li S, Zhang L, Zou C, Yu D (2012) The role of WRKY transcription factors in plant abiotic stresses. Biochem Biophys Acta 1819:120–128

Choi YJ, Hur YY, Jung SM, Roh JH, Nam JC, Park SJ (2012) Changes in gene expression of native Korean Vitis flexousa during flood and its subsequent recovery. In: Conference on mechanisms of abiotic stress tolerance in plants 8–9 June, Gaesin Culture Centre, Chungbuk National University, South Korea, p 194

Choi YJ, Hur YY, Jung SM, Kim SH, Noh JHM, Park SJ, Park KS, Yun HK (2013) Transcriptional analysis of *Dehydrin1* genes responsive to dehydrating stress in grapevines. Hortic Environ Biotech 54:272–279

Colmer TD, Voesenek LACJ (2009) Flooding tolerance: suites of plant traits in variable environments. Funct Plant Biol 36:665–681

Colmer TD, Gibberd MR, Wiengweera A, Tinh TK (1998) The barrier to radial oxygen loss from roots of rice (*Oryza sativa* L.) is induced by growth in stagnant solution. J Exp Bot 49:1431–1436

Comas LH, Bauerle TL, Essenstat DM (2010) Biological and environmental factors controlling root dynamics and function: effects of root ageing and soil moisture. Aust J Grape Wine Res 16:131–137

Cramer GR, Ergul A, Grimplet J (2007) Water and salinity stress in grapevines: early and late changes in transcript and metabolite profiles. Funct Integr Genom 7:111–134

Cramer GR, Van- Sluyfer SC, Hopper DW, Pascovici C, Keighley T, Haynes PA (2013) Proteomic analysis indicates massive changes in metabolism prior to the inhibition of growth and photosynthesis of grapevine (*Vitis vinifera* L) in response to water deficit. BMC Plant Biol 13:49

Dardeniz A, Muftuoglu NM, Altay H (2006) Determination of salt tolerance of some American grape rootstocks. Bangladesh J Bot 35:143–150

Davies FS, Lakso AN (1978) Water relations in apple seedlings changes in water potential components, Abscisic acid levels and stomatal conductances under irrigated and non irrigated conditions. J Am Soc Hortic Sci 103:310–313

Delaunois B, Colby T, Belloy N, Conreux A, Harzen A, Baillieul F, Clement C, Schmidt J, Jeandet P, Cordelier S (2013) Large-scale proteomic analysis of the grapevine leaf apoplastic fluid reveals mainly stress-related proteins and cell wall modifying enzymes. BMC Plant Biol 13:24

Deluc LG, Grimplet J, Matt M, Craig O, Karen AS, Cramer GR, Cushman JC (2006) Transcriptional profiling throughout grape berry development under well watered and water deficit stress conditions. International plant and animal genome conference, San Diego, 2006, W171

Deuschle K, Funck D, Forlani G, Stransky H, Biehl A, Leister D, van der Graaff E, Kunze R, Frommer WB (2004) The role of Δ1-pyrroline-5-carboxylate dehydrogenase in proline degradation. Plant Cell 16:3413–3425

Downton WJS, Loweys BR, Grant WJR (1990) Salinity effects on the stomatal behavior of grapevine. New Phytol 16:499–503

Dry PR, Loveys BR, During H (2000) Partial drying of the root zones of grape vines. 1. Transient change in shoot growth and gas exchange. Vitis 39:3–7

Dubrovina AS (2012) Expression of calcium dependent protein kinase (CDPK) genes in Vitis amurensis under abiotic stress conditions. J Stress Physiol Biochem 8:19

During H (1984) Evidence for osmotic adjustment to drought in grapevines (*Vitis vinifera*). Vitis 32:1–10

During H (1988) CO_2 assimilation and photorespiration of grapevine leaves: responses to light and drought. Vitis 27:199–208

During H, Dry PR (1995) Osmoregulation in water stressed roots: response of leaf conductance and photosynthesis. Vitis 34:15–17

During H, Loveys BR (1982) Diurnal changes in water relations and abscisic acid in field grown *Vitis vinifera* cvs. I. Leaf water potential components and leaf conductance under humid temperate and semiarid conditions. Vitis 21:223–232

During H, Loveys BR, Dry PR (1997) Root signals affect water use efficiency and shoot growth. Acta Horticult 427:1–14

Farooq M, Wahid A, Lee DJ (2009) Exogenously applied polyamines increase drought tolerance of rice by improving leaf water status, photosynthesis and membrane properties. Acta Physiol Plant 31:937–945

Fiedler S, Vepraskas MJ, Richardson JL (2007) Soil redox potential: importance, field measurements, and observations. Adv Agron 94:2–56

Fisarakis I, Chartzoulakis J, Stavrakas D (2001) Response of Sultana vines (*Vitis vinifera* L.) on six rootstocks to NaCl salinity exposure and recovery. Agric Water Manag 51:13–27

Fisarakis I, Nikolaou N, Tsikalas P, Therios I, Stavrakas D (2004) Effect of salinity and rootstock on concentration of potassium, calcium, magnesium, phosphorus and nitrate-nitrogen in Thompson seedless grapevine. J Plant Nutr 27:2117–2134

Foyer CH, Lelandais M, Kunert KJ (1994) Photooxidative stress in plants. Physiol Plant 92:696–717

Fozouni M, Abbaspour N, Doulati Baneh H (2012) Short term response of grapevine grown hydroponically to salinity: mineral composition and growth parameters. Vitis 51:95–101

Gibberd MR, Walker RR, Condon AG (2003) Whole-plant transpiration efficiency of 'Sultana' grapevine grown under saline conditions is increased through the use of a Cl-excluding rootstock. Funct Plant 30:643–652

Glissant D, Dédaldéchamp F, Delrot S (2008) Transcriptomic analysis of grape berry softening during ripening. J Int Sci Vigneet Vin 42:1–13

Greenway H, Munns R (1980) Mechanisms of salt tolerance in nonhalophytes. Ann Rev Plant Physiol 31:149–190

Grimplet J, Laurent D, Wheatley M, Osborne C, Schalauch K, Carmer G, Cushman J (2006) Tissue-specific mRNA profiling within the grape berry under well watered and water deficit stress conditions. International plant and animal genome conference, San Diego

Grumet R, Isleib TG, Hanson AD (1985) Genetic control of glycine betaine level in barley. Crop Sci 25:618–622

Hanana M, Daldoul S, Fouquet R, Deluc L, Leon C, Hoefer M, Barrieu F, Ghorbel A (2014) Identification and characterization of a seed-specific grapevine dehydrin involved in abiotic stress response within tolerant varieties. Turk J Bot 38:1157–1168

Hanin M, Brini F, Ebel C, Toda Y, Takeda S, Masmoudi K (2011) Plant dehydrins and stress tolerance – versatile proteins for complex mechanisms. Plant Signal Behav 6:1503–1509

Hardie WJ, Martin SR (2000) Shoot growth on de-fruited grapevines: a physiological indicator for irrigation scheduling. Aust J Grape Wine Res 6:52–58

Heidari Sharif Abad H (2002) Plant and salinity. The Publications of Research Institution for Forests and Rangelands. Tehran, Iran, p 199

Houimli SIM, Denden M, Mouhandes BD (2010) Effects of 24-pibrassinolide on growth, chlorophyll, electrolyte leakage and proline by pepper plants under NaCl-stress. Eur Asian J Bio Sci 4:96–104

Howarth CJ, Ashraf M, Harris PJC (2005) Genetic improvement of tolerance to high temperature. In: Ashraf M, Harris PJC (eds) Abiotic stresses: plants resistance through breeding and molecular approaches. Howarth Press, New York, pp 277–300

Ismail A, Riemann M, Nick P (2012) The jasmonate pathway mediates salt tolerance in grapevines. J Exp Bot 63:2127–2139. doi:10.1093/jxb/err426

Jackson MB (2004) The impact of flooding stress on plants and crops. http://www.plantstress.com/Articles/waterlogging_i/waterlog_i.htm

Jalili Marandi R, Jalil Doostali P, Hassani R (2009) Studying the tolerance of two apple roots to different concentrations of sodium chloride inside the glass. Mag Hortic Sci Iran 40:2

Jogaiah S, Oulkar DP, Banerjee K, Sharma J, Patil AS, Maske SR, Somkuwar RG (2013) Biochemically induced variations during some phenological stages in Thompson Seedless grapevines grafted on different rootstocks. S Afr J Enol Vitic 34:36–45

Jogaiah S, Ramteke SD, Sharma J, Upadhyay AK (2014) Moisture and salinity stress induced changes in biochemical constituents and water relations of different grape rootstock cultivars. Int J Agronom. Article ID 789087, 8 pp. doi:http://dx.doi.org/10.1155/2014/789087

Jorrey JG (1976) Root hormones and plant growth. Annu Rev Plant Physiol 27:435–439

Kaldenhoff R, Ribas-Carbo M, Sans JF, Lovisolo C, Heckwolf M, Uehlein N (2008) Aquaporins and plant water balance. Plant Cell Environ 31:658–666

Katam R, Saurez J, Williams S, Matta F, Gottschalk V (2013) Differential expression of transcripts to water deficit stress in Florida hybrid bunch grape. Proc Fla State Hortic Soc 126:8–13

Keller M, Kummer M, Vasconcelos MC (2001) Soil nitrogen utilization for growth and gas exchange by grapevines in response to nitrogen supply and rootstock. Aust J Grape Wine Res 7:2–11

Klingler JP, Balelli G, Zhu JK (2010) ABA-receptors. The start of a new paradigm in phytohormone signalling. J Exp Bot 61:3199–3210

Kodur S, Tisdall M, Tang C, Walker RR (2010) Accumulation of potassium in grapevine rootstocks (vitis) grafted to Shiraz as affected by growth, root-traits and transpiration. Vitis 49:7–13

Kozlowski TT, Pallardy SG (1984) Effects of flooding on water, carbohydrate and mineral relations. In: Kozlowski TT (ed) Flooding and plant growth. Academic, Orlando, pp 165–193

Kramer PJ, Boyer JS (1995) Stomata and gas exchange. In: Kramer PJ, Boyer JS (eds) Water relations of plants and soils. Academic, London, pp 257–282

Kuiper PJC (1968) Lipids in grape roots in relation to chloride transport. Plant Physiol 43:1367–1371

Kumar M, Busch W, Birke H, Kemmerling B, Nurnberger T (2009) Heat shock factor HsfB1 and HsfB2b are involved in the regulation of Pdf1.2 expression and pathogen resistance in Arabidopsis. Mol Plants 2:152–165

Kumar AS, Hima Kumari P, Sharna Kumar G, Mohanalatha C, Ravi Kishor PB (2015) Osmotin: a plant sentinel and a possible agonist of mammalian adiponectin. Front Plant Sci 6:163. doi:3389/fpls.2015.00163

Levitt J (1980) Responses of plants to environmental stresses, vol II. Academic, New York

Liu GT, Jun-Fang Wang JF, Cramer G, Dai ZW, Duan W, Xu HG, Wu BH, Fan PG, Wang LJ, Li SH (2012) Transcriptomic analysis of grape (*Vitis vinifera* L.) leaves during and after recovery from heat stress. BMC Plant Biol 12:174

Loulakakis LKA (1997a) Genomic organization and expression of an osmotin-like gene in *Vitis vinifera*. Vitis 36:157–158

Loulakakis LKA (1997b) Nucleotide sequence of a *Vitis vinifera* L. cDNA (Accession No. Y10992) encoding for osmotin-like protein (PGR97-064). Plant Physiol 113:1464

Lovisolo C, Schubert A (2006) Mercury hinders recovery of shoot hydraulic conductivity during grapevine rehydration: evidence from a whole plant approach. New Phytol 172:469–478

Lovisolo C, Hartung W, Schubert A (2002) Whole-plant hydraulic conductance and root-to-shoot flow of abscisic acid independently affected by water stress in grapevines. Funct Plant Biol 29:1349–135

Lovisolo C, Perrone I, Hartung W, Schubert A (2008) An abscisic acid-related reduced transpiration promotes gradual embolism repair when grapevines are rehydrated after drought. New Phytol 180:642–651

Lovisolo C, Perrone A, Carra A, Ferrandino, Flexas J, Medrano H, Schubert A (2010) Drought-induced changes in development and function of grapevine (*Vitis* spp.) organs and in their hydraulic and non hydraulic interactions at the whole plant level: a physiological and molecular update. Funct Plant Biol 37:98–116

Mangelsen E, Wanke D, Kilina J, Sundberg E, Harter K, Jaasson C (2010) Significance of light, sugar and amino acids supply for diurnal gene regulation in developing Barley caryopsis. Plant Physiol 153:14–33

Marguerit E, Brendel O, Leben E, Van Leeuwen C, Ollat N (2012) Rootstock control of scion transpiration and its acclimation to water deficit are controlled by different genes. New Phytol 194:416–429

Maurel C, Verdoucq L, Luu DT, Santoni V (2008) Plant aquaporins: membrane channels with multiple integrated functions. Ann Rev Plant Biol 59:595–624

Mazzucotelli E, Mastrangelo AM, Crosatti C et al (2008) Abiotic stress response in plants: when post-transcriptional and post-translational regulations control transcription. Plant Sci 174:420–431

Menendez AB, Rodriguez AA, Maiale SJ, Rodriguez-Kessler M, Jimenez-Bremont JF, Ruiz OA (2012) Polyamines contribution to the improvement of crop plants tolerance to abiotic stress. In: Tuteja N, Gill SS (eds) Crop improvement under adverse conditions. Springer, Morlenbach, pp 113–137

Mohammadkhani N, Heidari R, Abbaspour N, Rahmani F (2012) Growth responses and aquaporin expressions in grape genotypes under salinity. Iran J Plant Physiol 2:497–507

Morgan JM (1984) Osmoregulation and water stress in higher plants. Ann Rev Plant Physiol 35:299–319

Morlat R, Jacquet A (1993) The soil effects on the grapevine root system in several vineyards of the Loire valley (France). Vitis 32:35–42

Moschou PN, Wu J, Cona A, Tavladoraki P, Angelini R, Roubelakis- Angelakis KA (2012) The polyamines and their catabolic products are significant players in the turnover of nitrogenous molecules in plants. J Exp Bot 63:5003–5015

Mullins MG, Bouquet A, Williams LE (1996) Biology of grapevine. Press Syndicate of the University of Cambridge, Cambridge, p 239

Munn R (2002) Comparative physiology of salt and water stress. Plant Cell Environ 25:239–250

Nauriyal JP, Gupta OP (1967) Studies on salt tolerance of grape. I-Effect of total salt concentration. J Res Ludhiana 4:197–205

Nelson DE, Raghothama KG, Singh NK, Hasegawa PM, Bressan RA (1992) Analysis of structure and transcriptional activation of an osmotin gene. Plant Mol Biol 19:577–588

Neumann PM, Volkenburgh EV, Cleland RE (1988) Salinity stress inhibits bean leaf expansion by reducing turgor, not wall extensibility. Plant Physiol 88:233–237

Palaniappan R (1986) Salt tolerance studies in fruit crops. Annual report, IIHR Bangalore (India), pp 84–85

Palliotti A, Cartechini A, Nasoini L (2001) Grapevine adaptation to continuous water limitation during the season. Adv Hortic Sci 15:39–45

Paltronieri P, Bonsegna S, Domenico S, Santino A (2011) Molecular mechanisms in plant abiotic stress response. Field Veg Crop Res 48:15–24

Pan HQ, Zhan JC, Liu HT, Zhang JH, Chen JY, Wen PF, Huang WD (2006) Salicylic acid synthesized by benzoic acid 2-hydroxylase participate in the development of thermotolerance in pea plants. Plant Sci 171:226–233

Pandey RM, Divate MR (1976) Salt tolerance in grapes. I. Effect of sodium salts singly and in combination on some of the morphological characters of grape varieties. Indian J Plant Physiol 19:230–239

Parage C, Tavares R, Réty S et al (2012) Structural, functional, and evolutionary analysis of the unusually large stilbene synthase gene family in grapevine. Plant Physiol 160:1407–1419

Paranychianakis NV, Angelakis AN (2007) The effect of water stress and rootstock on the development of leaf injuries in grapevines irrigated with saline effluent. Agril Water Manag 2531:1–8

Parida AK, Das AB (2005) Salt tolerance and salinity effects on plants: a review. Ecotoxicol Environ Saf 60:324–349

Picaud S, Becq F, Dedaldechamp F, Ageorges A, Delrot S (2003) Cloning and expression of two plasma membrane aquaporins expressed during the ripening of grape berry. Funct Plant Biol 30:621–630

Prior LD, Grieve AM, Slavish PG, Gullis PR (1992) Sodium chloride and soil texture interactions in irrigated field grown Sultana grapevines. II. Plant mineral content, growth and physiology. Aust J Agric Res 43:1067–1084

Puhakainen T, Hess MW, Makela P, Svensson J, Heino P, Palva ET (2004) Overexpression of multiple dehydrin genes enhances tolerance to freezing stress in *Arabidopsis*. Plant Mol Biol 54:743–753

Raskin I (1992) Role of salicylic acid in plants. Annu Rev Plant Physiol Plant Mol Biol 43:439–463

Regina MA, Carbonneau A (1997) Gas exchanges in *Vitis vinifera* under water stress regime III. Abscisic acid and varietal behavior. Perquisa Agropecuaria Braseleira 32:579–584

Reinth M, Terregrosa L, Luchaire N, Chatbanyong R, Lecourieux D, Kelly MT, Romieu C (2014) Day and night heat stress trigger different transcriptomic responses in green and ripening grapevine (*Vitis vinifera*) fruit. BMC Plant Biol 14:108

Rizhsky L, Liang HJ, Shuman J, Shulaev V, Davletova S, Mittler R (2004) When defense pathways collide. The response of Arabidopsis to a combination of drought and heat stress. Plant Physiol 134:1683–1696

Rocheta M, Becker JD, Coito JL, Carvalho L, Amancio S (2013) Heat and water stress induce unique transcriptional signatures of heat-shock proteins and transcription factors in grapevine. Funct Integr Genomics. doi:10.1007/s10142-013-0338-z

Rodrigues ML, Chaves MM, Wendler R (1993) Osmotic adjustment in water stressed grapevine leaves in relation to carbon assimilation. Aust J Plant Physiol 20:309–321

Rorat D (2006) Plant dehydrins – tissue location, structure and function. Cell Mol Biol Lett 11:536–556

Rorat T, Szabala BM, Grygorowicz B, Yin Z, Rey P (2006) Expression of SK3-typed dehydrin in transporting organs is associated with cold acclimation in Solanum species. Planta 224:205–221

Roy SJ, Tucker EJ, Tester M (2011) Genetic analysis of abiotic stress tolerance in crops. Curr Opin Plant Biol 14:232–239

Ruggiero C, Di Lorenzo R, Agelino G, Scaglione G, Gambino C, Do Vaio C (2012) Root hydraulic conductivity in three self-rooted and grafted table grape cultivars. J Int Sci Vigne Vine 46:177–183

Ruhl EH (2000) Effect of rootstock and K^+ supply on pH and acidity of grape juice. Acta Horticult 512:31–37

Sadder MT, Doss AC (2014) Characterization of dehydrins ahDHN from Mediterranean saltbush (*Atriplex halimus*). Turk J Biol 38:469–477

Salisbury FB, Ross CW (1992) Plant physiology, 4th edn. Wadsworth Publishing Company, Belmont

Salleo S, Lo Gullo MA (1993) Drought resistance strategies and vulnerability to cavitation of some Mediterranean sclerophyllous trees. In: Borghetti M, Grace J, Raschi A (eds) Water transport in plants under climatic stress. Cambridge University Press, Cambridge, pp 99–113

Satisha J, Prakash GS (2006) The influence of water and gas exchange parameters on grafted grapevines under conditions of moisture stress. S Afr J Enol Vitic 27:40–45

Satisha J, Prakash GS, Murti GSR, Upreti KK (2006) Response of grape rootstocks to soil moisture stress. J Hortic Sci 1:19–23

Satisha J, Prakash GS, Murti GSR, Upreti KK (2007) Water stress and rootstocks influences hormonal status of grafted grapevines. Eur J Hortic Sci 72:202–205

Satisha J, Somkuwar RG, Sharma J, Upadhyay AK, Adsule PG (2010) Influence of rootstock on growth, yield and fruit composition of Thompson Seedless grown in the Pune region of India. S Afr J Enol Vitic 31:1–8

Sauer MR (1968) Effect of vine rootstocks on chloride concentration in Sultana scions. Vitis 7:223–226

Schultz HR (2007) Climate change and world viticulture. Cost action 858 workshop: vineyard under environmental constraints: adaptations to climate change. Abiotic stress ecophysiology and grape functional genomics. University of Lodz, Poland

Schultz HR, Matthews MA (1988) Resistance to water transport in shoots of *Vitis vinifera* L. Plant Physiol 88:718–724

Scienza A, Boselli M (1981) Frequency and biometric characteristics of stomata in some grapevine rootstocks. Vitis 20:281–292

Seguin B, de Cortazar IG (2005) Climate warming: consequences for viticulture and the notion of 'terroirs' in Europe. Acta Hortic 689:61–71

Serra I, Strever A, Myburgh P, Deloire A (2013) Review: the interaction between rootstocks and cultivars (*Vitis vinifera* L.) to enhance drought tolerance in grapevine. Aust J Grape Wine Res. doi:10.1111/ajgw.12054

Shang Y, Yan L, Liu IQ, Zheng C, Chao M, Xin Q, Wu FQ, Wang XF, Du SY, Jiang J, Zhang XF, Zhao R, Sun HI, Liu R, Yi YT, Zhang DP (2010) The Mg-chelatase h subunit of Arabidopsis antagonizes a group of transcription repressors to relieve ABA responsive genes of inhibition. Plant Cell 22:1909–1935

Sharma J, Upadhya AK (2008) Rootstock effect on Tas-A-Ganesh (*Vitis vinifera* L.) for sodium and chloride uptake. Acta Hortic 785:113–116

Sharma J, Upadhyay AK (2004) Effect of moisture stress on performance of own rooted and grafted vines of Tas-A-Ganesh (*Vitis vinifera* L.). Acta Hortic 662:253–257

Sharma J, Shikhamany SD, Satisha J, Raghupathi B (2006) Diagnosis of nutrient imbalances in bunch stem necrosis affected Thompson Seedless grapes grafted on Dogridge rootstocks using DRIS. Indian J Hortic 63:139–144

Sharma J, Upadhyay AK, Band D, Patil SD (2010) Studies on black leaf symptom development and its impact on nutrient profile and fruitfulness in Thompson Seedless grapevines grafted on Dogridge rootstock. Indian J Hortic 67:156–160

Sharma J, Upadhyay AK, Bande D, Patil SD (2011) Susceptibility of Thompson Seedless grapevines raised on different rootstocks to leaf blackening and necrosis under saline irrigation. J Plant Nutr 34:1711–1722

Shi H, Shen Q, Qi Y, Yan H, Nie H, Chen Y, Zhao T, Katagiri F, Tang D (2013) BR-Signaling KinaSE1 physically associates with Flagellin Sensing2 and regulates plant innate immunity in Arabidopsis. Plant Cell 25:1143–1157

Skene KGM, Barlass M (1988) Response to NaCI of grapevines regenerated from multiple shoot culture exhibiting mild salt tolerance in vitro. Am J Enol Vitic 39:125–128

Skibbe M, Qu N, Galis I, Baldwin IT (2008) Induced plant defenses in the natural environment Nicotiana attenuata WRKY3 and WRKY6 coordinate responses to herbivory. Plant Cell 20:1984–2000

Smirnoff N, Crawford RMM (1983) Variation in the structure and response to flooding of root aerenchyma in some wetland plants. Ann Bot 51:237–249

Song Y, Jing SJ, Yu DQ (2009) Overexpression of the stress induced OSWRKY08 improves the osmotic stress tolerance in Arabidopsis. China Sci Bull 54:4671–4678

Storey D, Schachtman P, Thomas MR (2003) Root structure and cellular chloride, sodium and potassium distribution in salinized grapevines. Plant Cell Environ 26:789–800

Striegler RK, Howell GS, Flore JA (1993) Influence of rootstock on the response of Seyval grapevines to flooding stress. Am J Enol Vitic 44:313–319

Sykes SR, Newman S (1987) Apparent variation in chloride accumulation between vines of cultivars Italia and Mataro grown under furrow irrigation. Aust Salinity Newsl 15:71

Takahashi T, Kakehi JI (2010) Polyamines: ubiquitous poly cations with unique roles in growth and stress responses. Ann Bot 105:1–6. doi:10.1093/aob/mcp259

Tavladoraki P, Alessandra Cona A, Federico R, Tempera G, Viceconte N, Saccoccio S, Battaglia V, Toninello A, Agostinelli E (2012) Polyamine catabolism: target for antiproliferative therapies in animals and stress tolerance strategies in plants. Amino Acids 42:411–426

Tesniere CM, Romeiu C, Vayda ME (1993) Changes in the gene expression of grapes in response to hypoxia. Am J Enol Vitic 44:445–451

Tisi A, Federico R, Moreno S, Lucretti S, Moschou PN, Roubelakis- Angelakis KA, Angelini R, Cona A (2011) Perturbation of polyamine catabolism can strongly affect root development and xylem differentiation. Plant Physiol 157:200–215

Toumi I, Moschou PN, Paschalidis KA, Bouamama B, Salem-fnayou AB, Ghorbel AW, Mliki A, Roubelakis-Angelakis KA (2010) Abscisic acid signals reorientation of polyamine metabolism to orchestrate stress responses via the polyamine exodus pathway in grapevine. J Plant Physiol 167:519–525

Tregeagle JM (2007) Sustainable salt exclusion by salt excluding rootstocks of grapevine (*Vitis*). PhD thesis, La Trobe University, Bundoora

Tregeagle JM, Tisdall JM, Blackmore DH, Walker RR (2006) A diminished capacity for chloride exclusion by grapevine rootstocks following long term saline irrigation in an inland versus a coastal region of Australia. Aust J Grape Wine Res 12:178–191

Troncoso A, Matte C, Cantos M, Lavee S (1999) Evaluation of salt tolerance of in vitro-grown grapevine rootstock varieties. Vitis 38:55–60

Turner NC, Jones MM (1980) Turgor maintenance by osmotic adjustment: a review and evaluation. In: Turner NC, Kramer PJ (eds) Adaptation of plants to water and high temperature stress. Wiley, New York, pp 78–103

Upadhyay A, Upadhyay AK, Bhirangi R (2012) Expression of Na^+/H^+ antiporter gene in response to water and salinity stress in salt tolerant and sensitive grape rootstocks. Biol Plant 56:762–766

Upadhyay AK, Sharma J, Satisha J (2013) Influence of rootstocks on salinity tolerance of Thompson Seedless grapevines. J Appl Hortic 15:173–177

Upreti KK, Murti GSR (2010) Response of grape rootstocks to salinity: changes in root growth, polyamines and abscisic acid. Biol Plant 54:730–734

Upreti KK, Varalakshmi LR, Jayaram HL (2012) Influence rootstocks on salinity tolerance of grapevine: changes in biomass, photosynthesis, abscisic acid and glycine betaine. Indian J Plant Physiol 17:128–136

Vandeleur KR, Mayo G, Shelden CM, Gilliham M, Kaiser NB, Tyerman DS (2009) The role of plasma membrane intrinsic protein aquaporins in water transport through roots: diurnal and drought stress responses reveal different strategies between isohydric and anisohydric cultivars of grapevine. Plant Physiol 149:445–460

Vanholme B, Grunewald W, Bateman A, Kohchi T, Gheysen G (2007) The tify family previously known as ZIM. Trends Plant Sci 12:239–244

Vaseva II, Anders J, Feller U (2014) Identification and expression of different dehydrin subclasses involved in drought response of *Trifolium repens*. J Plant Physiol 171:213–224

Vidigal P, Carvalho R, Amancio S, Carvalho L (2013) Peroxiredoxins are involved in two independent signalling pathways in the abiotic stress protection in *Vitis vinifera*. Biol Plant 57:675–683

Walker RB, Blackmore DH, Clingeleffer RP, Ray CL (2002) Rootstock effects on salt tolerance of irrigated field-grown grapevines (*Vitis vinifera*L. cv. Sultana). I. Yield and vigor inter-relationships. Aust J Grape Wine Res 8:3–14

Wan SB, Wang W, Wen PG, Chen JY, Kong WF, Pan QH, Zhan JC, Tian LI, Liu HT, Huang WD (2007) Cloning of phospholipase D from grape berry and its expression under heat acclimation. J Biochem Mol Biol 40:593–603

Wang L, Li SH (2006) Salicylic acid induced heat or cold tolerance in relation to Ca2+ homoeostasis and antioxidant systems in young grape plants. Plant Sci 170:685–694

Wang LJ, Li SH (2007) The effects of salicylic acid on distribution of ^{14}C assimilation and photosynthesis in young grape plants under heat stress. Acta Hortic 738:779–7851

Wang Y, Ying J, Kuzma M, Chalifoux M, Sample A, McArthur C, Uchacz T, Sarvas C, Wan J, Dennis DT et al (2005) Molecular tailoring of farnesylation for plant drought tolerance and yield protection. Plant J 43:413–424

Wang LJ, Chen ST, Kun WF, Li SH, Archbold DD (2006) Salicylic acid pre-treatment alleviates chilling injury and affect the antioxidant systems and HSP of peach during cold storage. Postharvest Biol Tech 41:244–251

Wang D, Pajerowska-Mukhtar K, Hendrickson Culler A, Dong X (2007) Salicylic acid inhibits pathogen growth in plants through repression of the auxin signaling pathway. Curr Biol 17:1784–1790

Wang LJ, Fan L, Loescher W, Duan W, Liu G, Cheng J, Luo H, Li S (2010a) Salicylic acid alleviates decreases in photosynthesis under heat stress and accelerates recovery in grapevine leaves. BMC Plant Biol 10:34

Wang W, Jang K, Yang HR, Wen PF, Zhang P, Wan HL, Huang WD (2010b) Distribution of resveratrol and stilbene synthase in young grapevines (*Vitis vinifera* L. Cv. Cabernet Sauvignon) and the effect of UV-C on its accumulation. Plant Physiol Biochem 48:142–152

Wang M, Vannozzi A, Wang G, Liang YH, Tonielli GB, Zenoni S, Cavallini E, Pezzotti M, Cheng ZM (2014) Genome and transcription analysis of the grapevine (*Vitis vinifera* L) WRKY gene family. Hortic Res 1: Article No. 14016. doi: 10.1038/hortres.2014.16

Wegner LH (2010) Oxygen transport in waterlogged plants. In: Mancuso S, Shabala S (eds) Waterlogging signalling and tolerance in plants. Springer, Berlin, pp 3–22

Wi SJ, Kim WT, Park KY (2006) Over expression of carnation S-adenosyl methionine decarboxylase gene generates a broad-spectrum tolerance to abiotic stresses in transgenic tobacco plants. Plant Cell Rep 25:1111–1121

Wimalasekera R, Tebartz F, Scherer GF (2011) Polyamines, polyamine oxidases and nitric oxide in development, abiotic and biotic stresses. Plant Sci 181:593–603

Xu Z, Zhou G (2008) Rcsponsc of lcaf stomatal density to water status and its relationship with photosynthesis in a grass. J Exp Bot 59:3317–3325

Yamaguchi K, Takahashi Y, Berberich T, Imai A, Miyazaki A, Takahashi T et al (2006) The polyamine spermine protects against high salt stress in *Arabidopsis thaliana*. FEBS Lett 580:6783–6788. doi:10.1016/j.febslet.2006.10.078

Yamaguchi-Shinozaki K, Shinozaki K (2006) Transcriptional regulatory networks in cellular responses and tolerance to dehydration and cold stresses. Annu Rev Plant Biol 57:781–803

Yan L (2013) Studies on water logging tolerance of grape rootstocks. Master of Horticulture (Pomology) thesis submitted to Shadong Agri Univ, Shadong, China. http://thebestthesis.com/?doc/2149903

Yan Y, He M, Zhu Z, Li S, Xu Y, Zhang C, Singer S, Wang Y (2012) Identification of the dehydrin gene family from grapevine species and analysis of their responsiveness of various forms of abiotic and biotic stress. BMC Plant Biol 54:743–753

Yildirim O, Aras A, Ergul A (2004) Response of antioxidant system to short term salinity stress in grape vine rootstocks-1616C and *Vitis vinifera* L. cv. Razaki. Acta Biol Cracov Ser Bot 46:151–158

Yohannes DB (2006) Studies on salt tolerance of Vitis spp. Ph.D thesis submitted to University of Agricultural Sciences, Dharwad, India, p 132

Yokoi S, Quintero FJ, Cubero B, Ruiz MT, Bressan RA, Hasegawa PM, Pardo JM (2002) Differential expression and function of Arabidopsis thaliana NHX Na+/H+ antiporters in the salt stress response. Plant J 30:529–539

Zamboni M, Fregoni M, Iacono F (1986) Compaortamento di specie edibridi di vite in condizioni di siccita. Attidel IV simpocio. Internazionale GeneticadellaVite Vignevini 12:119–122

Zawoznik MS, Ameneiros M, Benavides MP, Vasquez S, Groppa MD (2011) Response to saline stress and aquaporin expression in *Azospirillum*-inoculated barley seedlings. Appl Microbiol Biotechnol 90:1389–1397

Zhu B, Chen THH, Li PH (1995) Activation of two osmotin-like protein genes by abiotic stimuli and fungal pathogen in transgenic potato plants. Plant Physiol 108:929–937

Abiotic Stress Tolerance in Banana

12

Iyyakutty Ravi and M. Mayil Vaganan

Abstract

The key issues that are to be addressed to sustain agriculture in the twenty-first century are water availability, salinity and extreme temperatures. Bananas are staple food for hundreds of millions of people in Africa, and it is cultivated mostly in tropical and subtropical countries. In India, bananas are grown in the regions of humid tropics, humid subtropics and semiarid tropics, and some banana genotypes can grow up to an elevation of 2000 MSL. Suitable climate for banana growth is warm moist weather without strong winds throughout the year. The major constraints for banana production are largely dominated by biotic and abiotic stresses. Though research on biotic stresses in banana has drawn sufficient attention worldwide, abiotic stresses are not. The successful cultivation of water-loving Cavendish clones, in drought-prone areas with protected irrigation, has provided the sufficient momentum to research on drought in bananas. Banana plants, especially the Cavendish clones, are sensitive to abiotic stresses like soil moisture deficit, salinity, extreme temperatures and strong winds. As bananas are recalcitrant to conventional breeding approaches, clonal selection and tolerant genotype selections from germplasm are the major sources to address the abiotic stresses. In the recent past, a lot of attention has been given to study the basic mechanisms involved in tolerance and management of the abiotic stresses in banana, and these aspects are discussed in this chapter.

12.1 Introduction

Bananas (referring to banana, plantain and cooking bananas) are economically important high-value commercial as well as subsistence food crop. It is one of the earliest crop plants to have been domesticated. Originally, they have

I. Ravi (✉) • M.M. Vaganan
ICAR-National Research Center for Banana, Thogamalai Road, Trichy 620102, Tamil Nadu, India
e-mail: Ravi.I@icar.gov.in; iravi24@gmail.com

N.K.S. Rao et al. (eds.), *Abiotic Stress Physiology of Horticultural Crops*,
DOI 10.1007/978-81-322-2725-0_12

been adapted from the humid tropics to broad subtropics climatic conditions. With the exception of a few large companies, traditional banana growers are responsible for most production worldwide. Bananas are grown in about 130 countries around the world (FAOSTAT 2012), exhibiting a spectacular production of 122.85 million tons. India produces 30.25 million tons on an area of 0.811 million ha (NHB 2014).

Bananas originated in Indochina and Southeast Asia, where the greatest diversities of wild *Musa* species (*Musa acuminata* AA and *Musa balbisiana* BB) are found today and the earlier domestication occurred (Simmonds 1962). *Musa acuminata* gave rise to different varieties of AA diploid cultivars through natural hybridizations between various subspecies of the polymorphic species of AA. The diploid AA gave rise to triploid AAA through chromosome restitution during meiosis (Stover and Simmonds 1987; Price 1995).

In bananas, the rate of appearance of new leaves and rate of fruit growth are largely governed by temperature. Broadly speaking, banana cultivation is widely practised in areas with the 102 cm isohyet and the 15.5 °C (60 °F) winter isotherms which roughly coincides with 30° latitude N and S of the equator. Deep, well-drained loamy soil with adequate organic matter is ideal for its cultivation. The most suitable climate is with warm moist weather without strong winds throughout the year. The major abiotic factors which influence the banana growth and productions are water, temperature, wind and mineral nutrients. The banana is basically a plant of humid tropics but adapted to a wide range of climatic conditions from wet tropical to dry subtropical areas.

Abiotic stresses are caused by nonliving factors, light (high light, UV, darkness), water (deficit, desiccation, flooding), salt, temperature (frost, low, heat), nutrients (nutrient imbalance), oxidation stress, hypoxia and physical (wind, touch). These environmental factors cause disorders that are non-infectious. The extremes of water, temperature, salts, nutrients and light cause stress to the plants. Abiotic stresses have the potential to cause yield loss up to 65–87 %. Tolerance to these stresses depends on species, genotype and developmental stage of the plant. Stress resistance mechanism are avoidance (preventing exposure of plant to the stress), tolerance (allowing the plant to withstand against the stress) and acclimation (altering the physiology in response to the stress). Duration and rate of stress onset are the factors that can affect the response of plants to acclimatize against various types of stresses. The current technologies that are available to increase banana and plantain production have neglected the abiotic stress constraints (Wairegi et al. 2010).

12.2 Growth and Development

12.2.1 Vegetative Phase

The banana aerial shoot is comprised of the pseudostem (encircled leaf sheath) and leaves. This aerial shoot is borne on a subterranean rhizome, which grows to a length of about 0.3 m. Suckers or followers arise from lateral buds on the corm (Carr 2009). Banana shoot is a determinate type (Norman et al. 1984). Mechanically the aerial pseudostem is completely dependent on the support leaf sheaths reaching to a height of 2–8 m in cultivars. The leaf area of each cultivar can reach up to 2–3 m^2 and the leaf area of a plant may go up to 17–25 m^2 (Stover and Simmonds 1987). The leaf area index (L) may vary from 2 to 6 or more depending on the banana cultivar, season and plant density. The maximum use of photosynthetically active radiation is associated with 4.5 L (Turner 1990), and the leaf canopy with 4.5 L can intercept about 90 % of incoming solar radiation (Turner et al. 2007). The leaf primordia initiated at the shoot apex develop into leaves that rapidly encircle the shoot apex, and because of their condensation, it is difficult to identify the successive internodes (Carr 2009). However, on flowering, the successive internodes are visible, when a true erect stem appears (Barker and Steward 1962). For banana normal plant growth and leaf production, the optimum temperature is around 31–32 °C and leaf emergence stops below 9–10 °C. In summer, banana plant may produce a leaf per week, but half a leaf per month is produced during winter (<15 °C) (Turner 2003). Plant health and favourable climate rate influence leaf production. In a healthy banana plant, under favourable tropical conditions, a leaf is produced at 7–10 days' interval in "dwarf Cavendish" banana during summer. But in winter, it takes 15–20 days for a leaf production depending on the temperature. Banana plant leaf elongation rate is sensitive to soil moisture deficit stress with a range of −10 to −80 KPa (Hoffman and Turner 1993).

12.2.2 Root Development

Like many other monocotyledons, bananas follow the similar arrangement of roots. The root axes (cord roots), which produce primary laterals, arise from the upper part of corm and secondary laterals produced from the primary. Generally a root is considered, when it is having an axis with laterals. A banana plant (single shoot) may have several hundreds of roots (Price 1995). Root hairs are present just behind the root apex, and it measures up to 2 mm in length. The root axes can grow up to lengths of 3–5 and diameters of 4–10 mm, and besides they are relatively straight and cylindrical, (Draye et al. 2005).

Robinson (1987) recorded that primary and secondary laterals and root hairs remained functional (based on their external morphology) for about 8 weeks, 5 weeks and 3 weeks, respectively. In bananas, the number of roots and extension reach the maximum at flowering, and the root length can spread laterally up to 5 m, but mostly it is found within the radius of 0.6 m. Most of the roots can reach vertically up to 0.4 and sometimes reach 0.6 m (Carr 2009). On the size and distribution of the root system, irrigation plays small role but, cultivar differences may exist non-commercial cultivars and it needs to be studied which are grown in subsistence agriculture' (Price 1995).

12.2.3 Flower Development

The Cavendish (AAA) group of banana produces 40–50 leaves before flowering, and it may vary with genotypes. After emergence of 30–40 leaves, floral initiation occurs, and then the aerial stem starts to elongate through pseudostem till flowering. Floral initiation to inflorescence emergence from the throat of the pseudostem is called the floral phase. The type of inflorescence in a banana is called spike. The flowers are arranged in nodal clusters in two rows on transverse cushions (crown) (Carr 2009). Without pollination, the fruits grow from the female flowers (Purseglove 1972). The inflorescence on emergence behaves as positively geotropic in cultivated bananas and results in hanging of banana bunch. Irrigation to maintain optimum soil moisture during dry season produces more number of female flowers in cv. Robusta (AAA) (Holder and Gumbs 1982).

12.2.4 Fruit Development

In banana, flowering to harvest is seasonally variable. It ranges from 90 to 110 days in tropics and subtropics (summer) to 200 days in subtropics' autumn season. Flower initiation at winter season in subtropics produces small and malformed bunches (Robinson 1982). A banana bunch is one which contains hands (cluster of fruits), which is composed of fingers (individual fruit). The drought effect on bunch development was studied in cv. 'Grande Naine' AAA at the Canary Islands (28°37′N) by withholding irrigation for 63 days from flower emergence. The water stress affected finger length and reduced diameter by 9 % and bunch fresh weight by 41 % besides delaying fruit maturity (Mahouachi 2007).

12.3 Drought Stress Tolerance

Drought is a global problem for agriculture; it affects 28 % of the world's soils. Less water holding capacity and shallowness of soils are affecting half of soils intermittently in the agriculture land (Dudal 1976). 'Shortage of water in the root zone, resulting in decreased crop yield' is defined as drought in agriculture (Salekdeh et al. 2009). Drought avoidance and/or dehydration tolerance constitutes drought tolerance, which is finally considered by the successful reproductive behaviour of the species (Taylor et al. 2007). The plants having inherent mechanism for drought avoidence strategies may have deep roots, less use of available and conserve for reproductive organ development and maturation and modification of life cycle to match water availability. The plant's ability to dehydrate partially but remain alive and grow again as soon as water is available to them is called dehydration tolerance.

The plant response to drought is a complex process as it manifests with many protective adaptations. Drought triggers overheating of plant cells and tissues and dehydration, but the drought-resistant plants withstand against dehydration and overheating. There is no set of universal mechanism existing in plants to adapt to drought, and high temperature resistance and drought resistance are not always linked. During the process of ontogeny and whole plant preceding phylogeny, the drought resistance property is formed. Henckel (1964) defined that "drought-resistant plants are those which in the process of ontogenesis are able to adapt to the effect of drought and which can normally grow, develop and reproduce under drought conditions because of a number of properties acquired in the process of evolution under the influence of environment". Among the abiotic stress constraints, drought is the major stress affecting banana cultivation and its expansion into new growing areas. In the light of changing climate, the need of the hour is to have improvement and management strategies for existing commercial banana cultivars to combat abiotic stresses.

12.3.1 Plant-Water Relations

In banana, the presence of laticifers (containing latex) in leaves, fruits and corms and large air pockets in the leaves pose problems to use standard methods to measure water relations in water-deficit banana plants (Turner and Thomas 1998). However, the refractive index of exuded latex was correlated with water potential (Milburn et al. 1990) and further confirmed by Thomas and Turner (2001). The well-watered banana plants may have the water potential of 0 to −0.35 MPa and follow a diurnal cycle pattern within this remarkably narrow range. The plant-water status can be seen from the rate of extension of the youngest banana leaf as it is the most sensitive to plant-water status (Kallarackal et al. 1990), provided when the air temperature is not very hot (Thomas and Turner 1998). Therefore, leaf folding cannot be considered as a reliable plant trait for banana crop irrigation schedule (Thomas and Turner 1998).

Abiotic and biotic stresses are the dominant factors which affect banana production in a larger way. However, much effort has not been taken for research on abiotic stresses in banana than biotic stresses, where worldwide research is on bigger way. The most important abiotic stresses are drought, salinity and extreme temperatures. Among the abiotic stresses, drought, salinity and heat are the most important. In banana, drought has been addressed scarcely in early but in the face of changing climate and dwindling of natural. The Cavendish clones are the major commercially cultivated banana and traded in large scale throughout the world. These plants are generally grown in areas where assured water supply to the plants is available throughout the year. Now the successful cultivation of Cavendish clones under protective irrigation has given sufficient momentum to take up research in bananas on drought. In subtropical and semiarid areas, several improved management techniques have been adopted, especially in the areas of improvement of physical properties of the soil, enrichment of nutrients and water management through drip irrigation and fertigation. However, evaluation of banana germplasm in large scale in the field is limited, because the creation of logistics for the long-duration crop hinders and poses problem for the research in drought.

The IITA, Nigeria; ICAR-National Research Centre for Banana (NRCB), India; and CICY, Mexico, are engaged in research on screening and evaluating germplasm and landraces against drought tolerance. At ICAR-NRCB, 112 banana genotypes, from core germplasm (340), were screened against drought with leaf water retention capacity and identified a few drought-tolerant and moderately tolerant genotypes (Ravi and Uma 2009). Bananas are prone to water stress in subtropical and semiarid zones. Since banana plants are long-duration crop, it is prone to water stress any time in the crop cycle as in change in rainfall pattern, shortage of rainfall or decrease of groundwater. During this time supplementary irrigation is essential to shield the negative impacts of water deficit. Besides, higher leaf area index and shallow root system make banana susceptible to soil water stress (Robinson 1996).

12.3.2 Evapotranspiration from Banana Plantations

The amount of water to be applied based on cultivar, stage and season to all commonly grown cultivars in all banana-growing places is lacking, although some information is available for Cavendish clones. Robinson and Alberts (1989) estimated that, depending on the climatic conditions, the annual evapotranspiration of 1200– 2690 mm occurs in a banana plantation. Under semiarid condition, the water requirements for a drip-irrigated bananas on Mollisol or on an Ultisol with transient dry periods may range from 0.25 to 1.25 of class A pan factors, and irrigating bananas over and above the class A pan factor may significantly increase yield components of the plant crop and ratoon crops (Goenaga and Irizarry 1998; Young et al. 1985). Under humid tropics, the water requirement of bananas s about 1–1.4 times of class A pan evaporation (Stover and Simmonds 1987).

12.3.3 Genotypic Variation for Drought Resistance

Field experiments and research reports established that bananas are sensitive to soil moisture stress (Robinson 1996). Banana plants close the stomata under soil moisture stress condition, and it is linked to root signal passed on to leaves (Turner 2003; Thomas 1995); as a result the plant remains hydrated (Shamueli 1953; Turner and Thomas 1998). Though it conserves water in the leaves, due to closure of stomata, the plant carbon assimilation reduced (Thomas 1995). Besides, when the evaporation exceeds water absorption by roots, we may find the decrease in transpiration and stomatal conductance (Robinson and Bower 1988). However, the architecture of the root system, i.e. root length density (Ld) and specific root length (Lw), may have been linked to drought stress in bananas, and it is worth exploring (Turner 2005; Blomme et al. 2005).

Not much research reports are available on the effect of soil moisture deficit stress on bananas. In a field-grown banana, water stress was imposed at flowering which decreased the bunch weight to the tune of 42.07 %, 25.0 % and 18.83 % in cvs. Robusta, Karpuravalli and Rasthali, respectively (Ravi et al. 2013). In this experiment, cv. Robusta was found to be sensitive. The fruit length (11–14 %) and circumference (5.75–16 %) were decreased when soil moisture stress was effected at flowering (Anon 2008).

In bananas, the yield may drastically get reduced, when the irrigation interval became long and available and goes <66 % (Robinson and Bower 1988). Also, occurrence of drought at bunch emergence period may cause shorter fruits, duration of fruit filling may get extended, shorter green life and maturity bronzing in the fruits (Daniells et al. 1987). It is said that, because of the herbaceous nature and highly hydrated pseudostem to keep the plant upright, bananas require a considerable amount of available soil water for normal growth and development. Banana plant transpires *ca.* 40–50 mg H_2O

$dm^{-2} min^{-1}$ under full sunshine (Shamueli 1953; Morello 1954; Tai 1977). In a banana plantation with a density of 1500 plants/ha with LAI of 2.1, water requirements may range from 1170 m^3, 765 m^3 and 450 m^3 in sunny, semi-cloudy and cloudy weather, respectively (Ravi et al. 2013). The pulse system of irrigation practice is important for humid tropics and semiarid areas as it reduces the soil temperature in the upper soil layer and decreases NO_3 leaching loss (Lahav and Kalmar 1981).

Musa balbisiana genotypes might have been domesticated under extreme weather conditions and are often reported as resistant to abiotic stresses. Banana genotypes with the B genome are considered resistant to abiotic stresses than the A genome. In Egypt, the cultivars with AAB and ABB genome makeup cultivated in rural areas are recorded more resistant to drought than other Cavendish clones (De Langhe 2002), and 'Sugar' and ABB 'Pisang Awak' genotypes in Oman grow well to dry environment at the Agriculture research station of Salalah (De Langhe 2002). Thomas et al. (1998) reported that the B genome makes the plant hardy to dry environment as the plant with more B genome is recorded better gas exchange parameters and it is insensitive to air dryness.

In bananas, there is no single set of mechanism operating to resist or tolerate against drought or soil moisture stress. Reactions of banana genotypes towards drought and its physiological, biochemical and agronomic expression are reported by a few researchers (Garcia and Manzanilla 1994; Cayón et al. 1998; Bananuka et al. 1999; Ismail et al. 2000; Shamsuddin et al. 2000; Wagner et al. 2000; Abeywickrama and Weerasinghe 2002; Chai et al. 2005; Anon 2008; Ravi and Uma 2009; Ravi and Uma 2011, 2012; Ravi et al. 2013). The drought-resistant genotypes manifest less reductions in photosynthetic parameters and leaf area and maintain higher leaf water content (Bananuka et al. 1999). Banana plants selected for higher transpiration efficiency (TE) may perform well under well-watered as well as soil water-deficit conditions (Kissela et al. 2015)

12.3.4 Water Requirement and Management of Water Stress

The water requirement for total growth period is high, as it is an herbaceous and long-duration crop. The water requirements for banana in the humid tropics are 1200 mm and 2200 mm per year in the dry tropics. Any water shortage for this high water-loving crop may reduce the yield (www.fao.org). Adequate water supply and sufficient nutrients during early and late vegetative phase are crucial, which determine the plant growth potential and yield parameters (Turner 1995). The rate of leaf appearance gets affected by the occurrence of water deficits during these vegetative phases and results in negative effects on yield and yield parameters. Uniform rainfall distribution throughout the year at the rate of >100 mm per month with a temperature range of 25–30 °C favours banana growth and production. The major abiotic factors that hinder the expansion of the cultivation of banana are limited water supply with long dry season, extreme temperatures and stormy or cyclonic winds.

12.3.5 Management of Drought Stress

12.3.5.1 General External Symptoms

The general symptoms due to loss of water include reduction in leaf emergence rate, loss of turgor in cigar leaves, drooping of leaves, premature senescence, yellowing and drying of young leaves, bunch choking (Fig. 12.1) and reduction in pseudostem size. Sometimes, combined effects of soil moisture deficit and high temperature stresses damage plants (Fig. 12.2) more than their occurrence in isolation. In severe stress conditions the plant may topple down due to weakness of the pseudostem.

12.3.5.2 Critical Stages Susceptible to Soil Moisture Stress

Occurrence of soil moisture deficit stress at the third month after planting may not affect the bunch size but it delays shooting by

Fig. 12.1 Choked banana bunch due to soil moisture deficit stress

Fig. 12.2 Soil moisture deficit with higher atmospheric temperature affected banana plant

15–20 days. Soil moisture deficit stress, in Grand Nain, at the fifth month after planting may result in development of malformed bunches or reduction in number of hands and fingers. The occurrence of soil moisture deficit stress during the time of flowering and the first 1 month after flowering may affect the normal fruit development, i.e. reduced fruit length and girth and reduction in bunch weight to the tune of 42.07 % in susceptible banana cv. Robusta.

In bananas, the floral primordia start differentiating around the fifth month after planting in plants with a duration of 12 months, but the vegetative organs continue to grow. Drought stress, i.e. soil moisture stress, occurs during this period that may affect the number of hands, fruits per hand and size of fruits. Soil moisture deficit occurs during the bunch development which leads to poor fruit filling (Fig. 12.3) and leads to reduction in the fruit size and quality. A reduced number of leaf at flowering may extend the fruit maturity duration and some time the fruit may start ripening while in the plant itself and it leads to poor fruit quality and storage period.

12.3.5.3 Mitigation of Soil Moisture Deficit Stress

In banana cultivation, the possibility of soil moisture deficit stress occurrence is more, as the crop duration is of the minimum of 1 year and it passes through all the seasons during this growing period. Therefore, we must give due attention towards maintenance of optimum soil moisture during critical stages of banana plant growth, and any lapse would cause yield loss.

The practices to be followed in banana cultivation to tackle the soil moisture deficit stress are as follows:

Fig. 12.3 Plant affected with soil moisture deficit at the time of bunch development

(i) **General Agronomic Practices**
  - Deep planting of suckers (placing of sucker at 1.5′ depth), so that the growing roots go deep to extract the soil moisture
  - Installation of drip system, as it delivers water in the root zone and saves water by 25–30 %, and application of optimum quantity of water through drip at right stages
  - Mulching the area between banana plants with dead leaves/trashes or black polyethylene sheets to reduce evaporation loss

(ii) **Irrigation at Critical Growth Stages**

Since the depletion of total available soil water about 35 % during the total growing period is harmful to growth and fruit production, frequent irrigation is important. The irrigation interval depends on evapotranspiration and the soil water holding capacity in the rooting depth which may vary from 3 days under high evaporative conditions in light soils and up to 15 days under low evaporative conditions in high water-retaining soils. Under limited rainfall and irrigation water, it is advantageous to reduce the depth of each water application rather than to extend the irrigation interval. Regular water supply through irrigation over the total growing season as compared to rain-fed production, with seasonal differences in water supply, produces taller plants with greater leaf area and results in earlier shooting and higher yields. The interval between irrigations has a pronounced effect on yields, with higher yields being achieved when intervals are kept short. Under conditions of limited water supply, total production will be higher when full crop water requirements are met over a limited area than when crop water requirements are partially met over an extended area. The banana plant has sparse and shallow root system (Krishnana and Shanmugavelu 1980). Most feeding roots are spread laterally near the surface up to 3–4 m. The rooting depth generally does not exceed 0.75 m. In general 100 % of the water is obtained from the first 0.5 to 0.8 m soil depth with 60 % from the first 0.3 m.

(iii) Whenever drip irrigation is used under conditions of high evaporation, of low rainfall and particularly when irrigation water contains even a small amount of salt, accumulation of salts at the boundary of wet and dry soil area will occur. Under such conditions leaching will often be needed since banana plants are highly salt sensitive and damage to the crop can otherwise easily occur. Overhead sprinkler systems with small application at frequent intervals are commonly used in commercial banana plantations in many countries. To maintain the soil moisture, mulching with dry leaves is also followed.

Regular irrigation needs to be provided to maintain optimum soil moisture (80–90 % of available soil moisture) at the fifth to sixth months after planting, which coincides with floral primordial initiation and development. Irrigation is also crucial during flowering and subsequently 1 month after flowering. Adopting these practices would ensure the chances of getting normal bunch emergence and fruit development.

(iv) **Drought-Tolerant Cultivars**

In conditions where farmers do not get sufficient irrigation water for providing regular irrigation, growing 'Saba' (ABB), Monthan (ABB), Karpuravalli (ABB) and Poovan (AAB) which are relatively drought tolerant than other cavendish clones like 'Grand Naine' and 'Robusta' could be an option.

(v) **Maintenance of Optimum Number of Leaves**

There is a general belief that retaining more number of leaves at flowering would lead to development of big bunches. When unlimited water supply situation exists, maintenance of more leaves is not a harm. But when the farmer is having only limited water supply, they can maintain eight to ten top leaves at flowering. Thereby water loss through transpiration can be reduced and cut leaves can be used for mulching to cover the soil surface. Maintaining eight to ten numbers of top green leaves at flowering makes fruit development and quality as normal. Besides, at optimum number of leaves paves the way for light to reach the lower leaves, reduces transpirational water loss and decreases leaf spot disease inoculum to spread further (www.nrcb.res.in).

(vi) **Bunch Spray**

After flowering, opening of all hands and de-naveling of male bud, application of 2 % potassium sulphate as a bunch spray in the first instance and second bunch spray after 15–20 days of first spray increases finger length and girth and bunch weight. Since plant is having sufficient moisture during this stage, it hastens the fruit development (both length and girth). By adopting this practice, the bunch weight could be increased up to 1.5 kg as compared to non-sprayed bunches, besides producing fruits with glossy appearance which can fetch good market price (Ravi et al. 2014b).

12.4 Salt Stress Tolerance

In India, salinity and alkalinity is one of the major factors affecting agriculture productivity, 6.73 mha of land has been already affected by salts, and groundwater (25 %) used for agriculture is either saline or brackish water and it is predicted that by 2015, 11.7 mha land will likely to get affected by salts (www.cssri.org). The salt-affected soil problem is further invigorated by bringing large-scale additional areas under irrigation without an adequate development of drainage (FAO 1988).

Salt-affected soils limit agriculture production, because of toxic levels of salts or deficiency of plant essential nutrients. These soils have low productivity and have great potential for agriculture production if managed effectively. Problem soils primarily affect plant growth through root either directly or indirectly with combination of osmotic and ionic stress.

Salinity-related problems arise under a dry climate on saline soils and/or because of the use of low-quality irrigation water. Excessive salinity increases the sodium content in roots with marked depression in potassium uptake, reduces the growth rate, delays flowering and decreases crop yield. The optimal value for the soil K/Na ratio is 2.5 and banana yields decrease when sodium proportion is excess by 8 % of the sum of exchangeable cations. Since the adverse effects of salinity cannot be avoided completely, the cooking banana 'Saba' can be grown. There must be frequent irrigation with good-quality water with adequate drainage. Traditional banana-growing areas, especially in command areas, are becoming more saline due to secondary salinization, i.e. due to regular flood irrigation and poor drainage. Bananas are generally sensitive to salt stress, especially the cultivars like Nendran and Robusta. The salt stress leads to poor plant growth and yield. Banana, *acuminata* clones are considered sensitive to salinity (Israeli et al. 1986). The salt stress injuries are initially expressed as leaf margin necrosis and later progress towards the centre part of leaf. The symptoms in leaves appear in an acropetal manner. Salt effects manifest into relatively reduced stem thickness and prolonged duration to flowering, sometime, more than 2–3 months. Susceptible cultivars also fail to develop normal finger (fruit) size and produce low-quality marketable bunch.

Many plant species thrive well in the saline environment despite the existence of negative effects of salinity which affects physiology of plants, such as CO_2 assimilation, protein synthesis, respiration or phytohormone turnover (Mengel and Kirkby 1987). Therefore, the mechanisms to combat salt effect in the plant system operate from the whole plant to cellular level. Generally, in salinity salt-tolerant plants may manifest one or combination mechanisms, i.e. succulence of plant organs, salt exclusion, excretion of salts, osmotic adjustment through synthesis of combatable solutes and accumulation of ions in the tonoplasts and with more unsaturated fatty acids in the membrane composition. In India, banana is cultivated in more than 30 % of the area under adverse soil condition with pH >8.5, ECe >4 dSm^{-1} and ESP >15 (www.nrcb.res.in). Banana, acuminata clones are considered sensitive to salinity (Fig. 12.4). The lower photosynthetic activity in such leaves leads to yield loss due to reduced finger weight. The salt affects the fruit development in susceptible varieties especially in the Cavendish group of banana (Fig. 12.5), although not in tolerant genotype 'Saba' (Fig. 12.6).

Saba, a tolerant cooking banana, develops bunch normally under salinity conditions in affected fields (Fig. 12.6).

Plant growth is inhibited by salts present in soil water due to a more negative osmotic potential in the soil solution than in the plant water; thereby, it reduces the plant-water uptake and results in a reduced rate of plant growth. This phenomenon is called as salinity-induced water deficit. The excess salts entered into the plant system, by the force of transpiration stream, may cause ion toxicity and cause cell injuries in the transpiring organs, and thereby it further reduces the growth in plants, and it is called as ion-excess effect of salinity (Greenway and Munns 1980). In bananas, salinity influences the fruit physical parameters like fruit length, circumference, fruit pulp, peel weight, volume and density (Ravi et al. 2014). These are important commercial criteria which determine marketing quality and price of banana.

12.5 Temperature Stress

Temperature has a major influence on banana plant growth and development. As it has originated from tropical origin, the extreme temperatures, higher and lower, have a negative impact on the overall development. When the temperature is optimum (31–32 °C), each plant may produce four or five leaves a month, provided all other factors, viz., water, minerals, etc., are not limiting. In such an optimum environment, the banana plant may grow faster and comes to flowering even earlier than normal. But

Fig. 12.4 Banana plants grown in salt-affected field showing necrotic symptoms in leaf margins

Fig. 12.5 Robusta (Cavendish group) grown in salt-affected field showing poor bunch development

Fig. 12.6 Saba (cooking banana) grown in salt-affected field showing normal bunch development

in winter (13–14 °C) the plant may produce a half leaf per month and even the leaf emergence may stop, when the temperature drops below 9–10 °C. Temperature influences the orientation and arrangement of the leaves. In plants grown under cool conditions, the leaves are more upright, but under warmer conditions, they are more horizontal (Ravi et al. 2013). Under subtropics, for any banana genotype, plant development is slower than in the tropics but not significant in the case of yield and bunch size. For bunch development, the optimum temperature is around 21–22 °C and stops above 38 or 39 °C and below 9 or 10 °C. Under subtropical condition, a bunch emerging during winter may take about 5–6 months from flowering to harvest and 3 months for one emerging in summer. Therefore, temperature has a major influence on banana fruit growth rate and bunch sleeves may help the fruit grow faster during winter as it warms the fruit and maintains the temperature gradient uniformly. At higher temperatures, i.e. above the optimum, a sizeable amount of carbohydrates are used up for respiration by all other organs, i.e. leaves, pseudostem, corm and roots, and much less resource is available for growth.

12.5.1 Effect of High Temperature

Sunburn high temperature (>38 °C) associated with bright sunshine causes sunburn in the banana leaf (Fig. 12.7), and when the temperature is high, the peduncle and top hands of the bunch may also get affected by the sunburn.

Protective bunch sleeve (non-woven fibres) can be used to avoid sunburn. Care is needed if bunch covers made of polyethylene and we must place paper between fruit and bunch sleeve or the bunch sleeves with reflective coating outside are to be used. Bunch sleeves reduce the temperature gradient across the bunch. Bunches covered with bunch sleeves fruits are more uniform than the uncovered one.

12.5.2 Effect of Low Temperature

In the tropics the low temperature rarely causes the direct damage except in higher altitudes. Chilling causes the latex in the peel to coagulate and the latex ducts become discoloured. This gives dull yellow or even a greyish colour (severe chilling) to the peel of the ripened fruit. In bananas, the inflorescence impedance and bunch malformation occurs, when temperature drops below 10 °C. Chilling of fruit shows up as water-soaked patches underneath the skin surface, and the ripening becomes more difficult and the chilled skin turns black as the fruit ripens (www.niam.res.in). Chilling symptoms on leaves are seen after 2–4 days of stress and the leaves turns yellow. In the recently emerged leaf, the midrib may appear brown with water-soaked symptoms underneath (Turner and Lahav 1983). The potassium deficiency symptoms can be seen in older leaves. The older leaves break at the petiole and three or four leaves may hang like a skirt around the pseudostem. These leaves quickly turn yellow and die. Under the cooler conditions the banana cannot maintain all its leaves (www.niam.res.in).

Choking in choked banana plants that appear, the closeness of petioles looks like 'rosette' or 'choked' due to compactness. Banana choking is generally seen in plants suffering from soil moisture deficit stress, when the temperature is above 36 °C and during winter months. In these circumstances, the bunches may not emerge fully. This symptom is commonly seen in Cavendish bananas. 'Choked throat' is a kind of physiological disorder of bananas and here the bunch may get trapped at different stages of emergence. These bunches are highly susceptible to sunburn and produce low marketable valued fruits.

Fig. 12.7 Sunburn damage caused leaf blade

12.5.3 Frost

Once temperatures drop below freezing, damage occurs quickly, and a few hours later the leaves and bunches become water soaked, then blacken, and die. Frost rarely kills the whole plant but it usually damages only exposed tissues. Within a few days of frost exposure, the youngest leaf begins to emerge again and growth resumes. When plants are completely defoliated by frost, the formation of the bunch can be delayed, and, if a bunch is already present in the pseudostem, its yield will be reduced. These effects are not seen until some months after the frost has occurred.

12.5.4 Remedial Measures

- To avoid the adverse effects of cold weather, maintain good leaf area without disease to promote rapid filling of the bunch.
- Prune hands in the bunches to allow rapid filling of few hands.
- Plant in warm localities and provide shelter from cold winds.
- Avoid planting in areas where frost occurrence is frequent.
- Use bunch covers to protect fruits.
- Badly affected plants by frost are to be removed and immediately allow the next ratoon crop.

12.6 Water Logging

Oxygen is required for root respiration and decrease in availability of oxygen at root zone leads to anoxia (anaerobic condition). The anoxic condition could occur due to soil compaction and water logging. Therefore, air and water balance and water circulation are important in the rhizosphere region. Optimum air ratio in soil space is required for normal and good banana root growth. The girth of pseudostem is directly related to soil porosity. Continuous submergence of banana root system under water for more than 48 h leads to anoxic situation for root and impairs the normal respiratory metabolism and results in yellowing of leaves and plant death. Therefore, good drainage is required for normal root growth and metabolism. The periodic loosening of surface soil without affecting the root system is always beneficial for banana plant growth and production.

12.7 Conclusions

The key issues that are to be addressed to sustain agriculture in the twenty-first century are water availability, salinity, extreme temperatures and nutrient use efficiency. Many attempts have been made to increase productivity of agriculturally important crops under different types of abiotic stresses. Natural selection of plants against abiotic stresses made them to survive, but only the artificial selection and hybridization processes increased the crop yield. A lot of basic researches in cereal crops have provided significant knowledge gain on physiological and molecular responses towards abiotic stresses. But there are many research gaps in most of the plantation crops and banana is not an exception. The complex mechanism of abiotic stress tolerance in plants makes the research progress in slow pace for increasing yield in agriculturally important crops. Of late, many basic researches in the areas of crop physiology and genomics shed new lights into many abiotic stresses, viz., drought, salt, temperature, UV light, etc. Banana crop can withstand a moderate level of abiotic stress. But there is a definite trade-off between the stress tolerance and yield. Instead of losing the major yield, it can be reduced through earlier correction of abiotic stresses. The research findings reveal that the effect of abiotic factors greatly influence the banana production with a lot of limitations on control measures. Therefore, planning is very important before taking plantation in areas where abiotic stresses are a real concern. Thus, we need continuous research support on various aspects of abiotic stresses in the present climate changing scenario to sustain the productivity of economically important crops.

References

Abeywickrama G, Weerasinghe SS (2002) Physiological disorders of banana in the southern dry region. Ann Sri Lanka Dep Agric 4:431

Anon (2008) Annual report. National Research Centre for Banana (ICAR). Trichy

Bananuka JA, Rbaihayo PR, Tenywa MM (1999) Reactions of Musa genotypes to drought stress. Afr Crop Sci J 7:333–339

Barker WG, Steward FC (1962) Growth and development of the banana plant 1. The growing regions of the vegetative shoot. Ann Bot 26:389–411

Blomme G, Teugels K, Blanckaert I, Sebuwufu G, Swennen R, Tenkouano A (2005) Methodologies for root system assessment in bananas and plantains (*Musa* spp). In: Turner DW, Rosales FE (eds) Banana root system: towards a better understanding for its productive management. Montpellier, France, Proceedings of an International symposium, INIBAP, pp 43–57

Carr MKV (2009) The water relations and irrigation requirements of banana (*Musa* spp.). Expl Agr (UK) 50:487–497

Cayón MG, El-Sharkawy MA, Mejía de Tafur S (1998) Physiological responses to water stress in the Dominico-Harton plantain clone (Musa AAB Simmonds). Info Musa 7:12–14

Chai TT, Fadzillah NM, Kusnan M, Mahmood M (2005) Water stress-induced oxidative damage and antioxidant responses in micropropagated banana plantlets. Biol Planta 49:153–156

Daniells JW, Watson BJ, O'Farrell PJ, Mulder JC (1987) Soil water stress at bunch emergence increases maturity bronzing of banana fruit. Qld J Agric Anim Sci 44:97–100

De Langhe (2002) Diversity in Egypt. In: Banana fiversity in the Middle East. Jordan, Egypt, Oman (eds). INIBAP Publisher, Montpellier, pp 14–20

Draye X, Lecompte F, Pages L (2005) Distribution of banana roots in time and space: new tools for an old science. In: Turner DW, Rosales FE (eds) Banana root system: towards a better understanding for its productive management. International Network for the Improvement of Banana and Plantain, Montpellier, pp 58–74

Dudal R (1976) Inventory of the major soils of the world with special reference to mineral stress hazards. In: Wright MJ (ed) Plant adaptation to mineral stress in problem soils. Ithaca, New York, pp 3–13

FAO (1988) Soil bulletins. 39. FAO, Rome

FAO STAT (2012) http://faostat3.fao.org/home/index.html#SEARCII_DATA

Garcia OH, Manzanilla E (1994). Respuesta de clones de Musa a condiciones severas de estrés hídrico. INIBAP-LAC, Santo Domingo, Dominican Republic, pp 17

Goenaga R, Irizarry H (1998) Yield of banana growth with supplemental drip irrigation on an Ultisol. Exp Agric 34:439–448

Greenway H, Munns R (1980) Mechanisms of salt tolerance in nonhalophytes. Ann Rev Plant Physiol 31:149–190

Henckel PA (1964) Physiology of plants under drought. Ann Rev Plant Physiol 15:364–384

Hoffman HP, Turner DW (1993) Soil water deficits reduce the elongation rate of emerging banana leaves but the night/day elongation ratio remains the same. Sci Hortic 54:1–12

Holder GD, Gumbs FA (1982) Effects of irrigation at critical stages of ontogeny of the banana cv. 'Robusta' on growth and yield. Trop Agric (Trinidad) 59:221–226

Ismail MR, Yusoff MK, Awad MH, Marziah M, Shamsuddin ZH (2000) Growth and physiological processes of banana (*Musa* spp.) plants to soil water deficit. In: Wahab Z, Mahmud TMM, SitiKhalijah D, NorAiniMohd F, Mahmood M (eds) Proceedings of the first national banana seminar. Universiti Putra Malaysia, Serdang, pp 225–234

Israeli Y, Lahav E, Nameri N (1986) The effect of salinity and sodium adsorption ratio in the irrigation water, on growth and productivity of bananas under irrigation conditions. Fruits 41:297–302

Kallarackal J, Milburn JA, Baker DA (1990) Water relations of the banana. III. Effects of controlled water stress on water potential, transpiration, photosynthesis and leaf growth. Aust J Plant Physiol 17:79–90

Kissela E, van Astenb P, Swennena R, Lorenzenb J, Carpentier SC (2015) Transpiration efficiency versus growth: exploring the banana biodiversity for drought tolerance. Sci Hortic 185:175–182

Krishnan BM, Shanmugavelu KG (1980) Effect of different soil moisture depletion levels on the root distribution of banana cv. 'Robusta'. South Indian Hortic 28:24–25

Lahav E, Kalmar D (1981) Shortening the irrigation interval as a means of saving water in a banana plantation. Aust J Plant Physiol 32:465–477

Mahouachi J (2007) Growth and mineral nutrient content of developing fruit on banana plants (*Musa acuminata* AAA, 'Grand Nain') subject to later stress and recovery. J Hortic Sci Biotechnol 82:839–844

Mengel K, Kirkby EA (1987) Principles of plant nutrition, 4th ed. International Potash Institute, IPI, Bern, p 685

Milburn JA, Kallarackal J, Baker DA (1990) Water relations of the banana. I. Predicting the water relations of the field grown banana using the exuding latex. Aust J Plant Physiol 17:57–68

Morello J (1954) Transpiraçâo e balanço da agua da bananeira nas condiçôes do Estado de Sao Paulo, Univ. de Sao Paolo. Botánica 10:27–97

NHB (2014) http://nhb.gov.in/

Norman MJT, Pearson CJ, Searle PGE (1984) Bananas (*Musa* spp). In: The ecology of tropical food crops. Cambridge University Press, Cambridge, pp 271–285

Price NS (1995) Banana morphology. Part 1: roots and rhizomes. In: Gowen S (ed) Bananas and plantains. Chapman and Hall, London, pp 179–189

Purseglove JW (1972) Tropical crops: monocotyledons. Longman, London

Ravi I, Uma S (2009) Screening banana germplasm for drought tolerance. In: Singh HP, Mustaffa MM (eds) Banana new innovations. Westville Publishing House, New Delhi, pp 109–111

Ravi I, Uma S (2011) Phenotyping bananas and plantains for adaptation to drought. In: Philippe M, Jean-Marcel R (eds) Drought phenotyping in crops: from theory to practice. CGIAR Generation Challenge Programme/ CIMMYT, Texcoco

Ravi I, Uma S (2012) Progress report of the project on 'Screening Musa Germplasm for Drought Tolerance for the Resource Poor Farmers' submitted to Global Crop Diversity Trust. FAO, Rome, p 68

Ravi I, Uma S, Vaganan MM, Mustaffa MM (2013) Phenotyping bananas for drought resistance. Front Physiol 4: Article 9, doi:10.3389/fphys.2013.00009

Ravi I, Mayil Vaganan M, Mustaffa MM (2014) Management of drought and salt stresses in banana. Tech Folder No. 6. National Research Centre for Banana, Thayanur Post, Tiruchirappalli, Tamil Nadu, India, p 6

Ravi I, Mayilvaganan M, Mustaffa MM (2014) Bananas grown in salt affected soil impairs fruit development in susceptible cultivars. Andhra Agric J 61 (3):638–642

Robinson JC (1982) The problem of November-dump fruit with Williams banana in the subtropics. Subtropica 3:11–16

Robinson JC (1987) Root growth characteristics in banana. Inst Trop Subtrop Crops S Afr Inf Bull 183:7–9

Robinson JC (1996) Banana and plantains. CAB International, Wallingford

Robinson JC, Alberts AJ (1989) Seasonal variations in crop water-use coefficient of banana (cultivar 'Williams') in the subtropics. Sci Hortic 40:215–225

Robinson JC, Bower JP (1988) Transpiration from banana leaves in the subtropics in response to diurnal and seasonal factors and high evaporative demand. Sci Hortic 37:129–143

Salekdeh GH, Reynolds M, Bennett J, Boyer J (2009) Conceptual framework for drought phenotyping during molecular breeding. Trends Plant Sci 14:488–496

Shamsuddin ZH, Suzita S, Marziah M (2000) Effect of aluminium toxicity, drought, flooding, acid and salt stress on growth, protein and chlorophyll content of in vitro plantlets of banana cv 'Berangan'. In: Wahab Z, Mahmud TMM, SitiKhalijah D, Nor' AiniMohd F, Mahmood H (eds) Proceedings of the first national banana seminar. Universiti Putra Malaysia, Serdang, pp 250–254

Shamueli E (1953) Irrigation studies in the Jordan valley. Physiological activity of the banana in relation to soil moisture. Bull Res Coun Israel 3:228–247

Simmonds NW (1962) The evolution of the bananas. Longman, London, p 170

Stover RH, Simmonds NW (1987) Bananas. Longman, London, p 468

Tai EA (1977) Banana. In: Alvim P de T, Kozlowski TT (eds) Ecophysiology of tropical crops. Academic Press, New York, pp 441–460

Taylor CF, Paton NW, Lilley KS, Binz PA, Julian RK, Jones AR, Zhu W, Apweiler R, Aebersold R, Deutsch EW, Dunn MJ, Heck AJ, Leitner A, Macht M, Mann M, Martens L, Neubert TA, Patterson SD, Ping P, Seymour SL, Souda P, Tsugita A, Vandekerckhove J, Vondriska TM, Whitelegge JP, Wilkins MR, Xenarios I, Yates JR, Hermjakob H (2007) The minimum information about a proteomics experiment (MIAPE). Nat Biotechnol 25:887–893

Thomas DS (1995) The influence of the atmospheric environment and soil drought on the leaf gas exchange of banana (*Musa* spp). PhD thesis, The University of Western Australia

Thomas DS, Turner DW (1998) Leaf gas exchange of droughted and irrigated banana cv. Williams (*Musa* spp.) growing in hot, arid conditions. J Hortic Sci Biotechnol 73:419–429

Thomas DS, Turner DW (2001) Banana (*Musa* sp.) leaf gas exchange and chlorophyll fluorescence in response to soil drought, shading and lamina folding. Sci Hortic 90:93–108

Thomas DS, Turner DW, Eamus D (1998) Independent effects of the environment on the leaf gas exchange of three banana 8 *Musa* sp cultivars of different genomic constitution. Sci Hortic 75:41–57

Thomas DS, Turner DW, Eamus D (1998a) Independent effects of the environment on the leaf gas exchange of three banana 8 *Musa* sp cultivars of different genomic constitution. Sci Hortic 75:41–57

Thomas DS, Turner DW, Eamus D (1998b) Independent effects of the environment on the leaf gas exchange of three banana (*Musa* spp.) cultivars of different genomic constitution. Sci Hortic 75:41–57

Turner DW (1990) Modelling demand for nitrogen in the banana. Acta Hortic 275:497–503

Turner DW (1995) The response of banana plant to the environment. In: Gowen S (ed) Bananas and plantains. Chapman and Hall, London, pp 206–229

Turner DW (2003) Bananas-response to temperature. Agfact (Australia: NSW Agriculture), H6.2.6. 3

Turner DW (2005) Factors affecting the physiology of the banana root system. In: Turner DW, Rosales FE (eds) Banana root system: towards a better understanding for its productive management. INIBAP, Montpellier, pp 107–113

Turner DW, Lahav E (1983) The growth of banana plants in relation to temperature. Aust J Plant Physiol 10:43–53

Turner DW, Thomas DS (1998) Measurement of plant and soil water status and their association with leaf gas exchange in banana (*Musa* spp): a laticiferous plant. Sci Hortic 77:177–193

Turner DW, Fortescue JA, Thomas DS (2007) Environmental physiology of the bananas (*Musa* spp.). Braz J Plant Physiol 19:463–484

Wagner M, Surga Rivas JG, Marín C, Medina G, Belloso M, Delgado A, Dorantes (2000) Adaptabilidad de variedades FHIA (*Musa*) y su relación con índices morfométricos bajo condiciones de riego In: Cortés M, Dávila R, Chavarria Carvajal JA, Ortiz F, Acin Diaz NM, González A, Franqui RA, Lugo W, Guzmán Chaves JA (eds) Memorias XIV de la reunión de la associación para la cooperación en investigación de banano en el Caribe y en América tropical. ACORBAT, San Juan, Puerto Rico, pp 12

Wairegi LWI, Van Asten PJA, Tenywa MM, Bekunda MA (2010) Abiotic constraints override biotic constraints in East African highland banana systems. Field Crop Res 117:146–153

www.cssri.org

www.fao.org

www.niam.res.in

www.nrcb.res.in

Young SCH, Sammis TW, Wu IP (1985) Banana yield as affected by deficit irrigation and pattern of lateral layouts. Trans ASABA 28:507–510

13 Arid Zone Fruit Crops

N.K. Srinivasa Rao

Abstract

Arid lands are among the world's most brittle ecosystems, made more so by periodic droughts and increasing overexploitation of meagre resources (FAO (1989) Arid zone forestry. FAO conservation guide No. 20. FAO, Rome, p 143). Arid zone is characterised by extremes of temperature, low and erratic rainfall, high wind velocity, high evapotranspirational losses of water and light sandy soil with very low water holding capacity and poor fertility status. Looking to the very dismal future scenario of water in arid zone, growing of less water-requiring crops with high application efficiency is necessary. Drought-hardy crops especially perennial fruits with deep root systems are capable of surviving extreme radiation and temperatures and provide income security, nutritional security and food security. Ber, pomegranate, custard apple, aonla and date palm are the major fruit crops which are commercially grown in arid regions. In this chapter, major findings of research on abiotic stresses on major arid zone fruit crops are summarised.

13.1 Introduction

The arid region has extreme climatic conditions. In arid zones rain is very greatly low, and its distribution is erratic leading to low plant stand and productivity. The water storage capacity of soils in these areas is very low, being mainly poor textured and shallow soils, and is compared with high evaporative loss. In such regions the ability to produce agricultural crops is restricted. Usually on arid lands the potential evaporation of water from the land exceeds the rainfall. The major effects of extreme temperature (heat) and wind are to increase the rate of evaporation of water and thus to enhance the effects of aridity. Mechanical damage to crops may occur due to high wind velocity. Both are combated by changing the microclimate. Wind damage can be reduced by planting rows of trees perpendicular to the direction of prevailing winds which acts as windbreaks. Tall species like tamarind, *Casuarina* and *Eucalyptus* can act as windbreaks. As a

N.K.S. Rao (✉)
Division of Plant Physiology and Biochemistry, ICAR-Indian Institute of Horticultural Research, Hessaraghatta Lake Post, Bengaluru 560089, Karnataka, India
e-mail: nksraoiihr@gmail.com

N.K.S. Rao et al. (eds.), *Abiotic Stress Physiology of Horticultural Crops*,
DOI 10.1007/978-81-322-2725-0_13

general rule, a windbreak is effective over an area two and a half times the height of the tree. One must remember, however, that a windbreak may also rob crops of light water and nutrients. The advantages of a windbreak must be weighed against the disadvantages in any particular environment. Windbreaks can also be constructed of nonliving materials, which are likely to be expensive (Creswell and Martin 1993).

All living things need water to stay alive. Plants, however, need much more water than many living things. The amount of water a plant needs depends on the type of plant, how much light the plant gets and age of the plant. In arid zones vegetation is sparse. The native plant species have adaptations that are capable of growing, reproducing and surviving in semiarid, arid and even desert regions. Plants are adapted to aridity by several mechanisms. Some plants have evolved special root systems, while others have unique leaf characters that allow them to survive drought periods. In general plant species of arid zones have xerophytic leaf structures, physiological control of transpiration and assimilation (wax coating on leaf surfaces to check water loss and plants with very small or narrow leaves, thus reducing water loss) and can withstand much desiccation without dying. The number of tree species is very limited in arid zones and is slow growing due to environmental constraints (Sharma and Tewari 2009). Crop plants in arid regions may have any or a combination of such mechanisms. In all arid regions a major challenge is to manage water appropriately. The purpose of such management is to obtain water, to conserve it, to use it efficiently and to avoid damage to the soil (Creswell and Martin 1993). Drought-hardy crops especially perennial fruits with deep root systems are capable of surviving extreme radiation and temperatures and provide income security, nutritional security and food security. Amla, jamun, ber, pomegranate, custard apple, etc., are the fruit crops suitable for dry lands. The conventional system of irrigation especially in fruit crops is not congenial under such harsh agro climatic situations. Studies have shown that more than 70 % of applied water is lost during conveyance. Looking to the very dismal future scenario of water in arid zone, growing of less water-requiring crops with high application efficiency is necessary. In this chapter, major findings of research on abiotic stresses on major arid zone fruit crops are summarised.

13.2 Pomegranate

Pomegranate (*Punica granatum* L.) is native to hot dry regions of Afghanistan and Baluchistan (De Candolle 1967), but with time it has adapted to a wide range of environmental conditions. It belongs to the family Punicaceae and an important arid zone fruit from a commercial point of view. It can be grown in tropical and subtropical regions (Trapaidze and Abuladze 1998) and temperate zones (Purohit 1982, 1986). Sharma and Sharma (1990) have reported that it can fruit in hilly areas up to 1800 m altitude. Flowering and fruiting physiology are altered with the habitat (Pareek and Godara 1993). Pomegranate can be commercially produced in areas where the summer is warm and too hot and where rainfall is minimal during late summer/autumn. Water should be available for irrigation. Very high temperatures can cause sunburn injury on the fruit. Deep, loamy, well-drained soils are preferred, but the pomegranate has some tolerance to less than ideal drainage and to mild alkaline conditions (Johnson 1983). As a consequence, irrigation means have become a major factor for a successful plant production in dry climate zones. Not only is the absolute lack of water a problem but also the poor quality of irrigation water, due to excessive ion concentrations. Numerous attempts have been made to traditional breeding programmes to improve the drought tolerance of pomegranate, but commercial success has been limited thus far. Pomegranate is often cultivated in arid zones where water is scarce; therefore, germplasm needs to be exploited in breeding varieties tolerant to moisture stress (Jalikop 2010). Tree crops that can tolerate drought develop a morphological and/or physiological mechanism which enables them to avoid or postpone desiccation (Save et al. 1995;

Torrecillas et al. 1996). In perennial crops grown under arid and semiarid conditions, it is relevant to assess the extent of resumption of assimilation and growth rate after relieving water stress for resuming productivity (Rodrígueza et al. 2012). Pomegranate is considered to be drought tolerant and winter hardy and can thrive well under desert conditions (Aseri et al. 2008). Pomegranates can tolerate high temperatures and survive in arid and semiarid areas, but they need regular irrigation throughout the dry season to reach optimal yield and fruit quality (Sulochanamma et al. 2005; Levin 2006; Holland et al. 2006). Sulochanamma et al. (2005) found that drip irrigation had positive effects on pomegranate growth parameters such as tree height, stem diameter and plant spread. Positive effect was also noted on fruit yield and fruit weight (Prasad et al. 2003; Shailendra and Narendra 2005; Sulochanamma et al. 2005). Timing of irrigation and seasonal application at correct stages are important not only for better growth and yield of the pomegranate trees but also to monitor time of ripening. For example, in India, timing of irrigation is used to control and optimise the yielding season of evergreen pomegranates (Sonawane and Desai 1989). By applying different irrigation regimes, it was possible to direct the required time of fruit yield in Indian pomegranates. Despite the growing importance of pomegranate and the ever more frequent water shortages experienced in the most suitable regions for its growth, no information exists on the mechanisms developed by pomegranate trees to overcome drought. Pomegranate plants confront water stress by developing stress avoidance and stress tolerance mechanisms (Rodrígueza et al. 2012). These mechanisms appear to be complementary and take place gradually. They further observed that from the beginning of imposing water stress, leaf conductance decreased in order to control water loss via transpiration and avoid leaf turgor loss (stress avoidance mechanism). When, close to the end of the stress period, maximum stress levels were developed, active osmotic adjustment was triggered, which could have contributed to the maintenance of leaf turgor (stress tolerance mechanism). High relative apoplastic water content (42–58 %), would contribute to the retention of water at low leaf water potentials (Rodrígueza et al. 2012).

Recycled water and saline water are being used as an alternative water resource for irrigating pomegranate (Holland et al. 2009). They are amenable to irrigation with saline water. Many desert orchards in the Negev Highlands and in the southern Arava of Israel are irrigated with saline water. Salinity levels in the water of these orchards range between 2.5 and 4.0 dS/m. Salinity tolerance mechanism in pomegranate is not yet fully understood. Pomegranate tissues accumulate sodium, chlorine and potassium in response to irrigation with saline water, and the concentration of these ions increased with increased concentrations of salt in the irrigation water. Pomegranate can tolerate saline water up to concentration levels of 40 mM NaCl in the water (Doring and Luddar 1987; Naeini et al. 2004, 2006). Length of the main stem, length and number of internodes and the area of leaf surface are severely affected if the NaCl concentration in water exceeded 40 mM (Naeini et al. 2006). Soluble sugar content in the plant tissues was reported to decrease with increasing NaCl concentrations (Naeini et al. 2004). The use of saline water for irrigation can improve pomegranate fruit qualities such as acidity, sugar content, TSS, antioxidant value and medicinal properties. Polyamines are being used to enhance the drought and salinity tolerance of plants (Chattopadhyay et al. 2002; Verma and Mishra 2005). Polyamines are organic compounds found in all living organisms. In higher plants putrescence, spermidine and serine are the most abundant polyamines and are implicated in various developmental processes (Ton on et al. 2004). The exogenous application of polyamines to stress-affected cells or tissues could lead to injury alleviation and growth promotion in most cases, although the effects vary between polyamines and among plant species (Shen et al. 2000; Boucher et al. 1999; Carpel et al. 2004). Increase in the endogenous polyamines due to water stress has also been reported in various plants (Das et al. 1995; Chattopadhyay et al. 2002).

Enhanced polyamine biosynthesis can protect plants from drought by scavenging free radicals, stabilising membrane and cellular structures and maintaining a cation-anion balance (Boucher et al. 1999). Foliar application of polyamine ameliorated the harmful effects of drought stress on growth of pomegranate, and the use of putrescence and spermidine increased the proline concentration in the plant (Amri and Mohammadi 2012). An increase in the growth rate of pomegranate grown at different salinity levels with a foliar application of polyamines spermidine and putrescine was observed, although a decline in the growth rate occurred at salinity levels higher than 70 mM (Amri et al. 2011). Elevated levels of polyamines were observed in pomegranate cultivars grown under saline conditions. Polyamines might have prepared the cell to meet and combat stress by stabilising membranes and forming a potential of higher buffering and antioxidant capacity (Roy and Sengupta 2014). Accumulation of polyamines results in presumed protective effects, acting as free radical scavengers, stabilising cellular membranes and maintaining cellular ionic balance under salinity (Duan et al. 2008).

13.3 Custard Apple

Annonas are tropical in origin and grow well in hot and dry climate. An annual rainfall of 500–750 mm is adequate for growth and fruiting. Custard apple requires hot and dry climate during flowering and high humidity at fruit setting. The optimum growing conditions for custard apple production appear to be at temperatures ranging from 23 to 34 °C, 70 % or higher relative humidity (RH) during flowering and fruit set and non-limiting soil moisture from flowering to harvest. Low humidity is harmful for pollination and fertilisation. They can tolerate extreme temperatures (below freezing and up to 40 °C), and such extremes won't be ideal for production of flowers and fruits. They can be grown from sea level up to 1000 m. Where the summer temperature is very high (more than 40 °C) and humidity is low, there will be no fruit set though the flowering will be profuse. Prolonged temperatures below 13 °C can cause skin discolouration and splitting (Deuter 2011). Low soil temperatures (5–10 °C) caused a reduction in the leaf chlorophyll concentration. Soil temperatures below 10 °C may cause severe chilling injury leading to loss of cell membrane stability (Maritza et al. 2004). Fruit production of custard apple trees may benefit from wind protection. Properly placed and managed windbreaks may enhance tree growth, decrease the drying effect of winds and increase the relative humidity around trees. Dry, windy conditions during flowering may reduce fruit set, and percent fruit drop may be high by strong winds during fruit development. Strong winds along with heavy crop loads may result in breakage of branches. Custard apple trees have a relatively shallow and weak root system. Tree toppling and uprooting have been observed after hurricane-force winds. Even trees not toppled or uprooted appear to be damaged by very strong winds, resulting in loss of tree vigour and unsuccessful recovery.

The custard apple is not very particular about soil conditions and flourishes in all types of soils like shallow and sandy but fails to grow if the subsoil is ill drained. It can grow well in deep black soils provided they are well drained. A little salinity or acidity does not affect it, but alkalinity, chlorine, poor drainage or marshy wetlands hamper the growth and fruiting. Custard apple trees may withstand prolonged dry soil conditions but do so by reduced growth and defoliate leaves affecting fruit development. Prolonged drought stress may reduce the percent fruit set, fruit size and crop yields. Mild to severe drought stress reduced the fruit size by 10–50 % in *Annona atemoya* (a relative of custard apple). Therefore, periodic watering of sugar apple trees is recommended from flowering through fruit development to enhance fruit quality and production (Crane et al. 1994). Soil moisture stress (midday leaf water potential −2.0 MPa) reduced the flowering and fruit set. High vapour pressure deficit (1–2 Kpa) severely reduced both flowering and fruit set (George and Nissen

1988). In areas with good summer rainfall, a water reserve of 5 megalitres per hectare of orchard is necessary to maintain production in a dry year (George et al. 1998). Thirty-five germplasms of custard apple were 'evaluated in zones where rainfall was inadequate under rain-fed conditions for 3 years in 10-year-old trees for their growth and yield (Dhanumjaya Rao and Subramanyam 2011). Varieties, viz. Atemoya X Balanagar (highest fruit weight), Y. Palli (12; highest pulp weight), NLD (8; maximum T.S. S), Balanagar SR (more number of fruits/tree) and Ramphal (highest yield/tree), were found to be better and can be recommended for scarce rainfall zone under rain-fed conditions in poor fertile soils. Highest fruit yield was recorded in Ramphal (15.7 Kg/tree) followed by MMR-10 (15kg/tree) Highest fruit weight per tree was recorded (278.9 g) in Atemoya X Balanagar followed by Y. Palli (12; 270.6 g). High temperature, low atmospheric humidity, lack of irrigation water and natural stress result in less number of flowers, poor fruit set and low yield and degraded quality of fruit (Patel et al. 2010). These problems can be controlled by mulching which improves the moisture within the periphery of the plant. Maximum flowering duration and fruit retention were observed with wheat straw mulch + GA_3 (5 t/ha + 50 ppm) treatments. Highest number of fruits, fruit yield, fruit diameter and fruit pulp were also recorded under the same treatments. Wheat straw mulch gave 21 % higher fruit yield of custard apple. However, application of 20 ppm NAA was at par with GA_3 (50 ppm). On economic basis, 50 ppm GA_3 + wheat straw mulch followed by 20 ppm NAA + wheat straw mulch gave highest net income (Patel et al. 2010).

Custard apples are sensitive to salt in irrigation water. Water salinity should not exceed an electrical conductivity of 800 μS/cm (George et al. 1998). Symptoms of salt stress include marginal and tip necrosis (death) of leaves, leaf browning and drop, stem dieback and tree death. Custard apple trees are not tolerant to excess moisture or flooded soil conditions. Flooding for as little as 7–10 days may result in plant death. Symptoms of flooding stress include leaf chlorosis (yellowing), stunted leaf and shoot growth, leaf wilting and browning, leaf drop, stem dieback and tree death (Crane et al. 1994). Flooding of *Annona squamosa* even for short periods reduced growth, caused defoliation and severely reduced flowering and fruit set (Marler et al. 1994). Pond apple (*Annona glabra* L.), a species of *Annona* which has no commercial value (Popeno 1920), native to the tropical Americas, including south Florida, is extremely flood tolerant. Studies have suggested that grafting commercial *Annona* species and cultivars onto *A. glabra* rootstock may increase flood tolerance. Atemoya trees, cultivar '49–11' (a hybrid of *Annona* spp. *Annona reticulata* L.), grafted on *A. glabra* rootstock, had a significantly higher net CO_2 assimilation rate after flooding compared to the same plants grafted on *Annona reticulata* rootstock (Nunez-Elisea et al. 1997).

13.4 Ber

Ber is a tropical fruit (also known as Indian jujube), *Ziziphus mauritiana* Lam., which belongs to the family Rhamnaceae. The tree grows in most dry tropical and subtropical regions of the world. It is regarded as the king of arid zone fruits and also as poor man's apple. Ber provides livelihood support to farmers in arid zones, who often face the crop failures due to frequent drought and erratic rainfall. It grows in neutral or slightly alkaline, deep, sandy loams, and it can tolerate a range of soil types including those exhibiting moderate salinity. Soils should be well drained, but it can withstand temporary flooding. The tree has a deep and extensive root system (Anonymous 2001). Ber can tolerate very high summer temperatures (49–50 °C); fruit set can be affected at temperatures above 35 °C. The trees shed their leaves and enter dormancy in extremely hot summers. New leaves emerge, and with the onset of the rains, new foliage emerges. Summer dormancy may not occur or be very short in areas where temperatures are not so extreme. Growth, flowering and fruiting development phases may vary depending on the temperature conditions. Ber trees can withstand very short periods of freezing temperatures; however, frost will damage the young twigs and developing fruits and may kill the tree. Ber trees should not be planted in areas where minimum temperatures are lower than 4 °C for extended periods. For rain-fed production of ber, a minimum average yearly rainfall of 400 mm is required. Performance of the ber tree is adversely affected in humid areas having more than 1500 mm annual rainfall. Fruit yields are higher in high rainfall years and lower in low rainfall years (Pareek 2001). Meghwal et al. (2014) have studied the response of 22 jujube cultivars to natural climatic variability under rain-fed conditions in arid zone. The fruit yield was found to be positively correlated with annual rainfall irrespective of varieties. The differential dependence of fruit yield in different varieties of ber on total annual rainfall can be helpful in selecting varieties that are comparatively more resilient to climatic variability in Indian arid zone.

In *Z. mauritiana*, the abaxial leaf surface is densely pubescent, with numerous stomata arranged in the interveinal regions, whereas the adaxial surface is glabrous, with comparatively few, sunken stomata. Leaf hairiness, hypostomatous distribution and sunken stomata are all characteristic features of species that exist in drought-prone regions (Clifford et al. 2002). Research into the mechanisms for drought tolerance and avoidance in *Ziziphus* species has shown that *Z. mauritiana* and *Z. rotundifolia* exhibit a range of drought avoidance adaptations to progressive drought stress. Under well-irrigated conditions, *Ziziphus mauritiana* exhibited high rates of net photosynthesis and transpiration compared to other fruit tree species (Clifford et al. 1997), but during early stages of drought stress, the stomata were very sensitive to water deficit, and reduced stomatal conductance effectively increased intrinsic water use efficiency in the short term (Arndt et al. 2001). However, as water deficits increased, osmotic adjustment occurred concomitant with an increased root/shoot ratio to access deep soil moisture reserves, followed by leaf loss and ultimately drought-enforced dormancy (Arndt et al. 2001; Clifford et al. 1998). Under field conditions neither d^{13} C values nor soluble sugar concentrations changed markedly during the study period in *Z. mauritiana* leaves, and *Z. mauritiana* plants showed no osmotic adjustment and were observed during the dry season. Although *Z. mauritiana* exhibited the capacity for osmotic adjustment in glasshouse experiments (Arndt et al. 2001), the trees avoided drought stress under field conditions indicating that the root system has access to deeper moist soil layers (Arndt et al. 2000).

Metcalf and Chalk (1950) have reported abundant mucilaginous substances in plants of Rhamnaceae, including the genus *Ziziphus*. High mucilage concentrations of 7–10 % (dry weight basis) in the leaves, with large numbers of mucilage-containing cells in the upper epidermis and extracellular mucilage-containing cavities in the leaf veins and stem cortex, were observed in the drought-tolerant tree species *Ziziphus mauritiana* Lam. and *Z. rotundifolia* Lam. (Clifford et al. 2002). The main sugar constituents of the water-soluble mucilage extract were rhamnose, glucose and galactose. Glucose and starch contents declined significantly in crude mucilage extracts from droughted leaves. They concluded that in *Ziziphus* leaves, intracellular mucilages play no part in buffering leaf water status during progressive drought. In *Ziziphus* species, growing in environments with erratic rainfall, the primary role of foliar mucilage and glucans may be as sources for the remobilisation of solutes for osmotic adjustment, thus enabling more effective water uptake and assimilating redistribution into roots and stems prior to defoliation as the drought stress intensified (Clifford et al. 2002).

Drought and high temperature injury in ber can be ameliorated by soil application of paclobutrazol (Sankhla et al. 1989). Elongation growth of branches was inhibited, and the green colour of the foliage was intensified by soil application of paclobutrazol (8 mg a.i./tree). Fruits from treated plants had higher contents of sugars and ascorbic acid than the control fruits. Fruits from paclobutrazol-treated trees exhibited higher activities of scavenger enzymes like catalase, peroxidase and superoxide dismutase and decreased lipid peroxidation under moisture stress. Paclobutrazol also minimised moisture loss from the fruits and greatly reduced cracking of fruits caused by excessive moisture and thermal stress. Paclobutrazol can effectively protect ber fruits from stress injury in the Indian Desert (Sankhla et al. 1989).

13.5 Aonla/Amla

The aonla or Indian gooseberry (*Emblica officinalis*) tree is native to tropical Southeast Asia, particularly central or southern India, Pakistan, Bangladesh, Sri Lanka, Malaya, Southern China and Mascarene Islands (Pathak 2003; Anonymous 2012). A rich genetic diversity of aonla exists in northeastern region of India, particularly in Lower Assam, Meghalaya, Mizoram and Tripura (Yadav et al. 2001). Aonla is becoming more and more commercially important, owing to its hardy nature, suitability to various wastelands, high productivity/unit area (15–20 tons/ha) and nutritive and therapeutic value. Aonla fruits are a very rich source of vitamin C having an ascorbic acid content varying from 0.9 to 1.3 %. Aonla grows well in sandy loam to clay soils in India. It has a great tolerance to salinity and sodicity and is cultivated successfully in pH range of 6.0–8.0. However, production shall be highly benefited in deep and fertile soils. Heavy soils or high water table areas are not suited for cultivation.

The tree is not much affected either by hot wind or frost. The mature trees can tolerate freezing temperature as well as a temperature as high as 46 °C, but the temperature should not be high at the time of flowering (Pathak 2003). It affects fruit setting and may sometime lead to complete unfruitfulness, if there are hot and dry winds too. Dry spells result in heavy dropping and delay in initiation of fruit growth. Soon after the fruit set in spring, the fruits remain dormant through summer without any growth, making it a highly suitable fruit crop for arid region.

Aonla trees are hardy and stand very well against drought. Therefore, hardly any irrigation is practised. The young plants require watering during summer months at fortnightly intervals till they have fully established (Bajpai and Shukla 1985). Aonla orchards in general do not require irrigation particularly in the normal soil situations. Irrigation is not required during rainy and winter season. However, irrigation at an interval of 15–20 days is desirable in dry summer corresponding with period of fruit set and active plant growth particularly under wasteland situations (Pathak 2003). Brackish water should not be used for irrigation purposes. Irrigation of young aonla plantation is preferable at 10-day intervals during summer months. In the bearing plantation, first irrigation should be just after applying manures and fertilisers before flowering. Irrigation should be avoided during flowering period. Irrigation at 15-day interval should be given after fruit set particularly in the salt-affected soils. Aonla plants are highly susceptible to flooding; hence, care should be taken to save the plants from excess water during rainy season. Proper drainage and inter-cultivation of dhaincha (*Sesbania aculeata*) during rainy season has been observed as a possibility of saving aonla plants from water logging (Singh et al. 1993). Moisture stress in lean monsoon years restricts the fruit growth and also leads to heavy fruit drop. Since aonla cultivation is becoming popular in arid and semiarid regions, water harvesting and its recycling with efficient irrigation methods, viz. drip and pitcher coupled with mulching, will in a long way harness potentiality of aonla, particularly on the marginal lands.

13.6 Date Palm

Date palm (*Phoenix dactylifera*) originated in the ancient Mesopotamia region or in Western India (Wrigley 1995). Currently, the date palm is the primary crop in several arid and semiarid countries in North Africa, the Middle East and Central America where long and hot summers prevail with no (or at most low) rainfall and very low relative humidity level during the ripening period (FAO 2006). Date palms require well-drained deep sandy loam soils with pH of 8.0–10.0. The soil should have the ability to hold the moisture. The date palm tree has evolved through natural selection to be a drought- and salt-tolerant plant (Zaid and de Wet 2002a). The date palm can adapt to extreme drought, to heat and to relatively high levels of soil salinity (Yaish and Kumar 2015). It prefers constant moisture in the soil but sensitive to water logging.

Date palms can endure high temperatures of ±56 °C for several days under irrigation. During winter, it can tolerate temperatures below 0 °C. The zero vegetation point of a date palm is 7 °C, and below 7 °C growth stops, and this stage is called a resting period. Growth is active above 7 °C and reaches its optimum at about 32 °C. The growth will continue at a stable rate until the temperature reaches 38 °C/40 °C when it starts decreasing. When the temperature falls below 0 °C, it causes metabolic disorders which lead to partial or total damage of leaves. Pinnate margins turn yellow and dry out at −6 °C. Inflorescences are also heavily damaged by frost. When frost periods are forecasted, inflorescences should be protected with craft paper bags immediately after pollination. Leaves of medium and outside canopy will be damaged and dry out between −9 and −15 °C. If these low temperatures persist for a long period (12 h to 5 days), all leaves will show frost damage, and the palm will look as if it was burnt (Mason 1925a; Nixon 1937). The possibilities of harvesting poor fruit quality will be greater if more leaves are damaged. Poor flowering could also be expected. Nixon (1937) reported that date gardens that were irrigated during frost periods were less damaged than the ones that were not irrigated. Air humidity also affects the date quality during the maturation process. At high humidity, fruits become soft and sticky, while at low humidity they become very dry.

Soil salinity has become a global agricultural constraint especially true in arid and semiarid regions where a considerable amount of agricultural land area has been affected (Pitman and Läuchli 2002) and has led to significant economic losses in date palms and other crops (Cookson and Lepiece 2001). The major causes of increased soil salinity in arid and semiarid regions are insufficient rainfall coupled with over-irrigation using brackish or saline groundwater (Hillel 2000; Pitman and Läuchli 2002; Malash et al. 2008).

The date palm can adapt to extreme drought, to heat and to relatively high levels of soil salinity. The date palm tree has evolved through natural selection to be a drought- and salt-tolerant plant (Zaid and deWet 2002a) with an adaptation capacity exceeding barley, which is widely considered to be a salt-tolerant crop (Furr and Armstrong 1975). However, insufficient rainfall coupled with over-irrigation using brackish or saline groundwater has significantly reduced the productivity of the fruits as well as marked decrease in the viable numbers of the date palm trees (Yaish and Kumar 2015). Some date palm varieties have the ability to grow in close proximity to those ashore where they are often exposed to sea water during tidal currents. Therefore, the date palm could be considered one of the exceptional halophytic plants and may possess a suite of mechanisms for salinity tolerance. Plant species differ in their responses to saline conditions. Plants adapted to salinity have the ability to tolerate relatively high levels of soil salinity, while susceptible plants are not able to grow under the same salinity conditions (Munns and Tester 2008). Ionic toxicity and osmotic stress reduce the growth rate in salt-susceptible plants and may lead to plant death. Not much research work has been done on salt tolerance in the date palm. Ramoliya and Pandey (2003) screened particular date palm varieties for tolerance to salinity and found that some cultivars can endure high soil salinity levels of 12.8 dS m^{-1}(1 dS m^{-1} = 640 mg/1) with no visible effect on the seedling phenotype. Alrasbi et al. (2010) found that some other date palm varieties can tolerate up to 9 dS m^{-1} soil salinity and an excess of Na^+ ions accumulated in the leaves of the plants treated with high salt concentrations. Further work is needed at the physiological and molecular levels to fully understand the salt adaptation mechanisms in the date palm. Local varieties growing in relatively saline soils and expressing a normal phenotype may possess genes responsible for efficient salinity adaption mechanism than those growing in less saline soils. How these plants are able to survive and grow in such a condition remains to be studied. Date palms growing around the world should be characterised for phenotypic traits, for understanding the differential tolerance mechanism in date palm. Highly salt-tolerant and salt-susceptible varieties should be studied in depth. It is known that some other salt-tolerant species, such as mangroves, develop specific apoplastic barriers that can act as filters against NaCl uptake in their roots, minimising the salt movement into the plant (Krishnamurthy et al. 2011, 2014a). Kinetics and dynamics of salt uptake or salt exclusion in roots as well as the histology of the roots in both tolerant and susceptible date palm varieties should also be studied (Yaish and Kumar 2015). The physiological basis of salt tolerance in date palm was found as a strict control on Na^+ and Cl^- concentration in leaves and keeping up the K^+ content (Alrasbi et al. 2010). Differences in salt tolerance between date palm cultivars appear to be related to the salt exclusion mechanisms by the root parts (Greenway and Munus 1980), resulting in reduced Na^+ translocation to the shoots. Efforts should be made to compare the relative sensitivity of various cultivars to salt uptake and transport of NaCl and their interactions with nutrients. In addition, it is required to identify differences in salinity tolerance between date palm cultivars and thus start new breeding programmes to improve salinity tolerance. In order to bring arid and semiarid regions into production, future research should focus on using halophytes as an alternative crop and sea water for irrigation. Remote sensing as a tool to detect salinity stress should be effectively applied in future date palm development programmes (Yaish and Kumar 2015).

13.7 Future Strategies

The rains are infrequent and irregular and often come in a few heavy storms of short duration resulting in high run-off, instead of replenishing the groundwater, sometimes with no rain during long periods in arid and semiarid regions. The existing problems in both arid and semiarid regions are further aggravating because of climate change (Anonymous 2010). There is an expected precipitation decrease over the next century of 20 % or more in many arid zones. It is estimated that the arid and semiarid regions are set to suffer further from water shortages. Water will be the primary constraint for productivity and sustainability throughout the semiarid tropics in the coming years. High wind and solar regimes further increase the effect of rainfall variability, and the whole complex makes a fragile ecosystem in which small disturbances may cause great loss to the sustainability, which are sometimes irreversible. The hot arid zones of the world are economically and environmentally disadvantaged, with unique problems. The number of fruit tree species is very limited in arid zones, and in general, they are very slow growing due to prevailing environmental constraints. Agri-horticulture system has been found to give better and earlier production and resilience to erratic rainfall. Focus should be on the evaluation, improvement and encouragement of sustainable traditional agri-horticultural systems. The possibilities of natural regeneration/rehabilitation of arid zone ecosystems should be explored. Emphasis should be on conservation of rainwater, development of in situ rainwater conservation techniques that require less labour and maintenance and widespread adoption of rainwater conservation techniques. Plant-based water-absorbing/water-retaining materials should be developed. Genes that can improve water use efficiency and confer drought tolerance should be identified. For managing limited water supply, some concepts like crop response factor as a tool to crop sensitivity rating, optimal leaf area concept, optimal yield concept, optimal irrigation scheduling, water use efficiency maximising concept (protected agriculture) and partial root zone wetting concept may be adopted. In some of the crops where rootstocks are used, salinity and water stress-tolerant rootstocks need to be explored for better survival and productivity. Appropriate research in arid horticulture will not only provide nutritious fruits but also vegetation cover, shade and ever-scarce nutritious fodder for sheep, goats and camels under harsh conditions. The local germplasm found in arid areas is valuable, noble and distinct and exhibits a lot of genetic variability in colour, size, shape, productivity and quality. The maintenance and preservation of such biodiversity is of global significance. Adoption of these approaches will certainly lead to efficient management of scarce water in arid and semiarid region.

References

Alrasbi SAR, Hussain N, Schmeisky H (2010) Evaluation of the growth of date palm seedlings irrigated with saline water in the sultanate of Oman. In: Proceedings of the IV International Date Palm Conference, Vol. 882, Abu Dhabi, pp 233–246

Amri E, Mohammadi MJ (2012) Effects of timing of drought stress on pomegranate seedlings (*Punica granatum* L. cv 'Atabaki') to exogenous spermidine and putrescine polyamines. Afr J Microbiol Res 6:5294–5300

Amri E, Mirzaei M, Moradi M, Zare K (2011) The effects of spermidine and putrescine polyamines on growth of pomegranate (*Punica granatum* L. cv 'Rabbab') in salinity circumstance. Int J Plant Physiol Biochem 3:43–49

Anonymous (2001) Fruits for the future ber international centre for underutilized crops, Institute of Irrigation and Development Studies, University of Southampton, United Kingdom Department of International Development (DFID) Southampton, SO17 1BJ

Anonymous (2010) Semi-arid and arid ecosystems change: key issues and challenges National policy dialogue on climate change action. 8–9 Nov 2010, Chennai

Anonymous (2012) Cultivation of Aonla Agri hunt, Pakistan

Arndt SK, Wanek W, Clifford SC, Popp M (2000) Contrasting adaptations to drought stress in field-grown Ziziphus mauritiana and Prunus persica trees: water relations, osmotic adjustment and carbon isotope composition. Aust J Plant Physiol 27:985–996

Arndt SK, Clifford SC, Wanek W, Jones HG, Popp M (2001) Physiological and morphological adaptations of the fruit tree *Ziziphus rotundifolia* in response to progressive drought stress. Tree Physiol 21:705–715

Aseri GK, Jain N, Panwar J, Rao AV, Meghwal PR (2008) Biofertilizers improve plant growth, fruit yield, nutrition, metabolism and rhizosphere enzyme activities of Pomegranate (*Punica granatum* L.) in Indian Thar Desert. Sci Hortic 117:130–135

Bajpai PN, Shukla HS (1985) Aonla. In: Bose TK (ed) Fruits of India – tropical and sub-tropical. Naya Prokash, Calcutta, pp 591–600

Bhantana P, Lazarovitch N (2010) Evapotranspiration, crop coefficient and growth of two young pomegranate (Punica granatum L.) varieties under salt stress. Agric Water Manag 97:715–722

Boucher A, Aziz A, Larher F, Martin-Tanguy J (1999) Polyamines and environmental challenges: recent development. Plant Sci 140:103–125

Carpel T, Basie L, Christos P (2004) Modulation of the polyamine biosynthetic pathway in transgenic rice confers tolerance to drought stress. Proc Natl Acad Sci U S A 101:9909–9914

Chattopadhyay MK, Tiwari BS, Chattopadhyay G, Bose A, Sengupta DN, Ghosh B (2002) Protective role of exogenous polyamines on salinity-stressed rice (*Oryza sativa*) plants. Physiol Plant 116:192–199

Clifford SC, Kadzere I, Jones HG, Jackson JE (1997) Field comparisons of photosynthesis and leaf conductance in Ziziphus mauritiana and other fruit tree species in Zimbabwe. Trees 11:449–454

Clifford SC, Arndt SK, Corlett JE, Joshi S, Sankhla N, Popp M, Jones HG (1998) The role of solute accumulation, osmotic adjustment and changes in cell wall elasticity in drought tolerance in *Ziziphus mauritiana* (Lamk.). J Exp Bot 49:967–977

Clifford SC, Arndt SK, Popp M, Jones HG (2002) Mucilages and polysaccharides in *Ziziphus* species (Rhamnaceae): localization, composition and physiological roles during drought- stress. J Exp Bot 53:131–138

Cookson P, Lepiece A (2001) Could date palms ever disappear from the Batinah? Salination of a coastal plain in the sultanate of Oman. In: Mahdi KA (ed) Water in the Arabian Peninsula: problems and policies. Ithaca Press, Garnet Publishing Ltd., South Street, Reading, RG1 4QS, UK, pp 221–235

Crane JH, Balerdi CF, Ian Maguire I (1994) Sugar apple growing in the Florida home landscape. Fact sheet HS38, Horticultural Sciences Department, Florida Cooperative Extension Service, Institute of Food and Agricultural Sciences, University of Florida. ECHO, Myers, pp 1–23

Creswell R, Martin FW (1993) Dryland farming: crops & techniques for arid regions. ECHO, Myers, pp 1–23

Das S, Bose A, Gosh B (1995) Effect of salt stress on polyamine metabolism in *Brassica campestris*. Phytochemistry 39:283–285

De Candolle A (1967) Origin of cultivated plants. Hafner Publication Co/Distribution Co, New York/London/ Lucknow, pp 237–240

Deuter P (2011) Climate change-risks and opportunities of the custard apple industry. A paper presented at the National Australian custard apple conference 2011 -quality and quantity-finding and understanding the balance

Dhanumjaya Rao K, Subramanyam K (2011) Growth and yield performance of custard apple germplasm under scarce rainfall zone. Indian J Agric Res 45:156–160

Doring J, Luddar P (1987) Influence of sodium salts on the Na, Cl and SO4 contents in leaves, shoots and roots of *Punica granatum* L. Gartenbauwissenschaft 52:26–31

Duan JJ, Li J, Guo SR, Kang YY (2008) Exogenous spermidine affects polyamine metabolism in salinity-stressed *Cucumis sativus* roots and enhances short-term salinity tolerance. J Plant Physiol 165:1620–1635

FAO (1989) Arid zone forestry, vol 20, FAO Conservation Guide. FAO, Rome, p 143

FAO (2006/2005) Worldwide dates production statistics. Food and Agriculture Organization of the United Nations, Rome

Furr JR, Armstrong WW (1975) Water and salinity problems of Abadan Island date gardens. Ann Date Grow Inst 52:14–17

George AP, Nissen RJ (1988) The effects of temperature, vapour pressure deficit and soil moisture stress on growth, flowering and fruit set of custard apple (*Annona cherimola* × *Annona squamosa*) 'African Pride'. Sci Hortic 34:183–191

George A, Campbell J, Nissen B, Smith L, Meiburg G, Broadley R, Vock N, Rigden P (1998) Custard apple information kit. Agrilink, your growing guide to better farming guide. Manual. Agrilink Series QAL9904. Department of Primary Industries, Queensland Horticulture Institute, Brisbane

Greenway HR, Munus (1980) Mechanism of salt tolerance in non-halophytes. Ann Rev Plant Physiol 31:149–190

Hillel D (2000) Salinity management for sustainable irrigation: integrating science, environment, and economics. World Bank Publications, Washington, DC

Holland D, Bar-Ya'akov I, Hatib K (2006) Pomegranate research at Newe Ya'ar. Alon Hanotea 60:411–413

Holland DK, Hatib, Bar-Ya'akov I (2009) Pomegranate: botany, horticulture, breeding. Hortic Rev 35:127–191

Jalikop SH (2010) Pomegranate breeding. Fruit Veg Cereal Sci Biotechnol 4:26–34

Johnson JF (1983) Pomegranate growing. Agfact H3.1.42, New South Wales Department of Agriculture, Orange NSW 2800, pp 1–3

Krishnamurthy P, Ranathunge K, Nayak S, Schreiber L, Mathew MK (2011) Root apoplastic barriers block Na^+ transport to shoots in rice (*Oryza sativa* L.). J Exp Bot 62:4215–4228

Krishnamurthy P, Jyothi Prakash PA, Qin L, He J, Lin Q, Loh CS, Prakash PK (2014) Role of root hydrophobic barriers in salt exclusion of a mangrove plant *Avicennia officinalis*. Plant Cell Environ 37:1656–1671

Levin GM (2006) Pomegranate roads. In: Baer BL (ed) A Soviet botanist's exile from Eden. Floreat Press, Forestville, pp 15–183

Malash N, Flowers T, Ragab R (2008) Effect of irrigation methods, management and salinity of irrigation water on tomato yield, soil moisture and salinity distribution. Irrig Sci 26:313–323. doi:10.1007/s00271-007-0095-7

Maritza O, Schaffer B, Davies FS (2004) Soil temperature, physiology, and growth of containerized Annona species. Sci Hortic 102:243–255

Marler TE, George AP, Nissen RJ, Andersen PC (1994) Miscellaneous tropical fruits. In: Schaffer B, Andersen PC (eds) Handbook of environmental physiology of fruit crops, vol II, Subtropical and tropical crops. CRC Press, Boca Raton, pp 199–224

Mason SC (1925) The minimum temperature for growth of the date palm and the absence of a resting period. J Agric Res 31:401–414

Meghwal PR, Suresk Kumar, Bhatt RK, Akat Singh (2014) Response of Jujube cultivars to natural climatic variability under rain fed condition in arid zone. Paper presented at the conference: new dimension in agro meteorology for sustainable agriculture, At GB Pant University of Agriculture and Technology, Pantnagar, Uttrakhand, October 2014, India

Metcalf CR, Chalk L (1950) Anatomy of the dicotyledons, vol 1. Claredon Press, Oxford

Munns R, Tester M (2008) Annu Rev Plant Biol 59:651–681

Naeini MR, Khoshgoftarmanesh AH, Lessani H, Fallahi E (2004) Effects of sodium chloride-induced salinity on mineral nutrients and soluble sugars in three commercial cultivars of pomegranate. J Plant Nutr 27:1319–1326

Naeini MR, Khoshgoftarmanesh AH, Fallahi E (2006) Partitioning of chlorine, sodium, and potassium and shoot growth of three pomegranate cultivars under different levels of salinity. J Plant Nutr 29:1835–1843

Nixon RW (1937) The freeze of January 1937 – a discussion. Ann Rep Date Grow Inst 14:19–23

Nunez-Elisea R, Schaffer B, Crane JH, Colls AM (1997) Effects of flooding on growth and leaf gas exchange of *Annona* species. In: Proceedings Congresso Internacional de Anonaceas, University of Chapingo, Texcoco, Mexico, 12–14 November pp 124–132

Pareek OP (2001) Ber. International Centre for Underutilized Crops, Southampton

Pareek OP, Godara AK (1993) Crop regulation in pomegranate. In: Chadha KL, Pareek OP (eds) Advances in horticulture, vol 3, Fruit Crops. Malhotra Publishing House, New Delhi, pp 1229–1235

Patel NM, Patel DK, Verma LR, Patel MM (2010) Effect of cultural and chemical treatments on fruit set and fruit yield of custard apple (*Annona squamosa* Linn.) cv. Sindhanur. Asian J Hortic 5:498–502

Pathak RK (2003) Status report on genetic resources of Indian gooseberry-aonla (*Emblica officinalis* Gaertn.) in South and Southeast Asia. IPGRI Office for South Asia National Agriculture Science Centre (NASC), DPS Marg, Pusa Campus, New Delhi

Pitman MG, Läuchli A (2002) Global impact of salinity and agricultural ecosystems. In: Läuchli A, Lüttge U (eds) Salinity: environment – plants – molecules. Kluwer, Dordrecht, pp 3–20

Popeno W (1920) Manual of tropical and subtropical fruits. Hafner Press, Mcmillan Publishers, London

Prasad N, Bankar GJ, Vasistha BB (2003) Effect of drip irrigation on growth, yield and quality of pomegranate in arid region. Indian J Hortic 60:140–142

Purohit AG (1982) Flower induction in deciduous pomegranate in tropics. Sci Cult 48:146–47

Purohit AG (1986) Quantitative methods for estimation of soft-seededness of pomegranate. J Maharashtra Agric Univ 11:116–117

Ramoliya PJ, Pandey AN (2003) Effect of salinisation of soil on emergence, growth and survival of seedlings of Cordia rothii. Forest Ecol Manage 176:185–194

Rodrígueza P, Mellishob CD, Conejerob W, Cruza ZN, Ortu˜nob MF, Galindo A, Torrecillas (2012) Plant water relations of leaves of pomegranate trees under different irrigation conditions. Environ Exp Bot 77:19–24

Roy C, Sengupta DN (2014) Effect of short term NaCl stress on cultivars of *S. lycopersicum*: a comparative biochemical approach. J Stress Physiol Biochem 10:59–81

Sankhla N, Sankhla D, Upadhyay A, Davis TD (1989) Amelioration of drought and high temperature injury in fruits of ber by paclobutrazol. Acta Hortic 239:197–202

Save R, Biel C, Domingo R, Ruiz-Sanchez MC, Torrecillas A (1995) Some physiological and morphological characteristics of citrus plants for drought resistance. Plant Sci 110:167–172

Shailendra A, Narendra A (2005) The effect of trickle irrigation on growth, yield and quality of pomegranate (*Punica granatum*) cv. Ganesh in Chhattisgarh region. Mysore J Agric Sci 39:175–181

Sharma SD, Sharma VK (1990) Variation for chemical characters in some promising strains of wild pomegranate (Punica granatum L.). Euphytica 49:131–133

Sharma AK, Tewari JC (2009) Arid zone forestry with special reference to Indian hot arid zone. In: John N, Owens H, Lunday G (eds) Forests and Forests Plants. Eolss Publishers Company, UK, pp 90–130

Shen W, Nada K, Tachibana S (2000) Involvement of polyamines in the chilling tolerance of cucumber cultivars. Plant Physiol 124:431–439

Singh IS, Pathak RK, Diwedi R, Singh HK (1993) Aonla production and post harvest technology. Tech Bull, Department of Horticulture, N D U A T, Faizabad

Sonawane PC, Desai UT (1989) Performance of staggered cropping in pomegranate. J Maharashtra Agric Univ 14:341–342

Sulochanamma BN, Yellamanda Reddy T, Subbi Reddy G (2005) Effect of basin and drip irrigation on growth, yield and water use efficiency in pomegranate cv. Ganesh. Acta Hortic 696:277–279

Tonon G, Kevers C, Faivre-Rampant O, Grazianil M, Gaspar T (2004) Effect of nail and mannitol iso-osmotic stresses on proline and free polyamine levels in embryogenic Fraxinus angustifolia callus. J Plant Physiol 161:701–708

Torrecillas A, Alarcon JJ, Domingo R, Planes J, Sánchez-Blanco MJ (1996) Strategies for drought resistance in leaves of two almond cultivars. Plant Sci 118:135–143

Trapaidz TG, Abuladze LS (1998) Pomegranate cultivars resistant to cracking. Subtropicheskie Kul'tury 2:95–97

Verma S, Mishra SN (2005) Putrescine alleviation of growth in salt stressed *Brassica juncea* by inducing antioxidative defense system. J Plant Physiol 162:669–677

Wrigley G (1995) Date palm. In: Smartt J Simmonds NW (eds) Evolution of crop plants, 2nd ed. Longman Group, Essex, England, pp 99–403

Yadav DS, Mishra MM, Nath A (2001) Horticulture research: an overview. In: Verma ND, Bhatt BP (eds) Steps towards modernization of NEH agriculture. ICAR Research Complex For NEH Region, Umiam, pp 93–120

Yaish MW, Kumar PP (2015) Salt tolerance research in date palm tree (*Phoenix dactylifera* L.), past, present, and future perspectives. Front Plant Sci 6:348. doi:10.3389/fpls.2015.00348

Zaid A, de Wet PF (2002) Climatic requirements of date palm. In: Zaid A (ed) Date palm cultivation. Food and Agriculture Organization of the United Nations, Rome, pp 57–72

Litchi

14

Rajesh Kumar

Abstract

Litchi is an important commercial fruit crop providing nutritional security to millions of people of Southeast Asia. It belongs to the Sapindaceae family and is popularly known as the queen of fruits due to its attractive colour, taste, and quality. It is very fastidious in its climatic requirements and due to this specific characteristic, performance is highly influenced by the growing environment and nutrition. The responses and reactions may be due to various abiotic stresses leading to a series of morphological, physiological, and biochemical changes that adversely affect plant growth and development. The complexion further increases with the present scenario of a changing climate, which is causing a wide array of interacting stress factors that increase atmospheric temperature and decrease soil osmotic potential caused by uneven, irregular, and unpredictable rainfall patterns. The scarcity of resources and adverse climatic conditions, singularly or in combination, induce cellular damage and cause a reduction in the physiological growth processes in the plant body. The plant architecture physiology above the ground and the root system below can be adjusted with various biotechnological and other management practices to cope with the adverse effect of the abiotic stresses to achieve the production potential in an economically viable manner. The mechanisms of the physiological basis of abiotic stress tolerance/resistance or to avoid/escape the stress need to be understood using biotechnological tools. Consistent efforts for enhancing the accumulation of specific compounds in response to a stimulus, with compatible solutes, have resulted in some degree of tolerance in litchi plants. The

R. Kumar (✉)
ICAR-National Research Centre for Litchi, Muzaffarpur, Bihar 842002, India
e-mail: rajeshkr_5@yahoo.com

N.K.S. Rao et al. (eds.), *Abiotic Stress Physiology of Horticultural Crops*,
DOI 10.1007/978-81-322-2725-0_14

ability to redirect nutrients to imperative processes and the induction of adequate metabolic processes and adjustments are crucial for plant survival in conditions of stress. Proper agronomical management and good agricultural practices with favourable rhizosphere and manipulation have been the important adaptation and mitigation strategies for developing abiotic stress tolerance mechanisms in litchi, although efforts for breeding varieties and use of biotechnological approaches will also extend stress resistance characteristics.

14.1 Introduction

Litchi (*Litchi chinensis* Sonn.), a member of the Sapindaceae family, originated in China. This is an arillate fruit, widely cultivated in subtropical regions of the world which provides livelihood security to millions of people in Southeast Asia (Menzel and Waite 2005). Its production has undergone substantial expansion in the past 50 years with plantations increasing from 9400 ha in 1949 to 82,000 ha, in 2013 in India (Rai et al. 2001; NHB Databook 2013). Fruits are available from early May to early July in different areas, with only a small quantity being exported in the last few years. Attributes such as bright red colour, excellent flavour, super taste, and rich in nutrients, make it a very attractive and popular fruit crop in international markets (Wall 2006). In India, this is popularly known as the 'queen' of fruits (Fig. 14.2a, b). However, this crop has been meagrely exploited at its R&D level and till date, there exists a wide gap between actual production and potential production (Kumar 2015). As commercial viability lies in its production of an enhanced quantum of quality fruits, its commercialisation even around the world has also been slow due to specific climatic requirements for successful cultivation. The low production of this crop is the result of the interaction of many interdependent factors. The scenario of its production is very poor as well as the erratic cropping pattern in many areas, but there is an increasing trend of area and production under litchi in India (Fig. 14.1). Among the various factors, the cumulative effects of abiotic stresses significantly affect the production and productivity of litchi fruits.

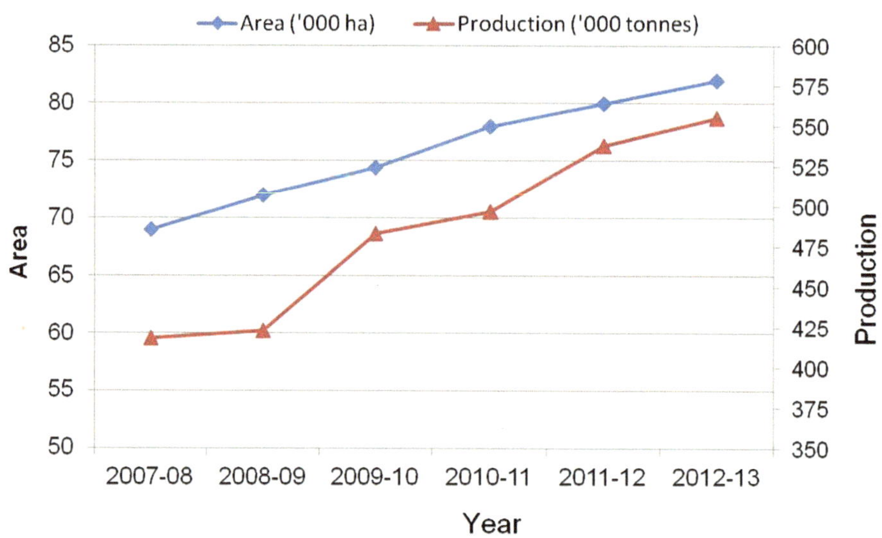

Fig. 14.1 Area and production of litchi at the national level

Fig. 14.2 (**a**, **b**) Healthy bunch of litchi fruits

14.2 Tree Architecture and Production Constraints

The litchi is very specific in its climatic requirement and is adapted to the warm subtropics, cropping best in regions with brief, dry, frost-free winters and long hot summers with high rainfall and humidity. The growth responses are accordingly influenced with the environmental variation and seasonal fluctuation. The annual life cycle of litchi, like other perennial fruit plants, has two distinct phases, the juvenile vegetative phase and the adult reproductive phase. With panicle emergence leading to transition to flowering, the plant enters the reproductive phase. The phases of increased plagiotropism, changes in reproductive competence, and branching characteristics after foliar morphology, in addition to the changes that accompany the transition (vegetative to reproductive) phase, have the distinct influence of abiotic factors (Chadha and Rajpoot 1969; Husen and Pal 2006).

The growth physiology and correlations between roots and shoot systems are important aspects to understanding the role of these systems in the health of litchi trees and the desirable performance with the interaction effects of environmental factors (Kumar and Nath 2013).

The root system of the litchi plant varies depending on the method of propagation used (Singh et al. 2012). Litchi seedlings are used for grafting on to a desirable variety and the resultant grafted trees that have reached maturity have well-developed vertical downward going tap roots due to seedling rootstock that can extend a few meters down into the ground towards the water table. These tap roots have some lateral branching. Most commercial plantations develop by the air-layered technique, which has no main tap root and entirely shallow lateral branching root systems often have their entire root distribution spread in the top layer of soil, extending to a depth of about 1–1.5 m. The root growth and distribution in the cultivar Tai So has been found growing down to 1 m in sandy soils, whereas trees growing in clay soils in South Africa have a shallow root system. Most of the roots of 8-year-old plants of the cultivar Tai So tree are growing in a sandy clay loam overlying a heavy clay in the top 30–40 cm in Queensland (Menzel and Waite 2005).

Furthermore other studies have shown that soil type influenced total root density and feeder root distribution (depth of the soil where 80 % of roots are located). There were more roots in a sandy soil than in a clay soil, but a smaller proportion was found at depth (feeder root distributions of 0–20 cm and to 0–60 cm, respectively). About 90 % of the roots were less than 2 mm in diameter, with no effect of soil type or depth (Menzel et al. 1990).

The lateral absorptive roots of a litchi tree extend out to about the drip line of the tree canopy. The diameter of the canopy is a good indicator of the extent of the root distribution system (Egea 2009). The root system of a litchi tree typically experiences different variable climaxes of growth each year: April to June (summer), July to August (rainy) after harvesting, September to November (pre-cool winter) when new shoots mature, and December to March (phase change from vegetative to reproductive). The root distribution pattern or the root system growth events seem to occur or remain in active phase when the rhizospheric soil is warmer (Agnihotri et al. 2009).

During late spring, summer, and fall seasons in the subtropics the shallow root system of a litchi tree is exposed to the daily cyclical extremes of heating and cooling. This stress is especially pronounced during the mid-summer postharvest months, that is, the growth and developmental phases. The soil type, soil tilth, inherent nutrient status, and microbial population (mainly the mycorrhizal fungi) aid in the growth and development of their host tree by significantly increasing the absorptive surface area of the roots, as well as assisting in the decomposition of organic matter. These factors decide the tolerance of plants to drought, seasonal temperature fluctuations, and pH balance, as well as minimise the negative effects of transplant shock, apart from protecting the plant from harmful diseases and pathogens and need or dependency of plants on external or supplemental nutrients and other input applications (Kumar and Nath 2013). If these natural soil components are missing, they should be replaced for healthy plant growth. The hemispherical habit of growth and shape of the litchi tree naturally keep the root system shaded under the canopy.

The root system in plants is a structure of dynamic nature, which modulates primary root growth and root branching by continuous integration of environmental inputs such as nutrition availability, soil aeration, soil moisture, and salinity. In the case of artificial stress conditions, the energy level and vigour decrease and cause a decline in growth and productivity. The efficiency of soil water absorption in trees depends on both spatial extension and density of their root system. Water uptake by individual trees from their spatial extension portion depends on the fine root exchange surface, that is, on their cumulated length or biomass. Both vertical root distribution and seasonal root growth dynamics depend closely on physical soil properties (mainly texture such as clay content, bulk density, content of coarse elements, etc.) and the physiological constraints on root survival and development (water table, oxygen supply, nutrients, aluminium or manganese toxicity, soil pH). Climate itself could also influence fine root dynamics.

Litchi plantations have been made with various spacings throughout the country. Traditional orchards were planted at low density and the trees were left to develop into large specimens. However, most of the new orchards are planted in high densities and have adopted techniques such as pruning and training. These technologies need to have been skilfully and widely adopted but there is still little understanding of tree physiology and its actual effect as the relationship between yield and the interception of light by canopy, which will lead to the development of new management techniques and increased productivity. Canopy geometry should be managed to achieve the maximum efficiency for trapping sun energy through foliage and channelling metabolites for quality fruit production; this operation also manipulates the tree vigour and use of maximum available light and temperature to increase productivity and quality as well as minimises the adverse effect of the weather (Monteith 1977; Srivastava 2007). The mechanism by which shoot reorientation favours flower production has also been investigated (Ito et al. 2004; Yang et al. 1992). Floral development progresses more rapidly on horizontal than on vertical shoots (Ito et al. 2004). Nutrition in conjunction with climate conditions has a significant influence on the tree vigour and canopy development in perennial fruit trees (Jinag et al. 1999; Shikhamany 2001).

The aspects of nutritional and hormonal imbalances in litchi are also causing a significant decline in fruit quality and productivity. Prolonged periods of overcast weather, extremes of temperature, and droughts have been found to reduce photosynthesis and productivity in most of its growing areas due to physiological disorders, mainly fruit drop and fruit cracking (Kumar 2015). In the litchi production system, erratic bearing behaviour and poor production with inferior fruit quality is experienced in most countries (Menzel and Waite 2005). This is often because winters are not cool or dry enough to induce a growth check prior to flowering. Irregular flowering in litchi is related to variations in the timing of flushing. The current shoot and its maturity coinciding with winter's cool temperature, flower panicles will be initiated or it will give rise to vegetative flushes (Batten and McConchie 1995; Menzel and Simpson 1994). Litchi growing in some parts of the world has shown that winter frosts damage the panicle and developing flowers and limit fruit production in that season (Chadha and Rajpoot 1969). In China, conditions during flowering are often cool, which can prevent pollination and fruit set. More frequently, the variations in climatic conditions (temperature extremes, drought, and low irradiance) cause seasonal variations influencing the entire litchi-growing areas of the world (Kumar 2015). Although this fruit crop is evergreen perennial in nature, it has a low ability to transport water from roots to leaves; this is why wet soil alone cannot prevent the development of tree water stress.

In the case of litchi achieving its potential production performance, it can be just as productive as any other highly productive subtropical fruit enterprise. In India, the major producing state is Bihar, where a longer cool period before flowering is available in most years. In West Bengal, occasional hot and dry conditions prior to flowering restrict production. In some other parts of the country, off-season production is also possible, which requires proper exploitation. Different types of abiotic stress adversely affect plant growth, quality production, and productivity of litchi (Nakata 1955). The present scenario of changing climate causes a wide array of stress factors interacting with each other to increase atmospheric temperature and decrease soil osmotic potential due to an uneven, irregular, and unpredictable rainfall pattern (Kumar 2015). Another influencing factor is the quality of water, that is, with a quantity of soluble salts in the root-growing zone. However, various abiotic stress conditions are often interrelated; these conditions singularly or in combination induce cellular damage affecting overall growth and performance of litchi plants.

In one projection report it has been estimated that more than 50 % of crop yield reduction is caused due to the effect of abiotic stresses (Acquaah 2007; Nath et al. 2012). Many factors act independently or in combination with the environment and its occurrence, influencing plant growth and physiology, and are difficult to distinguish (Wilhite 2000). To mitigate the ill effect of abiotic stresses, knowledge of effective plant responses that disturb the homeostatic equilibrium at the cellular and molecular level is essential in order to identify a common mechanism for multiple stress tolerance. Stresses also affect the biosynthesis, concentration, transport, and storage of primary and secondary metabolites. Abiotic stresses also cause protein dysfunction, which adversely affects growth and productivity and triggers a series of intrinsic changes in plants. The stresses affect the whole plant resulting in a stress response by diverting their scarce resources towards withstanding the changed environment. This entails a major change in their gene expression, proteins, metabolites, and so on (Mou 2011). Hence, adequate attention is required to study the various abiotic stresses affecting the physiology and biochemistry of litchi and also to develop some sort of predictive and protective strategies to develop tolerance/resistance mechanisms by this crop.

14.3 Major Abiotic Stresses

The various stresses caused by drought, salinity, extreme temperatures (cold and heat), nutrition, and irradiance on the litchi crop are of major significance. Droughts in the Indian condition are mainly caused by failure of the rainy season due to the southwest monsoon. It is believed to be an association between El Niño and La Niña events and weak monsoons (Samra and Singh 2002).

14.3.1 Drought

Litchi production is very much influenced by drought, which is a complex physical process of widespread significance. The impact of drought varies from crop to crop even at the intervarietal level and the severity can cause a wide variety of impacts due to many factors related to the underlying vulnerability of a region and is visible in terms of poor quality fruits and loss of fruit yield. Some of the marked responses to this stress are turgour loss, which leads to inhibition of growth. The output of litchi crop cultivation mainly depends upon rainfall and soil moisture (Menzel and Waite 2005). The water deficit due to short-term drought affects many aspects of plant physiology. In litchi and longan crops, drought closes the stomata and reduces CO_2 assimilation, stem extension, leaf expansion (Fig. 14.3), and fruit growth (Menzel 2005a; Smrintontip et al. 2013). The information required to address the dynamic nature of drought in terms of its onset, progression, intensity, and impact in litchi has been least developed, but with the available information efforts have been integrated for further critical analysis of this climatic anomaly causing abiotic stress to this perennial fruit crop.

14.3.1.1 Water Potential

The reduction in soil water content in the rhizosphere below the level required for sustaining normal plant growth and developmental procedures results in the immediate decline in leaf relative water content causing stomatal closure, which in turn renders a decrease in many physiological processes in the crop.

Litchi is also found to have been affected by seasonal drought in areas wherever it is grown. Most of the information on water relations of litchi has been in determining the impacts on plant–water relations and gas exchange under laboratory, glass house, and field conditions through attempts to model the opening and closing of stomata and some attempts to study the

Fig. 14.3 Leaves showing water deficiency

effect of drought on shoot growth, flowering, fruit production, and quality (Menzel 2005a). An attempt made to see the effect of drought influence on plant growth revealed the best growth results with regular weekly irrigation in summer and monthly in winter (Singh and Pathak 1983). The volume of water available for plant growth varies with soil type and rooting depth. The majority of the nutrient-absorbing roots are found in a 0.0–0.6 m horizon but the trees extract water from lower depths. The root water extraction pattern from 0.0 to 1.50 m depth varies from 28 to 13 %, that is, in descending order w.r.t. depth (Menzel et al. 1995). It is obvious that mature trees are able to extract soil water from considerable depths and that short-term drought may not necessarily reduce growth and other physiological processes. But there must not be any water stress even for a short period at critical stages such as phase change and fruit development. The large evergreen dense leafy canopy of this perennial fruit tree has a marked capacity for passive and active adjustment of cell solute concentration in response to varying water status in the soil and air.

Leaf water potential is a good measure indicating the exact nature and degree of stress experienced by the plants. Yet not enough studies have been done for proper recommendations required to control water status under drought (Menzel and Simpson 1994) to exhibit physiological responses up to optimum level (Chen and Liu 2001a, b; Menzel 2005b, c). The physiological and morphological basis of drought effects with reference to resistance/tolerance have been investigated in some greenhouse experiments, which indicated that as drought conditions occurred, leaf osmotic potential at full turgour decreased and total solute concentrations increased in leaf and stem (Menzel 2005b). Leaf water potential (ΨL) in litchi cv. Tai So declined from predawn to early afternoon. It was also observed that there was full recovery in late afternoon in the wet treated (control) plants but not in the dry treated ones. At each given time of a day, ΨL was lower in dry treated plants than those of the control (Menzel and Simpson 1994).

The other study was conducted with litchi cvs. Mauritius and Floridian, by Stern et al. (1998) and showed that mid-day stem water potential (Ψs) differed more markedly than ΨL and Ψs clearly indicated irrigation regimes regulating drought. The litchi cultivar Dongliu No. 1 showing drought-resistance nature had thicker leaves and higher specific leaf weight compared with the cultivar Chenzi showing attributes of a drought-sensitive nature. Under drought, relative water content (RWC) decreased, electrolyte and inorganic phosphate (Pi) leakages increased, and chlorophyll and carotenoid content were reduced, especially in Chenzi, indicating that Dongliu No. 1 had a stronger ability to maintain cell membrane stability (Chen and Liu 2001a). As the water deficit increased, carbon was remobilised from leaves and preferentially redistributed to stems and roots, and leaves were shed, resulting in reduced whole-plant transpiration and enforced dormancy. Thus, litchi showed a range of responses to different drought intensities indicating a low degree of plasticity in response to water deficits.

Understanding the physiological responses of drought in the field is also critical. A short drought in winter has given encouraging results in phase change from the vegetative to the reproductive stage and enhanced panicle emergence in litchi. Differences in the sensitivity of the shoots and fruit to the effects of drought have been studied regarding shoot growth, flowering, fruit production, and quality and showed a period of drought can assist flowering, at least in litchi, but is not essential (Chadha and Rajpoot 1969; Nakata and Suehisa 1969). They also studied the effects of irrigation in 8-year-old Tai So trees in Hawaii, where it is generally dry and recorded that the severe drought in October inhibited leaf growth in November and increased flowering and yield, indicating that drought can induce flowering, but the response is probably related to a shift in the timing of shoot growth. Several glasshouse experiments in Australia showed that drought had no direct effect on flowering but shoot growth was found very sensitive to changes in tree water status. Menzel et al. (1989) examined the vegetative flushing of Kwai May Pink under different irrigation regimes in a glass house. Growth decreased as the level and duration of drought increased, but none of the trees flowered at high temperatures. A period of drought before flower induction may assist flowering by delaying early shoot growth until winter. This can be used in areas that have dry winters.

Experiments have shown that soil water deficits can be used to manipulate flushing patterns and increase flowering and possibly fruit production, under some temporal circumstances (Menzel 2005a; Carr and Menzel 2014). Once panicles are initiated, developed and started flowering phases, the proper cross-pollination process during bloom may lead to best fruit set even when plants are well watered. However, a cyclic drought (ΨL −2.0 MPa) created by water regulation at every 4–7 days to field capacity reduced panicle growth and the numbers of flowers compared with plants watered daily (ΨL above −0.7 MPa). Most of the flowers abscised prematurely in water-stressed plants and the few flowers that reached anthesis were male. These results indicate that trees should be irrigated from panicle emergence to prevent water deficits reducing fruit set, although they do not indicate a threshold yL below which production is affected (Menzel et al. 1995). Experiments in small pots may not necessarily predict the response of mature trees in the field with a deep root system and slower development of drought, and hence proper field studies are required to have some sort of predictive options in litchi cultivation.

Drought reduced the number of fruit per tree, average fruit weight, flesh recovery, and yield (Menzel et al. 1995). If the drought is severe enough, fruit development will be affected, particularly the development of the fruit skin. Cell division is reduced and the fruit skin becomes inelastic, and often splits when the extreme dry spell comes. However, it is not a major problem in Vietnam, where the weather is less extreme. In the case of limited water supply, not all parts of the fruit develop at the same time (Menzel 2002). It has been reported that water deficits decreased total fresh fruit weight in cv. Bengal by 16 % compared with well-watered trees (Batten et al. 1994).

14.3.1.2 Photosynthesis

The inhibitory effect of drought on photosynthetic activity has been widely described and is mainly associated with stomatal conductance and metabolic limitations (Dash and Panda 2002; Chaves et al. 2009; Lawlor and Tezara 2009). Net photosynthesis rate is reduced as leaf water potential declines (Dash and Panda 2002) but in contrast some reports say that CO_2 assimilation is often independent of plant water status until a threshold level is reached and approaches zero when leaves wilt; this critical value can be used to estimate the required range in plant water status for production of maximum fruit yields (Roe et al. 1995; Menzel 2005a). Roe et al. (1995) also determined drought affected the biochemistry of CO_2 fixation. Loss of water from the leaf results in accumulation of solutes or nonmetabolite ions including magnesium (Mg) and potassium (K) which play a regulatory role in the calvin cycle (Kaiser et al. 1981). The accumulated solutes or ions also inhibit photosynthetic processes in the osmotic stressed leaf (Kaiser et al. 1981). In the case of peach, it has been reported that water scarcity affected tree carbon assimilation due to two interacting factors: (1) reduction in leaf photosynthesis (−23 and −50 % under moderate (MS) and severe (SS) water stress compared to low stress (LS) during growth season), and (2) reduction in total leaf area (−57 and −79 % under MS and SS compared to LS at harvest; Mitra et al. 2015). Similarly, the decrease in leaf water content (from 76.32 to 51.68 %) and the increase in leaf osmotic potential (Ψs) confirmed the deterioration of leaf water status in drought-stressed litchi plants. Some experiments in Israel have shown the correlation between gas exchange and leaf water potential (ΨL) with the indication that mid-day stem and leaf water potential were better indicators of tree water status and gas exchange in a range of fruit crops including litchi (Stern et al. 1998; Naor 2000). Studies to see the gas exchange of leaves growing in different environments are in progress, because of the opinion that the optimum conditions for photosynthesis may not necessarily reflect the environment where plant species evolved, inasmuch as survival may be more related to optimising leaf area or reproductive development rather than the rate of photosynthesis per unit leaf area (Atwell et al. 1999; Menzel 2005a).

14.3.1.3 Hormones

Hormonal assay of the plants influenced by drought also includes processes such as cell division and expansion, ABA synthesis, and sugar accumulation, which consequently reduce fruit yield and quality. Abscisic acid (ABA) content in litchi leaves increased with increasing drought, whereas indole acetic acid (IAA) and gibberellic acid contents decreased (Stern et al. 2003). As far as leaf photosynthesis is concerned, the evidence showed that drought inhibited CO_2 assimilation in litchi leaves (Menzel and Simpson 1994; Menzel 2005a; Song et al. 2006; Damour et al. 2008). It was accordingly reported that the ABA concentration in xylem sap from the three wettest treatments were similar, whereas cessation of irrigation (0 %) resulted in a steep decrease in the t-ZR level (zeatine riboside), further increase in dihydrozeatine riboside (DHZR), and a >25-fold increase in ABA level. ABA is considered to act as a hormone related to stress tolerance.

14.3.1.4 Metabolites

An important harmful effect of various abiotic stresses, mainly drought and other environmental odds at the cellular level, is the alteration of membrane architecture and permeability leading to the leakage of ions and decontrolled molecular trafficking. It is essential to understand about the compounds which are able to reduce the damaging effects of various stresses such as drought and the involved cellular processes that ameliorate the consequences of water loss (Stewart 1981; Neill et al. 2003; Krasensky and Jonak 2012). One of the biochemical changes occurring when plants are subjected to this harmful stress condition is the accumulation of reactive oxygen species (ROS; Smirnoff 1993). When plants are subjected to water deficit, a variety of ROS, such as superoxide ($O_2^{\cdot-}$), hydrogen peroxide (H_2O_2), and hydroxyl radicals ($OH^{\cdot}$), which cause oxidative damage in plants are generated (Fatemeh et al. 2011). Likewise, proline accumulation in leaves of low water potential is caused by increased biosynthesis, decreased oxidation in mitochondria, and decreased utilisation of protein synthesis (Raymond and Smirnoff 2002). Reactive oxygen species production is stimulated by environmental stresses, mainly drought (Sgherri and Navari-Izzo 1995; Smirnoff 1993; Chen and Liu 1998) but related study is very limited in the case of litchi. Chen (1997) reported an increase in relative permeability of plasma membranes of leaves for both litchi cultivars, viz. Dongliu No. 1 and Chenzi, and also observed it to be highly related to the increase in malondialdehyde (MDA) and H_2O_2 contents, indicating that the elevated level of H_2O_2 and the increased membrane-lipid peroxidation could be the cause for increased relative permeability of plasma membranes. Chen and Liu (2001b) reported that the extent of damage of chloroplasts and also the occurrence of plasmolysis in mesophyll cells was greater in the drought-sensitive litchi cultivar Chenzi. A decrease in leaf starch concentrations and a concomitant increase in the activities of amylase, invertase, and phenylalanine ammonia lyase (PAL) caused a shift in carbon partitioning in favour of soluble carbohydrates (Chen and Liu 1999; Damour et al. 2008). In severely drought-stressed leaves, high leaf nitrate reductase activities were paralleled by increases in proline concentration, suggesting an osmoprotective role for proline (Laura Fiasconaro et al. 2013). It was clearly indicated that the proline content of leaves increased with increasing drought, with a greater increase in Dongliu 1 than Chenzi. In contrast proline dehydrogenase activity in leaves decreased with a greater decrease in Dongliu 1 than that in Chenzi (Chen and Liu 1999).

14.3.1.5 Physiological Disorders

Litchi growing is threatened by many physiological disorders and the incidence and severity are increased with the occurrence of drought, as there may be different effects on fruit production depending on the level and timing of the water shortage (Li et al. 2000; Kumar and Kumar 2007;

Fig. 14.4 Fruit cracking in litchi

Kumar 2008). Batten et al. (1994) compared a set of unirrigated trees and trees irrigated weekly to replace 85 % of potential evapotranspiration in Australia (latitude 29° S) and the created drought stress condition from flowering to harvest resulted in smaller fruit size compared to control trees, but the number of fruit was more than double in the dry treatment. Greater fruit retention was attributed to less competition between leaf flushes and fruit.

The crop and its varieties receiving poor husbandry work have relatively low tree performance. In addition, the climatic conditions in the subtropics with relatively hot, dry prevailing winds during the initial fruit set period aggravate the problem of severe fruit abscission, commonly known as premature fruit drop. However, the exact physiological mechanism of this process is not so clear but the extent of premature fruit drop in litchi can be reduced with crop management practices to ensure economically sustainable cultivation of litchi in its growing regions (Kumar and Kumar 2007; Kumar 2008).

In general it has been observed that fruit skin cracking (Fig. 14.4) often occurs when trees are subject to drought soon after fruit setting (Kumar and Kumar 2007). Nevertheless, deficit irrigation can enhance fruit quality by raising dry matter percentage and sugar content. Moreover, controlled water deficit has been used as a technique to stimulate blossoming in litchi in the subtropics. Hence, regulated deficit irrigation and partial root zone drying techniques have been applied to withhold water during certain specific periods, thus producing moderate drought stress, which in turn results in improved yield, fruit quality, and water use efficiency (Rai et al. 2001). The role of hot dry conditions on fruit drop is not very clear. Higher rainfall would be expected to increase relative humidity and lead to reduction in fruit drop (Menzel and Waite 2005). Insect attacks are also factors in some areas, accounting for 25–99 % of green fruit drop in several locations. Throughout Asia and the Pacific, the problems of sunburning and skin-cracking (splitting) occurring are found, and are often associated with hot dry weather, drought, and low calcium concentrations (Kumar and Kumar 2007). However, the relationship between these disorders and tree management practices done at critical periods of fruit growth and development can have satisfactory outcomes/results. The flesh portion, that is, aril, grows rapidly at the fag end of the maturity period. The problem of skin-cracking was also found to occur after irrigation or pouring of heavy rain, or just an increase in relative humidity (Kumar and Kumar 2007).

14.4 Alkalinity/Salinity

Salt stress is causing varied effects on plant growth and metabolism, as well as alterations in physiological and biochemical processes (Bogges et al. 1976; Banu et al. 2010). However, the response to excess salts on litchi indicated that the litchi trees are susceptible to salt stress and not able to sustain high salt concentrations, hampering production and productivity. Litchi crops, more particularly under region-specific conditions of the country, have been observed as low and poor fruit (even quality) yields and are related to salinity/alkalinity hazards. No sincere efforts have been made for controlling the salinity problems and utilisation of the salt-affected soils (Singh 2003). The saline soil is characterised by poor physical properties such as low infiltration, low porosity, and high bulk density. The high concentrations of salts in sodic soil can not only injure the plant directly, but also degrade the soil structure by decreasing porosity and water permeability (Phang et al. 2008). Litchi trees should not be irrigated with water having an electrical conductivity greater than 0.5 dS per m or about 500 mg soluble salts per litre. Problems are intensified and cause damage during dry weather and excess fertilisation.

14.4.1 Cause and Effects

In the case of litchi, in response to excess salts the recommendation, as stated above, is that trees should not be irrigated with water having an electrical conductivity greater than 0.5 dS per m or about 500 mg soluble salts per litre. Damage sometimes occurs during dry weather, especially when young trees are overfertilised, as the tips and margins of the old leaves die (Samra and Singh 2002; Babita and Kumar 2008). Experiments were conducted in Australia to grow marcots in sand culture irrigated with two concentrations (i.e. 6 or 12 mM NaCl). It has been reported that at both concentrations, older leaves were shed with each new flush of growth. The cultivar Tai So was more sensitive than the cultivar Bengal, and this was reflected in greater uptake of salts. The concentrations of Na in the leaves of Tai So after 13 months in the control and 12 mM NaCl treatments were 240 and 22,000 ppm, respectively (FAO Lychee). Similarly, leaf Cl concentrations were 0.3 and 2.6 %. Salt or osmotic stresses affect the vegetative growth of perennial fruit plants, as observations recorded for both fresh weight and dry weight decreased significantly. It was also reported that the net photosynthesis declines with increase in salt as well as water stress (Hanson and Hitz 1982).

14.4.2 Photosynthesis

The effect of salt stress on plant performance and growth are mediated through decreases in stomatal conductance and photosynthesis and depend upon severity and duration of the stress period (Dash and Panda 2002). The accumulated solutes or ions also inhibit photosynthetic processes in the stressed leaves of some field crops (Dash and Panda 2002) due to decreased chlorophyll concentration (Boyer 1985; Kaiser et al. 1981). The net photosynthesis of the leaf declines with increase in salt as well as water stress (Hanson and Hitz 1982). The decline in total leaf area and net CO_2 fixation per unit leaf area are recorded, whereas dark respiration increases, leading to a drastic reduction in net CO_2 assimilation per unit leaf area per day. Low rates of net CO_2 fixation during the light period is caused by partial stomata closure, loss of turgour of mesophyll cells through salt accumulation in the apoplasm, or direct toxic effect due to ions (Hare Krishna 2012).

14.4.3 Metabolites and Physiology

There are a larger number of reports showing important metabolite regulations due to salt stress in field crops (Misra et al. 1987; Creus et al. 1997; Heikal et al. 1982). This osmotic adjustment in the cytoplasm is accomplished by means of dissolved substances that are compatible with the cellular enzyme systems and their metabolic processes. When plants are exposed to salinity mediated by high NaCl concentrations, steady-state transport of Na^+, Cl^-, K^+, and Ca^{++} are disturbed (Binzel et al. 1988). High levels of Na^+ and Cl^- in the apoplast alters aqueous and ionic thermodynamic equilibrium, resulting in hyperosmotic stress, ionic imbalances, and toxicity (Ehsanzadeh et al. 2009). In fruit plants, damages inflicted by salt stress may be due to disruption of ionic equilibrium, deleterious effect on the functioning of some enzymes, osmotic imbalance, membrane disorganisation, reduction in growth, inhibition of cell division–expansion, reduction in photosynthesis, and production of reactive oxygen species (Hare Krishna 2012). In saline conditions, uptake of abundantly available Na^+ and Cl^- therefore offers a comparatively cheap way to lower the tissue–osmotic potential. The adjustment of the cytoplasmic compartment (NaCl salts) is achieved via production of compatible osmolytes such as proline, mannitol, sorbitol, and glycine betaine. The latter also acts as an antioxidant and thus detoxifies reactive oxygen species. Ionic toxicity, osmotic stress, and nutritional defects under salinity lead to metabolic imbalances and oxidative stress (El-Shabrawi et al. 2010). From a practical point, salt stress can be imposed more easily and precisely in laboratory settings. Although the importance of salt and drought stress signalling was recognised long ago, few molecular components were known until recently. Both root and shoot growth with more shading of old leaves seems to have inhibition of enzyme reactions and inadequate compartmentalisation between cytoplasm and vacuoles results in salt accumulation in the leaf apoplasm leading to dehydration turgour loss, and a burning effect causing death of leaf cells and tissues.

Experimental attempts were made to grow marcots of two varieties in sand culture irrigated with 06 and 12 mM NaCl. At both concentrations, older leaves were shed with each new flush of growth. Tai So was more sensitive than Bengal, and this was reflected in greater uptake of salts. The concentrations of Na in the leaves of Tai So after 13 months in the control and 12 mM NaCl treatments were 240 and 22,000 ppm, whereas leaf Cl concentrations were 0.3 and 2.6 %, respectively. Salinity causes defoliation during fruit growth and development, and early ripening occurs, fruits are more exposed to the sun causing sunburn injury, fruit cracking, and ultimately poor quality having no market value. The chlorophyll content of leaves and rate of photosynthesis decrease. Salinity increases the rate of respiration of the roots. Protein synthesis declines (Batten et al. 1994).

A considerable part of litchi in India is grown in calcareous or alkaline soils, which favours the occurrence of Fe deficiency in trees, leading to Fe chlorosis. The deficiency occurs because of low availability Fe in high pH soils. The effect of salt stress on litchi tree performance and growth are mediated through decreases in stomatal conductance and photosynthesis (Hare Krishna 2012). Among the various mechanisms for adaptation to salt stress, one is osmoregulation that is at cellular adaptation, as evidence based on poor performance and attaining the declining stage of litchi plants in the case of saline/alkali conditions (environments) have been reported (Kumar and Nath 2013). Hence, remedial measures and suitable agronomic practices to reduce the concentrations to lower levels is the only viable option.

The inherent nutrient status of the soil and the internal nutrient requirement within the plant required within plant cells to sustain growth and metabolism, can be adjusted by appropriate management practices to avoid deficiencies and disorders (Atwell et al. 1999; Fernández-García et al. 2002). Agronomic means for controlling Fe chlorosis is adopted with alternative agricultural management techniques in litchi. For instance, in a soil with pH 7.6 and 8.0, the acidification of irrigation water to bring it down to normal pH 6.0–7.0 can solve the problem. In saline conditions, sodium chloride in higher concentrations reduces the growth of salt-sensitive plants subjected to salt stress and photosynthetic pigment composition changes due to salt stress. To combat this there is a need to develop the salt-tolerant cell lines of litchi. It has been reported earlier that an increase in the enzyme of the glyoxalase system imparts salinity tolerance in the model plant tobacco (Li 2011). This strategy by taking glyI + II genes together to manipulate the glyoxalase pathway for enhancing salinity tolerance in litchi has been found effective as the transgenic that developed showed higher salt tolerance as compared to the wild type, observed by less reduction in chlorophyll content in the leaf disc, which suggested an important role of glyI + II in conferring tolerance to plants under stress (salinity) conditions (Dash and Panda 2002; Sinha and Das 2013).

14.4.4 Hormones and Polyamines

A strict control of concentration and distribution of auxin level also may lead to aberrant growth as reported in many other plants (Sitbon et al. 1992; Sitbon and Perrot 1997). The balance of auxin levels (IAA) in the plant body may also be required for the adaptive recovery growth of litchi plants growing under such a stressed environment. Although the levels of cytokinins decreased, levels of ABA increased (Hare Krishna 2012). Elevated ABA levels are important for rapid osmotic adjustment to salinity. Polyamines such as putrecine accumulate in plants under salinity stress. Application of cytokinins or putrecine counteracts salinity-induced leaf senescence by stabilising plant cell membranes and enhancing protein synthesis.

14.5 Temperature

Both high and low temperatures can induce physiological and biochemical changes in the plant body, for altered crop growth and erratic bearing behaviour. The overview has clearly indicated that both extremely high and low temperatures have an adverse effect on the growth, production, and survival of plant species by causing chilling, freezing (frost), and heat injury (Singh 2003).

14.5.1 High Temperature

As an aggravation of global warming, high temperature would become one of the main nonbiological stress factors (Pasala et al. 2013) that restrict the growth and development of litchi plants. In general, stress due to increased temperature is a common agricultural problem (Fengxia et al. 2014) and it hampers plant growth right from its ontogeny, and at later stages may have an adverse effect on physiological processes such as photosynthesis, respiration, water relations, and membrane stability (Bhattacharya 2010; Hare Krishna 2012). Exposure of plant organs to severe temperature stress may lead to changes in phytohormones. The temperature range that supports litchi plant growth generally varies from 16 to 40 °C (Shao et al. 2010). The optimum temperatures for vegetative growth, transition into reproductive phase, and later for fruit growth and development, differ widely at the stages of development and are usually more responsive to night temperature (Ashraf and Hafeez 2004). As do other fruit crops, litchi has different tolerance levels for temperatures that vary with species, varieties, age, and stage of growth. Ripening is accompanied by changes in flavour, texture, colour, and aroma. In nonclimacteric fruits, the onset of ripening depends on a transition in ethylene production from a low basal rate associated with conducive expression of pigments and increased rate of respiration induced by a high temperature range (Hare Krishna 2012). Heat stress can be a major concern for litchi in its main growing regions, inasmuch as high temperature can cause significant damage to shoot and root growth inhibition, sunburn on leaves and maturing fruits, drying of branches and stems, anticipated leaf senescence and abscission, reproductive development, extent of pollination, fruit discolouration, and damage including fruit cracking, fruit drop, and retarded/underdeveloped fruits (Menzel and Waite 2005). Menzel and Paxton (1985) investigated the effects of temperature and found that stem extension, node production, and leaf production increased with increasing day/night regimes from 15/10 °C to 20/15 °C, 25/20 °C, and 30/25 °C, whereas the base temperature found for shoot growth was 11 °C and high temperatures increased the proportion of leaves (seedlings) and decreased the proportion of roots. The colour of leaves changed from red to pink to green with increasing temperature, and the authors concluded with the observation that the best vegetative growth in litchi occurs above 25 °C. Batten et al. (1992) indicated that the optimum temperature for stomatal opening and conductance was about 37 °C.

High temperatures reduced both the duration of flushing and the interval between flushes. The time of floral initiation in Calcuttia and Rose-Scented was studied at Kanpur, India and indicated that the first signs of floral differentiation occurred about 3–4 weeks after the minima fell below 10 °C and daily maxima at the start of these observations were as high as 30 °C (Menzel and Simpson 1988; Menzel et al. 1989). These studies highlight the complexities/difficulty of relating productivity of fruit trees to weather. The respiratory rate is disturbed with a change in temperature. Temperatures in the rhizosphere (root) also affect the rate of plant growth. Increasing root temperatures up to 26 °C may increase top growth and uptake of inorganic salts. High temperatures increase the rate of shoot elongation. In contrast, a few weeks of cool weather in winter favour flowering (Menzel et al. 1989).

Extended periods of temperatures above 30 °C during anthesis and fruit development can also reduce fruit set and possibly fruit quality. The effects of temperature on vegetative growth were initially studied in Australia using seedlings. High day/night of 30°/25°, 25°/20°, and 20°/15 °C compared with 15°/10 °C increased shoot growth in six selections, with a mean daily base temperature of 11 °C. In a later study with marcots, trees were flushed twice at 30°/25 °C and once at 25°/20 °C over 18 weeks (Menzel and Simpson 1992). In Australia, higher numbers of female flowers were associated with an average maximum temperature during early flower development of 18 °C, with lower numbers at 23 °C. In contrast, the rate of flower opening was related to the number of flowers per panicle and it was concluded that areas with winter maximum temperature above 25 °C were not well suited for litchi cultivation. Bagging can improve fruit quality, possibly due to cooler temperatures and higher humidity (Menzel and Simpson 1992).

Temperature has been shown to have strong effects on pollination, but these responses do not necessarily translate into better fruit production. Lack of proper pollination may lead to retarded fruits (Fig. 14.5; Kumar and Kumar 2007). Reports are available that optimum pollination occurred at 19–22 °C, with maximum fertilisation obtained after 7 days. At 15 °C, pollen tube elongation was strongly inhibited. However, from 15 to 27 °C, at least 10 % of ovules contained pollen tubes indicating that they were fertilised. Such a level of fertilisation appears sufficient for most cultivars to produce a high yield, although at 33 °C, all female flowers abscised, suggesting a limitation for good yields when days are above 30 °C for long periods. In southern Queensland, the proportion of female flowers that set was greater with later flowering when the maximum temperature was 30 °C than with earlier flowering when the maximum temperature was 24 °C. In contrast, fruit set or yield in northern New South Wales could not be attributed to differences in average or maximum temperatures during anthesis. It was proposed that fruit set failed because the male flowers failed to produce pollen. The other possibility was that the early female flowers were sterile (Menzel and Simpson 1988).

The average number of days from full bloom to harvest in Shahi in India was 68 days, equivalent to an average of 813 degree-days above 15 °C. In this study the base temperature data range was taken from data of Batten and Lahav (1994) that were based on stem growth not fruit development, although other workers reported that shoot growth still occurred with days of 15 °C (Shukla and Bajpai 1974). Rai et al. (2001) showed a strong correlation (r^2 = 99 %) between the number of days from full bloom to harvest and the number of degree-days above 15 °C, although there were 2 years out of five with the same number of days to harvest, but with different numbers of degree-days. This agrees with the more rapid fruit development in tropical areas.

Heat injury is common in litchi fruits and mostly occurs on the most exposed fruits on the south side of the tree. The heat-injured litchi

Fig. 14.5 Retarded litchi fruits

fruits show dark blotches on the skin without any damage to the pulp. Heat injuries to fruits are common where daily air temperatures of 40 °C and above are common in the hottest months and shortage of water aggravates the severity (Singh et al. 2012).

14.5.2 Low Temperature

Several subtropical and tropical tree species (mango, litchi, macadamia nut, avocado, and orange) are found to be induced to flower by exposure to low temperature (Mahajan and Tuteja 2005; Wilkie et al. 2008). Litchi also suffers from low temperature injury (mainly damage to foliage) which occurs below 0 °C and normal flowering is upset below 5 °C (Rai et al. 2001). It also seems to have the effect of lack of sufficient cool/low temperature during winters, resulting in poor growth and fruit yield in the successive growing season. In the case of litchi, the response to temperature has been found to be cultivar dependent. Temperature stress was found to influence negatively the panicle emergence, number and size of inflorescences, flowering phases, fruit set, fruit development, fruit yield, and quality because the several physiological and developmental processes including photosynthesis, respiration, morphogenesis, and fruit ripening, among others, are also sensitive to temperature.

However, the temperature has little effect on the rate of photosynthesis, a complex phenomenon, but the higher side of the optimum range has a marked increase in its rate but at third order inasmuch as effects are determined mainly by light intensity, CO_2 levels, and temperature. This fruit crop requires low temperature to induce flowering. Flower bud differentiation is greatly influenced by the temperature. Prevalence of an average temperature of 8–10 °C from January to February has been found conducive to floral bud differentiation or panicle emergence and subsequent morphogenesis in litchi (Indian condition). Similarly, it requires a mean temperature of 10.0 °C for rapid pollen tube growth. The direct physiological effects of temperature on litchi flowering and pollinating insect (mainly honeybees) responses for pollination success to warming, which include altered flower, nectar, and pollen production, could modify floral resource availability and reproductive output of pollinating insects. Similarly, pollinator responses, such as altered foraging activity, body size, and life span, could affect patterns of pollen flow and pollination success. As a result, the network structure could be altered as interactions are gained and lost or may not be effective in a beneficial way. Litchi and pollinator physiology are affected by warming and also affect complete pollination, leading to poor fruit set. The other critical part of the crop cycle is fruit set that is reduced when temperatures fall below 20 °C for extended periods during flowering.

Menzel and Simpson (1995) reported that flowering only occurs below 20 °C and this is easily or naturally achieved in subtropical conditions (Batten and Lahav 1994). The key factors to consider when assessing the potential of different areas for litchi are temperatures in winter which affect flower initiation, temperatures and light levels in spring which affect fruit set, and reliability of temperatures with rainfall which affect fruit development. The other critical part of the crop cycle is fruit set that is reduced when temperatures fall below 20 °C for extended periods during flowering.

14.5.3 Photosynthesis

Generally, in higher plants, photosynthetic rates are determined mainly by light intensity, CO_2 levels, and temperature. Temperature has little effect on the photosynthetic rate from 15 to 30 °C until light and CO_2 become saturated for the photosynthetic process. On warm days high mid-day leaf temperatures may inhibit photosynthetic activity. At high leaf temperatures, metabolic processes are significantly hampered. High temperature has been found influencing photosynthetic capacity and stomatal conductance by decreasing the activation state of rubisco. Furthermore, heat stress diminishes the amount of photosynthetic pigment (Hare Krishna 2012). The optimum temperature for efficient photosynthesis for litchi has been found to be 20 °C and in a study in Australia, whole tree gas exchange (gs) and photosynthesis (A) were greatest with photosynthetic photon flux density (PPFD) of 500 μ mol/m^2/s under a clear sky, but were still increasing with PPFD of 1000 μ mol/m^2/s under cloudy weather, possibly due to lower leaf temperature and vapour pressures (Lloyd et al. 1995).

14.5.4 Metabolites

High temperatures have been found to modulate the level of primary and secondary metabolites, in addition to the expression of a variety of heat shock proteins, other stress-related proteins, and production of reactive oxygen species which constitute major plant responses to heat stress (Hare Krishna 2012). Xu et al. (2006) reported that plants eliminated the ROS in higher plants resulting from heat stress by the antioxidant system of enzyme and nonenzyme. Ascorbate peroxidises (APX) is now one of the important antioxidant enzymes participating in the removal metabolism of active oxygen in higher plants. Moreover, the APX encoding apx genes in the family might be the functional genes affecting stress resistance in litchi (Shao et al. 2010). However, there are few reports on the relationships between APX and stress resistance in litchi at present. In order to have a basis for the heat-resistance molecular mechanism in litchi as well as the cultivation of a litchi variety with stronger heat resistance, the study on the embryogenic callus (EC) of litchi, the APX participated in response to high temperature stress, and cloning of the APX gene induced by high temperature stress, showed that the expression of apx genes was differently induced by high temperature stress. The level of APX activity was higher at 30 and 35 °C than at the control temperature, whereas the activity was lower at 40 °C than the control. The strength of APX isozyme bands was also elevated at 30–35 °C. A new band appeared at 35 °C. Both the strength and number of the bands decreased at 40 °C.

14.6 Nutritional Stresses

In nutritional jargon, nutrients are of five types, namely carbohydrates, proteins, fats, vitamins, and minerals as they are needed for providing energy and life processes through tissues in the plant body. Mineral nutrition affects metabolic activity, as mineral nutrients have essential and specific functions in cell metabolism (Singh

Fig. 14.6 Leaves showing nitrogen deficiency

2003). Nutritional stress due to problematic soils, that is, edaphic factors, as well as the mineral deficiency in the growth medium, are found to be stressful for the living system leading to various disorders and damage through reduction in growth, yield, and quality in fruit plants (Hare Krishna 2012). Likewise, the physiology of mineral nutrition in litchi is also vital to sustain productive life by completing the desired annual life cycle of the vegetative and reproductive phases. The nutrient disparity in the litchi plant body may result in manifestation of growth attributes and quality characteristics of fruits in the form of physiological disorders (Menzel and Simpson 1987). Many studies have established the fact that when litchi plants experience undernourishment, they develop symptoms of being unhealthy and affect their productive potential. Under field conditions, the levels of chemically available nutrients usually are much higher in the topsoil than in the subsoil (Singh 2003). The various physical and biological factors restricting the uptake of nutrients, both macro- and micronutrients, lead to a nutrient-deficient condition, causing deficiency symptoms on the plant body, forcing it to grow under a stressful environment.

14.6.1 Nutrient Deprivation/Deficiency

There is a need to study at length the aspect of essential versus beneficial mineral nutrition in the case of litchi. Deficiencies in nutrients can affect fruit development in litchi (Joubert 1970). Based on the references, it has been found under macronutrients, optimum nitrogen status in the cell is responsible for the biosynthesis and breakdown of proteins simultaneously in the plant body (cells/tissues). When the nitrogen supply is suboptimal, growth is retarded and in general all the metabolic processes slow down and a severe deficient condition causes necrosis and bleaching of pigments (Fig. 14.6) (Menzel et al. 1994). More or less similar is the case with sulphur deficiency, as it also inhibits protein synthesis and causes chlorosis. The phosphate is not reduced by the plants but remains in highly oxidised form. The phosphorus-deficient litchi plants show dark green colouration of leaves as compared to normal along with a wavy margin, twisting leaf lamina, and irregular shape (Fig. 14.7; Babita and Kumar 2008). The metabolic processes such as respiration, cell division, photosynthesis, and energy generation are severely affected due to phosphorus deficiency.

Fig. 14.7 Leaf (wavy margin and smaller size) due to phosphorus deficiency

The regulatory function of phosphorus in photosynthesis is largely responsible for the decrease in the rate of carbohydrate synthesis and reduced cell size. In contrast to other macronutrients, the Ca^{2+} deficiency causes in general cell wall disintegration apart from a reduction in cell division and cell elongation, hence affecting growth and vigour of the plant body and reduced fruit production, as its deficiency also leads to auxin deficiency (Brett and Waldron 1990). An excess (higher concentration) of calcium as a mineral nutrient is mostly found to be nontoxic in the case of litchi too. As magnesium deficiency impairs the degradation of starch it leads to an abnormal increase in the carbohydrate content of the leaves, also causing a drastic reduction in the rate of photosynthesis. Potassium-deficient litchi plants exhibit retarded growth, impaired lignification of vascular bundles, and chlorotic or necrotic leaf margins or even new twigs (Babita and Kumar 2008). The plants receiving an inadequate supply of K^+ are often found to be more susceptible to frost damage, which in turn is related to K^+ deficiency-induced less water intake at the cellular level.

In the case of litchi, it is likely that accumulation of phenolics in boron-deficient plants per se is responsible for cell damage. The boron-deficient plants have an overall reduction in growth and development of fruits, leading to poor/inferior quality of fruit production. Copper deficiency in litchi has also been noticed as affecting leaf growth and development, causing a drastic reduction in photosynthetic area.

14.6.2 Nutrient Dynamics Under Problematic Soils

Salinity impairs mineral nutrient utilisation in higher plants through its effects on nutrient availability, transport, and partitioning in the plant. The stress caused due to salinity induces ion deficiency and imbalance due to competition of nutrients such as K^+, Ca^{++}, and NO^{-3} with the toxic ions Na^+ and Cl^-. Litchi appears to be less sensitive to salt but is still a low-tolerance class of plants (Whiley and Sarnah 1980). In general, reduced growth, leaf scorching, and defoliation result in reduced fruit yield and quality. In the case of alkali soils, being poor in organic matter they contain a very low amount of nitrogen. The symptoms of sodium injury include appearance of mottle or chlorotic areas in leaves which usually end up in necrotic areas on tips, margins, and even in the internal areas (leaf burns and necrotic symptoms) and behave as iron-inefficient plants. Little critical information exists regarding the low pH and availability of major and micronutrients, which is restricted (Menzel 1987; Menzel et al. 2002). In the case of acid soils the low status of bases especially of Ca, Mg, Mo, and K along with toxicity of Al and Mn make the soil less fertile and present poor physicochemical conditions. Litchi is found to be surviving and producing under slightly acidic soils due to the presence of B, Zn, Cu, and also S. In the case of drought, nutrient uptake efficiency is totally lost and nutrients become unavailable to the litchi plants which affects normal growth and development. Root activity also ceased in the case of water-logged conditions created due to flood, and restricted uptake of nutrients caused a drastic reduction in plant growth; decline may start if the condition prevails for a longer period.

14.7 Irradiance

The light of the radiant energy functions on the fruit plants in various ways depending on the duration, quality, and intensity of the light. Duration refers to hours of illumination, the quality refers to the nature of the visible wavelength, and the intensity refers to the dose of illumination (Chattopadhyay 1994). Light influences the growth, development, and production of quality fruits.

14.7.1 Photoperiods

Although litchi is a photoinsensitive evergreen plant, many of its biological systems are sensitive to incident light because of the presence of chromophores that absorb radiation. The sensitivity/injury leading from absorption of light by a biological system for a detectable response may take from a few seconds to several days. There is a need for exhaustive studies to know the precise physiological significance of radiation effects in litchi, which is supposed to be a climacteric fastidious fruit crop. Low light intensity and irradiance, having insufficient light penetration into the canopy influences the canopy energy requirement directly, by reducing photon energy utilisation, thus decreasing productivity (Singh et al. 2012). UV radiation acts as an abiotic physical elicitor of the mechanism causing resistance by a rapid increase of stress response compounds. The high influence of UV radiation causes cellular damage by generating photoproducts in the DNA. The most effective protection mechanism stimulated under such a light regime is the biosynthesis of flavonoids and other UV-absorbing phenolic components. This in turn results in production of more anthocynins and hence, improved colour in exposed fruits. More numbers of transcriptomic and genomic related studies are required for the molecular mechanisms underlying fruit set and fruit development in litchi. Shading during early fruit development decreases fruit growth and induces fruit abscission. One study, in which high-throughput RNA sequencing (RNA-Seq) was employed for the de novo assembly and characterisation of the fruit transcriptome in litchi, and differentially regulated genes, which are responsive to shading, has been investigated using digital transcript abundance (DTA) profiling (Caiqin et al. 2013).

The avoidance mechanism of radiation injury in litchi can be correlated with the other studies which are based on the lesser capture of irradiation by water and presence of a higher amount of radioprotectant. Because much of the water absorption ability of the plants is based on the cell wall composition and presence of osmoticum in the cell cytoplasm, those plants having some mechanism of less water absorption can protect themselves against the radiation effect, as it is believed that the majority of perennial plants in the course of hardening or adaptation acquire the specific ability to repair the radiation-induced damage. The growth cycle of perennial fruit plants under a marked diurnal or rhythmic variation with distinct season change (temperature) is due to the inherent radiation tolerance property, and attributable to the metabolic state of the plant body. The age and the stage of development can be important factors in affecting radiation tolerance and this property also varies from organ to organ in the plant body. Aspects of other studies indicate the results that there is some role for intercellular salts and other compounds, which are stored in large quantities to neutralise the radiation-induced injury to cell and tissues, and may be exploited in the case of litchi for radiation stress injury avoidance and/or tolerance properties. Among the morphological characteristics, it has been observed that fruits exposed to sunlight develop better colour than those developing with canopy shade or slightly inside the canopy. The bright crimson red (mostly) of litchi fruits is chiefly due to the presence of flavonoids (mainly anthocynins).

14.8 Harvesting Problems

Litchi is categorised under the nonclimacteric group of fruits and does not ripen off the tree. Akamine and Goo (1973) monitored litchi during development and found that respiration and ethylene production declined continuously, typical of nonclimacteric fruit. Visual/commercial maturity has been found to differ from physiological maturity. Selective harvesting is frequently carried out, inasmuch as all fruits even on a tree do not ripen at one time. Harvesting at the optimum stage of maturity reduces different kinds of pre- and postharvest losses. Litchi fruit deteriorate rapidly after harvest because of water loss, pericarp browning, and rot development (Jiang et al. 2006). One of the challenges facing litchi fruit production globally is that regional climate regimes are becoming more unpredictable from year to year. Hence it is important to understand the effects of field level pre- and postharvest stress susceptibility for better storage and shelf-life potential of fruits (Toivonen 2005). The nature of pre- and postharvest abiotic stress events, is important for product quality and marketing, hence the technologies available to manage the sensitivity of fruits to stresses during handling and the distribution chain. However, at the preharvest level in litchi fruit production, the low relative humidity of the atmosphere for a longer period, alkalinity of the soil, salinity of the water used for irrigation, lack of organic matter in the soil, lack of mycorrhizal fungi in the soil, low soil temperatures, poor soil aeration, lack of mulching at the proper time, and/or a deficiency in one or more minor elements are responsible for poor quality of fruit and insignificant harvest (Storey 1973).

Pericarp/peel browning is one of the most important problems associated with this crop (Jiang et al. 2006). The structure and colour of anthocynins, the cause of redness of the litchi fruit pericarp, is dependent on the environment. At the molecular level, it has been found that an exogenous ATP supply enhanced antioxidant systems and maintained membrane integrity to delay browning and senescence of litchi fruit (Yi et al. 2008, 2009, 2010). Fruit ripening and senescence are active processes initiated by internal and environmental factors. The studies suggest that a cellular energy supply is a key factor controlling ripening and senescence events, and that ageing and browning of postharvest litchi fruits may be related to inadequate supplies and reduced efficiency of cellular energy generation (Azad et al. 2008; Saquet et al. 2000; Trippi et al. 1998) due to various stress problems. The ageing process in developing fruits is a decline in the adenosine triphosphate (ATP) level, as increased membrane permeability and enhanced reactive oxygen species production by harvested fruit are related to low ATP and energy charge (EC) levels (Duan et al. 2004; Saquet et al. 2003). Preharvest application of boron and calcium (Xuan et al. 2005), and postharvest treatments including ethylene inhibitors (Qu et al. 2006), high oxygen (Duan et al. 2004), providing anaerobic conditions (Jiang et al. 2007), controlled atmosphere (CA) storage (Saquet et al. 2000; Trippi et al. 1998; Saquet et al. 2003; Veltman et al. 2003), and application of an exogenous carbon source (Azad et al. 2008) and ATP (Song et al. 2006; Yi et al. 2010), can maintain tissue levels of ATP and EC, thereby delaying pericarp browning of litchi. In addition, exogenous ATP elevates ATP levels, inhibits ROS accumulation, and maintains unsaturated fatty acid levels and membrane integrity, thus delaying senescence and deterioration of harvested fruits (Jiang et al. 2003; Saquet et al. 2003; Qu et al. 2006; Jiang et al. 2007; Veltman et al. 2003; Yi et al. 2008, 2009, 2010). Therefore, browning and senescence are closely related to cellular energy status, and measures to maintain energy levels can delay ageing and inhibit occurrence of browning and deterioration to a certain extent. At the production level, abiotic stresses occurring can either be the primary cause (direct) for disorders that exhibit themselves during postharvest handling and storage practices or they can influence the susceptibility of a fruit to postharvest conditions that cause abiotic stresses resulting in disorders (indirect; Ferguson et al. 1999).

It is important to characterise the relationship between preharvest abiotic stresses occurring during production and postharvest abiotic stresses to which the fruit is exposed after harvest and during storage and distribution, because the solution to these different problems will be best resolved by focusing on preharvest or postharvest abiotic stress amelioration, respectively. Harvested fruits can be potentially exposed to numerous abiotic stresses during production, handling, storage, and distribution (Hodges 2003). Some of these stresses can be minor in nature, resulting in no quality loss or, in some cases, in quality improvement (Hodges et al. 2005) during distribution. However, when the abiotic stress is moderate or severe, quality losses almost always are incurred at market (Toivonen 2003a, b). As a consequence it is important to understand the nature and sources of abiotic stresses that affect fruits. In addition, with improved understanding, options for better management or resistance become available (Toivonen 2003b; Toivonen 2005). The occurrence of drought conditions during production of fruit crops is becoming more frequent with climate change patterns (Whitmore 2000). Although much work has been devoted to understanding drought effects on production and productivity of these crops (Whitmore 2000), there is limited published literature on the effects of preharvest water stress on responses to postharvest stresses and hence on subsequent quality and shelf life.

Fruits grown under high salinity will produce smaller, poor quality fruits with only higher total soluble solids. At excessive Na^+ and Cl^- rhizosphere concentration (activity), there are competitive interactions with other nutrient ions (e.g. K^+, NO_3^-, $H_2PO_4^-$) for binding sites and transport proteins in root cells, and thereafter for (re)translocation, deposition, and partitioning within the plant (White and Broadley 2001). Direct information is required to confirm that production of smaller litchi fruits and later postharvest desiccation and firmness decline problems, deteriorating the quality of the harvested fruits, are due to saline-stressed conditions.

It would be considered logical to assume that effects of exposure to high light are difficult to dissociate from effects exposure to high temperatures. However, research has shown that low light in the preharvest interval reduced colour development and other qualities of litchi fruits, whereas intense light increased the susceptibility of litchi fruits to developing skin cracking and sunburn patches. When litchi fruit is developed in ambient low light, which is suboptimal for photosynthetic activity, colour, taste, and shelf life of fresh fruit are much inferior. Very little efforts are being made to deal with postharvest stresses inasmuch as metabolite contents are general and directly proportional to relative levels or stress tolerance.

The role of temperature extremes causing quick deterioration in harvested fruit during postharvest handling and storage is important. Harvesting of litchi during mid-day and scorching sunlight exposure lead to quick loss of quality and marketing value. Postharvest temperature stress during distribution, transportation, and storage is an ongoing challenge for litchi fruit, particularly distant markets necessitating shipping by air or ocean container (East et al. 2008). The frequent jerk and breaks in cool chain temperatures can result in desiccation, browning, and softening (i.e. reduction in shelf potential) in litchi fruits. Chilling injury due to extreme low temperature effect also causes browning and yield loss in the litchi crop.

14.9 Adaptation and Mitigation Strategies

It is now established that development and productivity of perennial fruit plants are significantly and profoundly (mostly negatively) affected by environmental stresses, although they possess efficient defence mechanisms to cope with the adversities (Fujita et al. 2006). Abiotic stresses are now common threats: they affect the optimal function of the metabolism, growth, and production potential of fruit plants (including litchi). With increased understanding of the mechanisms of the physiological aspects of plants, advances have been made in increasing or developing the tolerance/resistance mechanisms to avoid/escape the different types of abiotic stresses (through biotechnological approaches at the molecular level) (Jiang et al. 2003). The traditional approach of enhancing the accumulation of specific compounds in response to a determined stimulus, as done with compatible solutes, has resulted in some degree of tolerance in litchi plants, and also demonstrates that the ability to redirect nutrients to imperative processes and the induction of adequate metabolic adjustments are crucial for plant survival during conditions of stress. Hence, agronomical management and good agricultural practices are also important adaptation strategies for developing abiotic stress tolerance mechanisms.

14.9.1 Biotechnological Approaches

The biotechnological approaches for mitigating the ill effects of abiotic stress conditions require clear understanding about the physiology involved as to how plants recognise external conditions and initiate protective reactions such as mechanisms through which the protein function is protected and maintained (Wang et al. 2004). Under abiotic stress adaptation strategies, protein folding stability and dynamics (PFSD) directly interact with the protein metabolic, as the altered protein surface properties of changing the water structure and hydrophobic interactions, thereby stabilising the folded states relative to unfolded states; these characteristics show the ability to maintain protein performance under abiotic stress, which mainly depends on intrinsic stability, chaperone activity, protein turnover, and extrinsic stabilisation through cosolvents (compatible solutes; Naganathan et al. 2006). Protein conformation and stability are dramatically affected by sudden changes in the environment, giving rise to protein unfolding and misfolding, and aggregation under stress conditions has to take protein dynamics, meaning conformational changes of proteins with time, into consideration (Ortbuauer 2013). Molecular motion as well as protein flexibility and dynamics are highly linked to enzymatic activity, which is clearly dependent on the particular environment of a protein (Daniel et al. 2003).

The beneficial effects of compatible solutes on plant tolerance to environmental stress through the conformational exchange in proteins and their folding stability under varying experimental conditions can be planned and studied accordingly for litchi. High osmolyte accumulation in plants seems to be a very effective strategy to adapt plants quickly and with a remarkable plasticity to various changes in their environment, as it (high osmolyte accumulation) serves to suppress protein unfolding and misfolding, enhances protein folding stability, and facilitates the protein refolding process after complete denaturation. The leaf cell structure and several physiological characteristics of 14 litchi (*Litchi chinensis* Sonn.) cultivars showed that the cultivar with large leaf cell tissue ratio had strong cold resistance; conversely the cold resistance of the cultivar with a low leaf cell tissue ratio was weak, therefore the leaf cell tissue ratio may be taken as an index to detect cold tolerance of litchi, where the contents of free proline, soluble sugar, and soluble protein, and the activity of superoxide dismutase (SOD) that were positively correlated with the leaf cell tissue ratio, can reflect differences in cold tolerance among various litchi cultivars, and, therefore, they could effectively be used as methods to determine the cold resistance of cultivars (She and Li 2010). A plant responds to a lack of water by halting growth and reducing photosynthesis, leaves may appear to change colour – usually to blue-green – foliage begins to wilt, and, if the plant is not irrigated, leaves will fall off and the plant will eventually die. The leaf surface having a thicker cuticle layer will have a better drought tolerance capacity compared to curving and folding (Blum 2011).

14.9.2 Plant Metabolomics and Applications

The pericarp of litchi has been the focus of studies associated with fruit size, coloration, cracking, and shelf life. Numerous postharvest treatments have been evaluated for enhancing abiotic stress resistance of litchi fruits, generally, temperature modulation, modified atmospheres (high O_2, CO_2, and low O_2), growth regulators, antitranspirants, antioxidant dips, growth regulators, nitric oxide, and ethanol have been tested (Menzel and Waite 2005). Application of treatments in combination can often improve the stress resistance level to the fruits by enhancing the shelf life. In an experimental attempt, application of tea polyphenols has been found to delay pericarp browning markedly, and alleviate the decreases in contents of total soluble solids (TSS) and ascorbic acid, as well as maintain relatively high levels of total phenolics and anthocyanins of litchi fruit after 30 days of cold storage. Meanwhile, it also reduced the increases in relative leakage rate and lipid peroxidation content, delayed the increases in both $O_2^{\cdot -}$ production rate and H_2O_2 contents, and increased SOD activity but reduced POD activity throughout this storage period. It also enhanced antioxidant capability, reduced accumulations of reactive oxygen species and lipid peroxidation, and improved membrane integrity. An analysis of the transcriptome of litchi pericarp showed that a large collection of transcripts and expression profiles associated with litchi fruit maturation processes, including coloration as most of the unigenes were annotated, provided a platform for litchi functional genomic research within this species (Lai et al. 2015).

Atmospheric treatments are quite often beneficial to controlling response to postharvest stresses. Modified or controlled atmospheres have been shown to help minimise chilling injury in a number of fruits. There have been some reports on the selection of germplasm and cultivars from breeding programs that will have greater (postharvest) stress resistance and hence better storage capability. Rai et al. (2001) were able to evaluate and select a cracking-resistant cultivar. Litchi cultivars which have various problems of physiological disorders should have measures of levels of apoplastic antioxidant enzymes and ROS in the apoplast, which can be instructive to understand the basis for difference in stress tolerance in selections of a breeding population; they are too labour intensive to consider for incorporation as selection tools in a breeding program. Microorganisms also change the pH conditions of their surroundings and act as effective buffering agents, developing the tolerance mechanism affecting the biological systems. Proteins, amino acids, fatty acids, and various organic acids also serve as cellular buffers and play a significant role in adaptation mechanisms (Liu et al. 2000; Liu et al. 2007).

14.9.3 Breeding Approaches

Climate change and weather extremes of temperature, and water status as well as deteriorating soil conditions, pose continuous problems in litchi production systems (Kumar 2015). Attempts at breeding abiotic stress resistance/tolerance in litchi are based on the plant's response to types of stresses with the aim to manipulate characteristic plant performance that will be better suited to withstand these stresses. The identification and characterisation of the mechanisms in plants and the physiological traits considered important for evaluating abiotic stress tolerance are (1) osmotic adjustment, protoplast tolerance, memebrane integrity; (2) hormonal regulation, ABA, GA, and ethylene; (3) leaf characteristics of cuticular thickness, reflectance, and movement; (4) gas exchange characteristics; (5) development of nutrient uptake efficiency and water use efficiency (WUE), component root traits including root length density, deep root to shallow root ratio, root to shoot ratio; (6) canopy temperature; (7) leaf initiation, expansion, survival rate, and water tolerance; (8) stay green and senescence; (9) fruit number, weight, pericarp colour, maturity time; and (10) harvest index, yield, and its components (Srinivasa Rao and Laxman 2013). The effective management, enhancement, evaluation, and valuation of genetic resources and development of improved cultivars with high-quality characteristics, productivity, tolerant to abiotic stresses, and resistant to pests and disease are in progress at ICAR-NRC on Litchi, Bihar (ICAR-NRCL Ann. Rep. 2013, 2014, 2015).

Early bearing is an important and desirable tree character for litchi. The commercial varieties should give abundant and regular harvests. It is very important to obtain cultivars which flower and bear abundantly without the presence of special temperature and humid conditions. This characteristic will make the variety more widely adaptable and extend cultivation to other areas. The selected variety must have branches with a wide crotch angle preferably 120–130° that can bear the heavy bearing load of fruits. The plant type or variety should be resistant to extremes of temperature which will enhance production in relatively disadvantaged areas. Although the species is relatively resistant, still, it is better to find a more resistant plant type. Saline soils and irrigation water with a high salt concentration are common in many parts of the litchi growing belts. Therefore, salt resistance or tolerance is a desirable characteristic. The litchi plant is moisture stress tolerant, however, the fruits suffer a lot due to moisture stress. Some varieties are relatively more tolerant to this and therefore these should be selected for cultivation. Cultivars which present a wide range of adaptability are to be preferred over those requiring light soils for commercial cultivation. Multilocation trials are a safe and efficient way to select a variety which can withstand adverse soil conditions. It has been noted that moderately to slightly vigourous varieties contribute to some extent to early age flowering and should generally be preferred. The spreading growth habit makes harvesting easier and increases wind resistance, hence, should be preferred.

14.9.4 Agrotechnological Mitigation and Adaptation Measures

The adaptation strategies to address abiotic stresses in litchi are as follows.

14.9.4.1 Orchard Establishment

New plantations with a highly specialised and recommended way will lead to a healthy orchard having proper growth and development and enough energy to endure abiotic stress conditions. The considerations and activities required for new plantations are proper planning, selection of site, land preparation, layout, planting of saplings, as well as orchard protection and management, incorporation of (about two baskets) soil from the root zone of old litchi trees encourages the mycorrhiza growth.

14.9.4.2 Balanced Nutrition

Balanced nutrition is a must for growing and fruit-bearing plants/trees which ensures better productivity and higher quality even withstanding adverse climatic conditions, inasmuch as litchi responds significantly to exogenously applied manures and fertilisers depending upon age and cultivar (Kotur and Singh 1992). The effects of a graded dose of NPK on growth, yield, and quantity of fruits have been found encouraging (Rai et al. 2002) for sustainable production. The litchi orchards regularly fed with higher doses of organic manure have better yield and quality as compared to orchards maintained with chemical fertilisers. Additional applications of micronutrients (Ca, Zn, B, Cu, and Mn) are always recommended on a per soil and leaf nutrient analysis basis. Zinc is applied in the form of 0.5 % zinc sulphate hydrated with lime, which helps in reducing fruit drop and enhancing fruit yield and quality. In acidic soil application of 10–15 kg lime/tree once in 3 years has been found to increase the fruit yield and quality in perennial fruit crops. Foliar application of zinc sulphate (0.1 %) is done twice, 10–15 days before flowering for improving the sex ratio and to reduce fruit drop. If a deficiency of zinc and magnesium is observed, application of 150–200 g $ZnSO_4$ and 150–200 g $MgSO_4$, respectively, per plant during September has been found to be beneficial. Foliar application of 0.1 % borax, 2–3 times during the period of fruit growth and development of the trees enhances fruit retention, minimises cracking, improves fruit colour and sweetness, and advances maturity. Other micronutrients such as Fe (ferrous sulphate), Cu (copper sulphate), and Mg (magnesium sulphate) are applied if deficiency symptoms are observed. Two to three sprays in a year are sufficient to maintain the trees in good health. Litchi exhibit a slow to moderate growth rate. This can be tremendously accelerated through proper and effective fertilisation. Litchi trees are more susceptible to pathogens and fungus when they are stressed.

14.9.4.3 Use of Organic Sources and Biofertilisers

The rhizosphere of litchi requires proper nourishment to benefit from nutrient–microbe synergy because of their perennial framework and root configuration. To improve the fertility of the well-balanced soil organic matter should be added regularly. A green manure crop is grown and incorporated into the soil, which improves its fertility, moisture-holding capacity, and physical condition at the initial level of developing orchards. In litchi, a conjoint use of half of the recommended dose of chemical fertilisers, organic manure, and biofertilisers has improved the soil health and enhanced the quantum of quality fruit production under Extra class category, that is, export quality (Kumar 2010). Microorganisms also change the pH conditions of their surroundings and act as effective buffering agents, developing the tolerance mechanism affecting biological systems (Rao and Pathak 1996). Proteins, amino acids, fatty acids, and various organic acids also serve as cellular buffers and play a significant role in adaptation mechanisms.

14.9.4.4 Use of Plant Growth Regulators (PGRs)

The exogenous application of plant growth regulators/hormones enabling plants to adapt to changing environments by mediating growth, development, nutrient allocation, and source/sink transitions, have been successfully employed in litchi too. Foliar application of the plant bioregulator, NAA (20 ppm) at an interval of 10–15 days during the period of fruit growth and development, twice or thrice, has been found to be useful for minimising fruit drop. IAA may be substituted for NAA if the latter is not available. Spraying plain water four to six times in the early morning hours of the day during the advanced stage of fruit growth and development has been found to be highly effective for achieving better growth with minimised fruit cracking, attributed to internal phytohormone regulation. Actually, plant growth regulators/hormones rapidly alter gene expression by inducing or preventing the degradation of transcriptional regulators via the ubiquitin–proteasome system (Santner and Estelle 2010). The response of plants to abiotic stress, especially water stress, is ABA signalling and ABA-responsive genes. ABA synthesis is one of the fastest responses of plants to abiotic stress, triggering ABA-inducible gene expression (Yamaguchi-Shinozaki and Shinozaki 2006) and causing stomatal closure, thereby reducing water loss via transpiration (Wilkinson and Davies 2010) and eventually restricting cellular growth.

14.9.4.5 Pruning and Canopy Management

Pruning and training to maintain the tree canopy of semicircular or open umbrella shape is a routine annual activity in litchi, aimed at increasing light interception, surface bearing area, productivity, stabilising yield, and improving fruit quality (Gardener et al. 1952). Litchi needs to have adaptive features to its light environment. More severe or corrective pruning to reduce effective height and spread may be effective in reducing water use because transpiration is related to canopy size and leaf area density. However, significant pruning may be needed to change transpiration by manipulating the canopy surface area. Summer pruning improves light penetration within the tree canopies and it should be followed just after the harvesting of fruits is over. It is beneficial in the restart of current healthy shoots and to some extent in discouraging the development of a microclimate for pest and disease growth in litchi. When the reduction in water supply is going to be drastic, a mature plantation can be saved for later years by heavy pruning (sometimes called 'dehorning' or 'stumping'). It also increases the input use efficiency and better regrowth. Proper pruning and training of the plant in the initial stage is essential to provide the required framework. Unwanted branches should be pruned skilfully under corrective pruning to provide definite shape and to promote growth of the trunk and crown of the tree. The branches not contributing for canopy development or thin nonfruiting unproductive branches inside the canopy in growing and mature trees should also be pruned. Dried, diseased, and scissors-shaped branches should also be periodically removed to maintain the desired canopy and healthy productive growth. Root pruning is also practiced in heavy soils to reduce the vegetative growth of the canopy surface area. The objective of pruning is to provide a good canopy for more input use efficiency, leading to enhanced flowering and quality fruit yield (Goren and Gazit 1992; Menzel et al. 2000). High-density planting adopting a double row system has also been found to be superior in terms of yield and quantity of fruits, but only feasible by adopting proper pruning and through appropriate canopy management from the initial stage, to develop a strong framework with enhanced surface bearing area and plant height at manageable (lower) height to make it responsive to intercultural operations and more responsive to chemical sprays (Kumar and Nath 2013).

14.9.4.5.1 Rejuvenation

Reiterative pruning and training for canopy rebuilding of old, senile, unproductive litchi trees for rejuvenation and transforming them into a young bearing orchard by restoration of youthful vigour is a medium-term strategy for enhanced quantum of quality fruit production in a sustainable manner and better withstand even adverse climatic conditions. Hard-core pruning or skeletonising the tree through timely careful and skilful operations, and proper scientific management to encourage replacement with new young branches can maintain tree health, quality fruit production, and longevity (Kumar and Kumar 2007; Kumar 2008; Kumar and Nath 2013).

14.9.4.5.2 Deficit Irrigation

A deficit irrigation (DI) situation is always correlated with the level of water supply in relation to maximum crop evapotranspiration (ET). Quantifying the ET reduction in the case of perennial fruit crops affected by DI (net water savings) as direct measurements with the models used to estimate the actual ET of stressed canopies is still quite empirical. But when the principle is applied as if the soil water supply were insufficient to meet the crop demand, and growth and transpiration are reduced, then DI induces an ET reduction, although below its maximum potential (Fereres et al. 2003). Deficit irrigation has shown great potential to increase irrigation water use efficiency of litchi production with a slight deviation in potential yield as reported by Mali et al. (2015). The water deficit was manifested in shrinkage of trunk diameter, depression of fruit growth, and overheating of the leaves, inasmuch as litchi is one of the most environmentally sensitive tropical tree fruit crops (Ghosh 2001). Plant growth and development are also sensitive to internal water deficit caused by lack of soil water or dry air conditions (Menzel et al. 1986). The use of a phytomonitoring technique to litchi growing in China has been found seriously profitable, because China is the largest litchi producer (Chen and Huang 2001). Because it is ranked from moderately sensitive to somewhat tolerant depending upon the specific location, this observation highlights a main concern about using deficit irrigation, which results in the reduced size of fruit from such treatments (Lopez et al. 2011). As among fruit quality characteristics, size of fruit is important, because larger fruit have lower surface area to volume ratios, which confers lower relative water loss (Shibairo et al. 1997). Presumably, deficit irrigation practices have been found to enhance adaption by developing a less water-permeable cuticle in fruit crops (Lopez et al. 2011).

14.9.4.5.3 Fruit Thinning

In cross-pollinated crop litchi, increased fruit set through efficient and effective pollination is always encouraged but for quality fruit production, the practice of fruit thinning is now being recommended to the growers to produce fruit of marketable sizes thus increasing crop value (Kumar and Kumar 2007). This practice is common and achieves higher grower revenue, even though yields are lower. Also, because the presence of a greater number of fruits per bunch/panicle requires more water for aril development, it enhances transpiration. Proper fruit thinning can reduce transpiration rates somewhat in many tree species (in litchi) and decrease the level of water stress in the trees, leading to an acceptable commercial size for the remaining fruit. Although not much experimental evidence is available, these measures appear to be very effective in litchi when compared to other commercial fruit crops of bunch-bearing behaviour. Furthermore, because the reduction in transpiration associated with fruit thinning is mediated by reduced stomatal conductance, it may also reduce heat damage where water deficit and high temperatures occur at the same time (Kumar and Kumar 2007).

14.10 Conclusion

Drought, heat, cold, salinity, and irradiance are among the major abiotic stresses that adversely affect plant growth, quality production, and productivity of litchi fruits causing an unfavourable series of morphological, physiological, biochemical, and molecular changes. Crop management can have a significant influence on susceptibility to stress. The physiology of plant architecture above the ground and the root system below can be adjusted with various biotechnological, technological, and other adaptive measures to cope with the adverse effect of the abiotic stresses to achieve the production potential in an economically viable manner. The traditional approach of enhancing the accumulation of specific compounds (i.e. metabolites) in response to a determined stimulus, as done with compatible solutes, has resulted in some degree of tolerance in litchi plants. The ability to redirect nutrients to imperative processes and the induction of adequate metabolic adjustments are crucial for plant survival during conditions of stress. Proper agronomical management and good agricultural practices with favourable rhizosphere and canopy manipulation have been found to be important adaptation and mitigation strategies for developing abiotic stress tolerance mechanisms in litchi, although efforts for breeding stress resistance in the field will also extend stress resistance characteristics and cultivars.

References

Acquaah G (2007) Principles of plant genetics and breeding. Blackwell, Oxford

Agnihotri PK, Lal RL, Mishra DS, Pavan S (2009) Root distribution pattern of litchi as influenced by seasonal variations. Indian J Hortic 66(2):257–260

Akamine EK, Goo T (1973) Respiration and ethylene production during ontogeny of fruit. J Am Soc Hortic Sci 93:381–383

Ashraf M, Hafeez M (2004) Thermotolerance of pearl millet and maize at early growth stages: growth and nutrient relations. Biol Plant 48:81–86

Atwell BJ, Kriedemann PE, Turnbull CA (1999) Plants in action. McMillan Education Australia, South Yarra, p 664

Azad AK, Ishikawa T, Sawa Y, Shibata H (2008) Intracellular energy depletion triggers programmed cell death during petal senescence in tulip. J Exp Bot 59 (8):2085–2095

Babita S, Kumar R (2008) Feed your litchi orchard properly. Indian Hortic 53(3):18–20

Banu MN, Hoque MA, Watanabe-Sugimoto M, Islam MM, Uraji M, Matsuoka K (2010) Proline and glycinebetaine ameliorated NaCl stress via scavenging of hydrogen peroxide and methylglyoxal but not superoxide or nitric oxide in tobacco cultured cells. Biosci Biotechnol Biochem (Research Support, Non-U.S. Gov't) 74(10):2043–2049

Batten DJ, Lahav E (1994) Base temperature for growth processes of lychee (*Litchi chinensis*), a recurrently

flushing tree, are similar but optima differ. Aust J Plant Physiol 21:589–602
Batten DJ, McConchie CA (1995) Floral induction in growing buds of lychee (*Litchi chinensis*) and mango (Mangifera indica). Aust J Plant Physiol 22:783–791
Batten DJ, Lloyd J, McConchie CA (1992) Seasonal variation in stomatal response of two cultivars of lychee (*Litchi chinensis* Sonn.). Aust J Plant Physiol 21:589–602
Batten DJ, McConchie CA, Lloyd J (1994) Effects of soil water deficit on gas exchange characteristics and water relations of orchard lychee (*Litchi chinensis* Sonn.) trees. Tree Physiol 14:1177–1189
Bhattacharya A, Vijayalaxmi (2010) Physiological responses of grain legumes to stress environments. In: Climate change and management of cool season grain legume crops. Springer, Dordrecht
Binzel ML, Hess FD, Bressan RA, Hasegawa PM (1988) Intracellular compartmentation of ions in salt adapted tobacco cells. Plant Physiol 86:607–614
Blum A (2011) Plant water relations, plant stress and plant production (Chap. 2). In: Plant breeding for water-limited environments. Springer, New York. doi:10.1007/978-1-4419-7491-4-2
Bogges SF, Aspinall D, Paleg LG (1976) Stress metabolism IX. The significance of end product inhibition of proline synthesis and of compartmentation in relation to stress induced proline accumulation. Aust J Plant Physiol 3:513–525
Boyer JS (1985) Water transport. Annu Rev Plant Physiol 36:473–516
Brett CT, Waldron K (1990) Physiology and biochemistry of plant cell walls. Vol. II. Plant physiology. Unwin Hyman Publisher, London, p 194
Caiqin L, Wang Y, Huang X, Li J, Wang H, Li J (2013) De novo assembly and characterization of fruit transcriptome in litchi chinensis Sonn., and analysis of differentially regulated genes in fruit in response to shading. BMC Genomics 14:552
Carr MKV, Menzel CM (2014) The water relations and irrigation requirements of lychee (*Litchi chinensis* Sonn.): a review. Exp Agric 50(04):481–497
Chadha KL, Rajpoot MS (1969) Studies on floral biology, fruit set and its retention and quality of some litchi varieties. Indian J Hortic 26:124–129
Chattopadhyay TK (1994) A textbook on pomology. Vol I, fundamentals of fruit growing. Kalyani publishers, New Delhi
Chaves MM, Flexas J, Pinheiro C (2009) Photosynthesis under drought and salt stress: regulation mechanisms from whole plant to cell. Ann Bot 103:551–560
Chen LS (1997) The physiological and biochemical responses and adaptability of litchi (*Litchi chinensis* Sonn.) to water stress. PhD dissertation, Fujian Agricultural University, Fuzhou, China
Chen HB, Huang HB (2001) China litchi industry: development, achievements and problems. Acta Horticult 558:31–39
Chen LS, Liu XH (1998) Effects of water stress on active oxygen metabolism in litchi leaves. Acta Hortic Sin 25:241–246
Chen LS, Liu XH (1999) Effects of water stress on nitrogen and nucleic acid metabolisms in litchi (*Litchi chinensis* Sonn.) leaves and their relation to drought resistance. Acta Physiol Sin 25:49–56
Chen LS, Liu XH (2001a) Physiology and biochemistry of litchi under drought. Acta Hortic 558:297–299
Chen LS, Liu XH (2001b) Effects of water stress on ultrastructure of litchi leaves. J Fujian Agric Univ 30:171–174
Creus CM, Sueldo RJ, Barssi CA (1997) Shoot growth water status in Azospirillum – inoculated wheat seedlings grown under osmotic and salt stresses. Plant Physiol Biochem 35:939–944
Damour G, Vandame M, Urban L (2008) Long-term drought modifies the fundamental relationships between light exposure, leaf nitrogen content and photosynthetic capacity in leaves of the lychee tree (*Litchi chinensis*). J Plant Physiol 165:1370–1378
Daniel RM, Dunn RV, Finney JL, Smith JC (2003) The role of dynamics in enzyme activity. Annu Rev Biophys Biomol Struct 32:69–92
Dash M, Panda SK (2002) Physiology of salt stress in plants-a review. In: Panda SK (ed) Advances in stress physiology of plants. Scientific Publishers (India), Jodhpur, pp 93–110
Duan XW, Jiang YM, Su XG, Liu H, Li YB, Zhang ZQ, Zheng YH, Jiang WB (2004) Role of pure oxygen treatment in browning of litchi fruit after harvest. Plant Sci 167:665–668
East AR, Tanner DJ, Maguire KM, Mawson AJ (2008) The influence of breaks in storage temperature on 'Cripps Pink' (pink lady) apple physiology and quality. HortSci 43:818–824
Egea G (2009) Usefulness of establishing trunk diameter based reference lines for irrigation scheduling in almond trees. Irrig Sci 27:431–441
Ehsanzadeh P, Nekoonam MS, Azhar JN, Pourhadian H, Shaydaee S (2009) Growth, chlorophyll and cation concentration of tetraploid wheat on a solution high in sodium chloride salt: hulled versus free – threshing genotypes. J Plant Nutr 32(1):57–80
El-Shabrawi H, Kumar B, Kaul T, Reddy MK, Singla-Pareek SL, Sopory SK (2010) Redox homeostasis, antioxidant defense, and methylglyoxal detoxification as markers for salt tolerance in Pokkali rice. Protoplasma (Research Support, Non-U.S. Gov't) 245(1–4):85–96
Fatemeh N, Yaghoobi MM, Kalantari KM (2011) Effect of exogenous arginine on alleviation of oxidative damage in tomato plant underwater stress. J Plant Interact 6(4):291–296
Fengxia T, Gong J, Zhang J, Feng Y, Wang G, Guo Q, Wang W (2014) Overexpression of monoubiquitin improves photosynthesis in transgenic tobacco plants following high temperature stress. Plant Sci 226:92
Fereres E, Goldhamer DA, Parsons LR (2003) Irrigation water management of horticultural crops. Historical review compiled for the American Society of Horticultural Science's 100th anniversary. HortSci 38:1036–1042

Ferguson IB, Volz R, Woolf A (1999) Preharvest factors affecting physiological disorders of fruit. Postharvest Biol Technol 15:255–262

Fernández-García N, Martínez V, Cerdá A, Carvajal M (2002) Water and nutrient uptake of grafted tomato plants grown under saline condtions. J Plant Physiol 159(8):899–905

Fujita M, Fujita Y, Noutoshi Y, Takahashi F, Narusaka Y, Yamaguchi-Shinozaki K, Shinozaki K (2006) Crosstalk between abiotic and biotic stress responses: a current view from the points of convergence in the stress signaling networks. Curr Opin Plant Biol 9:436–442

Gardener VR, Bradfor FC, Hooker HD (1952) Fundamental of fruit production. McGraw Hill Book Company Inc, New York

Ghosh SP (2001) World trade in litchi: past, present and future. Acta Horticult 558:23–30

Goren M, Gazit S (1992) Small stature litchi orchards: a new approach to growing of litchi. Acta Horticult 349:69–72

Hanson AD, Hitz WD (1982) Metabolic responses of mesophytes to plant water deficits. Ann Rev Plant Physiol 33:163–203

Heikal MM, Shaddad MA, Ahmed AM (1982) Effect of water stress and gibberellic acid on germination of flax, sesame and onion seeds. Biol Plant 24:124–129

Hodges DM (2003) Postharvest oxidative stress in horticultural crops. Food Products Press, New York. ISBN 1-56022-962-4

Hodges DM, Lester GE, Munro KD, Toivonen PMA (2005) Oxidative stress: importance for postharvest quality. HortSci 39(924):929

Husen A, Pal M (2006) Variation in shoot anatomy and rooting behaviour of stem cuttings in relation to age of donor plants in teak (*Tectona grandis* Linn.f.). New For 31(1):57–73

ICAR-NRCL Annual Report (2012–2013) Published from National Research Centre on Litchi, Muzaffarpur Bihar, India

ICAR-NRCL Annual Report (2013–2014) Published from National Research Centre on Litchi, Muzaffarpur Bihar, India

ICAR-NRCL Annual Report (2014–2015) Published from National Research Centre on Litchi, Muzaffarpur Bihar, India

Ito BA, Yoshiko H, Hayama H, Kashimura Y (2004) Reorientation of shoots to the horizontal position influences the sugar metabolism of laterals buds and shoot internodes in Japanese pear (*Pyrus pyriflora* (Burm.) Nak.). J Hortic Sci Biotech 79:416–422

Jiang RH, Mao YL, Chen TS, Yan L, Xing SH, Lin SQ (1999) Soil fertility of orchards and nutrient situation of fruit trees in Changtai town. J Agric 46:1541–1556

Jiang YM, Yan L, Lichter A, Li J (2003) Post harvest biology and technology of litchi fruit food. Agric Environ 2:76–81

Jiang YM, Wang Y, Song L, Liu H, Lichter A, Kerdchoechuen O, Joyce DC, Shi J (2006) Post harvest characteristics and handling of litchi fruit – an overview. Aust J Exp Agric 46:1541–1556

Jiang YM, Jiang YL, Qu HX, Duan XW, Luo YB, Jiang WB (2007) Energy aspects in ripening and senescence of harvested horticultural crops. Stewart Postharvest Rev 2:1–5

Joubert AJ (1970) The litchi. Bulletin of the Citrus and Subtropical Fruit Research Institute, Nelspruit, No. 389

Kaiser WM, Kaiser G, Prachuab PK, Wildman SG, Herber U (1981) Photosynthesisis under osmotic stress. Plants 153:416–422

Kotur SC, Singh HP (1992) Studies on leaf sampling of litchi (*Litchi chinensis*Sonn.). In: National seminar on recent developments in litchi cultivation, held in Rajendra Agricultural University, Pusa, Samastipur, Bihar, May 30–31, p 23

Krasensky J, Jonak C (2012) Drought, salt, and temperature stress-induced metabolic rearrangements and regulatory networks. J Exp Bot 63(4):1593–1608

Krishna H (2012) Physiology of fruit production. Stadium Press (India) Pvt. Ltd, New Delhi, p 101

Kumar R (2008) Rejuvenation of old senile litchi orchard. Technical bulletin 5. National Research Centre on Litchi (ICAR), Muzaffapur

Kumar R (2010) Evaluation of dynamic substrates under integrated plant nutrient management affecting growth, yield and quality of litchi (*Litchi chinensis* Sonn.). In: Proceedings third international symposium on Longan, Lychee and other fruit trees in Sapindaceae family. Acta Hortic 863:243–248

Kumar (2015) Climatic issues affecting sustainable Litchi (Litchi chinensis Sonn.). In: Choudhary ML, Patel VB, Siddiqui MW, Verma RB (eds) Climate dynamics in horticultural science-impact adaptation, and mitigation, vol 2. Apple Academic Press of CRC Press Group, pp 95–110

Kumar R, Kumar KK (2007) Managing physiological disorders in litchi. Indian Hortic 52(1):22–24

Kumar R, Nath V (2013) Climate resilient adaptation strategies for litchi production. In: Singh HP, Srinivasa Rao NK, Shivashankar KS (eds) Climate resilient horticulture: adaptation and mitigation strategies. Springer India, New Delhi, pp 81–88

Lai B, Hu B, Qin Y-H, Zhao J-T, Wang H-C, Hu G-B (2015) Transcriptomic analysis of *Litchi chinensis* pericarp during maturation with a focus on chlorophyll degradation and flavonoid biosynthesis. BMC Genomics 16:225

Laura Fiasconaro M, Sanchez-Diaz M, Carmen Antolin M (2013) Nitrogen metabolism is related to improved water use efficiency of nodu, ated alfaalfa grown with sewage sludge under drought. J Plant Nutr Soil Sci 176(1):110–117

Lawlor DW, Tezara W (2009) Causes of decreased photosynthetic rate and metabolic capacity in water-deficient leaf cells: a critical evaluation of mechanisms and integration of processes. Ann Bot 103:561–579

Li JG (2011) Litchi. In: Chen JZ (ed) Southern China fruits cultivation. Chinese Agriculture Press, Beijing

Li JG, Huang H, Gao F, Huang X, Wang H (2000) An overview of the research in litchi fruit cracking. In Abstract 1st International symposium on litchi and longan. Guangzhou, China, p 46

Liu K, Fu HH, Bei QX, Luan S (2000) Inward potassium channel in guard cells as a target for polyamine regulation of stomatal movements. Plant Physiol 124:1315–1325

Liu JH, Kitashiba H, Wang J, Ban Y, Moriguchi T (2007) Polyamines and their ability to provide environmental stress tolerance to plants. Plant Biotechnol 24:117–126

Lloyd J, Wong SC, Styles JM, Batten DJ, Priddle R, Turnbull C, McConchie CA (1995) Measuring and modelling whole – tree – exchange. Aust J Plant Physiol 22:987–1000

Lopez G, Larrigaudière C, Girona J, Behboudian MH, Marsal J (2011) Fruit thinning in 'Conference' pear grown under deficit irrigation: implications for fruit quality at harvest and after cold storage. Sci Hortic 129:64–70

Mahajan S, Tuteja N (2005) Cold, salinity and drought stresses: an overview. Arch Biochem Biophys 444:139–158

Mali SS, Das B, Bhatnagar PR (2015) Effect of water application method and deficit irrigation on yield, quality and irrigation water use efficiency of litchi (*Litchi Chinensis Sonn.*) cv Shahi. Int J Irrig Water Manage 2015:7 Pages. ISSN: 5423-5294© International Scholars Journals

Menzel CM (1987) Lychee nutrition: a review. Sci Hortic 31:195–224

Menzel CM (2002) The development of the lychee industry in Australia. Acta Hortic 575:391–396

Menzel CM (2005a) Plant water relations and irrigation. In: Menzel CM, Waite GK (eds) Litchi and Longan-botany production and uses. CABI Publishing, Wellingford, p 305

Menzel CM (2005b) Photosynthesis and productivity. In: Menzel CM, Waite CK (eds) Litchi and longan: botany, production and uses. CABI Publishing, CAB International, Wallingford, pp 153–182

Menzel CM (2005c) Plant water relations and irrigations. In: Menzel CM, Waite CK (eds) Litchi and longan: botany, production and uses. CABI Publishing, CAB International, Wallingford, pp 183–207

Menzel CM, Paxton BF (1985) The effect of temperature on growth and drymatter production of lychee seedling. Sci Hortic 26:17–23

Menzel CM, Simpson DR (1987) Lychee nutrition: a review. Sci Hortic 31:195–224

Menzel CM, Simpson DR (1988) Effect of temperature on growth and flowering of litchi (*Litchi chinensis* Sonn.) cultivars. J Hortic Sci 63:347–358

Menzel CM, Simpson DR (1992) Flowering and fruit set in lychee (*Litchi chinensis* Sonn.) in sub-tropical Queensland. Aust J Exp Agric 32:105–111

Menzel CM, Simpson DR (1994) Lychee. In: Schaffer B, Andersen PC (eds) The handbook of environmental physiology of fruit crops Vol. II. Sub-tropical and tropical. CRC Press, Boca Raton, pp 123–141

Menzel CM, Simpson DR (1995) Temperatures above 20°C reduce flowering in lychee (*Litchi chinensis* Sonn.). J Hortic Sci 70:981–987

Menzel CM, Waite GK (2005) Litchi and lonngan: botany, production and uses. CABI Publishing, Wellingford, Oxfordshire, UK

Menzel CM, Banks AG, Simpson DR (1986) Effect of aerial and soil environment on leaf water potential of lychee: implications for irrigation scheduling and productivity. Acta Horticult 175:363–370

Menzel CM, Rasussen TS, Simpson DR (1989) Effects of temperature and leaf water stress on growth and flowering of litchi (*Litchi chinensis* Sonn.). J Hortic Sci 64:739–752

Menzel CM, Aitken RL, Dowling AW, Simson DR (1990) Root distribution of lychee trees growing in acid soils of subtropical Queensland. Aust J Exp Agric 30:699–705

Menzel CM, Haydon GF, Doogan VJ, Simpson DR (1994) Time of nitrogen application and yield of Bengal lychee on a sandy loam soil in subtropical Queensland. Aust J Exp Agric 34:803–811

Menzel CM, Oosthuizen JH, Roe DJ, Doogan VJ (1995) Water deficits at anthesis reduce CO_2 assimilation and yield of lychee (*Litchi chinensis* Sonn.) trees. Tree Physiol 5:611–617

Menzel CM, Olesen T, McConchie C, Wiltshire N, Diczabalis Y, Wicks C (2000) Lychee, Longan and Rambutan: optimum canopy management. Rural Industries Research and Development Corporation, Canberra, p 92

Menzel CM, Bagshaw J, Campbell T, Greer N, Noller J, Olesen T, Waite GK (2002) Lychee information kit, Queensland. Department of Primary Industries, Nambour

Misra AN, Sahu SM, Mishra M, Singh P, Meera I, Das N, Kar M, Sahu P (1987) Sodium chloride induced changes in leaf growth, pigment and protein content in two rice cultivars. Plant Biol 39:257–262

Mitra R, Davarynejad GH, Genard M, Bannayan M, Azizi M, Vercambre G (2015) Peach water relations, gas exchange, growth and shoot mortality under water deficit in semi arid weather conditions. PLoS ONE 10: e0120246

Monteith JL (1977) Climate and the efficiency of crop production in Britain. Phil Trans R Soc Lond B 281:277–294

Mou B (2011) Improvement of horticultural crops for abiotic stress tolerance: an introduction. HortSci 46:1068–1069

Naganathan AN, Doshi U, Fung A, Sadqi M, Munoz V (2006) Dynamics, energetics, and structure in protein folding. Biochemistry 45(28):8466–8475

Nakata S (1955) Floral initiation and fruit set lychee, with special reference to the effects of sodium nephthalene acetate. Bot Gaz 117:126–134

Nakata S, Suehisa R (1969) Growth and development of *Litchi chinensis* as affected by soil moisture stress. Am J Bot 56:1121–1126

Naor A (2000) Midday steam water potential as a plant water stress indicator for irrigation scheduling in fruit trees. Acta Horticult 537:447–454

Nath P, Sane VA, Asif MH, Sane AP, Trivedi PK (2012) Fruit crops: omic approaches towards elucidation of abiotic stress tolerance. In: Improving crop resistance to abiotic stress. Tuteja Plant Stress Omics O-BK, vol 1, Wiley-VCH Verlag GmbH & Co., KGaA, Weinheim, Germany, pp 1033–1048

Neill JS, Desikan R, Hancock J (2003) Nitric oxide as a mediator of ABA signaling instomatal guard cells. Bul J Plant Physiol (Special Issue):124–132

NHB Data Book (2013) National Horticulture Board, Gurgaon

Ortbuauer M (2013) Abiotic stress adaptation: protein folding stability and dynamics. In: Abiotic stress – plant responses and applications in Agriculture. University of Vienna, Vienna

Pasala KR, Raina SK, Kumar S, Bhagat KP, Singh Y, Bal SK (2013) Adaptation and mitigation strategies of plant under drought and high temperature stress. In: Climate change and plant abiotic stress tolerance. Wiley, Weinheim

Phang TH, Shao G, Lam HM (2008) Salt tolerance in soybean. J Integr Plant Biol (Research Support, Non-U.S. Gov't Review) 50(10):1196–1212

Qu H, Duan X, Su X, Liu H, Jiang Y (2006) Effects of anti-ethylene treatments on browning and energy metabolism of harvested litchi fruit. Aust J Exp Agric 46:1085–1090

Rai M, Nath V, Dey P, Kumar S, Das B (2001) Orchard managment and production. In: CHES (ed) Litchi: genetic resources, production, protection and post-harvest management. ICAR, Ranchi, pp 33–40

Rai M, Dey P, Gangopadhyaya KK, Das B, Nath V, Reddy NN, Singh HP (2002) Influence of nitrogen, phosphorus and potassium on growth parameters, leaf nutrient composition and yield of litchi (*Litchi chinensis*). Indian J Agric Sci 72:267–270

Rao DLN, Pathak H (1996) Ameliorative influence of organic matter on biological activity of salt affected soils. Arid Soil Res Rehabil 10:311–319

Raymond MJ, Smirnoff N (2002) Proline metabolism and transport in maize seedlings at low water potential. Ann Bot 89:813–823

Roe DJ, Oosthuizen JH, Menzel CM (1995) Rate of soil drying and previous water deficits influence the relationship between CO_2 assimilation and tree water status in potted lychee (*Litchi chinensis* Sonn.). J Hortic Sci 70:15–24

Samra JS, Singh G (2002) Drought management strategies. Indian Council of Agricultural Research, New Delhi, pp 1–45

Santner A, Estelle M (2010) The ubiquitin–proteasome system regulates plant hormone signaling. Plant J 61:1029–1040

Saquet AA, Streif J, Bangerth F (2000) Changes in ATP, ADP and pyridine nucleotide levels related to the incidence of physiological disorders in 'Conference' pears and 'Jonagold' apples during controlled atmosphere storage. J Hortic Sci Biotechnol 75(2):243–249

Saquet AA, Streif J, Bangerth F (2003) Energy metabolism and membrane lipid alterations in relation to brown heart development in 'Conference' pears during delayed controlled atmosphere storage. Postharvest Biol Technol 30:123–132

Sgherri CLM, Navari-Izzo F (1995) Sunflower seedlings subjected to increasing water deficit stress:oxidative stress and defence mechanisms. Plant Physiol 93:25–30

Shao W, Lai ZX, Huang Q, Guo YQ (2010) Changes of APX Isozymes under high temperature stress and cloning of an APX functional fragment in embyrogenic callus of litchi. Acta Horticult 863:183–187

She WQ, Li XH (2010) A difference in cold tolerance among litchi cultivars and its mechanism. Acta Horticult 863:299–304

Shibairo SI, Upadhyaya MK, Toivonen PMA (1997) Post-harvest moisture loss characteristics of carrot (*Daucus carota* L.) cultivars during short-term storage. Sci Hortic 71:1–12

Shikhamany SD (2001) Canopy management of the tropical and subtropical fruit crops. Indian J Hortic 58:28–32

Shukla RK, Bajpai PN (1974) Blossom bud differentiation and ontogeny in litchi (*Litchi chinensis* Sonn.). Indian J Hortic 31:224–228

Singh DP (2003) Stress physiology. New Age International (P) Limited Publishers, New Delhi, p 164

Singh SN, Pathak RA (1983) Effect of irrigation intensity, its frequency and manures on the growth of litchi, cv. Calcuttia. Punjab Hortic J 23:197–202

Singh G, Nath V, Pandey SD, Ray PK, Singh HS (2012) The litchi. FAO, New Delhi

Sinha S, Das DK (2013) Transformation of litchi (*Litchi chinensis* Sonn.) with GLY-I + II gene leads to enhanced salt tolerance. Int J Biotechnol Allied Fields 1:483–495

Sitbon F, Perrot-Rechenmann C (1997) Expression of auxin regulated genes. Plant Physiol 100:443–455

Sitbon F, Hennion S, Sunderg B, Little CHA, Olsson O, Sandberg G (1992) Transgenic tobacco plants coexpressing th Agrobacterium tumifaciens iaa M and Iaa H gene display altered growth and indoleacetic acid metabolism. Plant Physiol 99:1062–1069

Smirnoff N (1993) The role of active oxygen in response of plants to water deficit and desiccation. New Phytol 125:27–58

Smrintontip C, Tiyayon P, Sringam K, Pantachod S, Naphrom D, Ruamrungsri S, Sruamsiri P (2013) Influence of water regimes and potassium chlorate on floral induction, leaf photosynthesis and leaf water potential in longan. J Agric Sci 5(6):211–220

Song L, Liu H, Su X, You Y, Jiang Y (2006) Effects of adenosine triphosphate on the vase life of cut carnation flowers. Aust J Exp Agric 46:137–139

Srinivasa Rao NK, Laxman RH (2013) Phenotyping horticultural crops for abiotic stress tolerance. In: Singh HP, Srinivasa Rao NK, Shivashankar KS (eds) Climate resilient horticulture: adaptation and mitigation strategies. Springer India, New Delhi, pp 147–158

Srivastava KK (2007) Canopy management of fruit crops. International Book Distributing Company, Lucknow

Stern RA, Meron M, Naor A (1998) Effect of fall irrigation level in Mauritius and Floridian lychee on soil and plant water status, flowering intensity and yield. J Am Soc Hortic Sci 123:150–155

Stern RA, Naor A, Nehama B (2003) Xylem-sap zeatin-riboside and dihydrozeatin-riboside levels in relation to plant and soil water status and flowering Mauritius lychee. Sci Hortic 98(3):285–291

Stewart CR (1981) Proline accumulation: biochemical aspects. In: Paleg LG, Aspinall D (eds) Physiology and biochemistry of drought resistance in plants. Academic, Sydney, pp 243–259

Storey WB (1973) The lychee. Yearbook,-California-Avocado-Society 56:75–86

Toivonen PMA (2003a) Effects of storage conditions and postharvest procedures on oxidative stress in fruits and vegetables. In: Hodges DM (ed) Postharvest oxidative stress in horticultural crop. Food Products Press, New York, pp 69–90

Toivonen PMA (2003b) Postharvest treatments to control oxidative stress in fruits and vegetables. In: Hodges DM (ed) Postharvest oxidative stress in horticultural crop. Food Products Press, New York, pp 225–246

Toivonen PMA (2005) Postharvest storage procedures and oxidative stress. HortSci 39:938–942

Trippi VS, Paulin A, Pradet A (1998) Effect of oxygen concentration on the senescence and energy-metabolism of cut carnation flowers. Physiol Plant 73:374–379

Veltman RH, Lentheric I, Van der Plas LHW, Peppelenbos HW (2003) Internal browning in pear fruit (*Pyrus communis* L. cv Conference) may be a result of a limited availability of energy and antioxidants. Postharvest Biol Technol 28:295–302

Wall MM (2006) Ascorbic acid and mineral composition of longan (*Dimocarpus longan*), lychee (*Litchi chinensis*) and rambutan (*Nephelium lappaceum*) cultivars grown in Hawaii. J Food Compos Anal 19:655–663

Wang W, Vinocur B, Shoseyov O, Altman A (2004) Role of plant heat-shock proteins and molecular chaperones in the abiotic stress response. Trends Plant Sci 9(5):244–252

Whiley AW, Sarnah JB (1980) Salinity tolerance of some avocado, litchi and macadamia nut trees. Bienn Rep Maroochy Hortic Res Station Aust 2:16–17

White PJ, Broadley MR (2001) Chloride in soils and its uptake and movement within the plant: a review. Ann Bot 88:967–988

Whitmore JS (2000) Drought management on farmland. Kluwer Academic Publishers, Dordrecht

Wilhite DA (2000) Drought as a natural hazard: concepts and definitions. In: Wilhite DA (ed) Drought: a global assessment, vol 1. Routledge, New York, pp 1–18

Wilkie JD, Sedgley M, Olesen T (2008) Regulation of floral initiation in horticultural trees. J Exp Bot 59:3215–3228

Wilkinson S, Davies WJ (2010) Drought, ozone, ABA and ethylene new insights from cell to plant to community. Plant Cell Environ 33:510–525

Xu S, Li J, Zhang X (2006) Effects of heat acclimation pretreatment on changes on membrane lipid peroxidataion, antioxidant metabolites and ultrastructure of chloroplasts in two cool season turfgrass species under heat stress. J Environ Exp Bot 56:274–285

Xuan H, Streif J, Saquet A, Romheld V, Bangerth F (2005) Application of boron with calcium affects respiration and ATP/ADP ratio in 'Conference' pears during controlled atmosphere storage. J Hortic Sci Biotechnol 80:633–637

Yamaguchi-Shinozaki K, Shinozaki K (2006) Transcriptional regulatory networks in cellular responses and toletance to dehydration and cold stresses. Annu Rev Plant Biol 57:781–803

Yang HM, Ozaki T, Ichii T, Nakanishi T, Kawai Y (1992) Diffusible and extractable auxins in young Japanese pear trees. Sci Hortic 51:97–106

Yi C, Qu HX, Jiang YM, Shi J, Duan XW, Joyce DC, Li YB (2008) ATP-induced changes in energy status and membrane integrity of harvested litchi fruit and its relation to pathogen resistance. J Phytopathol 156:365–371

Yi C, Jiang YM, Shi J, Qu HX, Duan XW, Yang B, Prasad NK, Liu T (2009) Effect of adenosine triphosphate on changes of fatty acids in harvested litchi fruit infected by *Peronophythora litchii*. Postharvest Biol Technol 54:159–164

Yi C, Jiang YM, Shi J, Qu HX, Xue S, Duan XW, Shi JY, Prasad NK (2010) ATP-regulation of antioxidant properties and phenolics in litchi fruit during browning and pathogen infection process. Food Chem 118:42–47

Part IV

Abiotic Stress Tolerance in Horticultural Crops: Plantation and Tuber Crops

15 Coconut and Areca Nut

S. Naresh Kumar, V. Rajagopal, and K.V. Kasturi Bai

Abstract

The plantation crops, owing to their perennial nature, live through the cycles of several types of stresses. Abiotic stresses such as droughts, dry spells, high and low temperatures and floods significantly affect the growth, development and yield of the plantations. At seedling stage, the high light intensity stress causes seedling mortality if no protective measurements are taken. Studies indicated that plantation crops such as coconut impart drought tolerance through morphological, anatomical, physiological, biochemical and molecular mechanisms. The tolerant genotypes with better revival capacity are also identified. Though genotypic improvement through breeding is important, it is more important to improve the population by incorporating the progeny of identified in situ drought-tolerant and high-yielding palms. This should hasten the selection process for abiotic stress tolerance as well as for high yield as the in situ drought-tolerant palms have withstood the naturally occurring stresses in their life cycle. Apart from genetic improvement, the agronomic management becomes very important in plantation crops for sustaining the yield. Therefore, the soil moisture conservation measures and drip irrigation become very important. In this chapter, these issues are discussed in detail with research evidences with special reference to coconut.

S.N. Kumar (✉)
Centre for Environment Science and Climate Resilient Agriculture, ICAR-Indian Agricultural Research Institute, NRL Building, New Delhi 110012, India
e-mail: nareshkumar.soora@gmail.com

V. Rajagopal • K.V. Kasturi Bai
ICAR-Central Plantation Crops Research Institute, Kasaragod 671124, Kerala, India

N.K.S. Rao et al. (eds.), *Abiotic Stress Physiology of Horticultural Crops*, DOI 10.1007/978-81-322-2725-0_15

15.1 Introduction

Growth and productivity of coconut and areca nut palms are influenced by the external factors such as rainfall, temperature, relative humidity and sunshine duration apart from the soil characteristics and management practices. The optimum weather conditions for good growth and nut yield in coconut are well-distributed annual rainfall between 130 and 230 cm, mean annual temperature of 27 °C and sunlight ranging from 250 to 350 Wm^{-2} with at least 120 h per month of sunshine period (Child 1974; Murray 1977). Application of NPK is at 500N:320P_2O_5:1200 K_2O per palm/year and optimal irrigation is at 200 L/palm once in 4 days or at 66 % E as drip irrigation. Similarly, areca nut requires a temperature range of 14–36 °C, though it can grow in temperatures ranging from 5 to 40 °C. Also, it requires a well-distributed annual rainfall ranging above 1,200 mm with relative humidity neither very high nor low. Any deviations from the optimal conditions cause the palms to experience the stress conditions. The wide variation in nut yield implies that there is a wide gap in potential and realized yield showing the poor exploitation of production potential. Coconut plantations are seldom grown under ideal management conditions as well as they are continuously exposed to vagaries of nature as perennial crops. While coconut plantations face abiotic and biotic stresses, areca nut plantations in India are generally well managed but have major issues of biotic stresses.

Amount and distribution of rainfall have significant influence on nut yield in coconut (Lakshmanachar 1963; Abeywardena 1968). The relationship between rainfall and nut yield varied (Mathai and Panicker 1979) due to magnitude of rainfall and other factors (Vijayakumar et al. 1988). Weather data-based descriptive models for yield forecasting are developed for different agroclimatic zones (Naresh Kumar et al. 2008). These models had high R^2 with 4-year lag, suggesting the possibility of forecasting first year nut yield using weather data of first to fourth year. However, with the development and validation of the process-based InfoCrop coconut simulation model (Naresh Kumar et al. 2008), now it is not only possible for yield forecast but also the model is applied for various studies such as climate change impacts, adaptation gains (Naresh Kumar and Aggarwal 2013) and also for yield gap analysis (Naresh Kumar et al. 2008) as well as for assessing the carbon sequestration potential with high reliability and also under multifactor influence. In this chapter, major findings of research on abiotic stresses in coconut are summarized. Since the information on areca nut is not as exhaustive and scientifically analysed as in the case of coconut, relevant available information is provided.

15.2 Coconut

In general, coconut plantations are exposed to abiotic stresses such as drought, temperature (high and low), light intensity (high and low), cyclonic wind and flooding stress. The environmental stresses affect coconut yield in almost all coconut-growing areas (Coomans 1975; Mathes 1988; Bhaskara Rao et al. 1991).

15.2.1 Drought Stress: Extent of the Problem

Drought is the major constraint for coconut productivity in rainfed plantations in India as well as in other coconut-growing countries. Hence, the research efforts and the depth of availability of information are more. Coconut is mainly grown as a rainfed crop and the productivity is about 50 % less as compared to irrigated gardens. As a rainfed crop, it faces summer dry spells each year apart from the frequent occurrence of drought years. Water deficit at seedling stage could lead to seedling death. Though coconut seedlings are generally irrigated for their establishment, aftercare is not optimal. Tall palms flower in about 6 years after planting while the dwarfs and hybrids flower by about fourth year. The duration from the initiation of inflorescence primordium to nut maturity is about 44 months. Out of this, the prefertilization period is about 32 months and nut development phase (postfertilization) is 12 months. The primordial initiation, ovary development and button size nut stage are very sensitive to water stress. Coincidence of drought with critical stages affects nut yield (Rajagopal et al. 1996, 2000a) not only in the current year but also in the subsequent 3 years to follow, thus making the problem more severe across all agroclimatic regions of India (Naresh Kumar et al. 2007b). Severely affected palms take at least 4 years to recover after going through stress period. Thus the economic loss to the growers is perennial and significant mainly in rainfed zones of Kerala, Karnataka, Orissa, the northeastern region and Lakshadweep Islands. Average productivity of these regions is significantly low (5,000–8,000 nuts/ha/year) compared to the predominantly irrigated regions (13,000–20,000 nuts/ha/year). The rainfed coconut plantations in intermediate and dry zones of Sri Lanka (Peiris and Peries 1993; Peiris et al. 1995; Peiris and Thattil 1988), Zanzibar (Juma and Fordham 1998) and Indonesia (Bonneau and Subagio 1999) are affected by droughts. Thus, drought is the major constraint for productivity across coconut-growing countries.

In order to sustain the productivity, planting drought-tolerant cultivars with faster recovery potential and agronomical management becomes utmost important. Hence, the research efforts were focused on development of drought-tolerant cultivars as well as agronomic management strategies. In nature, the physiology of a plant's response to drought stress is rather complex. Because of the dynamics of soil water depletion and atmospheric water demand, the length and severity of stress have spatio-temporal variations. Moreover, the response of a plant to abiotic stress is a multidimensional one with responses starting at the cellular and intercellular level to the organ and phenological stage level varying with space and time. Over generations, plants developed adaptive mechanisms, i.e. molecular, biochemical, physiological, anatomical, morphological and phenological, to overcome drought stress.

The length and intensity of dry spell and influence of rainfall and dry spell on the nut yield in major coconut-growing areas in different agroclimatic zones, viz., in the western coastal area, hot, subhumid and per-humid (Kasaragod, Kerala; Ratnagiri, Maharashtra); in Western Ghats, hot, subhumid and per-humid (Kidu, Karnataka) and hot and semiarid (Arsikere, Karnataka); and in eastern coastal plains, hot and subhumid (Veppankulam, Tamil Nadu; Ambajipeta, Andhra Pradesh), in India has indicated that the dry spell was longer in Ratnagiri (216 days) and Arsikere (202 days) and shorter at Kidu (146 days), which differentially affected the nut yield (Naresh Kumar et al. 2008). The annual nut yield under rainfed conditions varied at different zones. Fluctuations in coconut yield during different years could be explained on the basis of rainfall distribution. However, the length and number of dry spells are more important than the total rainfall per se which influences the nut yield (Rajagopal et al. 2000b; Naresh Kumar et al. 2008). In view of the long duration (44 months) between the inflorescence initiation to nut maturation, the occurrence of dry spell in any year would affect the yield for the subsequent 3–4 years, depending on the critical stages. Consecutive droughts in Coimbatore District (Tamil Nadu) during 1998–2002 reduced the coconut production by about 3 lakh nuts/year for 4 years. The productivity loss was to the tune of about 3500 nuts/ha/year (Naresh Kumar et al. 2008) (Fig. 15.1) (Photo 15.1).

15.2.2 Nut Yield in Relation to Coincidence, Intensity and Length of Drought Stress

The coconut palm is influenced considerably by the environmental variables in its productive features especially under rainfed condition. Of all the climatic factors, rainfall has the maximum influence on the seasonal fluctuation in yield (Abeywardena 1968). In coconut palm, there is a long duration (~44 months) between inflorescence primordial initiations to nut maturity with the prefertilization period (~32 months) being longer than the postfertilization (12 months) period. Hence, the impact of drought occurring during any of the critical stages of the development of inflorescence affects nut yield (Rajagopal et al. 1996). The impact of drought on the ontogeny of coconut inflorescence integrating the overall occurrence of dry spell

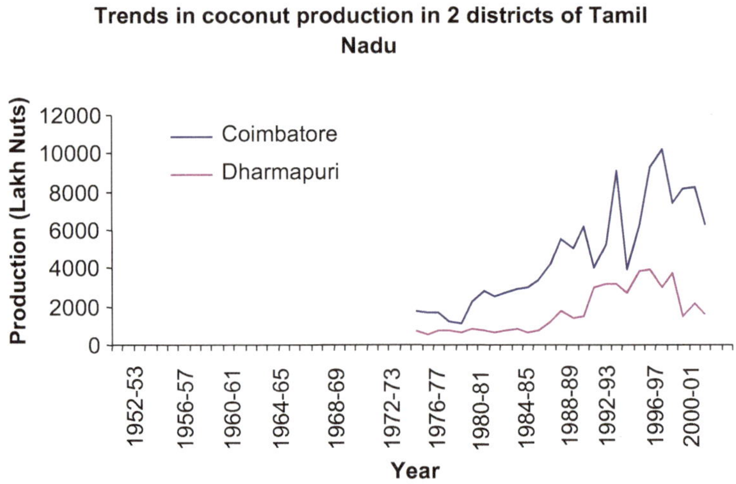

Fig. 15.1 Trends in coconut production in two districts of Tamil Nadu

Photo 15.1 Drought affected coconut garden

and growth stages of the developing nut has been delineated (Rajagopal et al. 1996). The intricate relationship between dry spell and stages of nut development right from inflorescence initiation to the nut maturity as well as annual nut yield in different agroclimatic zones has been well described (Rajagopal et al. 1996, 2000a; Naresh Kumar et al. 2008). Physiological traits responsible for drought tolerance correlated with yield performance under stress conditions, and some of the cultivars identified as drought tolerant also proved to be good yielders (Bhaskara Rao et al. 1991; Rajagopal et al. 1992). There were genotypic variations for the drought index in coconut (Pomier and de Taffin 1982). By providing life-saving irrigation during summer months, the adverse effects of dry spells especially on the development of the inflorescence primordium can be reduced (Naresh Kumar et al. 2003).

The influence of drought on nut yield becomes manifest in the subsequent year (Bhaskara Rao et al. 1991). Seven lag periods were worked out to relate 11 weather variables with coconut yield (Vijayakumar et al. 1988). Among the variables tested, rainfall had positive influence on five of the seven lag periods while temperature and relative humidity on two or three lag periods. The weather-based descriptive models for different agroclimatic zones were developed (Naresh Kumar et al. 2009). The prediction models with 3- and 4-year lag had high R^2 values. The models differed for usage of parameters in different agroclimatic zones, indicating the relative importance of these parameters in respective conditions for realizing the nut yield in coconut. Interestingly, the parameters used in models for the western coastal area (hot, subhumid and per-humid) are temperature and relative humidity, as indicated even in the classification of these areas. Models were verified for 2 years and prediction of yield during 1998–1999 and 1999–2000 within 10 % confidence level validated these models. The study indicates that the relative humidity and temperature play an important role during the ontogeny of inflorescence and nut development. The descriptive models, developed based on weather data, can be used for the prediction of coconut yield 2–4 years in advance within acceptable range of accuracy. The yields to be realized can be bracketed within the predicted range obtained from models using 1-, 2-, 3- and 4-year lags (Naresh Kumar et al. 2009; Fig. 15.2).

Among the hybrid combinations in India, two of them performed well under drought conditions in which yield stability was found maintained through lower reduction in bunch production, pistillate flower production, nut setting and yield (Bhaskara Rao et al. 1991; Table 15.1).

Among the three hybrid combinations, viz., Chandrasankara (COD × WCT), Chandralaksha (LCT × COD) and Lakshaganga (LCT × GBGD), Chandrasankara outyielded both but Lakshaganga and Chandralaksha performed well under drought situations than Chandrasankara (Table 15.1). Physiological traits responsible for drought tolerance correlated with yield performance under stress conditions and some of the cultivars identified as drought tolerant also proved to be good yielders (Rajagopal et al. 1992). However, stability analysis needs to be done assessing the suitability of cultivars across the agroclimatic zones (Siju Thomas et al. 2005).

15.2.3 Button Shedding

Flowering in coconut is a continuous process and the development and growth of inflorescence is intimately associated with the development and growth of the leaf (Patel 1938). Once the inflorescence (spadix) opens, it takes about 12 months for full maturity of the buttons and harvest of nuts. Being a tree crop of indeterminate growth habit, the partitioning of metabolites towards vegetative and reproductive growth is a continuous process and interlinked during the entire span of its life. Shedding of buttons is one of the major constraints in coconut production. Under normal conditions, within 3 months of the opening of the inflorescence, about 80 % of the pistillate flowers are shed. Extensive reports are available on the impact of various factors on the shedding of buttons (Sudhakara 1990; Ramadasan et al. 1991; Kasturi Bai et al. 2003).

During summer months under unirrigated condition, soil water deficit is the major cause for the shedding of buttons, which gets aggravated with the changes in the micrometeorological variables. Although high rainfall is not harmful to the growth and productivity of palms, shedding can be observed due to the impairment of pollination and fertilization (Menon and Pandalai 1958). Significant differences have been observed in pistillate flower production, shedding of buttons and nut production between the cultivars (Kasturi Bai et al. 2003). High correlation has been observed between the pistillate flower production and shedding of buttons. This implies that the increase or decrease in the button shedding depends on the number of pistillate flowers in the spadix which show variation between cultivars.

There are also seasonal variations in pistillate flower production and button shedding under irrigated and unirrigated conditions (Table 15.2). In WCT and its two hybrids, WCT × COD and COD × WCT, pistillate flower production and percentage button shedding did not differ significantly under irrigated and rainfed conditions. However, increasing trend in pistillate flower production was observed under irrigated conditions. In general, in all the varieties, lower pistillate flower production is observed in the post-monsoon season as compared to summer (February to May) and monsoon (June to September) months.

Irrespective of irrigation or rainfed condition, button shedding occurred in two peaks in all the

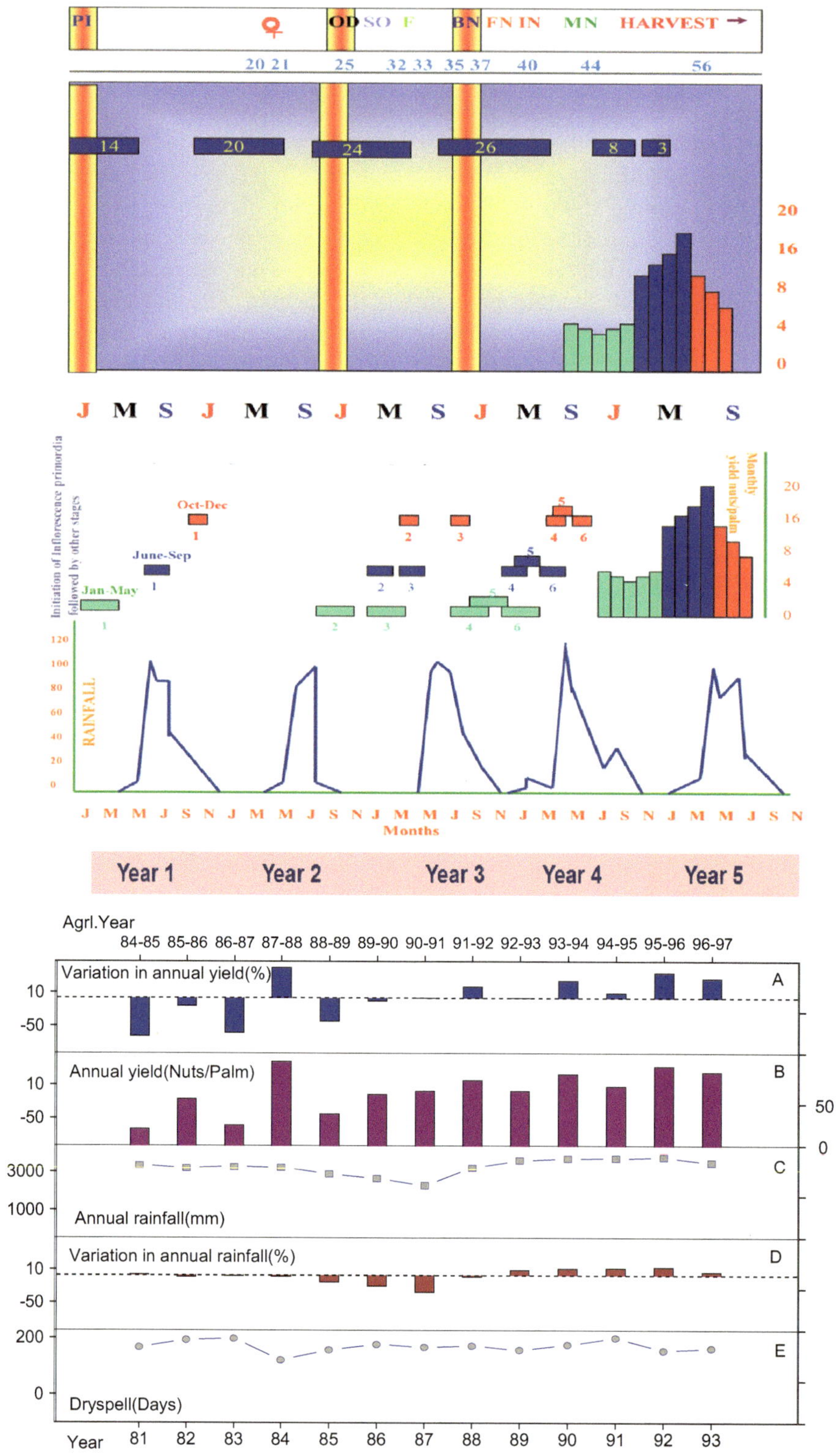

Fig. 15.2 Annual nut yield and variation in annual yield as influenced by annual rainfall and dry spell over a period of time at Kasaragod

Table 15.1 Performance of hybrids during good and drought-influenced years

Condition	COD × WCT	WCT × COD	WCT × GBGD	LCT × COD	LCT × GBGD
Number of bunches					
Good years	10.5	10.4	9.9	9.5	9.8
DI years	6.1	7.2	5.1	7.7	7.7
% reduction	42	31	45	19	21
Number of pistillate flowers/bunch					
Good years	35.0	32.5	33.3	26.8	30.1
DI years	27.4	24.5	24.5	29.9	26.7
% reduction	22	29	30	5	22
Setting %					
Yield, nuts/palm					
Good years	145	117	98	95	108
DI years	36	43	28	81	60
% reduction	75	64	72	15	44

Source: Bhaskara Rao et al. (1991)
DI years, drought-influenced years

Table 15.2 Seasonal variation in pistillate flower production (PFP-No.) and button shedding under irrigated (I) and rainfed (RF) conditions (mean values)

		WCT		COD × WCT		WCT × COD	
Period	Parameters	RF	I	RF	I	RF	I
Pre-monsoon (Feb–May)	PFP	29.0	30.0	24	29	22	32
	%drop	64.2	54.3	44.3	47.7	47.2	56.7
Monsoon (June–Sept.)	PFP	28	20	29	24	23	25.0
	%drop	76.6	63.4	56.6	64.9	60.8	70.9
Post-monsoon (Oct–Jan)	PFP	17	19	16	16.0	15	20.0
	% drop	38.3	45	32.1	39.6	33.0	50.0
Mean	PFP	25.0	23	2.3	23	20.0	26.0
	% drop	59.1	54.2	44.3	50.7	47.0	59.2

Source: Kasturi Bai et al. (2002b)

varieties, one during summer months and the other during monsoon. During post-monsoon season, the number of pistillate flower production is low; consequently, the drop percentage is also low in all the varieties. Significant differences in pistillate flower production and percentage drop were observed only between different months (Kasturi Bai et al. 2003).

15.2.4 Response of Palms to Field Drought Stress

Exposure to excess radiation (>265 Wm^{-2}), temperature (>33 °C) and vapour pressure deficit (>26 m bar) cause stress in coconut palms (Kasturi Bai et al. 1988) and are further aggravated by water deficit. The duration of dry spell during initiation of inflorescence primordium, ovary development and button size nut stages in that order has greater influence on nut yield than other stages (Rajagopal et al. 1996). Critical levels of water deficit vary with soil type. Soil types and compaction levels influenced the water stress in seedlings (Nainanayake and Bandara 1998). A water deficit of 110 mm is critical in sandy loam soil, at which the stomata close (Rajagopal et al. 1989). The photosynthetic rates and dry matter production and its partitioning are influenced by the soil water status (Kasturi Bai 1993; Naresh Kumar et al. 2002a). Soil water deficit coupled with atmospheric evaporative demand during dry months affected coconut palms more in red

Table 15.3 Relationship between aridity index and symptoms of drought in coconut

Aridity index (%)	Morphological symptoms	Drought intensity
65	Nil	Slight
65–85	Drooping of lower leaves	Moderate
85–100	Beginning of drying of drooped leaves and button shedding	Moderately severe
Prolonged with 100 for 5 weeks	Drying of drooped leaves falling of tender and immature nuts, burning trace on nuts due to the high intensity of radiation	Severe
Prolonged with 100 for > 5 weeks	Palms show the death symptoms, drying of the spindle leaf under poor management conditions	Critically severe

Source: Rao (1985)

sandy loam than laterite soil (Voleti et al. 1993a). Hybrids had higher stomatal resistance resulting in maintenance of high water potential during stress in laterite soil than in sandy loam. However, the hybrid, COD × WCT, was sensitive to water stress in sandy loam soil. Exposure of palms to moisture stress for 3–4 weeks led to reduction in the vegetative dry matter by 15–18 % and reproductive dry matter by 20–22 %, as compared to irrigated palms (Rajagopal et al. 1989). Soil water deficit in unirrigated conditions reduced the leaf water potential and enhanced the activity of stress-sensitive enzymes (Shivashankar et al. 1991). Four general types of leaves were found in coconut canopies, viz., (i) leaves with higher photosynthesis rate (Pn) and water use efficiency (WUE) than mean performance of canopy leaves, (ii) leaves with higher Pn and lower WUE, (iii) leaves with lower Pn and higher WUE and (iv) leaves with lower Pn and lower WUE (lower leaves) (Naresh Kumar and Kasturi Bai 2009a). An oval-shaped canopy is more suitable for higher photosynthesis efficiency, WUE and productivity as compared to X-shaped and semicircle-shaped canopies. These results indicate that the canopy shape plays a role in the overall performance of photosynthesis and water use efficiencies and productivity in coconut. Results also indicate coconut as a source-limited plant (Naresh Kumar and Kasturi Bai 2009a).

15.2.4.1 Morphological Symptoms

In rainfed conditions, a prolonged dry spell affects the palm. An 'aridity index' of 100 % for a prolonged period of 5–10 weeks severely affects the productivity of coconut palms (Table 15.3). When exposed to such severe moisture stress, adverse effects such as bending and breaking of dry leaves, poor spathe development and most bunches with only one or two nuts are seen (Rao 1985).

The drought index, calculated based on the morphological symptoms, is related to nut yield (Ramadasan et al. 1991). The drought index was found to be the lowest in the hybrids WCT × PHOT and WCT × GBGD and the highest in WCT × MOD among ten hybrids. The 'index to drought tolerance', calculated based on the percentage of dry leaves (n), compared to the number of living ones (N), i.e. (n/N) X 100, was higher for the hybrid PB-121 and least for Rennell Tall × WAT among five hybrids (Pomier and de Taffin 1982).

The characterization of drought, in different coconut-growing areas, in six states of India falling under different agroclimatic zones, revealed that the length and intensity of dry spell adversely affect the source-sink relationship and consequently the nut yield (Naresh Kumar et al. 2003; 2006a). These agroclimatic zones varied for annual rainfall from 718 to 3,338 mm. Similarly dry spell varied from 146 to 216 days and consequently annual nut yield varied from 30 to 68. In view of the long duration (44 months) between the inflorescence initiation to nut maturation, the occurrence of dry spell in any 1 year would affect the yield for the subsequent 3–4 years. The longer dry spell affects the nut yield for next 4 years to follow with stronger impact on fourth year, irrespective of the total rainfall. Apart from this, current year

Table 15.4 Correlations between anatomical features and morpho-physiological parameters

Parameters		r value
Leaflet thickness	Palisade parenchyma cell size	0.73
	Spongy parenchyma cell size	0.82
	Stomatal frequency	−0.67
	Photosynthetic rate	−0.41
	Transpiration rate	−0.12
Cuticle thickness	Epicuticular wax content (ECW)	0.86
Adaxial	Transpiration rate	−0.69
	Photosynthetic rate	0.13
	Leaf water potential	0.64
Abaxial	ECW	0.72
	Transpiration rate	−0.79
	Photosynthetic rate	−0.59
	Leaf water potential	0.50
Sub-stomatal cavity size	Stomatal frequency	−0.54
	ECW	0.61
	Transpiration rate	−0.15
	Photosynthetic rate	−0.11
Palisade parenchyma size	Photosynthetic rate	−0.77
	Transpiration rate	−0.38
	Leaf water potential	0.11
Spongy parenchyma size	Photosynthetic rate	−0.67
	Transpiration rate	−0.05
	Leaf water potential	0.10

Source: Naresh Kumar et al. (2000)

nut yield is also affected (Naresh Kumar et al. 2007).

15.2.4.2 Leaflet Anatomical Adaptations in Relation to Drought Stress

Anatomical basis of physiological efficiency for drought tolerance in coconut is delineated. The increase in leaflet thickness is mainly due to increase in parenchyma cell size. It is also associated with lowered stomatal frequency, an indication of adaptation to drought. Increased leaf thickness and thick cuticle are some of the xeromorphic characteristics as observed in WCT and FMS cultivars. An increase in thickness of the leaflet causes a decrease in the ratio of the external surface to its volume. Water in leaves is conducted not only by the veins and bundle sheath extensions but also by the mesophyll cells, epidermis and intercellular spaces. Water transport towards the epidermis is much higher through the palisade tissue than the spongy parenchyma. Increased parenchyma cell size, as observed in WCT and FMS, means less intercellular space/unit area. This may help in reducing the water conductance towards epidermis thus reducing the transpiration rates and maintaining high water potentials, a characteristic feature of drought tolerance. The volume of intercellular spaces in xeromorphic leaves is low thus reducing the water transport to epidermal cells. The structure favourable to high photosynthetic rates (large palisade parenchyma tissue surface due to small parenchyma cells, Table 15.4) induces at the same time high transpiration rates because of higher intercellular space (Naresh Kumar et al. 2000). Stomatal frequency (SF) and index (SI) also play a major role in plant water relations. Variations in SF and SI were observed among the talls, dwarfs and hybrids (Rajagopal et al. 1990).

The WCT and FMS have thick leaflets, thick cuticle on both sides and larger parenchyma, hypodermal and water cells compared to PHOT, WCT × COD, COD × WCT, GBGD and MYD (Naresh Kumar et al. 2000). This implies adaptability or tolerance of the former cultivars for drought condition than the latter. The water storage tissues supply water to other tissues when water is limited. Cultivars having thick cuticle are able to maintain higher leaf water potentials (Table 15.4). Drought-tolerant types also have more scalariform thickening on xylem tracheids in vascular bundles and large sub-stomatal cavities. These traits are less favourable in PHOT and WCT × COD implying moderate tolerance to drought. The values for these traits are the least in COD × WCT, GBGD and MYD indicating susceptibility to drought. However, the difference between moderately tolerant and susceptible cultivars for these traits is narrow. Large sub-stomatal cavities will help in maintaining enough internal CO_2 concentrations required for sustaining the photosynthetic rates during the stress period when the stomata are partially closed. High internal CO_2 concentrations may help in reducing the water loss through stomata. Epidermal cell size (upper and lower) and guard cell size are related to the drought tolerance characteristic of a cultivar. It is possible that the cumulative effect of all these traits contributes to the adaptation to drought stress (Naresh Kumar et al. 2000).

In coconut, the leaf (epicuticular wax) ECW, which influence the energy balance of leaf, did not differ significantly among cultivars and hybrids during favourable conditions. Thin-layer chromatographic analysis revealed the qualitative difference in the composition of wax during different stages of palm growth. Major components identified were hydrocarbons and esters while the alcohols were identified in the stress period. On the other hand, fatty acids occurred only during post-stress (Voleti and Rajagopal 1991). Almost a three to fourfold increase in ECW was noticed during dry season in some of the coconut hybrids, viz., WCT × GBGD, WCT × COD, LCT × COD and LCT × GBGD (Voleti and Rajagopal 1991; Kurup et al. 1993). The physiological age of palms and leaves influenced the formation of wax on leaf surface. Leaves of coconut seedlings have almost 50 % less ECW than those from adult palms even at same degree of stress.

15.2.4.3 Physiological Responses to Drought Stress

15.2.4.3.1 Stomatal Gas Exchange and Water Relations

Coconut responds to water stress by stomatal regulation and deposition of ECW to maintain leaf water potential. Osmotic adjustment is also found to be key in imparting tolerance to drought (Kasturi Bai and Rajagopal 2000). Stomatal regulation has been found to be a key factor in controlling the water balance in coconut (Milburn and Zimmermann 1977; Rajagopal et al. 1986, 1988), and it is used to monitor development of stress in different coconut cultivars (Kasturi Bai et al. 1988; Juma et al. 1997; Voleti et al. 1993a). Rainfed palms had higher leaf-to-air vapour pressure deficit (*LAVPD*), leaf temperature (T_{leaf}) and leaf-to-ambient temperature difference (ΔT), whereas the irrigated palms had higher photosynthetic rate (*Pn*), Ψ_{leaf} and transpiration rates (*E*) (Rajagopal et al. 2000a). A reduction in Fv/Fm (photochemical efficiency) and Fm/Fo (quantitative indicator of the Ψ_{leaf}) with decreasing leaf water potential during stress period indicated damage to photosynthetic apparatus (Kasturi Bai et al. 2006b). Based on principal component loadings, it is found that chlorophyll fluorescence transients, viz., F0 and t½ and Ψ_{leaf}, are important traits which can be used to differentiate and screen the coconut seedlings that can adapt to water stress condition (Kasturi Bai et al. 2008).

The leaf water potential (Ψ_{leaf}), an indicator of plant water status, showed a vertical gradation from middle leaf upwards, with magnitude being higher under rainfed condition (Voleti et al. 1993b). The spindle leaf maintains higher ψ_{leaf} throughout the day irrespective of rainfed or irrigated conditions (Voleti et al. 1993b). Seasonal variations in the ψ_{leaf} occur depending on the weather, type of soil and soil water availability (Shivashankar et al. 1991; Voleti et al. 1993a). A rapid screening method was developed based on Ψ_{leaf} in excised leaflets (Rajagopal et al. 1988) for easy handling of a large number of genotypes. The Ψ_{leaf} declined with time to different degrees of stress among the genotypes, indicating the degree of tolerance/ susceptibility to stress. In general, ψ_{leaf} is lower in palms grown in red sandy loam than those in laterite soil. In irrigated conditions, the ψ_{leaf} is maintained at relatively high level even during the non-rainy period. A diurnal fluctuation in the ψ_{leaf} has been observed in coconut (Rajagopal et al. 2000a). The daytime trends in Pn and gs rates show two peaks, one at 9:00 and another at 15:00 hours under rainfed condition. But, under irrigated condition, the peaks were less distinct. The inverse relationship between Ci/Ca and instantaneous WUE (Pn/E) indicates that during rainfed growth non-stomatal limitations reduce the Pn rates compared to irrigated conditions (Rajagopal et al. 2000a). The instantaneous WUE (*Pn/E*) also had two peaks during the day in rainfed palms. In irrigated palms, instantaneous WUE was significantly higher than rainfed ones. The intrinsic WUE (*Pn/gs*) also had almost similar trends (Rajagopal et al. 2000a). Stomatal regulation has been found to be the first line of adaptive mechanisms of coconut to withstand the water deficit conditions (Rajagopal et al. 2000a). In dwarfs, low *rs* resulted in high *E* in turn lowering ψ_{leaf}. On the other hand, in talls and hybrids like WCT × WCT, high *rs* is associated with high ψ_{leaf}. Among the cultivars and hybrids, ψ_{leaf} is relatively high (−1.20 MPa) in FMS, Andaman Giant and SSAT among the talls, in COD among the dwarfs and in WCT × GBGD and WCT × WCT among the hybrids. Thus, behaviour of palms in drought conditions depends on several factors, viz., water relation components like E, gs and ψ_{leaf} and agrometeorological factors like solar radiation, rainfall and humidity, as well as soil factors.

The photosynthetic rates (*Pn*) reduced due to water stress mainly because of increase in stomatal or mesophyll resistance, with higher reduction noticed in susceptible types than in tolerant cultivars (Kasturi Bai et al. 1998). The *Pn* rate is independent of ψ_{leaf} until a threshold ψ_{leaf} is reached. The Pn/gs increased while *Pn/E* (instantaneous WUE) decreased during stress period. Drought-tolerant hybrids such as WCT × COD, LCT × GBGD and LCT × COD exhibited higher increase in *Pn/gs* ratio as well as higher WUE than that of susceptible types during stress period. The relation between *Pn/gs* and WUE is linear in both tolerant and susceptible types. The relation between *Pn/Ci* and Pn is significant and linear in both drought-tolerant and susceptible types (Kasturi Bai et al. 1998).

The potential of palms for higher DM production is reflected in WUE. WUE, determined based on dry matter accumulation (g. DM mm^{-1} water used) as well as by gas exchange measurements (μmol CO_2. $mmol^{-1}$ H_2O), ranged between 28.8 and 69.3 gDM mm^{-1} water (Kasturi Bai et al. 1996a) among the cultivars/ hybrids. The three hybrids, viz., WCT × COD, LCT × GBGD and LCT × COD and WCT, had higher WUE than the others. WUE was higher in unirrigated palms than in irrigated ones (Rajagopal et al. 1989), and under mild stress conditions, the WUE is improved in coconut juvenile palms (Rajagopal et al. 2000a). Preliminary results from carbon and oxygen isotope discrimination studies at the CPCRI in India, Coconut Research Institute (CRI) in Sri Lanka and Essex University in the United Kingdom indicated strong negative relation between carbon isotope discrimination and WUE (Anonymous 2008).

When high evaporative demand in the atmosphere prevails, cultivars exhibit differential adaptability through stomatal regulation (Jayasekara et al. 1993), which is high in hybrids followed by talls. In dwarfs, it is almost 50 % less than that in hybrids. This indicates the higher transpiration loss of water in dwarfs than in talls and hybrids. Among the hybrids studied, COD × WCT had significantly low *rs*. The transpiration rates (E) are significantly lower in WCT × WCT, FMST, JVGT, LCT × COD, LCT × GBGD, WCT × COD and WCT than in other cultivars and hybrids. Based on leaf water status, gas exchange and membrane lipids at the nursery stage, PB-121 was found to be tolerant and WAT moderately tolerant to drought among five coconut varieties (Repellin et al. 1994). Similarly,

Kasturi Bai et al. 2001 reported the extent of variation in physiological characters among the parents and hybrids at the nursery stage.

15.2.4.3.2 Membrane Stability and Osmotic Adjustment

Solute leakage differed among the cultivars and with the maturity of leaf. WCT and hybrid WCT × COD had higher stability of membranes indicating relative tolerance, while the hybrid COD × WCT had higher leakage indicating susceptibility. The membrane stability is also influenced by the potassium (K+) and phenol levels. Less K+ leakage was seen in WCT × COD as compared with the COD × WCT hybrid. Higher phenol content in leachate is seen also due to heat stress which varies among cultivars and leaf positions. Coconut palms accumulate organic solutes such as sugars and amino acid during stress period. However, during moderate stress period, solute accumulation did not differ significantly between the drought tolerant and susceptible. Accumulation of these solutes was more in WCT × WCT than COD × WCT during severe stress condition (Kasturi Bai and Rajagopal 2000).

15.2.4.3.3 Root-Shoot Signals

Roots in drying soil are known to overproduce abscisic acid (ABA) thus providing signals to shoot for closure of stomata for water regulation in plants (Zhang and Davies 1989). In coconut also, root-shoot relationship was reported to be an effective indicator of soil compaction and water stress for coconut seedlings (Nainanayake et al. 2000). A high ABA/cytokinin ratio in the leaf has positive influence on water use efficiency (WUE), whereas a high ABA/cytokinin ratio in the root has a negative influence on WUE in coconut seedlings (Kasturi Bai et al. unpublished data).

15.2.4.3.4 Role of K^+ and Cl^- Nutrition in Relation to Drought Tolerance

The K^+ and Cl^- nutrition impart drought tolerance to coconut through stomatal regulation (Braconnier and D'Auzac 1990; Braconnier and Bonneau 1998). Unlike in most of the crops where malate serves as a balancing ion for K^+, in coconut, the absence of chloroplasts in the guard cells deprives the availability of malate (Braconnier and D 'Auzac 1985). Probably Cl^- replaces malate as an osmoticum for maintaining the turgidity and thereby resisting the effects of water stress. Hence it becomes the most essential ion required for coconut growth, particularly under dry conditions. The Cl^- ion has been shown to increase water absorption and reduce transpiration by stepping up osmotic pressure within the cells in coconut (Ollagnier et al. 1983). Chlorine is important in coconut nutrition and for resistance to water stress; the critical level of Cl was identified as 0.7 % in 14th leaf (Bonneau et al. 1993, 1997). Potassium nutrition also plays important role in drought tolerance in coconut (Quencez and de Taffin 1981; Rajagopal et al. 2000b; Rajagopal and Naresh Kumar 2001). Application of KCl increased the drought tolerance of palms under dry conditions (Ollagnier et al. 1983; Rajagopal and Naresh Kumar 2001). Palms fertilized with higher levels K_2O under rainfed condition shown higher stomatal regulation (Lubina 1990). An inadequate supply of K^+ and Cl^- results in symptoms like yellowing and drying of leaves caused by an imbalance in the water relations of palms.

15.2.4.4 Biochemical Responses to Drought Stress

The biochemical response of coconut palm to drought stress includes up-regulation or synthesis of scavenging enzymes to maintain cell membrane integrity thus enabling cells to tolerate stress. Concentrations of leaf epicuticular wax, proline, reducing sugars and amino acids increased during summer and tall cultivars, viz., WCT and LCT, exhibited more accumulation of these compounds. This resulted in wider adaptability of tall cultivars to different environments (Siju Thomas et al. 2006a, b).

15.2.4.4.1 Activities of Enzymes

Decrease in leaf water status during drought stress or during prolonged dry spell caused an increase in the activities of the stress-sensitive enzymes, viz., peroxidase (PO), polyphenol oxidase (PPO), superoxide dismutase (SOD), acid phosphatase (APh) and L-aspartate: 2-oxoglutarate amino transferase (AAT), in WCT palms, while activities of malic dehydrogenase (MDH) and nitrate reductase (NR) were decreased (Shivashankar et al. 1991; Kasturi Bai et al. 1996b, 2003). Among five hybrids, WCT × SSGT was more tolerant to drought in terms of water relation components and activities of enzymes such as malic dehydrogenase, peroxidase, SOD and catalase (Kasturi Bai et al. 2005a). Thus, regulation of enzyme activities imparted tolerance to drought stress. Similar responses were observed with PEG-induced osmotic stress, except that the intensity of change was much higher (Shivashankar 1988) due to the rapid induction of stress. Increase of APh and AAT activities in coconut leaves subjected to osmotic stress is correlated with decline in ψ_{leaf} (Rajagopal et al. 1988). The threshold values of leaf water potential for sudden increase in activities of APh and AAT are −1.6 M Pa and −1.8 M Pa, respectively, indicating that APh is affected earlier in the sequence. Changes in the activity or appearance of isozymes represent the relative tolerance of coconut cultivars to water stress (Shivashankar 1988). Increased intensity of APh isozyme II shows the susceptibility of the cultivar to water deficits since APh is a hydrolytic enzyme (Shivashankar and Nagaraja 1996). In osmotically stressed coconut leaves, peroxidase isozymes change quantitatively, while PPO (Shivashankar 1988) and APh isozymes change both quantitatively and qualitatively. During stress, one of the isozymes of APh (APh II) undergoes alterations in its kinetic parameters, especially in its Km value towards PNPP and other naturally occurring phosphate compounds like ATP, NADP and glucose-6-phosphate (Shivashankar and Nagaraja 1996). The variability in the isozyme patterns of enzymes like esterase, peroxidase, phosphoglucoisomerase, alcohol dehydrogenase, glutamate oxaloacetate transaminase and acid phosphatase was also reported in coconut germplasm (Fernando and Gajanayake 1997). Soluble protein levels also increased during stress indicating increased solubilization of enzyme protein or additional synthesis in response to stress.

The NR activity decreases even when there is no perceptible change in relative water content (RWC) of leaves and the fall in activity is higher in the drought-susceptible cultivars than in the drought-tolerant cultivars. Thermal stability of NR in vivo in several coconut cultivars was found to be correlated with drought tolerance (Shivashankar 1992). An increase in the intensity of peroxidase bands could be considered as an adaptive mechanism. Two additional fast-moving bands of PPO were located in the drought-susceptible varieties under stress, while the drought-tolerant cultivars showed no change (Shivashankar 1988). It is clear that the drought-tolerant varieties are endowed with a biochemical mechanism to prevent the adverse effects of drought by appropriate regulation of enzyme activities.

15.2.4.4.2 Membrane Stability in Relation to Drought Stress

At the cellular level, the integrity of membranes is affected, and the extent of solute leakage is regulated by the membrane stability. Normal cell functions are affected due to changes in peroxidation of cell wall lipids (LP) during stress resulting in increased cell permeability and solute leakage. In coconut, lipid peroxidation was high in drought-susceptible cultivars as compared to tolerant ones (Chempakam et al. 1993). An increase in the activities of peroxidase and SOD during stress appears to be essential for maintaining the integrity of the cell, since they along with catalase protect the cell against oxidative damage. Higher levels of PPO in susceptible cultivars are due to a higher rate of cell injury, resulting in the release of PPO from plastids to the cytoplasm under induced stress (Chempakam et al. 1993; Kasturi Bai et al. 1996b) and also under water stress (Shivashankar 1988). Thus, tissue peroxidation levels are negatively correlated with activities of SOD, catalase and peroxidase, while PPO is positively correlated. In seedlings, a negative relationship between ψ_{leaf} and peroxidation of cell wall lipids has been found. As ψ_{leaf} decreased, an increase in lipid peroxidation has been observed during stress period, while relief of stress reversed this indicating recovery (Kasturi Bai et al. 2001).

Drought tolerance is thus characterized by higher activities of the protective enzymes like SOD, catalase and peroxidase and consequently coupled with lower levels of lipid peroxidation and higher membrane integrity (Shivashankar 1988; Chempakam et al. 1993; Kasturi Bai et al. 1996b). Coconut seedlings of the tolerant group maintained lower water loss and lipid peroxidation than the susceptible group and a negative correlation between leaf water potential and lipid peroxidation was observed (Kasturi Bai et al. 2001). Total leaf lipid and chloroplastic major lipid (monogalactosyldiacylglycerol) contents reduced in drought-susceptible cultivars (Repellin et al. 1997).

15.2.4.5 Molecular Markers for Drought Tolerance Traits

Studies on stress-responsive proteins in coconut indicated expression of low-, medium- and high-molecular-weight stress proteins in seedlings exposed to drought stress (Naresh Kumar et al. 2007a). Studies using RAPD and ISSR analysis indicated correlation between leaf water potentials and molecular markers in coconut cultivars (Manimekalai et al. 2004). Water stress-related MAPK genes were sequenced in coconut (Bobby et al. 2010). However, development of molecular markers for drought tolerance is still in progress. Efforts are on to link the drought tolerance to molecular diversity to find putative molecular markers, which can be useful for marker-assisted selection (MAS). Even though lack of a viable regeneration technique is a bottleneck for genetic engineering of coconut palm, the molecular markers should be identified for use in large-scale rapid screening of germplasm. This will not only increase the efficiency for selection of parental material but also will reduce the gestation period for breeding improved varieties with drought tolerance (Batugal 1999). The RFLP analysis indicated that tall and dwarf ecotypes from Pacific and Far East Asia were different from those from India, Sri Lanka and West Africa (Lebrun et al. 1998, 1999). An in vitro screening technique was developed using NaCl as the osmoticum at different concentrations in coconut embryo culture medium (Karunaratne et al. 1991).

Plants have earliest signalling pathways after perception of stimuli. The evolutionary conserved mitogen-activated protein kinase (MAPK) cascade is a major pathway by which extracellular signals are transduced into internal responses. A cDNA clone of 467 bp encoding an MAPK (designated CnMAPK1) was isolated from leaves of coconut plantlets subjected to water stress using polyethylene glycol (PEG). CnMAPK1 was significantly homologous to other plant MAPKs, viz., MPK7 and MPK5 from *Zea mays*, SIPK and MAPK6 from *Oryza sativa* and FLRS from *Triticum aestivum* (93 % identity). Coconut MAPK belongs to the serine/threonine kinases (STKs), plant TEY MAPK subfamily group A (Bobby et al. 2012) (Fig. 15.3 and Photo 15.2).

A total of 129 transcripts are differentially expressed or up-regulated during water stress in coconut seedlings. Out of ten amplified water stress-responsive candidate genes, five, viz., AP2, CBF, MAPK, NAC and 14-3-3, were up-regulated during water stress. Many of these have significant homology with closely related species in Arecaceae. From the overall experimentation, 50 highly significant gene sequences based on the similarity, identity and E value were chosen for designing functional markers, which will be used for future studies to screen coconut germplasm collection for tolerance/susceptibility to water stress. From these, 40 markers were designed from DDRT-PCR analysis and 10 markers were RT-PCR amplified. These functional markers may be used for screening germplasm and in marker-assisted selection for breeding drought-tolerant coconut cultivars (Bobby et al. 2012).

It is possible to link the in vitro and nursery screening techniques to molecular techniques for the development of molecular markers. Once the

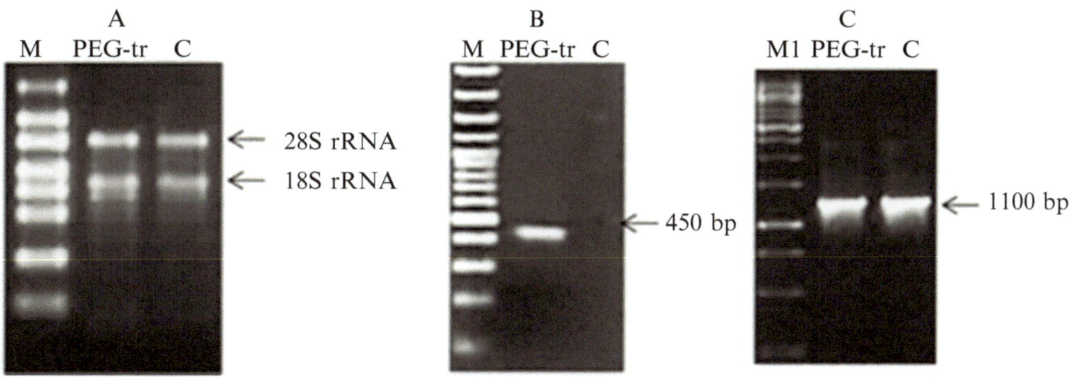

Fig. 15.3 (**a**) RNA isolation. (**b**) RT-PCR amplification of coconut transcript for MAPK. (**c**) RT-PCR amplification of coconut transcript for rbcL. M, 100 bp DNA ladder; M1, 1 kb DNA ladder; *PEG-tr*, 20 % PEG treated; *C*, control (Bobby et al. 2012)

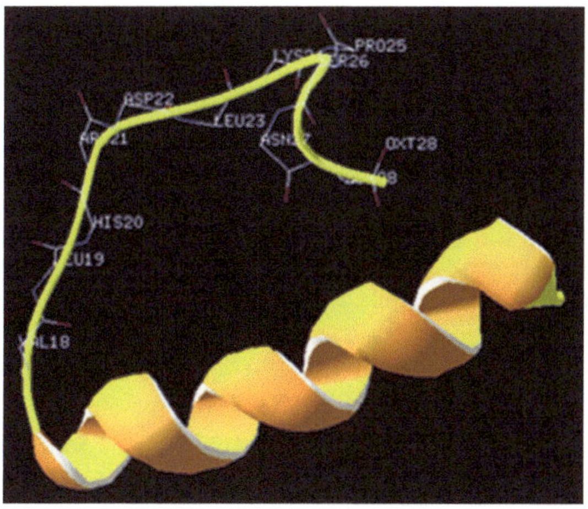

Photo 15.2 Predicted 3-D structure of coconut CnMAPK1 protein generated using MODELLER program (Bobby et al. 2012)

markers are established, they will be of prime importance to identify the parental material in breeding for drought tolerance. At the same time, it is essential that the stability of drought tolerance through pheno-phases is also to be established. Thus development of molecular markers and application of biotechnological tools for development of drought-tolerant coconut varieties needs more emphasis and concerted efforts. The future challenge is in overcoming the bottlenecks in implementation of genetic engineering for development of drought-tolerant coconut varieties.

15.2.5 Drought Tolerance Mechanism in Coconut

Based on all the above, the mechanism of drought tolerance and stability in yield of coconut under water stress conditions has been deciphered (Schematic Diagram 15.1). Drought tolerance in coconut is the cumulative effect of several inductive morphological, anatomical, physiological and biochemical mechanism (Rajagopal and Kasturi Bai 2002; Naresh Kumar et al. 2000). Genotypes possessing the above traits of drought tolerance can be used in breeding programmes. Further, the genetics of these important traits are being looked into for developing future coconut improvement strategies.

15.2.6 Coconut and Other Abiotic Stress

Apart from drought, coconut is affected by other stress like flooding, cyclone, etc. Loss due to 1996 cyclone in Konaseema led to a reduction in coconut yields by about 3350 lakh nuts/year for 6 years. The loss in East Godavari district alone was to the tune of about 2200 lakh nuts/year in 6 years. The productivity was reduced by 6200 nuts/ha/year in E. Godavari District and by ~4100 nuts/ha/year in AP. Similarly, a supercyclone in Orissa severely damaged the

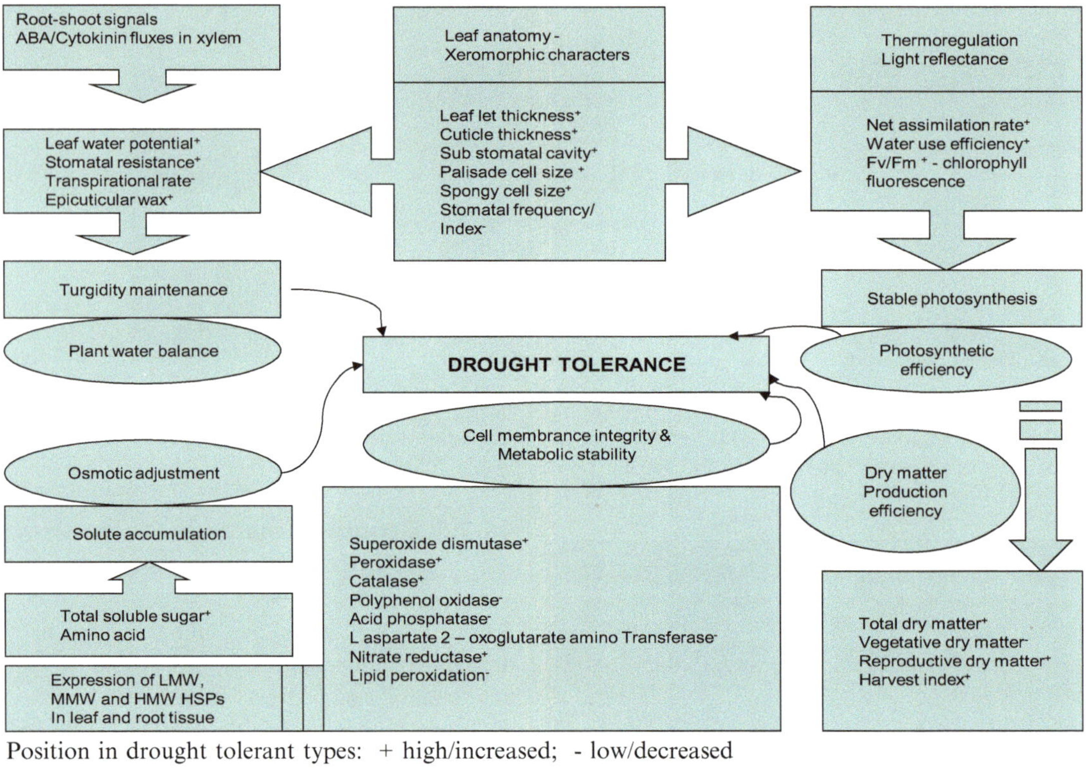

Schematic Diagram 15.1 Adaptive strategies of coconut palm under stressful conditions

yield (Naresh Kumar et al. 2008) (Fig. 15.4 and Photos 15.3 and 15.4, Table 15.5).

15.2.6.1 Flood-Affected Palms

Coconut palms generally are not affected by short-term (10–15 days) flooding but if water stagnation prolonged, palms suffer physiological drought, a condition where palms will not be able to uptake the water or nutrients due to hampered root activity as a consequence of lack of oxygen for root respiration.

15.2.6.2 Abiotic Stress Responsive Protiens in Coconut

Coconut seedlings of different cultivars and cross combinations subjected to stresses like water, high temperature and flooding stresses independently showed an increase in the concentration of heat-stable protein fraction (HSPF) in leaf tissue even when the total protein concentration reduced due to stress (Naresh Kumar et al. 2007). The percentage of HSPF in leaf tissue increased with the decrease in leaf water potentials (Table 15.6). Quantitative changes in proteins also were observed in leaf and root tissue due to temperature and flooding. In HSPF, proteins of ~66 KDa are found in root and leaf tissue. Leaflet tissue have specific proteins in the range of ~10 and ~14 KDa which are not present in the root tissue. Two extra proteins of 66 KDa and ~76 KDa appeared in water-stressed WCT seedlings. They were also present in seedlings exposed to high light intensity (~1,500 mmol/m^2/s) apart from an extra LMW HSP of 14.4 KDa. Among the MMW proteins, a protein of 53 KDa is present in all WCT samples. The LMW protein of 20.1 KDa present in non-stress seedlings disappeared during water stress period. In temperature-induced and flooded leaflet tissue, new proteins (LMW range ~17 KDa) were observed. In temperature-induced root tissue, a protein band in the range of 30 KDa was observed but not in flooded root or in control root. Results indicate quantitative and qualitative variations in stress proteins in coconut seedlings

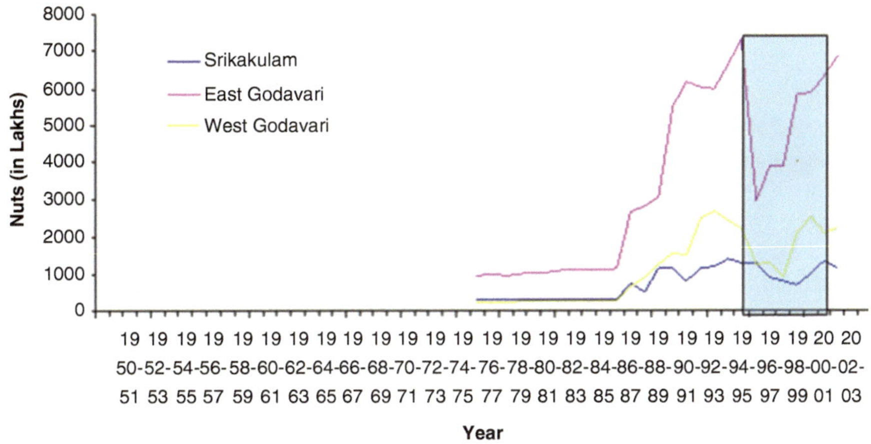

Fig. 15.4 Trends in coconut production in three districts of AP

Photo 15.3 Cyclone affected palms

subjected to different abiotic stresses (Naresh Kumar et al. 2007).

15.2.6.3 Response of Coconut to Elevated CO_2 and Temperature

Elevated CO_2 (550 and 700 umol/mol) and temperature conditons caused an increase in the activities of superoxide dismutase and catalase, whereas they reduced polyphenol oxidase activity. On the other hand, peroxidase (POX) activity decreased in elevated temperature, while it increased under elevated CO_2 conditions. The POX activity and membrane stability index (MSI) were positively correlated. By virtue of greater MSI and lower MDA content, WCT and COD $\times$ WCT were rated to be tolerant to oxidative stress among the seedlings of three cultivars, viz., WCT, LCT and COD, and two

Photo 15.4 Flood affected palms

Table 15.5 Loss in nut production in AP due to 1996 cyclone

Year	Srikakulam	E. Godavari	W. Godavari	AP state
1996–1997	−28.9	−4,245.0	−1,203.7	−5,753.4
1997–1998	−456.3	−3,400.0	−1,089.2	−4,943.8
1998–1999	−604.5	−3,313.4	−1,332.0	−5,726.3
1999–2000	−706.9	−1,289.2	17.9	−2,019.1
2000–2001	−406.0	−963.6		−1,232.4
2001–2002	−71.4	−208.0		−394.3
Loss in 6 years lakh nuts	−2,274.17	−13,419.3	−3,607.06	−20,069.5
Loss lakh nuts/year	−379.0	−2,236.5	−901.7	−3,344.9

hybrids, viz., WCT × COD and COD × WCT, under elevated CO_2 and temperature conditions. These cultivars may adapt better to changing climates (Naresh Kumar et al. 2008; Sunoj et al. 2014). After exposure to elevated CO_2 and temperature treatments for about 2 years, the elevated CO_2 increased the accumulation of proline in leaf tissue in spite of maintenance of high leaf water potentials (Fig. 15.5). On the other hand, elevated temperature caused only a slight increase in proline concentration, despite the reduced leaf water potentials (Fig. 15.6). These results suggest a greater role for proline in coconut adaptation to high CO_2 concentrations (Sunoj et al. 2009a; 2013). Elevated CO_2 and temperature reduced total phenolic concentration in majority of cultivars during both the seasons (Fig. 15.7). However, in WCT, a popular cultivar extensively grown in Kerala, India, the concentration of total phenolic compounds increased in elevated CO_2 and T conditions. On the other hand, reduction in concentration of phenolic compounds is more in COD, a dwarf cultivar. The effect of elevated temperature in reducing the concentration of phenolic compounds is found to be more during the post-monsoon period than in the pre-monsoon period. Overall results indicate that in climate change scenarios, WCT is likely to remain relatively more tolerant while COD, LCT and COD × WCT may become more predisposed to pests and

Table 15.6 Quantitative and qualitative changes of proteins in coconut seedlings subjected to different stresses

Type of stress on seedlings/tissue	Quantitative and qualitative changes in proteins		
	Protein concentration	HSPF	Extra proteins during stress (kDa)
Water stress: leaflet	Decrease	Increase	116, 76, 30, 14.4
Water stress + high light intensity: leaflet	Decrease	Increase	116, 76, 66, 20.1, 14.4
Temperature: leaflet	Increase	Increase	17
Root	Decrease	Increase	30
High temperature:			
leaflet	Decrease	Decrease	–
Root	Decrease	Decrease	–
Flooding: leaflet	Increase	Increase	17
Root	Decrease	Decrease	17

Source: Naresh Kumar et al. (2007)

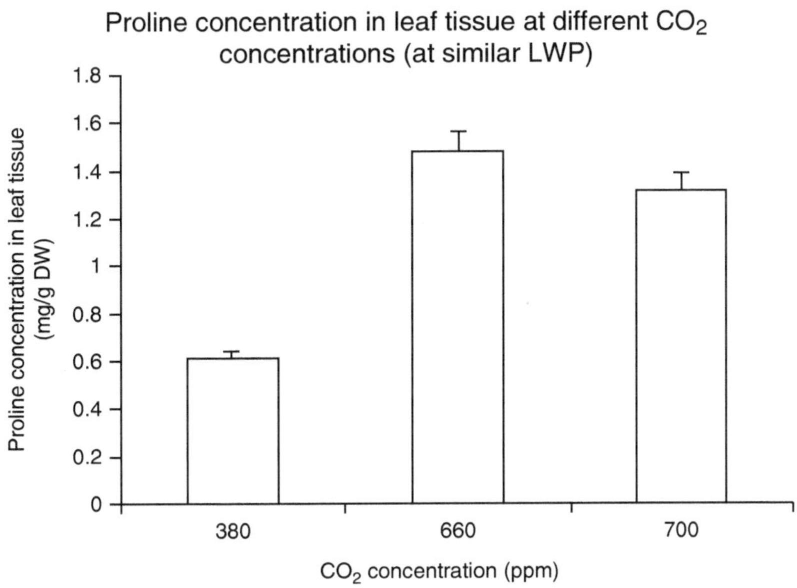

Fig. 15.5 Proline concentration (mg/g DW leaf tissue) in coconut seedlings grown at three atmospheric CO_2 concentrations. The leaf water potential of seedlings was similar at ~ −7.2 bars

diseases (Naresh Kumar et al. 2008; Sunoj et al. 2013). Rhizosphere enzymatic activity also varied in elevated CO_2 and temperature conditions (Sunoj et al. 2015).

Even though anatomical changes provide plants with long-duration or lifetime adaptation thereby bringing about stability to adaptation of plants, these parameters are less studied. Studies indicated that (1) elevated CO_2 and temperature reduced stomatal aperture area, stomatal complex area and guard cell area, (2) elevated T significantly reduced stomatal index and leaflet thickness, (3) elevated CO_2 and T increased the thickness of upper cuticle and diameter of meta xylem and (4) elevated CO_2 increased thickness of leaflets and of the upper epidermal layer (Table 15.7 and Photo 15.5). Some of the parameters such as stomatal index was not significantly influenced by elevated (CO_2), while thickness of lower cuticle and upper epidermal layer was not significantly influenced by elevated (T). However, there exists a variable response of anatomical features of cultivars and hybrids to these external factors. These anatomical modifications in coconut seedlings are likely to make them more adapted to climate change situations. Results indicate that coconut seedlings have the capacity to adapt to future

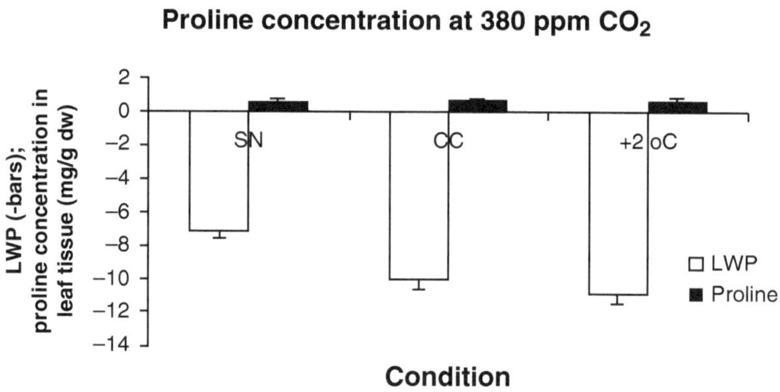

Fig. 15.6 Proline concentration (mg/g DW leaf tissue) and leaf water potential of coconut seedlings grown in different conditions (shade net (SN), control open top chamber (CC) and elevated temperature (+2 °C over ambient temperatures) at atmospheric CO_2 concentrations (380 ppm))

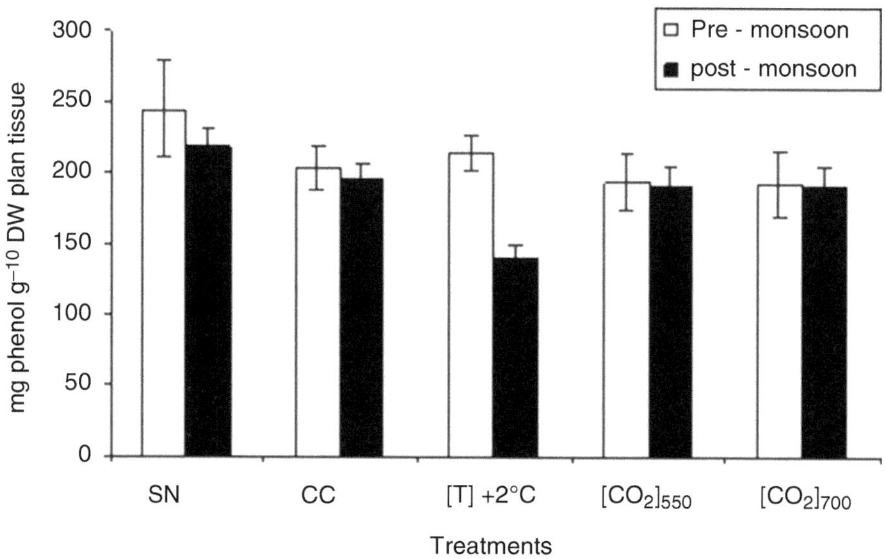

Fig. 15.7 Variation in total phenol concentration (mg phenol g^{-1} dry plant tissue) in coconut seedlings under elevated (CO_2) and (T) during pre- and post-monsoon seasons

climate by modifying their leaf anatomy (Naresh Kumar et al. 2008; Murali Krishna et al. 2009, 2013).

15.2.6.4 Photo-oxidative Stress in Coconut Seedlings

The main cause for damage due to high light intensity is photo-oxidative stress, and studies on chlorophyll fluorescence indicated a clear case of excess light energy under high light conditions causing stress to coconut seedlings (Naresh Kumar and Kasturi Bai 2009b). Quantum yield of photochemistry of leaflets exposed to high light was significantly less than those under shade. Seedlings exposed to high light and then shifted to shade have shown significant improvement in quantum yield. The Fv/Fm ratios indicated stress in seedlings under high light intensity. Excess light energy harvested by chlorophyll antenna caused high non-photochemical quenching resulting in production of biologically toxic super oxide, hydrogen peroxide and hydroxyl radicals, which damaged the cell membrane integrity as indicated by increased lipid peroxidation (Fig. 15.8) and chlorophyll bleaching. It is apparent that photoinhibition of

Table 15.7 Influence of elevated temperature and CO_2 on leaflet anatomical parameters of coconut seedlings. Mean values of five cultivars. Each datum is a mean of 100 observations

Parameter	Shade net	Chamber control	Elevated temperature (+2 °C)	Elevated (CO_2) 550 ppm	Elevated (CO_2) 700 ppm	CV (%)	CD at 5 %
Leaf stomata aperture area (μm^2)	41.6	37.7	31.5	23.5	17.8	18.5	0.63
Stomata index (%)	10.2	9.8	8.5	9.4	9.8	8.1	0.09
Leaflet thickness (μm)	178.0	147.9	141.9	173.9	165.6	6.1	1.10
Hypodermis cell length (μm)	17.9	15.0	16.1	17.3	17.0	18.7	1.73
Leaf lower cuticle thickness (μm)	2.9	2.2	2.3	2.6	2.6	15.1	0.04
Leaf upper cuticle thickness (μm)	3.4	2.2	2.8	3.3	2.8	12.7	0.04
Leaf lower epidermis cell thickness (μm)	5.8	5.0	5.1	5.4	5.1	13.3	0.08
Leaf upper epidermis cell thickness (μm)	6.2	5.2	5.3	6.3	5.9	15.2	0.10
Leaf metaxylem vessel lignification (μm)	1.9	1.6	1.7	1.7	1.6	15.1	0.03
Leaf metaxylem vessel diameter (μm)	32.1	26.5	31.0	30.2	28.8	8.2	0.46

photosynthesis takes place at (1) PSII down-regulation and (2) damage to PS II system in initial stages of light exposure, and under prolonged exposures, inhibition is caused due to (3) chlorophyll bleaching and (4) damage to chloroplast and cell membrane integrity, followed by reduction in photosynthetically active leaf area due to scorching thus reducing canopy photosynthesis (Fig. 15.9, Tables 15.8 and 15.9) (Naresh Kumar and Kasturi Bai 2009b). Protein concentration in leaf tissue was higher in seedlings under exposed conditions. Three distinct LMW proteins were found in seedlings exposed to high light intensities. These three LMW proteins have pI of 4.9, 8.4 and 10.15 with MW less than 20 kDa. Results clearly demonstrate the events that take place at early stage to subsequent cascading effects leading to the scorching and death of leaf under severe conditions (Naresh Kumar and Kasturi Bai 2009b). Coconut seedlings are sensitive to moisture stress and to high light intensity stresses. The seedlings when exposed to high light intensities experience photo-oxidative stress. Release of oxygen and superoxide radicals causes membrane damage thus causing the leaf scorching and, in severe cases, seedling death (Naresh Kumar et al. 2007). Studies indicated that the seedlings need to be supplied with adequate water and maintained under shade. Even field-planted seedlings need to be protected from high sunlight intensities, which may be by tying the leaves together or by providing a shield of dried coconut leaves around the coconut seedling canopy, which is tied. Further, field-planted seedlings need to be irrigated not only for better establishment but also to protect them from moisture stress. Moisture stress in combination with high light intensity causes the seedling death.

15.2.7 Approaches for Enhancing Drought Tolerance in Coconut

Drought is a major constraint for coconut productivity. Thus, screening of germplasm for drought tolerance attracted the research efforts and those led to generation of a lot of scientific knowledge. However, the breeding for drought

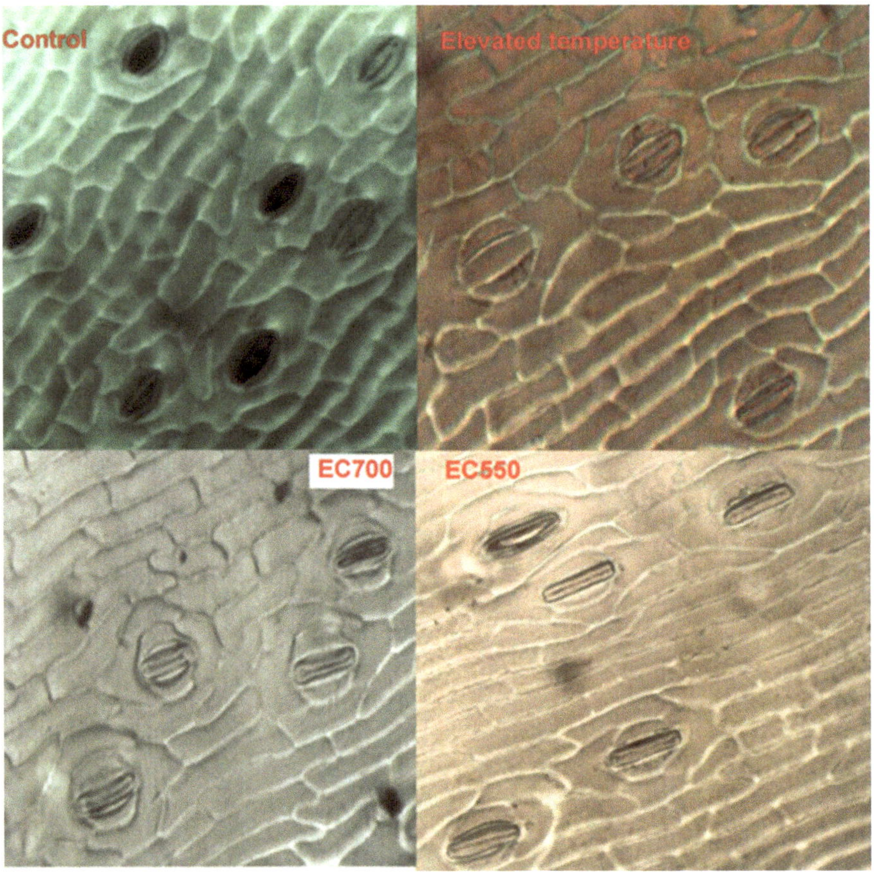

Photo 15.5 Epidermal impression showing the stomatal response of coconut leaf to elevated (CO_2) 550 and 700 and temperature (+2 °C over control OTC) conditions (Murali Krishna et al. 2013)

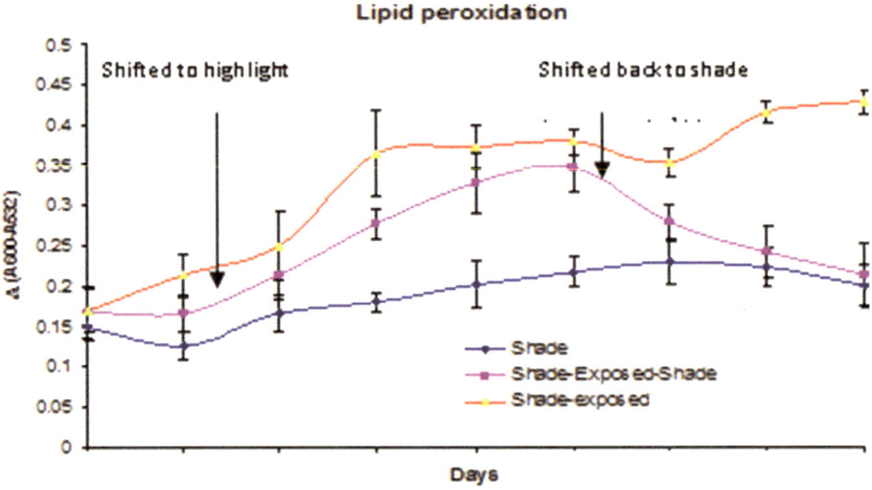

Fig. 15.8 Lipid peroxidation of leaflet under different light conditions

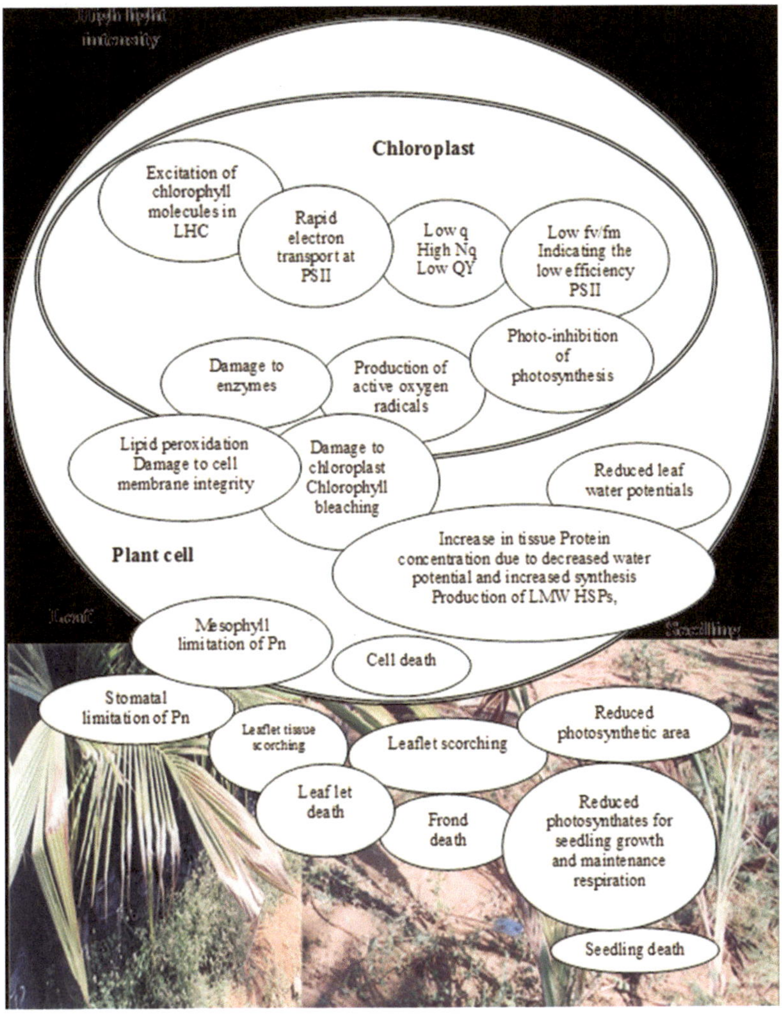

Fig. 15.9 Mechanism of photo-oxidative stress in coconut seedlings: early events to seedling death (Source: Naresh Kumar and Kasturi Bai 2009b)

tolerance in coconut started only recently. Recent encouraging results on regeneration techniques further the genetic engineering approach to impart drought tolerance in high-yielding cultivars.

Classical approach to improve crop performance in water-limited environments is to select the genotypes that survived and have an improved yield in these environments. Another approach is identification and selection of traits that contribute to drought tolerance and high water use efficiency. Identification of in situ tolerant palms and using their selected progeny for population improvement in plantations are possibly the most promising and stable method.

15.2.7.1 Screening for Drought Tolerance in Coconut

Drought-tolerant coconut palms can be selected at seedling stage in a nursery and at adult palm stage. Apart from these, one can use an in vitro screening technique as well. The screening of coconut germplasm can be done using morphological, anatomical, physiological and biochemical traits. Further, molecular marker-assisted selection criterion is one, which is to be developed. It is essential to note that one has to develop the threshold levels for development of stress in given climatic and soil conditions. The functional relationship in mechansim of

Table 15.8 Physiological and biochemical characteristics of leaflet under different light conditions

Parameter		Condition: Shade	Shade-exposed-shade	Shade-exposed	CD at $P = 0.05$
Specific leaf weight after treatments	FW (mg/cm^2)	20.57	20.03	20.91	NS
	DW (mg/cm^2)	6.77	6.69	7.84	0.23
Water in tissue (%)		67.07	66.66	62.39	2.33
Leaf water potential (bars)		−8.02	−9.91	−10.67	1.2
Chlorophyll components (mg/g FW)	Chl. a	1.00	0.91	0.58	0.11
	Chl. b	0.43	0.36	0.27	0.08
	Total chl.	1.43	1.26	0.85	0.14
	Chl. a/b ratio	2.54	2.31	2.12	0.09
Variation over shade (%)	Chl. a		−9.5	−42.2	5.21
	Chl. b		−17.0	−36.3	4.31
	Total chl.		−11.82	-40.30	4.02
	Chl a/b ratio		−8.9	−16.5	3.22
Leaf proteins (mg/g FW)		9.45	10.9	12.4	0.52
	Increase over shade (%)		15.34	31.22	5.32
Functional leaf area (m^2)		2.02	1.59	0.96	0.31
	Decrease over shade (%)		−21.29	−52.48	10.6
Rate of degradation of leaf area m^2/month		0.25	0.75	1.20	0.16
	Increase over shade (%)		200.00	380.00	56.00

Table 15.9 Chlorophyll fluorescence, gas exchange and microclimate parameters of leaflets of the same frond under different light conditions

Parameter		Condition: Shade	Exposed	CD at $P = 0.05$
Chlorophyll Fluorescence	Quantum yield	0.46	0.15	0.06
	ETR	24.37	79.11	12.3
	QP	0.42	0.15	0.92
	QN	0.86	0.94	0.03
Photosynthetic	Photo (μmol/m^2/s)	3.42	1.21	0.42
	Cond (mol/m^2/s)	0.05	0.06	NS
	Ci (Pa)	106	185	23.2
	E (mmol/m^2/s)	1.12	2.23	0.26
	Pn/E (μmol/mmol)	2.13	1.07	0.29
	Pn/gs (μmol/mol)	75.4	45.3	14.6
Microclimatic	PAR (μmol/m^2/s)	350	1,300	–
	VpdL (kPa)	2.41	2.78	0.11
	T_{leaf} (°C)	31.15	34.87	0.59
	T=leaf-air (°C)	2.962	4.652	0.35

drought tolerance should form the basic criteria for selecting screening method in coconut breeding strategy (Schematic Diagram 15.2).

15.2.7.2 Traits for Drought Tolerance

Cell size and number, sub-stomatal cavity size, stomatal frequency, epicuticular wax content and thickness, leaf thickness, stomatal resistance,

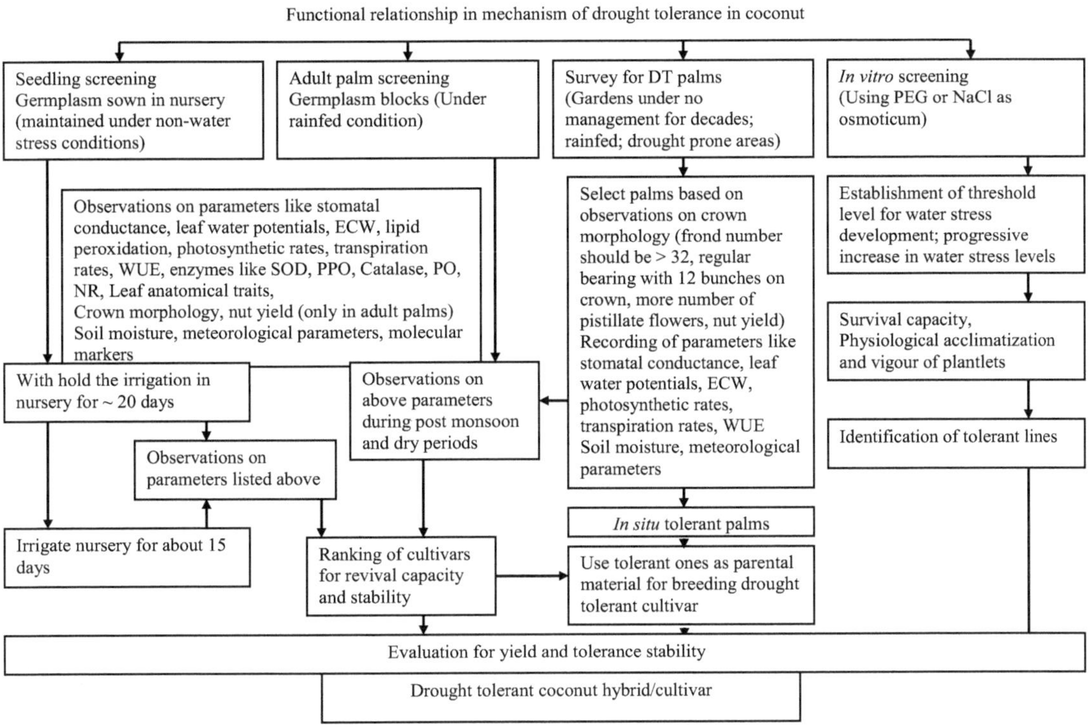

Schematic Diagram 15.2 Screening for drought tolerance in coconut

water potential components, cell membrane stability, water use efficiency and activity levels of scavenging enzymes are the essential anatomical, physiological and biochemical traits for assessing moisture stress tolerance in plants (Rajagopal et al. 1991; Kasturi Bai 1993; Champakam et al. 1993; Shivashankar et al. 1991; Naresh Kumar et al. 2000; Siju Thomas et al. 2006, 2008). Based on these, coconut germplasm collections comprising talls, dwarfs and hybrids were screened under field conditions for drought tolerance (Rajagopal et al. 1990).

15.2.7.3 Ranking of Cultivars for Drought Stress Tolerance

The ranking for drought tolerance was done based on all stress-sensitive parameters, viz., stomatal regulation, leaf water potential and ECW content (Rajagopal et al. 1990), and on biochemical traits, viz., lipid peroxidation, polyphenol oxidase, superoxide dismutase, catalase and peroxidase (Chempakam et al. 1993), as well as the anatomical characteristics (Naresh Kumar et al. 2000). All dwarf (s) performed badly, whereas all hybrids except COD × WCT and all tall(s) except the SSAT, ADOT and LMT were highly ranked. Based on anatomical features such as thicker leaflets, thick cuticle on sides, larger palisade and spongy parenchyma cells, larger hypodermal cells, water cells and sub-stomatal cavity, genotypes like WCT, FMS and PHOT and WCT × COD hybrid were identified as relatively tolerant to drought stress (Naresh Kumar et al. 2000). Two cultivars, viz., San Ramon and Ambakelle Special, were identified as drought tolerant in Sri Lanka (Wikremaratne 1987). In Ivory Coast, PB-121 was identified as tolerant while WAT was classified as moderately tolerant and Rennell Tall × WAT as the most sensitive to drought based on the drought tolerance index and effect of edaphic drought on the leaf water status, gas exchange and membrane lipids (Pomier and de Taffin 1982; de Nuce de Lamothe and Benard 1985; Repellin et al. 1994). Thus, coconut cultivars with different levels of drought tolerance could

Photo 15.6 In situ tolerant palm

be identified based on the desirable traits, which reflect on the overall water relations of palms. Presence or absence of desirable traits imparts higher degree of drought tolerance (e.g. WCT × WCT; FMST; LCT; WCT × COD, LCT × GBGD and LCT × COD) or drought susceptibility (e.g. MYD) (Rajagopal et al. 2000a). Although the expression of physiological traits are influenced by weather variables (Kasturi Bai et al. 1988; Rajagopal et al. 2000a; Gomes et al. 2002), diversity analysis in coconut varieties for drought-responsive physiological traits indicated that both *rs* and ECW are important variables, which contribute to diversity. This implies that in coconut, grouping could also be formed based on their tolerance to stress when appropriate traits are specifically used in diversity analysis Kasturi Bai et al. (2006a).

15.2.7.4 Identification and Characterization of In Situ Tolerant Palms

The plants that can withstand the natural occurrence of drought and other stresses and still produce good yield are of premium value, as they may possess the desirable genes. Surveys in hotspot areas were conducted to identify the palms yielding very high compared to others in their vicinity. Palms at different agroclimatic regions were identified in farmers' plots with desirable canopy shape and leaf number with better yield. The physiological water use efficiency of these palms was also found to be high (Photo 15.6, Table 15.10) (Naresh Kumar et al. 2002b). This type of in situ tolerant plants with desirable traits should be used in breeding programmes, which will help in reducing the time gap in breeding for drought-tolerant cultivars in coconut (Naresh Kumar et al. 2002b).

15.2.8 Genetics of Drought Tolerance-Related Physiological and Biochemical Traits

Coconut cultivars with desirable characters were selected and crossed in a 2 × 4 line x tester mating design to study the combining ability and gene action with respect to drought-responsive physiological traits (Rajagopal et al. 2007). Seedling transpiration rate and leaf water potential showed higher specific combining ability (sca) effects than general combining ability (gca) effects due to predominance of

Table 15.10. Physiological, biochemical and morphological characters of in situ drought-tolerant palms (mean of 2 years data collected on palms identified in farmers' plots under different agroclimatic zones – adopted from Naresh Kumar et al. 2002b)

Parameter	In situ tolerant	Susceptible palms
Physiological		
Net photosynthetic rates (μmol/m^2/s)	6.51 ± 0.39	4.68 ± 0.49
Transpiration rates (mmol/m^2/s)	2.43 ± 0.14	2.08 ± 0.18
Stomatal conductance (mol/m^2/s)	0.071 ± 0.005	0.056 ± 0.006
Instantaneous physiological water use efficiency (μmol $_{CO_2}$/mmol $_{H_2O}$)	2.67 ± 0.12	2.25 ± 0.11
Intrinsic water use efficiency (μmol $_{CO_2}$/mol $_{air}$)	0.10 ± 0.004	0.097 ± 0.005
Leaf-to-air temperature difference (°C)	3.43 ± 0.19	3.62 ± 0.24
Biochemical		
Total carbohydrates (mg/g dry tissue)	137.0 ± 6.4	140.5 ± 7.1
Reducing sugars (mg/g dry tissue)	20.7 ± 3.5	19.7 ± 2.7
Starch (mg/g dry tissue)	55.0 ± 2.86	53.2 ± 3.2
Total proteins (mg/g dry tissue)	211.7 ± 5.5	210.2 ± 8.2
Free amino acids (mg/g dry tissue)	12.24 ± 0.7	11.8 ± 0.5
Morphological and yield		
Number of leaves on crown	32 ± 1.6	27 ± 1.5
Number of bunches	13 ± 0.7	11 ± 1.0
Pistillate flower production/bunch	28 ± 7.5	18 ± 3.7
Mature nuts/bunch	10 ± 1.3	4 ± 1.2

nonadditive gene action indicating heterosis for this character. The *Pn* under stress was additive with good combining ability, while the *Pn* during non-stress and recovery were governed by non-additive gene action that could be exploited for heterosis. In case of lipid peroxidation, gene action was unpredictable in non-stress with additive gene action being nil with low dominance. These indicate that the nature of gene action governing drought-sensitive traits can be exploited by selecting proper breeding strategies for drought tolerance.

In changing climates, efficient management of available water is very important for sustainable crop production. Growing cultivars with high water use efficiency is even more important in the case of coconut, a perennial plantation crop mainly grown under rainfed and marginal conditions which faces annual summer stress. Drought tolerance research at the Central Plantation Crops Research Institute (CPCRI) in Kasaragod, India, indicated variability for WUE, dry matter production and yield in coconut cultivars (Rajagopal et al. 1989; Kasturi Bai et al. 1996a; Naresh Kuamr et al. 2000). Seedling stage screening of 75 cross combinations, reciprocal crosses, and their parents for drought tolerance and revival capacity at CPCRI indicated that in general, tall(s) and hybrids with tall(s) as mother palms had higher drought tolerance compared to dwarf(s) and dwarf as a mother palm. Heterosis was observed for some of the desirable characters for drought tolerance. Earlier studies (Rajagopal et al. 2000a) indicated the possibility of exploitation of heterosis of some of the drought-tolerant traits in evolving the drought-tolerant hybrids.

15.2.9 Drought Management

In the rainfed areas, drought is the major constraint for the crop productivity, more so in coconut since they are widely grown in different soil types such as sandy, sandy loam, laterite and forest soils in the states of Kerala, Karnataka, Tamil Nadu, Andhra Pradesh, West Bengal, Orissa, Maharashtra, northeastern states, Andaman and Nicobar Islands and Lakshadweep Islands of India. As this is mainly grown under rainfed condition, productivity is affected due to the dry summer months starting from December/

January to April/May. During the period, the soil water deficit coupled with increase in atmospheric water demand aggravates the situation leading to soil as well as atmospheric drought. Ideally, this is the time when they should receive adequate water supply in order to get better yields. Being perennial in nature, the water requirement is also fairly high and the approach then has to be to use the available water source with high productivity efficiency. Thus it is important not only to identify the varieties, which can withstand moisture stress conditions in the field, but also to evolve management strategies for conserving available water sources in order to mitigate adverse effects of drought. The drought management strategies and low-cost soil moisture conservation practices using in situ biowaste is recommended for each agroclimatic zone where coconut is grown (Naresh Kumar 2004; Naresh Kumar and Rajagopal 2005).

15.2.9.1 Water Management Through Irrigation Scheduling and Soil Moisture Conservation

The frequency and amount of irrigation influence the water relations and DM production of coconut palms (Rajagopal et al. 1989; Kasturi Bai et al. 1997). Summer irrigation (Nelliat and Padmaja 1978) and soil moisture conservation practices like husk burial in basins, leaf mulching, *Gliricidia* culture and application of compost and farm waste in basins increase nut yield in coconut (Rajagopal et al. 2000b; Rajagopal and Naresh Kumar 2001). The studies on the extent of influence of irrigation on coconut palms grown on sandy and laterite soils with different levels of irrigation through drip [in sandy, 66, 100 and 133 % of open pan evaporation (Eo); in laterite, 33, 66 and 100 % of Eo and basin (100 % of Eo)] indicated that the source parameters, viz., photosynthetic rate, Ψ_{leaf} and photosynthetic II (PS II) efficiency, varied with the irrigation level and soil type. Palms receiving irrigation also showed marked improvement in pistillate flower production and their retention to produce higher nut yields. Three types of physiological conditions in source and sink relationships were observed in the palms based on the type of irrigation treatment they were subjected to (Naresh Kumar et al. 2002a). The drip irrigation provided conditions for better physiological efficiency of source and sink for high WUE and yield. WUE was found to increase at field, plant and leaflet level (Naresh Kumar et al. 2002a).

The net photosynthetic rates were higher in the palms grown in laterite soils compared to those grown in sandy soils, and irrigation significantly increased the Pn rates and stomatal conductance. The chlorophyll fluorescence, PS II efficiency parameter Fv/Fm, an indicator of extent of physiological stress in leaf, has been found to be higher in irrigated palms compared to the rainfed palms (Naresh Kumar et al. 2002a). The pistillate flower production and setting percentage normally increase with irrigation. The percentage of nut retention is an indicator of efficiency of conversion of pistillate flowers into mature nuts. Percent increase in nut yield in irrigated palms over rainfed palms was higher in sandy soils compared to laterite soils (Naresh Kumar et al. 2002a). The consequent increase in yield could be related to the increases in source (the Pn rates) and sink (pistillate flower production) efficiency under irrigation.

The physiological conditions of source and sink in palms grown under different systems of irrigation are defined (Table 15.11) (Naresh Kumar et al. 2002a). The drip irrigation is a system where not only the available water is used to the optimum with negligible losses but also because of the presence of dry zones in the root system possibly acting as the stomatal regulation system to provide optimal physiological efficiency for higher WUE and better yields. Drip irrigation increases the WUE not only at a field level but at a plant and leaf level also. From the study, it is indicated that even in basin irrigation, by applying water in such a way that the dry pockets are created in the root system, it may be possible to increase WUE with high yields (Naresh Kumar et al. 2000) (Photo 15.7).

Soil moisture conservation also improved the yields by improving the source-sink efficiencies (Naresh Kumar et al. 2003, 2006). Results of a multilocation experiment indicated that the soil moisture conservation practices helped to retain and make soil moisture available for longer

Table 15.11 Summary of physiological conditions of source and sink as influenced by the type of irrigation

Source	Sink	Condition	Yield/ WUE	Remarks
Low Ψ_{leaf}, *E*, *gs* and *Pn*	Less pistillate flower production, nut retention	Rainfed	Low/ low	Less available water in root zone
Low Ψ_{leaf}, high *E*, *gs* and *Pn*	More pistillate flower production, nut retention	Basin irrigation	High/ low	Adequate availability of water; no dry pockets in root zone
High Ψ_{leaf}, medium *E* and *gs* and high *Pn*	More pistillate flower production, nut retention	Drip irrigation	High/ high	Availability of water in optimum; dry pockets in root zone

Source: Naresh Kumar et al. (2002a)

Photo 15.7 The plantations with drip irrigation (*left*) survived the drought (1998–2002) while those without timely irrigation failed to cope (*right*) with drought in Pollachi area of Tamil Nadu, India

periods during summer (Photos 15.8 and 15.9). This helped to maintain higher photosynthetic rates and water use efficiency thus helping improve pistillate flower production, nut retention and yield (Table 15.12; Naresh Kumar et al. 2003). The prolonged soil moisture availability due to soil moisture conservation treatments helped to increase the photosynthetic source number and efficiency as also the sink capacity and efficiency (Naresh Kumar et al. 2002b, 2006).

15.3 Areca Nut

Unlike scientific information on coconut, the information on areca nut is very meagre as far as the abiotic stress tolerance is concerned. The cultivation of areca nut is mostly confined to 28° north and south of the equator. In general, areca nut is mainly grown in low altitudes. This to some extent depends on the latitude. In the north-east region of India (Assam and West Bengal), the major areca nut-growing area is in plains, since at higher elevations the winter temperature would be too extreme for the crop. Although areca palms grow at an altitude of up to 1000 m above MSL, at higher altitudes, the quality of the fruits will not be good. In the high-altitude areas, the endosperm of the fruit does not develop sufficient hardiness. It is also reported that high altitudes affect the germination of seeds and quality of marketable dry kernel. The percentage

Photo 15.8 Soil moisture conservation practices in coconut. Polyphenol depeleted coirpith was used. Generally, one year Sun and rain exposed coirpith is phenol depleted and can be used for mulching

Photo 15.9 Soil moisture conservation practices in coconut using coconut husk

of germination of nut and the proportion of dry weight of kernel to whole fruits are less at altitudes above 850 m than in the lower altitudes.

The yield potential of areca palm mainly depends on climate. Generally more than 50 % of variations in yield are due to climatic differences. Though the areca nut is grown under different agroclimatic conditions, it is very sensitive to extreme climatic conditions. The most important climatic factors that influence the growth and development are altitude, relative humidity and rainfall. Temperature also

Table 15.12 Effect of drought management practices on nut yield at different centres (mean for 5 years)

Centre/treatment		Nut yield/palm/year	
		Pretreatment	Post-treatment
Arsikere	Rainfed	48	50
	Treatments[a]	55	66
	Irrigated	119	127
Kidu	Rainfed	60	62
	Treatments	62	69
	Irrigated	92	105
Veppankulam	Rainfed	52	70
	Treatments	51	89
	Irrigated	116	130
Ratnagiri	Rainfed	68	68
	Treatments	78	89
	Irrigated	112	115
Ambajipeta	Rainfed	100	97
	Treatments	105	117
	Irrigated	140	152

Source: Naresh Kumar et al. (2003)
[a]Soil moisture conservation treatments

influences the crop growth and yield to some extent. Regression analysis of weather variables of 12 years indicated that areca nut yield is influenced by relative humidity, evaporation and rainfall.

Areca palms grow well within the temperature range of 14–36 °C. However, the crop is being grown in temperatures ranging from 5 to 40 °C. But extreme temperatures and wide diurnal variations are not conducive for the healthy growth of the palms. Heavy damage to foliage and death of palms is reported when the minimum temperature was below −2.8 °C. Even temperature around 5 °C with low humidity causes severe foliage damage. Though areca nut flourishes well in tracts of heavy rainfall, it is grown in areas with wide variations in annual rainfall ranging from 750 mm to more than 4,500 mm. In low-rainfall areas, the palms are irrigated. Very high or low relative humidity is not conducive for growth and development of areca nut. Relative humidity directly influences the water relations of palm and indirectly affects leaf growth, photosynthesis, pollen dispersal, occurrence of diseases and finally economic yield. Relative humidity has considerable influence on evapotranspiration and hence on the water requirement of areca nut. Areca nut plantations also are affected by drought stress causing a loss ranging from 10 to 50 % depending on the duration and severity of stress. The loss due to drought was more severe in nontraditional areas of areca nut cultivation than in traditionally grown areas. Once affected, palms require 2–3 years to recover to normal yield condition. In severe cases, deaths of palms are reported from parts of Karnataka, India. Similarly, areca nut is affected by the high light intensities not only palms at seedling stage but also the grownup palms. Palm trunks get scorching damage due to high light intensities.

15.3.1 Management of Abiotic Stresses in Areca Nut Plantations

As in the case of coconut, providing drip irrigation, soil moisture conservation and raising the drought-tolerant areca nut varieties are the commonly recommended strategy for overcoming the drought stress in areca nut. To overcome the high light intensity, the seedlings are provided shades, and the trunks are covered with dry leaves to avoid direct incidence of sunlight (Photo 15.10).

Photo 15.10 Trunk protection against direct sun light in arecanut

15.4 Constraints and Opportunities

Abiotic stresses constrain the productivity of plantation crops. Drought is a major constraint for coconut productivity in entire coconut-growing areas at a global level. Realization of impact of drought on coconut yield forced increased attention towards this problem. A methodical research approach led to understanding the drought tolerance mechanism in coconut. So far, conventional breeding strategies were applied for development of drought-tolerant varieties and hybrids. However, this takes a lot of time and testing for yield stability under stressful conditions is time consuming. Lack of large-scale regeneration techniques handicapped the genetic engineering approach to impart drought tolerance in high-yielding cultivars. Hence it is very much important to globally co-ordinate the breeding for drought tolerance programmes as studies indicated that hybrids with talls as parents can perform better under water stress conditions. It is essential to conserve the natural desirable gene pools present in the in situ palms in farmers' fields before they become extinct. These materials are highly valuable for crop improvement programmes. Comprehensive molecular markers need to be developed for rapid screening of coconut germplasm for drought tolerance at a global level. Further, it is of importance to characterize the nature and intensity of drought in different coconut-growing areas in order to develop suitable drought management strategies. More importantly, available technologies need to be demonstrated and disseminated to the farmers for improving productivity in not only coconut but also areca nut.

15.5 Conclusions and Future Thrust

To sum up, a coconut responds to the stressful environments at morphological, anatomical, physiological and biochemical levels. Generally, drought coincides with high temperature and high light intensity stresses making the problem more complex. However, it is also important to note that the adaptive strategies of plants for different abiotic stresses are overlapped in most of the cases as the nature of abiotic stresses themselves. Further studies should be focused to develop the molecular markers linked to desirable traits and to understand the inheritance patterns of these traits. The genotypes with desirable traits for tolerance to stress conditions can be used in breeding strategies for abiotic stress tolerance in future crop improvement programmes. Identification and characterization of field-tolerant palms to abiotic stresses will be helpful in exploiting the natural tolerance to abiotic stresses. More basic research efforts are needed for the areca nut as is done for the coconut.

The results obtained so far indicate that variation exists among the tall(s), dwarf(s) and hybrids for drought-tolerant traits. Generally, tall(s) and hybrids with tall(s) as mother palms have higher drought tolerance compared to dwarf (s) and hybrids with dwarf(s) as mother palms. The heterosis for drought-tolerant traits can be exploited for breeding for drought tolerance. Further, in situ tolerant palms need to be identified and used in breeding programmes. Observations delineated the mechanism for drought tolerance (Schematic Diagram 15.1) and breeding for drought tolerance (Schematic Diagram 15.2) based on morphological, anatomical, physiological and biochemical traits. Germplasm should be screened for targeted traits under the targeted environments varying in time and space. In perennials, the stability of such tolerance over a period of time is the key factor for realizing stable yields even during stress years. Special emphasis should be given for field tolerance and in situ tolerant plants. The plants which can withstand the natural occurrence of drought and other stresses and still produce good yield are of premium in nature for their possessing desirable gene pool. Early vigour and high revival capacity are very significant factors that should take pivotal place. Selecting such genotypes should be based only on field stress. These experiments can be extrapolated to other germplasm sources, which were not studied so far, and for making a cross combination of potential success to come out with cultivars/hybrids with high drought tolerance and stable yield.

References

Abeywardena (1968) Forecasting coconut crops using rain fall data. III Session of FAO Tech. Working party on coconut production, protection and processing. Jakarta, Indonesia, PLCNP: 68(29)

Anonymous (2008) CPCRI, Annual report, 2007–2008, CPCRI Pub, Kasaragod, India, p 49

Batugal PA (1999) The role of international cooperation in the development of biotechnology in coconut. In: Oropeza C, Verdeil JL, Ashburner GR, Cardena R, Santamaria JM (eds) Current advances in coconut biotechnology. Kluwer Academic Publishers, Dordrecht, pp 19–30

Bhaskara Rao EVVB, Pillai RV, Mathew J (1991) Relative drought tolerance and productivity of released coconut hybrids. In: Silas EG, Aravindhakshan M, Jose AI (eds) Coconut breeding and management. KAU, Vellanikkara, pp 44–149

Bobby P, Rajesh MK, Kiran AG, George T, Naresh KS (2010) Differential expression of genes regulated in response to water stress in coconut In: International conference on coconut biodiversity for prosperity, 27–30 October 2010, Central Plantation Crops Research institute, Kasaragod, Kerala.TS3-PO2, pp 53

Bobby P, Rajesh MK, Naresh KS (2012) Molecular cloning and characterization of a partial mitogen-activated protein kinase cDNA, CnMAPK1 induced in response to water stress in coconut Indian J. Horticulture 69(2):145–150

Bonneau X, Subagio K (1999) Coconut growing in zones at risk of drought. Plant Rech Dev 6:432–442

Bonneau X, Ochs R, Kitu WT, Yuswohadi (1993) Le chlore: Un element essential de la nutrition minerale des cocotiers hybrids dans le Lampung (Indonesie). Oleagineux 48:179–189

Bonneau X, Boutin D, Bourgoing R, Sugarianto J (1997) Le chlorure de sodium, fertilisant ideal du cocotier en Indonesie. Plant Rech Dev 4:336–346

Braconnier S, Auzec JD (1985) Anatomical study and cytological demonstration of potassium and chlorine fluxes associated with oil palm and coconut stomatal opening. Oleagineux 40:5547–5551

Braconnier S, Auzec JD (1990) Chloride and stomatal conductance in coconut. Oleagineux 45:259–265

Braconnier S, Bonneau X (1998) Effects of chlorine deficiency in the field on leaf gas exchanges in the PB 121 coconut hybrid. Agronomie 18:563–572

Chempakam B, Kasturi Bai KV, Rajagopal V (1993) Lipid peroxidation in relation to drought tolerance in coconut (*Cocos nucifera* L.). Plant Physiol Biochem 20(1):5–10

Child R (1974) Coconut, 2nd edn. Longman, London, p 335

Coomans P (1975) Influence des facteurs climatiques sur les fluctuations saisonnieres et annuelles de la production du cocotier. Oleagineux 30:153–159

de Nuce de Lamothe M, Benard G (1985) L'hybride de cocotier PB 121 (ou MAWA) (NJM x GOA). Oleagineux 40:255–266

Fernando WMU, Gajanayake G (1997) Patterns of isozyme variations in coconut (*Cocos nucifera* L.) populations used for breeding improved varieties. Plant Rech Dev 4:256–261

Gomes FP, Mielke MS, Almed AAF (2002) Leaf gas exchange of green dwarf coconut (*Cocos nucifera* L. var. nana) in two contrasting environments of the Brazilian North-East region. J Hortic Sci Biotech 77 (6):766–772

Jayasekara C, Ranasinghe CS, Mathes DT (1993) Screening for high yield and drought tolerance in coconut. In: Nair MK, Khan HH, Gopalasundaram P, Bhaskara Rao EVV (eds) Advances in coconut research and development. Oxford & IBH Publishing Co. Pvt. Ltd., New Delhi, pp 209–218

John Sunoj VS, Naresh Kumar S, Muralikrishna KS (2009a) Proline may assume greater role in coconut adaptation to elevated CO_2 and temperature conditions. Proceedings of national workshop on climate and development. College of Agriculture, Kerala Agriculture University, Kerala, pp 159–166

John Sunoj VS, Naresh Kumar S, Muralikrishna KS, Kasturi Bai KV (2009b) Proline may assume greater role in coconut (*Cocos nucifera* L.) adaptation to elevated CO_2 and temperature conditions. In: Proceedings of the National workshop on Climate and Development. 29–30 June, 2009. KAU Pub, pp 159–166

John Sunoj VS, Naresh Kumar S, Muralikrishna KS (2013a) Variation in total phenols concentration in coconut (*Cocos nucifera* L.) seedlings under elevated CO_2 and temperature in different seasons. In: Sundaresan J, Sreekesh S, Ramanathan AL, Len S, Ram B (eds) Climate change and environment. Scientific Publishers, India, p 286

John Sunoj VS, Naresh Kumar S, Muralikrishna KS (2013b) Variation in total phenols concentration in coconut (*Cocos nucifera* L.) seedlings under elevated CO_2 and temperature in different seasons. In: Sundaresan J, Sreekesh S, Ramanathan AL, Len S, Ram B (eds) Climate change and environment. Scientific Publishers, India, pp 140–149

John Sunoj VS, Naresh Kumar S, Muralikrishna KS (2014) Effect of elevated CO2 and temperature on oxidative stress and antioxidant enzymes activity in coconut (*Cocos nucifera L.*) seedlings. Indian J Plant Physiol. doi:10.1007/s40502-014-0123-6

John Sunoj VS, Naresh Kumar S, Muralikrishna KS, Padmanabhan S (2015) Enzyme activities and nutrient status in Coconut (*Cocos nucifera* L.) seedling rhizosphere soil after exposure to elevated CO_2 and temperature. J Indian Soc Soil Sci 63(2):191–199

Juma M, Fordham R (1998) The effect of environmental stress on coconut (Cocos nucifera L.) growth in Zanzibar. In: Trees for life – the key to development. In: Topper CP, Caligari PDS, Kullaya AK, Shomari SH, Kasuga LJ, Masawe PAL, Mpunami AA (eds), Proceedings of the international cashew and coconut conference. Dar es Salaam, Tanzania, pp 342–347

Juma MA, Hornung RKW, Topper (1997) Effects of induced water stress on coconut leaf stomata. In: Trees for life – the key to development. In: Topper CP, Caligari PDS, Kullaya AK, Shomari SH, Kasuga LJ, Masawe PAL, Mpunami AA (eds), Proceedings of the international cashew and coconut conference. Dar es Salaam, Tanzania, pp 320–325

Karunaratne S, Santha S, Kovoor A (1991) An *in vitro* assay for drought tolerant coconut germplasm. Euphytica 53:25–30

Kasturi Bai KV (1993) Evaluation of coconut germplasm for drought tolerance. Ph.D. thesis, Mangalore University, Mangalore, India

Kasturi Bai KV, Rajagopal V (2000) Osmotic adjustment as a mechanism for drought tolerance in coconut (*Cocos nucifera* L.). Indian J Plant Physiol 5 (4):320–323

Kasturi Bai KV, Voleti SR, Rajagopal V (1988) Water relations of coconut palms as influenced by environmental variables. Agric For Meteorol 43:193–199

Kasturi Bai KV, Rajagopal V, Chempakam B, Prabha CD (1996a) Assay of enzymes in coconut cultivars and hybrids under non-stress and stress conditions. J Plant Crops 24:548–554

Kasturi Bai KV, Rajagopal V, Prabha CD, Ratnambal MJ, George MV (1996b) Evaluation of coconut cultivars and hybrids for dry matter production. J Plant Crops 24:23–28

Kasturi Bai KV, Rajagopal V, Balasimha D, Gopalasundaram P (1997) Water relation, gas exchange and dry matter production of coconut (Cocos nucifera L.) under un-irrigated and irrigated conditions. CORD XIII 2:45–58

Kasturi Bai KV, Rajagopal V, Balasimha D (1998) Variation in net carbon assimilation and related parameters in coconut (*Cocos nucifera* L.) under field conditions. Plant Physol Biochem 25(2):163–166

Kasturi BaI KV, Rajagopal V, Ratnambal MJ (2001) Relationship between leaf water potential and peroxidation of cell wall lipids in coconut. J Plant Biol 28 (2):161–164

Kasturi Bai KV, Srinivasa Reddy DV, Ratnambal MJ, Laxman RH (2003) Factors contributing to female flower production and button shedding in coconut (*Cocos nucifera* L). J Plant Crops 31(2):33–36

Kasturi Bai KV, Rajagopal V, Chempakam B, Naresh Kumar S (2005a) Inductive response of protective enzymes in coconut (*Cocos nucifera* L.) subjected to stress. Indian J Hortic 62(4):363–366

Kasturi Bai KV, Rajagopal V, Champakam B, Naresh Kumar S (2005b) Inductive response of protective enzymes in coconut palms subjected to moisture stress. Indian J Hortic 62(4):363–366

Kasturi Bai KV, Rajagopal V, Arunachalam V (2006a) Assessment of diversity in coconut varieties for drought responsive physiological traits. J Plant Crops 34(2):118–120

Kasturi Bai KV, Rajagopal V, Naresh Kumar S (2006b) Chlorophyll fluorescence transients with response to leaf water status in coconut. Indian J Plant Physiol 11(4):410–414

Kasturi Bai KV, Rajagopal V, Naresh Kumar S (2006c) Chlorophyll fluorescence traits with response to leaf water status in coconut. Indian J Plant Physiol 11(4):410–414

Kasturi Bai KV, Naresh Kumar S, Rajagopal V, Vijayakumar K (2008) Principal component analysis of chlorophyll fluorescence transients for tolerance to drought stress in coconut seedlings. Indian J Hort 65 (4):471–476

Kurup VVGK, Voleti SR, Rajagopal V (1993) Influence of weather variables on the content and composition of leaf surface wax in coconut. J Plant Crops 2:71–80

Lakshmanachar MS (1963) Studies on the effect of rain fall on coconut crops. Indian Coco Bull 16:370–372

Lebrun P, N'Cho YP, Seguin M, Grivet L, Baudouin L (1998) Genetic diversity in coconut (*Cocos nucifera* L.) revealed by restriction fragment length polymorphism (RFLP) markers. Euphytica 101:103–108

Lebrun P, Grivet L, Baudouin L (1999) Use of RFLP markers to study the diversity of the coconut palm. In: Oropeza C, Verdei JL, Ashburner GR, Cardena R, Santamaria JM (eds) Current advances in coconut biotechnology. Kluwer Academic Publishers, Dordrecht, pp 73–87

Lubina R (1990) Effect of irrigation and nutrient levels on water relations of coconut genotypes. M.Phil. thesis, Mangalore University, Mangalore, India, pp 70

Manimekalai R, Nagarajan R, Bharathi M, Naresh Kumar S (2004) DNA polymorphism among coconut (*Cocos nucifera* L.) cultivars and reciprocal derivatives differing in drought tolerance. J Plant Crops 32:117–122

Mathai G, Panicker KS (1979) Relationship between yield of coconut and rainfall pattern in the backwater region of Kerala. Agric Res J Kerala 16(2):253–25

Mathes DT (1988) Influence of weather and climate on coconut yield. Coco Bull 5(1):8–10

Menon KPV, Pandalai KM (1958) The coconut palm. A monograph. The times of India Press, Bombay, pp 293–303

Milburn JA, Zimmermann MH (1977) Preliminary studies on sap flow in *Cocos nucifera* L. 1. Water relations and xylem transport. New Phytol 79:535–541

Muralikrishna KS, Naresh Kumar S, John Sunoj VS (2009a) Elevated CO_2 and temperature reduce Stomata density in coconut. Proceedings of national workshop on climate and development. College of Agriculture, Kerala Agriculture University, Kerala, pp 153–158

Muralikrishna KS, Naresh Kumar S, John Sunoj VS, Kasturi Bai KV (2009b) Elevated carbon dioxide and temperature reduce stomatal density in coconut (*Cocos nucifera* L.). In: Proceedings of the national workshop on climate and development. KAU Pub, 29–30 June, 2009, pp 153–158

Muralikrishna KS, Naresh Kumar S, John Sunoj VS (2013a) Elevated CO_2 and temperature affect leaf anatomical characteristics in coconut (*Cocos nucifera* L.). In: Sundaresan J, Sreekesh S, Ramanathan AL, Sonnenschein L, Boojh R (eds) Climate change and island and coastal vulnerability. Springer, Dordrecht. doi:10.1007/978-94-007-6016-5

Muralikrishna KS, Naresh Kumar S, John Sunoj VS (2013b) Elevated CO_2 and temperature affect leaf anatomical characteristics in coconut (*Cocos nucifera* L.). In: Sundaresan J, Sreekesh S, Ramanathan AL, Sonnenschein L, Boojh R (eds) Climate change and island and coastal vulnerability. Springer, Dordrecht, pp 141–153. ISBN 978-94-007-6015-8

Murray DV (1977) Coconut palm. In: Alvim TA, Kozlowski TT (eds) Ecophysiology of tropical crops. Academic, New York, pp 384–407

Nainanayake NPAD, Bandara DC (1998) Effect of water stress on coconut (*Cocos nucifera* L.) seedlings under different soil types and compaction levels. Trop Agric Res 10:12–26

Nainanayake NPAD, Bandara DC, Nissanka SP (2000) Root shoot relationships: an effective indicator of soil compaction and water stress for coconut (*Cocos nucifera* L.) seedlings. Trop Agric Res 12:151–162

Naresh KS (2004) Drought management in coconut gardens. CPCRI Publication, 16p

Naresh KS, Rajagopal V (2005) Drought management in coconut gardens. In: Rajagopal V, Naresh Kumar S (eds) Value addition to weather data: advisory service to farmers. CPCRI Publication, pp 25–31

Naresh Kumar S, Aggarwal PK (2013) Climate change and coconut plantations in India: impacts and potential adaptation gains. Agril Syst http://dx.doi.org/10.1016/j.agsy.2013.01.001

Naresh Kumar S, Kasturi Bai KV (2009a) Photosynthetic characters in different shapes of coconut canopy under irrigated and rainfed conditions. Indian J Plant Physiol 14(3):215–223

Naresh Kumar S, Kasturi Bai KV (2009b) Photo-oxidative stress in coconut seedlings: early events to leaf scorching and seedling death. Braz J Plant Physiol 21(3):223–232

Naresh Kumar S, Rajagopal V, Anitha Karun (2000) Leaflet anatomical adaptations in coconut cultivars for drought tolerance. In: Muraleedharan N, Raj Kumar R (eds), Recent advances in plantation crops research. Proceedings of PLACROSYM XIII, 16–18 December 1998, Coimbatore, pp 225–229

Naresh Kumar S, Rajagopal V, Laxman RH, Dhanapal R, Maheswarappa HP (2002a) Photosynthetic characteristics and water relations in coconut palm under drip irrigation on sandy and laterite soils. In: Rethinam P, Khan HH, Reddy VM, Mandal PK, Suresh K (eds) Plantation crops research and development in the new millennium. Coconut Development Board Pub, Kochi, pp 116–120

Naresh Kumar S, Rajagopal V, Siju Thomas T, Cherian VK, Hanumanthappa M, Kumar A, Srinivasulu B, Nagvekar D (2002b) Identification and characterization of in situ drought tolerant coconut palms in farmers' fields in different agro-climatic zones. In: Sreedharan K, Vinod Kumar PK, Jayaram, Chulaki BM (eds) Proceedings of PLACROSYM XV. Code word process and Printers, Mangalore, pp 335–339

Naresh Kumar S, Rajagopal V, Siju Thomas T, Cherian VK (2003) Influence of soil moisture conservation on coconut (*Cocos nucifera* L.) under different agroclimatic conditions. In: Proceedings of the 2nd International Congress on Plant Physiology, 8–12 January 2003, IARI, New Delhi, India, pp 25

Naresh Kumar S, Rajagopal V, Siju Thomas T, Cherian VK (2006a) Effect of conserved soil moisture on the source-sink relationship in coconut (*Cocos nucifera* L.) under different agro-climatic conditions in India. Indian J Agric Sci 76(5):277–281

Naresh Kumar S, Kasturi Bai KV, George J, Balakrishnan A, Siju Thomas T (2007a) Stress responsive proteins in coconut seedlings subjected to water, high-light, and flooding and high-temperature stresses. Indian J Hortic 64(4):373–380

Naresh Kumar S, Rajagopal V, Siju Thomas T, Cherian VK, Ratheesh Narayanan MK, Ananda KS, Nagawekar DD, Hanumanthappa M, Vincent S, Srinivasulu B (2007b) Variations in nut yield of coconut (*Cocos nucifera* L.) and dry spell in different agroclimatic zones of India. Indian J Hortic 64 (3):309–313

Naresh Kumar S, Kasturi Bai KV, Rajagopal V, Aggarwal PK (2008) Simulating coconut growth, development and yield using InfoCrop-coconut model. Tree Physiol 28:1049–1058

Naresh Kumar S, Rajagopal V, Cherian VK, Siju Thomas T, Sreenivasulu B, Nagvekar DD, Hanumanthappa M, Bhaskaran R, Vijaya Kumar K, Ratheesh Narayanan MK, Amarnath CH (2009a) Weather data based descriptive models for prediction of coconut yield in different agro-climatic zones of India. Indian J Hortic 66(1):88–94

Naresh Kumar S, Rajeev MS, Vinayan, Nagvekar DD, Venkitaswmy R, Raghava Rao DV, Boraiah B, Gawankar MS, Dhanapal R, Patil DV, Katuri Bai KV (2009b) Trends in weather and yield changes in past in coconut growing areas in India. J Agrometeorol 11(1):15–18

Nelliat EV, Padmaja PK (1978) Irrigation requirement of coconut and response to levels of fertilizer under irrigated condition during the early bearing stage. In: Nelliat EV (ed) Proc. PLACROSYM-1. Sharada Press, Mangalore, pp 186–199

Ollagier M, Ochs R, Pomier M, de Taffin G (1983) Effect of chlorine on the hybrid coconut PB 121- in the Ivory Coast and Indonesia-Growth, tolerance to drought, yield. Oleagineux 38(5):309–321

Patel JS (1938) The coconut. A monograph Govt. Press, Madras, p 102

Peiris TSG, Peries RRA (1993) Effects of bimonthly rainfall on coconut yield in the low country intermediate zone (IL_1) of Sri Lanka. Cocos 9:1–11

Peiris TSG, Thattil RO (1988) The study of climate effects on the nut yield of coconut using parsimonious models. Exp Agric 34:189–206

Peiris TSG, Thattil RO, Mahindrapala R (1995) An analysis of effect of climate and weather on coconut (*Cocos nucifera*). J Exp Agric 31(4):451–460

Pomier M, de Taffin G (1982) The tolerance to drought of some coconut hybrids. Oleagineux 37(2):55–62

Quencez P, de Taffin G (1981) Relation between potassic nutrition and rainfall in oil palm and coconut growing. Oleagineux 36:1–7

Rajagopal V, Kasturi Bai KV (2002) Drought tolerance mechanism in coconut. Burotrop 17:21–22

Rajagopal V, Naresh Kumar S (2001) Avenues to improve productivity potential under drought condition – a case study on coconut. In: Chetti MB (ed) Souvenir on national seminar on role of plant physiology for sustaining quantity and quality of food production in relation to environment, pp 31–36

Rajagopal V, Patil KD, Sumathykutty Amma B (1986) Abnormal stomatal opening in coconut palms affected by root (wilt) disease. J Exp Bot 37:1398–1405

Rajagopal V, Shivashankar S, Kasturi Bai KV, Voleti SR (1988) Leaf water potential as an index of drought tolerance in coconut (*Cocos nucifera* L.). Plant Physiol Biochem 15(1):80–86

Rajagopal V, Ramadasan A, Kasturi Bai KV, Balasimha D (1989) Influence of irrigation on leaf water relations and dry matter production in coconut palms. Irrig Sci 10:73–81

Rajagopal V, Kasturi Bai KV, Voleti SR (1990) Screening of coconut genotypes for drought tolerance. Oleagineux 45(5):215–223

Rajagopal V, Shivashankar S, Kasturi Bai KV (1991) Physiological and biochemical basis of coconut production. CORD VII 2:12–30

Rajagopal V, Kasturi Bai KV, Pillai RV, Vijayakumar K (1992) Relationship between physiological characters and nut yield in coconut genotypes under rainfed condition. J Plant Crops 20:277–283

Rajagopal V, Shivashankar S, Mathew J (1996) Impact of dry spells on the ontogeny of coconut fruits and it's relation to yield. Plant Res Dev 3(4):251–255

Rajagopal V, Naresh Kumar S, Kasturi Bai KV, Laxman RH (2000a) Day time fluctuations in photosynthetic parameters and water relations in juvenile palms of coconut grown under rainfed and irrigated conditions. J Plant Biol 27(1):27–32

Rajagopal V, Kasturi Bai KV, Naresh Kumar S (2000b) Adaptive mechanism of coconut palms in changing environment conditions for higher production. In: Extended summaries vol 2 of international conference on managing natural resources for sustainable agricultural production in the 21st century, New Delhi, 14–18 Feb, 2000, vol 2. Agro-biodiversity, pp 778–779

Rajagopal V, Kasturi Bai KV, Naresh Kumar S, Niral V (2007) Genetic analysis of drought responsive physiological characters in coconut. Indian J Hortic 64 (2):181–189

Ramadasan A, Balakrishnan TK, Rajagopal V (1991). Response of coconut genotypes to drought. Ind Coc J 21: 2–5.

Rao GSLHVP (1985) Drought and coconut palm. Indian Coconut J 15:3–6

Repellin A, Daniel C, Zuily Fodil Y (1994a) Merits of physiological tests for characterizing the performance of different coconut varieties subjected to drought. Oleagineux 49(4):155–169

Repellin A, D'Arcy Lameta A, Pham Thi AT, Tashakorie A, Zuily Fodil Y (1994b) Physiological parameters as screening tools for drought-stress resistant varieties of coconut palm (*Cocos nucifera* L.) In: 2. Colloque. Sciences Vegetales, Saint-Malo, 12-- 14 October, 1994. Proceedings. Paris (France), Societe Francaise Physiologie Vegetale, p 299

Repellin A, Pham Thi AT, Tashakorie A, Sahsah Y, Daniel C, Zuily Fodil Y (1997) Leaf membrane lipids and drought tolerance in young coconut palms (*Cocos nucifera* L.). Eur J Agric 6:25–33

Shivashankar S (1988) Polyphenoloxidase isozymes in coconut genotypes under water stress. Plant Physiol Biochem 15:87–92

Shivashankar S (1992) Thermal stability of nitrate reductase in relation to drought tolerance in coconut. J Plant Crops 20:267–273

Shivashankar S, Nagaraja KV (1996) Water stress induces kinetic changes in the properties of an acid phosphatase isozyme from coconut leaves. Plant Physiol Biochem 23(1):21–26

Shivashankar S, Kasturi Bai KV, Rajagopal V (1991) Leaf water potential, stomatal resistance and activity of enzymes during the development of moisture stress in coconut palm. Trop Agric 68(2):106–110

Siju Thomas T, Rajagopal V, Naresh Kumar S, Arunachalam V, Cherian VK (2005) Stability analysis for dry matter production and yield components of coconut in two agro-climatic regions of India. Indian J Plant Physiol 10(1):1–8

Siju Thomas T, Naresh Kumar S, Cherian VK, Rajagopal V (2006a) Role of certain biochemical compounds in adaptation of coconut to different weather conditions- a study at two agro-climatic regions of India. Indian J Hortic 63(1):1–7

Siju Thomas T, Naresh Kumar S, Cherian VK, Kasturi Bai KV, Rajagopal V (2006b) Role of certain biochemical compounds in adaptation of coconut to different weather conditions- a study in two agro-climatic regions of India. Indian J Hortic 63(1):1–7

Siju Thomas T, Naresh Kumar S, Cherian VK, Rajagopal V (2008) Gas exchange parameters and canopy area in relation to coconut productivity in two-agro climatic regions of India. Trop Agric 85(1):24–35

Sudhakara K (1990) Button shedding and premature nut fall in coconut. J Plant Crops 18(2):66–77

Vijayakumar K, Mathew J, Amarnath CH, Nambiar PTN, Jose CT, Balakrishnan TK (1988) Influence of weather on coconut yield. In: Rao GSLHV, Nair RR (eds) Agrometeorology of plantation crops. Kerala Agricultural University, Trichur, pp 124–132

Voleti SR, Rajagopal V (1991) Extraction and identification of epicuticular wax in coconut. Pl Physiol Biochem 18(2):88–90

Voleti SR, Kasturi Bai KV, Nambiar CKB, Rajagopal V (1993a) Influence of soil type on the development of moisture stress in coconut (*Cocos nucifera* L.). Oleagineux 46:505–509

Voleti SR, Kasturi Bai KV, Rajagopal V (1993b) Water potential in the leaves of coconut (*Cocos nucifera* L.) under rainfed and irrigated conditions. In: Nair MK, Khan HH, Gopalasundaram P, Bhaskara Rao EVV (eds) Advances in coconut research and development. Oxford & IBH publishing Co Pvt. Ltd, New Delhi, pp 243–245

Wickremaratne MRT (1987) Breeding coconuts for adaptation to drought. Coco Bull 4:16–23

Zhang J, Davies WJ (1989) Abscisic acid produced in dehydrating roots may enable the plant to measure the water status of the soil. Plant Cell Environ 12:73–81

Cocoa and Cashew 16

D. Balasimha

Abstract

The tropical plant cocoa (*Theobroma cacao* L.), the source of chocolate, is endemic to Amazon basin. Its cultivation has subsequently extended to tropical and subtropical regions of South and Central America, West Africa, and Asia Pacific. In cocoa, an effective stomatal regulation facilitates the trees to withstand moisture stress during summer months. Morphological and anatomical adaptations also favored the accessions during drought conditions. Development of rapid screening methods for stress tolerance and identification of promising accessions for selective breeding has been done. Biochemical parameters like waxes that impart stress tolerance in some of the accessions have been well delineated. Carbon sequestration studies have indicated that cocoa has potential in climate change issues. Higher CO_2 and temperature levels influenced photosynthetic characteristics. Cashew has gained economic importance because of its export potential. Cashew (*Anacardium occidentale* L.) is a native of northern part of South America and introduced into India in the sixteenth century. In west and east coast regions, cashew is grown under rainfed conditions. The regions where cashew is grown generally receive 3000–4000 mm of rainfall. Screening of cashew germplasm at three different agroclimatic regions has resulted in identification of few drought-tolerant lines. Detailed studies on photosynthesis in relation to drought, irrigation, and fertilizers have been done and results discussed.

D. Balasimha (✉)
Central Plantation Crops Research Institute, Regional Station, Vittal, 574243 Karnataka, India

B-404, Mantri Alpyne, Banashankari 5th Stage, Bengaluru 560061, India

ICAR-Central Plantation Crops Research Institute, Kasaragod, Kerala, India
e-mail: dbsimha@gmail.com; balasimhad@rediffmail.com

16.1 Introduction

Over the past couple of decades, the global climate change has become a major concern and scientific studies to understand this complex phenomenon is given increasing attention. There are

N.K.S. Rao et al. (eds.), *Abiotic Stress Physiology of Horticultural Crops*,
DOI 10.1007/978-81-322-2725-0_16

far-reaching impacts on crop production and human society. Abiotic stresses like drought, temperature, salinity, and elevated CO_2 impact physiology and productivity of crop plants. Plantation crops like cocoa and cashew being perennial in nature have to undergo repeated cycles of stresses during lifetime. It is therefore imperative to understand the mechanism of responses and adaptation of these crops. The cocoa and cashew crops are of high economic value contributing substantially to the agricultural exports at global and at national levels of several countries.

The tropical plant cocoa (*Theobroma cacao* L.), the source of chocolate, is endemic to Amazon basin. Its cultivation has subsequently extended to tropical and subtropical regions of South and Central America, West Africa, and Asia Pacific. Cocoa, being perennial in nature, has to face the impact of climate change even during a single generation or in a standing plantation. In Southern India, cocoa is cultivated as a mixed crop with areca nut and coconut. During the summer months, the weather is dry for 3–5 months with high evaporative demand resulting in soil and atmospheric drought of varying degrees of severity. Areca nut to a considerable extent and coconut on a limited scale are raised as irrigated crops. However, nonavailability of water toward late summer season makes it difficult to cultivate mixed crops. Detailed studies have been carried out in order to develop drought-tolerant lines in cocoa and to understand basic mechanisms of tolerance.

Cashew (*Anacardium occidentale* L.) is a native of northern part of South America and introduced into India in the sixteenth century. It has gained economic importance because of its export potential. In west and east coast regions, cashew is grown under rainfed conditions. The regions where cashew is grown generally receive 3000–4000 mm of rainfall. There is a long spell of dry season during summer extending from 3 to 6 months. Cashew responds positively to irrigation (Yadukumar and Mandal 1994).

16.2 Effect of Weather Variables

An increase in evaporative demand owing to low humidity changes the stomatal resistance. Besides light and temperature also influence stomatal resistance. The stomatal resistance increased with decrease in relative humidity, which was magnified by soil moisture stress in field-grown cocoa trees (Balasimha and Rajagopal 1988). With higher relative humidity (RH), the leaf water potential was maintained at higher values. Stomata of intact leaves generally open in response to increase in light intensity. In cocoa leaves, however, the effects of RH and soil moisture stress decreased the effect of light. It indicates that RH has stronger influence than light. The correlation coefficients among various environmental factors and plant responses indicate that ambient temperature did not influence stomatal responses but showed highly significant correlation with leaf temperature. Stomatal regulation by internal (stress) and external factors (RH) leads to decrease in transpiration, which was a major adaptation for water conservation.

The climatic conditions in West African countries are similar to those prevailing in Southern India with two main rainy periods, viz., March to July and September to December (Asomaning 1976). In marked contrast to this, the rainfall in Brazil is very well distributed throughout the year. This difference in climate accounts for a bimodal crop pattern in Brazil in contrast to that of single main crop in Africa (Alvim 1977). The sustained water supply obviously helps in continuous flowering and fruiting throughout the year. Rainfall is also well distributed in other Latin American countries, Malaysia and Sri Lanka. High correlations between rainfall and yield have been reported (Gordon 1976). Other climatic factors such as temperature, light intensity, and day length normally are not limiting factors for cocoa yields except in Brazil which experiences nearly 4 months of low temperature during winter (Alvim 1977).

The cocoa tree needs a high and well-distributed rainfall, possibly with a short dry spell to stimulate flowering. Two dissimilar crop patterns are observed under rainfed and irrigated conditions in India. Correlations of yield of cocoa with various weather variables showed that numbers of rainy days of previous year, sunshine hour of current year, and maximum temperature (4th fortnight and 21st fortnight) of current year were found significant (Vijayakumar et al. 1991).

Cashew is grown in dry farming conditions and concentrated in intertropical regions. These areas are normally low in soil fertility with temperatures and humidity generally high. The rainfall occurs in certain periods and there is usually long dry periods affecting the yields of cashew (Bezerra et al. 2007).

16.3 Water Relations and Stomatal Regulation

Cocoa is very sensitive to water stress. Drought affects several physiological processes leading to a reduction in crop yield. Water potential (WP) of leaf is a major quantitative character used to assess water stress. Cocoa plants show changes in water relations when soil moisture drops to 60–70 % of available range. The onset of drought decreased water potential and relative water content (Balasimha 1982, 1983, 1987). Diurnal variations exist in water potential; however, a plateau WP is reached once the stomata close beyond – 1.5 MPa (Balasimha et al. 1991); this is the stage when turgor pressure falls and leaves start wilting. The changes in osmotic potential of cocoa during different months were not as marked as water potential (Balasimha 1987). Leaf water potential was appreciably lower in exposed leaves as compared to shaded ones, presumably due to higher temperature and lower RH in exposed area. The moisture stress results in accumulation of abscisic acid (ABA) in leaves (Alvim et al. 1974). The accumulation of ABA was related to drought tolerance among cocoa accessions (Balasimha and Anil Kumar 2000).

16.4 Seasonal Changes in Nitrate Reductase and Other Indicators of Water Stress

The relative water content (RWC) of leaves of rainfed cocoa plants was lower than those of irrigated plants. This tended to decrease in both conditions with progress of summer months. The content of proline in leaves was maintained at similar levels in irrigated plants, while there was considerable accumulation in rainfed plants during the dry period. With the onset of rains, proline decreased. The nitrate reductase (NR) activity was high during February–April in irrigated plants and low during rainy season, whereas in rainfed plants, the activity was low during the dry period. The decrease in NR in rainfed plants is an adaptation to stress conditions. Since NR is light inducible, the reduction during monsoon is due mainly to low light levels. When seasonal changes in NR were studied among tolerant and susceptible trees, the activity was higher in the former during summer months. However, the activities did not vary appreciably among the accession types (Balasimha 1982).

16.5 Morphological Adaptation

The reduction of cell growth is one of the earliest manifestations of water deficits. This is considered to be more sensitive to water stress than stomatal conductance and CO_2 assimilation. The reduction in leaf area helps plants to adapt to periods of drought. A reduction in leaf area might have an advantage in reducing transpiration. However, once leaf expansion is complete, an increase in epicuticular waxes (EW) of leaves, which can reduce cuticular transpiration, is beneficial.

In cocoa accessions there were no significant differences in leaf area under watered conditions (Balasimha et al. 1985). The specific leaf weight (SLW) and EW content were significantly different among cocoa accessions. Accessions having high specific leaf weight and higher EW were found to be drought tolerant (Balasimha 1987; Balasimha et al. 1985). Leaf anatomical characters of some cocoa accessions have also been studied. The SLW can be a good indicator of leaf anatomy and positive relation between leaf thickness and palisade layer thickness were obtained. Using some of the morphological attributes, it was possible to screen out 80 % of test material in the initial stages for drought tolerance.

16.6 Comparative Drought Tolerance

The water relation components and stomatal behavior were determined in 14 accessions of cocoa for detailed analysis after a preliminary screening. Differences in plant variables between drought-tolerant and susceptible accessions were studied (Balasimha et al. 1988). Based on parametric relationships with drought tolerance, the rank sums of these accessions led to a selection of NC23, NC29, NC31, NC39, and NC42. These drought-tolerant accessions had effective stomatal regulation resulting in decreased transpiration water loss. There was a decrease in osmotic potential and NR and an increase in soluble sugars in respect to stress. The stability of NR during summer season was higher in tolerance than susceptible trees both in leaf and root. The stomatal resistance (rs) was negatively correlated with transpiration. The cocoa accessions with higher drought tolerance characteristics and yield potential under drought should be useful material for selective breeding.

16.6.1 Screening Germplasm for Drought Tolerance

16.6.1.1 Cocoa

After evaluating the mechanism of drought tolerance in cocoa, it was found that as a result of reduced transpiration, leaf turgidity was maintained thereby giving higher water potential. A rapid screening method for drought tolerance in cocoa has been described which eliminated difficulty of using field plants. This method utilizes measurements of water potential in excised leaves. The decrease in water potential was more pronounced in susceptible as compared to tolerant accessions under induced stress (Balasimha and Daniel 1988). Sixty-five cocoa accessions have been so far screened for drought tolerance in the first phase. Five have been found to be drought tolerant as mentioned above. Another 52 accessions introduced to the germplasm have been screened in the second phase. Out of these another five accessions were found to be drought tolerant (Balasimha et al. 2013).

Four high-yielding trees and three drought-tolerant trees were used for crossing to combine these characters in the hybrids. High-yielding trees, viz., 1–14 (Red axil), I-21, I-29 (Amel x Na33), and II-67 (Landas 363), were used as female parents, while NC23/43 (P3 × P), NC 29/66 (P6 × P4), and NC42/94 (T86/2) were used as pollen parents. Nine F1 hybrids raised from these cross combinations were screened for drought tolerance by rapid screening technique. The parents were also screened for comparison. The seedlings were planted in the field on a replicated randomized block design. Measurements of leaf water potential during dry periods were made every year starting from 1992. The seedling screening and field evaluation showed that hybrids I-21 × NC 42/94 and I-29 × NC 23/43 retained higher water potential and stomatal resistance indicating their drought tolerance (Balasimha et al. 1999). Photosynthetic parameters were also measured in these hybrids and parents in the field.

Cashew is usually a rainfed crop and known to be generally tolerant to drought conditions. The variations in these physiological parameters will help in identifying high photosynthetically productive genotypes/varieties for high yield and drought tolerance. Photosynthetic characteristics of selected 23 accessions are given in Table 16.1.

Photosynthesis rates (Pn) ranged from 1.77 to 7.18 μmol/m^2/s, while transpiration ranged from 3.22 to 10.07 mmol/m^2/s. Transpiration rate had positive correlation with stomatal conductance (Ramalaxmi et al. 2011). Similar relationships have been reported in an earlier study (Balasimha

1991). Even under dry summer conditions when VPD of 1.94–4.10 kPa was prevalent, the stomatal conductance was not affected appreciably. Increasing stomatal opening allows greater CO_2 diffusion with enhanced photosynthesis. When accessions were compared, 1/4 Ullal, M 16/1, and A18/4 showed highest stomatal conductance, Pn, and transpiration rates. These also recorded highest WUE thus indicating these accessions to have very good photosynthetic efficiency. However, stomatal conductance did not show any trend directly with carbon isotope discrimination. The leaf water potential was generally high and did not show any effect of drought. These responses of cashew accessions demonstrate high adaptability to dry environments.

16.6.2 Photosynthesis

16.6.2.1 Cocoa

The efficient stomatal closure might hamper carbon assimilation. It was, therefore, necessary to study photosynthesis and yield performance of tolerant accessions vis-à-vis susceptible ones. Annual pod yield was higher in tolerant accessions than in susceptible ones studied over a period of 5 years. Since the stomata resume their original open condition immediately after stress is relieved, it is reasonable to assume that stomatal closure did not affect productivity. This is corroborated by the lack of correlation between yield and stomatal conductance, due to stress ($r = 0.04$). It is also of interest to note that there was a positive correlation to yield with stomatal conductance, at recovery ($r = 0.52$, $P \geq 0.05$). Therefore, accessions, which display these specific favorable attributes, can be used as source for breeding to bring desirable characters into a single ideotype with a good expectation of increasing drought tolerance (Balasimha et al. 1988). Thus, the stomatal responses are primary events in affecting the photosynthesis. Influences of environmental factors on photosynthesis in cocoa accessions were studied (Table 16.2; Balasimha et al. 1991). Net photosynthesis (Pn), transpiration (E), and stomatal conductance (gs) showed significant seasonal variations. During periods of low vapor pressure deficit (VPD), Pn was the highest. The drought-tolerant accessions maintained higher leaf WP. There was no significant difference in Pn between accession types. The depression in Pn was principally due to lowered stomatal

Table 16.1 Photosynthetic parameters of cashew accessions

Accessions	Pn (μmol/m^2/s)	Transpiration (mol/m^2/s)	Stomatal conductance (mmol/m^2/s)	Intercellular CO_2 (ppm)
Mean ($n = 23$)	3.73	6.79	393.86	309.9
F value	3.24	3.35	4.33	4.57

Light ranged from 101 to 709 μmol/m^2/s; VPD 1.94–4.10 kP

Table 16.2 Seasonal changes in P_n and other of tolerant and susceptible cocoa accessions

Parameters	Accession types	Month					Mean
		Feb	Apr	May	Jun	Oct	
P_n (μmol m^{-2}s^{-1})	Tolerant	3.47	1.98	2.16	3.50	4.17	3.06
	Susceptible	2.41	2.48	2.20	2.73	3.89	2.74
g_s (mol m^{-2}s^{-1})	Tolerant	0.14	0.07	0.07	0.18	0.27	0.15
	Susceptible	0.10	0.09	0.08	0.14	0.31	0.14
E (mmol m^{-2}s^{-1})	Tolerant	2.77	1.91	2.44	1.82	3.54	2.50
	Susceptible	1.86	2.39	2.45	1.84	5.37	2.78
WUE (μmol CO_2/mmol H_2O)	Tolerant	1.31	1.03	0.90	2.10	1.25	1.32
	Susceptible	1.31	1.03	0.91	1.76	0.88	1.18
ψ (-Mpa)	Tolerant	0.55	0.69	0.93	0.40	0.63	0.64
	Susceptible	0.54	0.91	1.03	0.43	0.70	0.73

conductance. Because of such linear relationships of Pn to gs, ratio of CO_2 internal/ CO_2 ambient remained relatively constant despite seasonal and accession-type variations (0.80–0.88). This type of relationship is rarely measured in the field. There was also no significant difference in internal CO_2 (Ci) showing that it did not limit net assimilation rate. The Pn/gs ratio increased during dry months accompanied by a decrease in internal CO_2. Similar decrease in CO_2 assimilation in seedlings has been reported (Joly and Hahn 1989). Daily carbon balances of individual leaves of cocoa seedlings were decreased with reduced leaf water potentials. The instantaneous water use efficiency (Pn/E) was positively correlated with P_N, and it was higher in drought-tolerant trees. The Pn/Ci ratio decreased in dry season and was positively correlated with Pn/E. Thus, high Pn/Ci with low gs and E during stress period may be an important contributing factor to drought tolerance in cocoa.

Diurnal patterns of Pn and related parameters were studied in the months of December and April. Pn and gs were highest at 10:30 h and declined toward midday and afternoon. The stressed plants had lower Pn and gs values. The lowest value of WP was recorded at midday. At midday highest levels of PAR and VPD were recorded. The decrease in gs was more pronounced in April when both soil and atmospheric moisture were low and demand for water was high. Thus, there was depression in Pn under transient water stress.

Stomata normally close in response to increasing VPD. The responses of Pn become even clearer when it is plotted against VPD. Maximum Pn was recorded up to a VPD of 1.8 kPa; thereafter Pn declined. Temperature also had an effect on Pn, being optimum at 31–33 °C, which is characteristic of tropical crops grown in warm, less arid tropics. Light saturation of Pn occurred at a PAR of 600 μmol $m^{-2}s^{-1}$ (Fig. 16.1). However, high light intensity decreased Pn under water stress conditions. The relationship of environmental variable and Pn in cocoa clearly demonstrates that the latter is influenced by both soil and atmospheric droughts (Fig. 16.2).

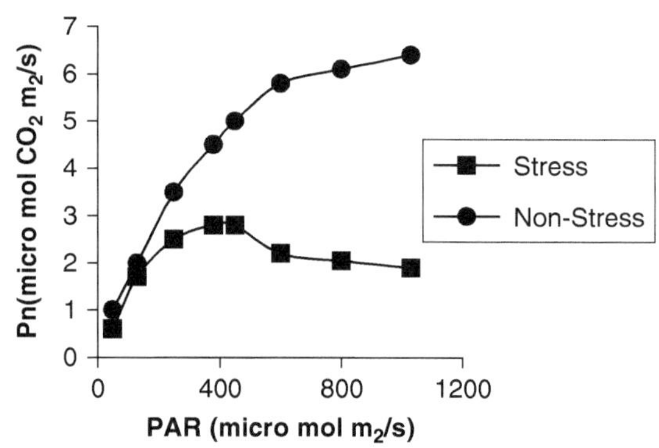

Fig. 16.1 Light saturation curve for Pn in cocoa under different conditions

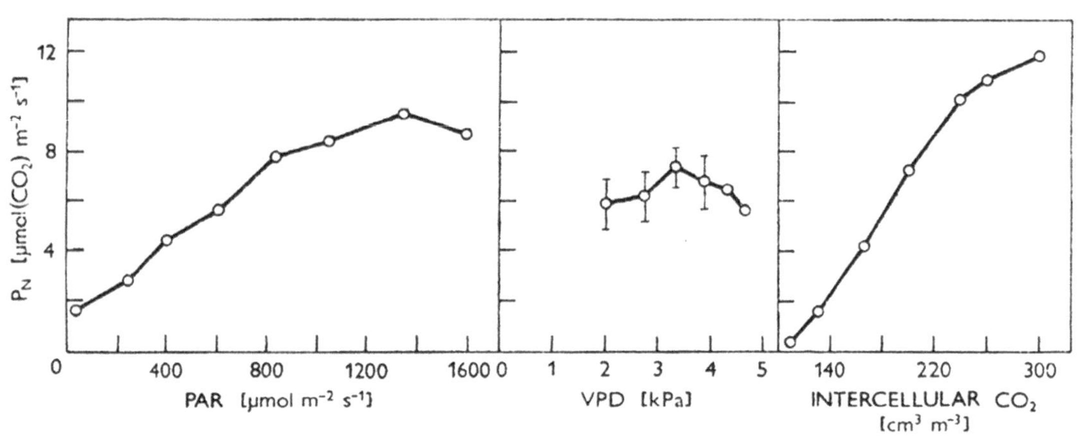

Fig. 16.2 Pn in cashew under different conditions of light, VPD and CO_2 (Int)

16.6.2.2 Cashew

Photosynthetic studies on cashew were first reported by Balasimha (1991). The effect of leaf position, irradiance, and environmental parameters were described. Pn was linearly related to stomatal conductance and transpiration. Light saturation was at 1,200–1,400 μmol/m^2/s. The CO_2 compensation was between 80 and 100 ppm (Fig. 16.3). The photosynthetic rate (Pn) was studied in cashew in relation to plant density (Balasimha and Yadukumar 1993), leaf position, and radiation (Palanisamy and Yadukumar 1993). Influence of leaf age and branch girdling (Schaper and Chacko 1993) and effects of drying and irrigation on gas exchange (Schaper et al. 1996; Blaikie et al. 2001) have been studied in Australia. In Brazil, photosynthetic responses to varying environmental conditions in young cashew plants are studied (Pereira de Souza et al. 2005).

Irrigation and fertilizer levels affected the photosynthetic characteristics in cashew (Yadukumar and Balasimha 2006). Photosynthesis, transpiration, and stomatal conductance were significantly higher in irrigated and fertilizer-applied trees than in unirrigated and unfertilized ones. The reduced levels in photosynthesis in unirrigated trees were associated with higher air/leaf temperature and lower relative humidity. Leaf water potential did not vary significantly due to irrigation and fertilizer treatments. Gas exchange, leaf water status, and yields of irrigated cashew were higher in Australia (Schaper et al. 1996). Thus for improving cashew performance under dry and high temperatures, regime irrigation would be suitable.

16.6.3 Chlorophyll Fluorescence as an Index of Stress Tolerance

16.6.3.1 Cocoa

Chlorophyll fluorescence parameters were studied in cocoa accessions belonging to drought-tolerant and susceptible types. Chlorophyll fluorescence showed significant differences at seasons and between accession types and leaf ontogeny (Balasimha 1992; Balasimha and Daniel 1995). The F_0 values were lower during drier months (March–May) as compared to other months. The F_0 was significantly higher in susceptible accessions showing that PSII was affected to a greater extent. However, F_M and F_V were lower in tolerant types. Correlations between Pn and fluorescence quenching parameters were done to know the nature of relationships. The fluorescence parameter, viz., F_V/F_M, can be used to estimate quantum yield of noncyclic electron transport and derived as F_M-F_S/F_M. These values were found to be directly proportional to fluorescence quenching in cocoa. There was also a linear correlation of Pn with PQ. Thus, the data demonstrate that the Pn shows linear relations with F_o/F_M and fluorescence quenching. These values can be used for routine measurement of photosynthetic performance of leaves.

The application of chlorophyll fluorescence as a tool to screen cocoa for drought tolerance was investigated. Excised leaves were incubated in PEG 4000 (−1.9 Mpa) solution either in low or high light conditions for 2 h. The incubation in low light (80 μmol $m^{-2}s^{-1}$) did not change F_v/F_M ratios in accession types. Under high light incubation (1690 μmol $m^{-2}s^{-1}$), F_0 value increased in susceptible leaves. There were also changes in F_v/F_M ratios which decreased when incubated with either PEG or air drying. These changes were more pronounced in susceptible leaves than in tolerant ones. Under PEG stress WP was higher in tolerant as compared to susceptible leaves. Due to heat stress (47 °C), treatments in excised leaves F_0 values increased, while F_v, F_M, and F_v/F_M decreased significantly. This increase was more in susceptible trees. The higher F_v/F_M ratio in tolerant cocoa under heat-stressed conditions indicates that reduction of quantum yield of PSII was to a lesser extent. The Hill activity in isolated chloroplasts showed a similar trend. Thus the overall efficiency of PSII was affected more in susceptible leaves. The results suggest that relative chlorophyll fluorescence can be used for screening cocoa drought tolerance (Balasimha and Namboothiri 1996).

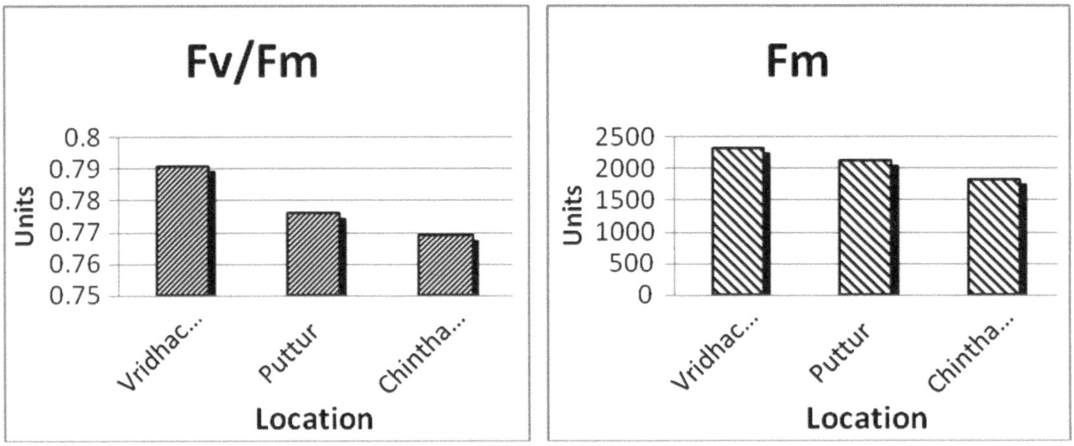

Fig.16.3 Chlorophyll fluorescence indices in cashew grown at different locations

Table 16.3 Changes in leaf characteristics of cocoa accessions

	Tolerant		Susceptible	
Parameter	Unstressed	Stressed	Unstressed	Stressed
Epicuticular wax (μg/cm^2)	24.78	38.86	18.01	31.66
Total lipid (mg/g FW)	10.46	7.39	9.97	6.27
Neutral lipid (%)	69.40	77.20	66.40	78.00
Glycolipid (%)	20.90	19.10	22.80	17.00
Phospholipid (%)	9.70	3.70	10.70	4.90
Total sterol (%)	3.50	4.40	3.60	4.80
Sterol/phospholipid ratio	0.36	1.19	0.34	0.98
Electrolytic leaching (%)	15.56	20.19	17.96	27.66

16.6.3.2 Cashew

The characters F_v, F_M, and F_v/F_M were studied in cashew germplasm at three different agroclimatic regions (Balasimha et al. 2011). There were significant variations in the germplasm collections screened. The chlorophyll fluorescence indices also showed different levels in the three locations (Fig. 16.3).

16.6.3.3 Epicuticular Waxes: Lipid End Membrane Stability

Epicuticular wax increases due to stress and helps the plants to withstand drought. Waxes accounted for 0.4–0.8 % as total leaf dry weight. The EW increased due to stress and tolerant secessions showed higher levels. Waxes contained hydrocarbons, esters, ketones, alcohols, free fatty acids, and two unidentified components. No qualitative differences were observed between secession types. Lipids constituted 0.62–1.04 % of total leaf fresh weight. Water stress in April increased total lipid content to an extent of 20 % in tolerant and 30 % in susceptible accessions. Polar lipid fraction decreased while neutral lipids increased due to stress (Table 16.3). The main fatty acids of lipid fractions were myristic, palmitoleic, oleic, linoleic, and linolenic acids (Bhat et al. 1990).

The activities of peroxidase and superoxide dismutase were studied in cocoa accession types. The peroxidase activity remained almost constant with decreasing WP in tolerant while it increased in susceptible trees. There was also a higher stability of SOD in tolerant trees. The lipid peroxidation was lower in tolerant trees both during prestress and stress conditions. These changes, along with lipid fraction ratios recorded and lower electrolytic leaching, demonstrate the increased membrane stability.

The epicuticular wax contents of cashew showed wide differences ranging from 65 to 202 μg/cm^2. There were variations among the accessions. These parameters are also contributing factors for drought tolerance and higher values are positive traits.

16.7 Water Use Efficiency Under High CO_2 and Temperature

Plantation crops such as cocoa, being perennial in nature, have to face the impact of climate change even during a single generation. In order to fill the gap in knowledge on response under long-term exposure at whole plant level to elevated CO_2 and temperature, cocoa plants were grown in open-top chamber (OTC) facility, wherein two elevated CO_2 levels (550 and 700 μmol CO_2 mol^{-1}) and elevated temperature (+2 °C above ambient OTC) were maintained apart from the chamber control (ambient temperature and CO_2) (Naresh Kumar et al. 2012). Results done over 2 years indicated that the elevated CO_2 significantly increased net photosynthetic rates (Pn) while the transpiration rates (E) remained almost similar to that in ambient CO_2 conditions (Fig. 16.4). As a consequence, the instantaneous water use efficiency (Pn/E-WUE) has increased by 51 % at 550 μmol CO_2 mol^{-1} and by 112 % at 700 μmol CO_2 mol^{-1} as compared to that in chamber control. Even though a 2 °C increase in temperature above control chamber temperature caused a slight (~4 %) reduction in the Pn rates, there was an overall increase in the instantaneous WUE by about 20 % indicating the ability of plants to adjust the stomatal conductance in such a way that they are able to efficiently utilize the water in case of increase in temperature. Elevated CO_2 and temperature increased the concentrations of chlorophyll relatively more than that of chlorophyll b and thus increased the chl a/b ratio. Elevated CO_2 and temperature also increased the concentration of total soluble sugars and starch in leaf tissue. The chlorophyll transients such as F_0, F_v, and F_M did not significantly differ due to growing condition. The F_v/F_M was more than 0.8, indicating that plants were not under stress in all the treatments.

Increase in photosynthesis under elevated CO_2 has been reported in several other C_3 crops (Baxter et al. 1995; Case et al. 1998; Osborne et al. 1997; Wang et al. 2000). Since cocoa is a C_3 plant, the photosynthetic rate accelerated due to the increased availability of CO_2 at the RUBISCO active site. Increased concentration of CO_2 in sub-stomatal cavity seems to have checked the transpiration loss of H_2O molecules from the stomata. This enhanced the water use efficiency as indicated by increase in

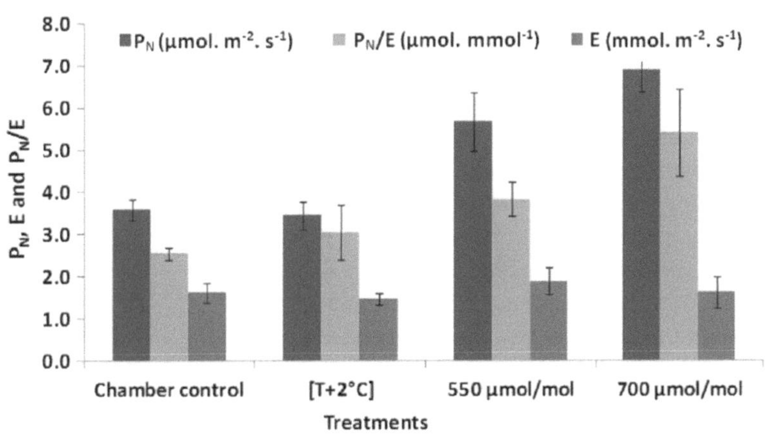

Fig. 16.4 Changes in net photosynthetic rate (Pn), instantaneous water use efficiency (Pn/E), and transpiration rate (E) in the physiologically mature leaves of cocoa plants grown in elevated CO_2 and temperature conditions for 2 years. Each bar represents the mean values of 525 readings each from seven cultivars/hybrids

Pn/E. Exposure to elevated CO_2 not only caused a significant increase in the soluble sugar and starch contents in plants but also caused an increase in the cellulose content (Teng et al. 2006). There is often an increase in non-structural carbohydrates in leaves (Farrar and Williams 1991; Stitt 1991). In cocoa, elevated temperature did not affect the PSII levels, showing relative tolerance among the cocoa genotypes to marginal increases in temperature. The genotypic differences in chlorophyll fluorescence in response to variation in temperature (Daymond and Hadley 2004) are reported, and chlorophyll fluorescence was used as the indicator of stress in cocoa plants and as one of the criteria for selecting drought-tolerant cocoa plants (Balasimha and Namboothiri 1996). The elevated temperature-induced reduction in transpiration rates is mostly mediated by stomatal regulation. Increase in WUE in plants grown in elevated temperature and CO_2 indicates that in future climates, where rise in atmospheric CO_2 accompanies rise in temperature, it is likely that cocoa plants stand net beneficiaries with respect to photosynthesis and water balance in leaves. This situation may lead to increased biomass accumulation in plants.

Results indicate that elevated CO_2 and temperature may prove to be beneficial for cocoa plantations in climate change scenarios, particularly in view of the fact that cocoa is grown as an intercrop in the palm gardens and mostly are maintained under water non-limiting conditions. In addition to increased physiological water use efficiency in elevated CO_2 and temperature conditions, there is a need to increase the field-level (agronomic) water use efficiency through drip irrigation or fertigation for harnessing the potential benefits due to elevated CO_2 and also to avoid or minimize the adverse effects, if any, of elevated temperature in future climates in cocoa-growing regions in India.

In the cashew-growing regions of Ghana and Côte d'Ivoire, yearly and monthly minimum and maximum temperatures are expected to increase by 2030, while rainfall changes may be trivial (CIAT 2011). It is predicted to reach a mean of 31.1 °C during this period. Increasing the CO_2 may increase photosynthetic assimilation rates and reduce stomatal conductance (Pereira de Souza et al. 2005).

16.8 Carbon sequestration

Mitigation of green house gas emission is one of the important aspects related to quest against climate change. Carbon sequestration by terrestrial biomass is one of the mitigation options used for the reduction of GHGs. Agroforestry systems provide opportunity for carbon sequestration under clean development mechanism. Arecanut cocoa based cropping systems produce abundant biomass to qualify for this. The standing biomass increased over time indicating accumulation of biomass in stem and also due to increase in yield by arecanut and cocoa plant with age up to 20th year of planting (Fig. 16.5). In case of cocoa yield stabilization takes place around five years of field planting (Balasimha 2001). Around 30–50 % of photosynthetically active radiation is transmitted trough arecanut canopy (Balasimha 1989; Balasimha and Subramonian 1984). Cocoa, with its compact and high leaf area, is able to intercept 90 % of the remaining PAR. Cocoa production under shade trees has been reported to be 1 t /ha/year in Costa Rica (Beer et al. 1990). However, biomass production under arecanut and cocoa agro forestry system under tropical conditions prevailing in India is higher. The microclimate especially shade, soil moisture and temperature in arecanut gardens were found to be ideal for cocoa growth and productivity (Shama Bhat and

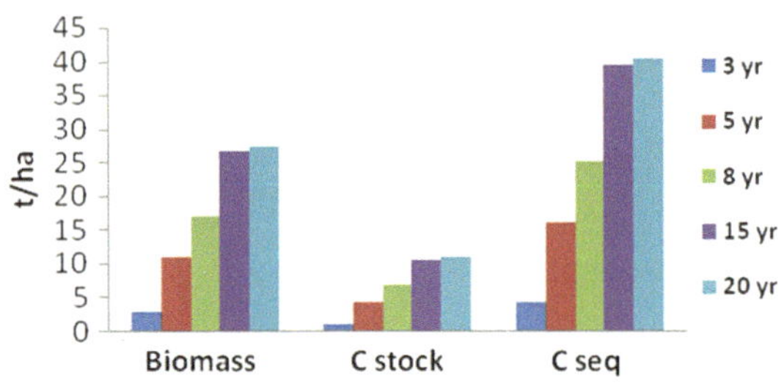

Fig. 16.5 Biomass, carbon stock and carbon sequestration in cocoa

Bavappa 1972). More over these systems are reported to be highly compatible at arecanut and cocoa population ratio at 1:2.

Carbon estimations, done on the basis of biomass and carbon percentage in tissue indicate net carbon sequestration by arecanut-cocoa system (Balasimha and Naresh Kumar 2010, 2013). The carbon sequestration increased considerably during the growth of these plants (Fig. 16.5). Positive carbon sequestration estimations in various land use systems in Southern Cameroon including several shaded cocoa based agroforestry systems were reported (Kotto-Same et al. 1997). However, these include the uncertainties related to future shifts in global climate, land-use and land cover, the poor performance of trees and crops on substandard soils and dry environments, pests and diseases. In addition, more efforts are needed to improve methods for estimating carbon stocks and trace gas balances such as nitrous oxide (N_2O) and methane (CH_4) to determine net benefits of agro forestry on the atmosphere.

High soil organic carbon levels indicate higher sequestration of carbon by soil as well as in the system. Temporal changes in soil organic carbon of cocoa-gliricidia agroforests in Indonesia has indicated that it remains fairly stable in different stratum (Smiley and Kroschel 2008). There is long spell of dry season during summer extending from 3 to 6 months. Therefore, it can be assumed that areca – cocoa system accumulates high amount of carbon in the soil.

The study has thus revealed that the biomass and primary productivity is considerable with areca-cocoa mixed crop and comparable to any agroforestry systems involving cocoa (Alpizar et al. 1986; Beer et al. 1990). However, in view of the fact that only afforestation and reforestration systems can be used for CDMs (Shrestha et al. 2005), the areca–cocoa system cannot be considered for carbon trade under clean development mechanism as of now. However, these methods on above ground biomass and carbon estimations provide useful basic information on these aspects and can be used as models in future, if there is any change in the policies when IPCC is to renegotiate the procedures.

References

Alpizar L, Fassbender HW, Heuveldop J, Folster J, Enriquez G (1986) Modelling agroforestry systems of cacao (*Theobroma cacao*) with *Cordia alliodora* and *Erythrina poeppigiana* in Costa Rica. I. Inventory of organic matter and nutrients. Agrofor Syst 4:175–189

Alvim P d T (1977) Cacao. In: Alvim P d T, Kozlowki TT (eds) Ecophysiology of tropical crops. Academic, London, pp 279–313

Alvim R, Alvim P d T, Loremi R, Saunders PF (1974) The possible role of abscisic acid and cytokinins in growth rhythms of *Theobroma cacao* L. Rev Theobroma 4:3–12

Asomaning EJA (1976) Cocoa research in Ghana. In: Simmons J (ed) Cocoa production. Praeger Publishers, New York, pp 168–201

Balasimha D (1982) Seasonal changes in nitrate reductase activity and other indicators of plant water stress in field cacao (*Theobroma cacao* L.) plants. Plant Physiol Biochem 9:74–79

Balasimha D (1983) Water relations and physiological responses to water stress in cacao. Plant Physiol Biochem 10(Special Vol):65–71

Balasimha D (1987) Cocoa. In: Sethuraj MR, Raghavendra AS (eds) Tree crop physiology. Elsevier Sci. Publ, Amsterdam, pp 263–285

Balasimha D (1989) Light penetration patterns through areca nut canopy and leaf physiological characteristics in intercrops. J Plant Crop 16(suppl):61–67

Balasimha D (1991) Photosynthetic characteristics of cashew trees. Photosynthetica 25:419–423

Balasimha D (1992) Net CO_2 assimilation and chlorophyll fluorescence in cocoa trees. Plant Physiol Biochem 19:23–26

Balasimha D (2001) Growth and yield of cocoa grown under areca in relation to plant density and anopy architecture. In: Proceedings of the 13th international cocoa research conference, Kota Kinabalu, 2000, pp 365–372

Balasimha D, Anil Kumar V (2000) Stomatal resistance and ABA concentration in cocoa plants due to drought. In: Recent advances in plantation crops research. UPASI, Valparai, pp 230–233

Balasimha D, Daniel EV (1988) A screening method for drought tolerance in cocoa. Curr Sci 57:395

Balasimha D, Daniel EV (1995) Changes in chlorophyll fluorescence during cocoa leaf development. J Plant Crop 23:64–65

Balasimha D, Namboothiri CGN (1996) Chlorophyll fluorescence of cocoa in relation to drought tolerance. Plant Physiol Biochem 23:64–66

Balasimha D, Naresh Kumar S (2010) Net primary productivity, carbon sequestration and carbon stocks in areca- cocoa mixed cropping system. In: Proceedings of the 16th international cocoa research conference, Bali, pp 215–226

Balasimha D, Naresh Kumar S (2013) Net primary productivity, carbon sequestration and carbon stocks in

areca- cocoa mixed cropping system. J Plant Crop 41:8–13
Balasimha D, Rajagopal V (1988) Stomatal responses of cocoa (*Theobroma cacao*) to climatic factors. Indian J Agric Sci 58:213–216
Balasimha D, Subramonian N (1984) Nitrate reductase activity and specific leaf weight of cocoa and light profile in areca nut-cocoa mixed cropping. In: Proceedings of the PLACROSYM VI, Kottayam, ISPC, Kasaragod. pp 83–88
Balasimha D, Yadukumar N (1993) Effect of plant density on photosynthesis in cashew. Indian J Plant Physiol 36:5–7
Balasimha D, Subramonian N, Subbaiah CC (1985) Leaf characteristics in cocoa (*Theobroma cacao* L.) accessions. Café Cacao The 29:95–98
Balasimha D, Rajagopal V, Daniel EV, Nair RV, Bhagavan S (1988) Comparative drought tolerance of cacao accessions. Trop Agric 65:271–274
Balasimha D, Daniel EV, Bhat PG (1991) Influence of environmental factors on photosynthesis in cocoa trees. Agric For Meteorol 55:15–21
Balasimha D, Anil Kumar V, Viraktamath BC, Ananda KS (1999) Leaf water potential and stomatal l resistance in cocoa hybrids and parents. Plant Rech Dev 6:116–120
Balasimha D, Ramalaxmi, Suchith N, Nayak MG, Jeeva S, Narasimha Reddy MN, Jose CT (2011) Chlorophyll fluorescence, stomatal conductance and yield of cashew germplasm from three agroclimatic regions of South India. J Plant Crop 39:232–235
Balasimha D, Elain Apshara S, Jose CT (2013) Genotypic variations in chlorophyll fluorescence and stomatal conductance of cocoa in relation to drought tolerance. J Plant Crop 41:40–45
Baxter R, Bell SA, Sparks TH, Ashenden TW, Farrar JF (1995) Effects of CO_2 concentrations on three montane grass species.III. Source leaf metabolism and whole plant carbon partitioning. J Exp Bot 46:917–929
Beer J, Bonnemann W, Chavez HW, Fassbender HW, Imbach AC, Martel I (1990) Modelling agroforestry systems of cacao (*Theobroma cacao*) with laurel (*Cordia alliodora* or poro (*Erythrina poeppigiana* in Costa Rica. V. Productivity indices, organic material models and sustainability over ten years. Agrofor Syst 12:229–249
Bezerra MA, De Lacerda CF, Gomes Filho E, De Abreu CEB, Prisco JT (2007) Physiology of cashew plants grown under adverse conditions. Braz J Plant Physiol 19:449–461
Bhat PG, Daniel EV, Balasimha D (1990) Epicuticular waxes, lipids and membrane stability of cocoa trees in relation to drought tolerance. Indian J Exp Biol 28:1171–1173
Blaikie SJ, Chacko EK, Lu P, Muller WJ (2001) Productivity and water relations of field-grown cashew: a comparison of sprinkler and drip irrigation. Aust J Exp Agric 41:663–673
Case AL, Curtis PS, Snow AA (1998) Heritable variation in stomatal responses to elevated CO_2 in wild radish, *Raphanus raphanistrum* (Brassicaceae). Am J Bot 85:253
CIAT (2011) Predicting the impact of climate change on cashew growing regions in Ghana and Cote d'Ivoire. International Center for Tropical Agriculture (CIAT), Managua. Final report September, 2011, pp 35
Daymond AJ, Hadley P (2004) The effects of temperature and light integral on early vegetative growth and chlorophyll fluorescence of four contrasting genotypes of cacao (*Theobroma cacao*). Ann Appl Biol 145:257–262
Farrar JF, Williams ML (1991) The effects of increased atmospheric carbon dioxide and temperature on carbon partitioning, source-sink relations and respiration. Plant Cell Environ 14:819–830
Gordon J (1976) Cocoa: its nature habitat and cultivation. In: Simmons J (ed) Cocoa production. Praeger Publishers, New York, pp 3–29
Joly RJ, Hahn DT (1989) Net CO_2 assimilation of cocoa seedlings during periods of plant water deficit. Photosynth Res 21:151–159
Kotto-Same J, Woomer P, Moukam A, Zapfack L (1997) Carbon dynamics in slash-and-burn agriculture and land use alternatives in humid forest zone of Cameroon. Agric Ecosyst Environ 65:245–256
Naresh Kumar S, Murali Krishna KS, John Sunoj, Balasimha D (2012) Effect of elevated CO_2 and temperature on photosynthesis and chlorophyll fluorescence of cocoa (*Theobroma cacao* L.) in open top chambers. 17th international cocoa research conference, 15–20 Oct 2012, Yaounde, pp 46
Osborne CP, Drake BG, Laroche J, Long SP (1997) Does long – term elevation of CO_2 concentration increase photosynthesis in forest floor vegetation/ (Indiana strawberry in a Maryland Forest). Plant Physiol 114:337–344
Palanisamy K, Yadukumar N (1993) Photosynthesis in relation to radiation and leaf position in cashew trees. Photosynthetica 29:113–116
Pereira de Souza R, Ribeiro RV, Machado EC, Oliveira RF, Gomes de Silveira JA (2005) Photosynthetic responses of young cashew plants to varying environmental conditions. Pesqisa Agropecuaria Bras 40:735–744
Ramalaxmi, Naik S, Balasimha D, Nayak MG (2011) Photosynthetic characteristics in cashew accessions. J Plant Crop 39:210–212
Schaper H, Chacko EK (1993) Effect of irradiance, leaf age, chlorophyll content and branch girdling on gas exchange of cashew (*Anacardium occidentale* L) leaves. J Hortic Sci 68:541–550
Schaper H, Chacko EK, Blaikie SJ (1996) Effect of irrigation on leaf gas exchange and yield of cashew in Northern Australia. Aust J Exp Agric 36:861–868
Shama Bhat K, Bavappa KVA (1972) Cocoa under palms. In: Proceedings of the conference on cocoa and coconuts in Malaysia, pp 116–121

Shrestha RM, Sharma S, Timilsina GR, Kumar S (2005) Baselines for afforestation and reforestation (A&R) projects. In: Lee MK (ed) Baseline methodologies for clean development mechanism projects. UNEP, Roskilde, pp 120–143

Smiley GL, Kroschel J (2008) Temporal change in carbon stocks of cocoa – gliricidia agroforests in Central Sulawesi, Indonesia. Agrofor Syst 73:219–231

Stitt M (1991) Rising CO_2 levels and their potential significance for carbon flow in photosynthetic cells. Plant Cell Environ 14:741–762

Teng N, Wang J, Chen T, Wu X, Wang Y, Lin J (2006) Elevated CO_2 induces physiological, biochemical and structural changes in leaves of *Arabidopsis thaliana*. New Phytol 172:92–103

Vijayakumar BG, Veerappa Devaru G, Balasimha D, Khader KBA, Ranganna G (1991) Influence of weather on areca nut and cocoa yield. J Plant Crop 19:33–36

Wang X, Curtis PS, Pregitzer KS, Zak DR (2000) Genotypic variation in physiological and growth responses of *Populus tremuloides* to elevated atmospheric CO_2 concentration. Tree Physiol 20:1019–1028

Yadukumar N, Balasimha D (2006) Effects of drip irrigation and fertilizer levels on photosynthesis in cashew. Indian J Hortic 63:310–315

Yadukumar N, Mandal RC (1994) Effect of supplementary irrigation on cashew nut yield. In: Problems and prospects. Centre for Water Resource Development and Management, Calicut, pp 79–84

Black Pepper and Water Stress

17

K.S. Krishnamurthy, S.J. Ankegowda, P. Umadevi, and Johnson K. George

Abstract

Black pepper is mainly grown as a rainfed crop. Total rainfall and its distribution influence pepper productivity. Heavy rains during flowering reduce the rate of pollination and continuous heavy rainfall promotes vegetative development and limits flowering. On the other hand, break in the rainfall experienced at different stages during critical period following flower initiation leads to severe reduction in yield. Black pepper needs to be irrigated in summer to harvest good crop. Low spike intensity in rainfed situation is due to staggered and delayed spiking, lower bisexual flowers, anthracnose incidence, and spike shedding. Both plant height and leaf area are affected by water stress. Tolerant genotypes maintain higher root growth, higher relative water content, and lower cell membrane leakage under stress condition. In general, tolerant genotypes accumulated all the amino acids in higher quantities compared to susceptible ones during water stress. A, g_s, and E decreased drastically, while T_{leaf} increased after 6 days of stress induction. Water stress reduced the activity of catalase and acid phosphatase and increased the activities of peroxidase, glutathione reductase, and polyphenol oxidase enzymes. Lipid peroxidation was high in drought-susceptible cultivars compared to tolerant cultivars. There was no difference in the expression profile of isoforms of catalase and peroxidase enzymes between control and stress, while super oxide dismutase showed an additional isoform in some genotypes under stress condition. Studies on 2D SDS gel electrophoresis indicated that the proteins relevant to photosynthesis were downregulated during drought suggesting reduced photo oxidative stress and protection of

K.S. Krishnamurthy (✉) • P. Umadevi • J.K. George
ICAR-Indian Institute of Spices Research, Kozhikode 673012, Kerala, India
e-mail: kskrishnamoorthy@spices.res.in

S.J. Ankegowda
ICAR-Indian Institute of Spices Research, Regional Station, Appangala, Madikeri, Karnataka, India

N.K.S. Rao et al. (eds.), *Abiotic Stress Physiology of Horticultural Crops*,
DOI 10.1007/978-81-322-2725-0_17

photosynthetic machinery during stress. Studies on exogenous application of ABA indicated a role for ABA in stress tolerance.

Gene expression studies showed higher expression levels of dehydrin, osmotin, and DREB genes in tolerant genotype, implicating their possible role in imparting drought tolerance.

17.1 Introduction

Black pepper is a crop of humid subtropics and is a shade-loving plant. It is mainly grown as a rainfed crop. Total rainfall and its distribution play an important role in black pepper cultivation and productivity. The rainfall requirement of the crop varies from 2000 to 3000 mm. Tropical temperature and high relative humidity with little variation in day length throughout the year is relished by the crop. A relative humidity of 60–95 % is optimum at various stages of growth. The crop tolerates temperature between 10 and 40 °C. The ideal temperature is 23–32 °C with an average of 28 °C. Optimum soil temperature for root growth is 26–28 °C (De waard 1969; Wahid and Sitepu 1987). It does not tolerate excessive heat and dryness (Sivaraman et al. 1999). It is a weather-sensitive crop and yield is influenced considerably by environmental factors. Crop yield is the function of many factors like weather, soil type and its nutrient status, management practices, and other inputs available. Of these, weather plays an important role, probably more so in India where aberrant weather such as drought, flood, etc. is a rule rather than an exception. Interannual yield variation in black pepper in India is mainly due to weather (Kandiannan et al. 2007). Vijayakumar et al. (1985) found that black pepper vines exposed to direct solar radiation developed physiological disorders. Even under favorable soil moisture conditions, black pepper vines kept under shade (7 % incident light) remained green and healthy, whereas those exposed to sunlight turned yellow and developed necrotic patches during summer (Vijayakumar and Mammen 1990). Shade to the extent of 50 % boosted the growth of black pepper cuttings in the nursery (Senanayake and Kirthisinghe 1983).

17.2 Black Pepper-Weather Relationship

Black pepper annual cycle is broadly divided into (1) lag period (March–May), it is interval between harvest and subsequent spike initiation; (2) spike emergence and flower bud differentiation (June–July), new leaf accompanied by spike; (3) spike enlargement and berry formation (August–October); (4) berry development (November–December); and (5) maturity and harvest (January–February). These phases may overlap and vary depending on rainfall distribution, variety, and location. Regression model for different weather parameters with black pepper fresh spike yield has been developed and brought out the magnitude of association, i.e., $RH_{Max} >$ RAIN $> T_{Min} > T_{Max} >$ Sunshine $>$ WIND $> RH_{Min} >$ EVPN (Kandiannan et al. 2011a). This indicates that RH_{Max} and rainfall have greater role in obtaining good yield in black pepper. One of the drawbacks of pepper production is the indiscriminate use of varieties irrespective of the prevailing agroclimatic conditions. In a state like Kerala with diverse climatic and soil conditions, identification and popularization of location-specific varieties are important steps in increasing the production and productivity (Nybe et al. 1999).

17.2.1 Rainfall and Pepper Productivity

Availability of water during critical crop growth stages such as during initial years of establishment, flowering. and berry filling periods is very important for obtaining good yields. A period of dry spell before reproductive primordial initiation helps in inducing profuse flowering (Ridley 1912; Pillay et al. 1988). Heavy rains during flowering reduce the rate of pollination and continuous heavy rainfall promotes vegetative development and limits flowering (Pillay et al. 1988). Pillay et al. (1988) compared the rainfall pattern and pepper yields during two extremely adverse years (1980–1981 and 1986–1987) with that of a favorable year (1981–1982) and noted that during both adverse years, there was a distinct break in the rainfall experienced at different stages during critical period following flower initiation which lead to 75 % reduction in yield in both the years. But during favorable year (1981–1982), precipitation remained steady without any break and the pepper yields were high. They observed that a low rainfall (70 mm in 20 days during May) is enough for triggering off flushing and flowering process. Significant correlations were obtained between rainfall received (amounts to 100 mm to attain field capacity during first half of May) with yield in black pepper ($r = 0.75$) and also with rainfall received during the second half of June and yield ($r = 0.90$) (if preceded by rainfall in the first half of May) (Pradeepkumar et al. 1999). Remold (1997) indicated that dry spell is essential during March–May months for better flowering. But once the process is set off, there should be continuous shower until fruit ripening, and any dry spell even for a few days, within this critical period of 16 weeks (flowering to fruit ripening), would result in low yield. It was also found that high total annual growth of fruiting branches was required in June–July, coinciding with peak period of monsoon for better yield. High dry matter accumulation was observed in branches just before shoot elongation and flowering during April–May (Mathai and Sastry 1988). It was also noted that the change in rainfall pattern during 1999–2000 crop season affected the flowering and yield of black pepper in Idukki, a predominant black pepper-growing zone (John et al. 1999). Similar observation was also noted in Bangka, Indonesia, by Suparman (1998). There is an increasing trend in rainfall during summer months (Kandiannan et al. 2011b) that could affect the flowering pattern of black pepper. The late commencement of Southwest monsoon delays flowering. Heavy Northeast monsoon showers after a spell of dry period after SW monsoon results in high spike drop.

In India years 1987 and 2002 were drought years as the rainfall during those years was less than 80 % of long-term average. So, the rainfall as well as pepper production data for these years were collected from black pepper-growing regions and compared with the long-term (1984–2004) average (Table 17.1). The rainfall deficit during 1987 was around 20 % for Wayanad region, 15 % for Kannur, 10 % for Trichur, and only around 7 % for Idukki region, while the deficit during 2002 was more than 30 % for black pepper-growing regions except Trichur (15.8 %). The black pepper productivity was also below normal (mean of 1984–2004) for all the places indicating the negative influence of rainfall deficit on black pepper productivity. Interestingly, though the rainfall deficit was more pronounced during 2002, productivity reduction was more during 1987 compared to 2002 for Wayanad, Idukki, and Trichur. A close look at the rainfall distribution pattern revealed that the rainfall was more evenly distributed during 2002. Also, the pre-monsoon (January–May) and early monsoon (June–July) rainfall was more during 2002 compared to 1987 which would have helped in better flowering and early fruit set during 2002 compared to 1987. Also it was found that the mean temperature for the year 2002 was less than that for the year 1987 for Idukki and Trichur which also would have contributed for better performance during 2002 (Krishnamurthy et al. 2015). In high elevation, in a year of pre-monsoon failure, black pepper needs to be irrigated in summer to harvest good crop in coffee-based cropping system where black pepper is a mixed crop. Low spike intensity in rainfed situation is due to staggered and delayed spiking, lower bisexual flowers, anthracnose incidence, and spike shedding. Earlier workers also have reported lower number of berries in rainfed situation and are attributed to delay spiking, lower number of bisexual flowers,

Table 17.1 Rainfall and productivity deficits during drought years

	Rainfall (mm)			% Deficit		Productivity (kg/ha)			% Reduction	
Place	Mean	1987	2002	1987	2002	Mean	1987	2002	1987	2002
Wayanad	1,931	1,574	1,194	18.5	38.2	402	222	387	44.8	3.90
Kannur	3,348	2,849	1,894	14.9	43.4	241	177	147	26.6	39.0
Idukki	1,902	1,772	1,313	6.8	31.0	327	138	238	57.8	27.2
Trichur	2,743	2,463	2,309	10.2	15.8	174	88	129	49.4	25.9

Table 17.2 Influence of pre-monsoon irrigation on pepper yield and quality

Treatment	Soil moisture (%)	No. of spikes per 0.5 m^2	Spike length (cm)	Dry yield (kg/vine)	Essential oil (%)	Oleoresin (%)
Irrigated	15–16	26.5	11.9	5.8	1.65	8.75
Rainfed	6–8	17.3	9.2	2.8	1.65	6.90
CD value ($p = 0.05$)		2.1	1.4	1.2	NS	1.1

and partial setting (Ravindran et al. 2000; Ankegowda et al. 2011).

17.2.2 Pre-monsoon Irrigation and Pepper Productivity

Correlation studies between rainfall and productivity and results of basin irrigation study by Ankegowda et al. (2011) suggested that pre-monsoon rain has an influence on pepper productivity. This leads to the development of summer irrigation package for black pepper which involves providing fortnightly irrigation (at 50 L $vine^{-1}$) during March to May months and shade regulation during May first fortnight for better productivity and quality (Table 17.2).

17.3 Water Stress

Intensive shedding of berries in the advanced stage of fruit development is thought to be due to physiological disturbance in the plant caused by prolonged spell of drought or heavy rains or the sharp and sudden alteration of the two. Intensive shedding occurs during years in which heavy Northeast monsoon showers are received after a spell of dry period followed by SW monsoon (Sukumara pillay et al. 1977).

17.3.1 Morphological Parameters as Affected by Water Stress

New growth put forth by the pepper plant under water stress is very limited. Both plant height and leaf area are affected by water stress, but leaf area development is affected to the maximum extent. During moisture stress, leaf expansion rate decreased before soil moisture reaching critical levels indicating that screening cultivars based on leaf expansion rate may not be useful in black pepper (Ramadasan and Vasantha 1994). Root shoot ratio and relative water content showed negative correlation with wilting. Genotypes (Acc 813, 931, and 1495) with least wilting symptoms after 2 months of water stress (14 % soil moisture) showed highest relative water content and root shoot ratio indicating tolerant genotypes maintain higher root growth under stress conditions (Table 17.3).

17.3.2 Physiological Parameters as Affected by Water Stress

Chlorophyll a, chlorophyll b, and total chlorophyll content decreased with severe water stress. Chlorophyll/carotenoid ratio can be used as an index for stress studies in black pepper (Vasantha et al. 1989). However, the wax content of the leaf increased under water stress (Thankamani and

Table 17.3 Root/shoot ratio and relative water content as affected by water stress

Genotype	Root/shoot	RWC (%)
Panniyur-1	0.058	75.4
Sreekara	0.051	72.8
4057	0.062	73.0
813	0.074	83.0
1216	0.062	75.1
931	0.082	80.6
1218	0.057	72.7
1368	0.066	77.9
1495	0.081	82.5
CD at 5 %	0.015	1.4

Ashokan 2002). During water stress, relative water content decreased and cell membrane leakage increased. The decrease in relative water content and increase in membrane leakage intensified with stress intensity. Relative water content was positively correlated, while membrane leakage was negatively correlated with days taken for wilting. Tolerant accessions had higher relative water content and lower membrane leakage values. These parameters can be used to screen black pepper germplasm for drought tolerance (Krishnamurthy et al. 1998). Moisture loss from the excised leaves was positively correlated with relative water content and negatively with membrane leakage. The accessions which lost less moisture had higher relative water content and lower membrane leakage. The physiological characters of high stomatal resistance, low rate of transpiration, and low leaf water potential (less negative) were found to influence the drought tolerance character of black pepper. Leaf shedding and deep root penetration were also found to influence tolerance (Vasantha et al. 1990).

17.3.3 Water Stress and Compatible Solutes

Water-stressed plants accumulated more reducing sugars compared to control plants (Krishnamurthy et al. 1998). Thankamani and Ashokan (2004) reported that the nitrogen content increased with moderate and severe stress and maximum N content was observed in Panniyur-5, a relatively drought-tolerant variety. Potassium, phosphorus, calcium, zinc, and copper contents were also high in this variety under water stress, while an increase in magnesium content was observed in varieties Uthirankotta-2, Kumbakodi, and Panniyur-5. Severe stress decreased manganese content in all the varieties except Poonjarmunda. Iron content increased due to severe stress in varieties such as Kalluvally-4, Kumbakodi, and Padarpan. Among the tested varieties, Panniyur-5 showed more solute accumulation than all the other tolerant varieties. Higher leakage of calcium, magnesium, and potassium were noticed from water-stressed plants when leaf bits from control and stressed plants were put in distilled water (unpublished).

17.3.4 Influence of Water Stress on Amino Acid Composition

Total free amino acids content increased during water stress in black pepper (Krishnamurthy et al. 1998). Quantification of individual amino acids in tolerant and susceptible accessions revealed that most of the amino acids increased during water stress in both the tolerant and susceptible genotypes, while alanine content decreased in all the genotypes. In general, tolerant genotypes accumulated all the amino acids in higher quantities compared to susceptible ones during water stress (unpublished, Table 17.4).

17.3.5 Water Stress Effects on Gas Exchange Parameters

Gas exchange parameters such as photosynthetic rate (A), stomatal conductance (g_s), leaf temperature (T_{leaf}), and transpiration rate were studied in both irrigated and water-stressed pepper plants. A, g_s, and E decreased drastically, while T_{leaf} increased after 6 days of stress induction. However, the rate of decrease in photosynthetic rate and transpiration rate was less in tolerant varieties than the susceptible ones. Gas exchange parameters were very low and could not be

Table 17.4 Amino acids as affected by water stress (↑/↓ over control)

Acc No Amino acid	4201	4208	4211	4212	4206	4195	4230	4216
Asp	↑	↑	↓	↓	↑	↓	↑	↑
Thr	↑	↑	↑	↓	↑	↓	↑	↑
Ser	↑	↑	↓	↓	↑	↓	↑	↑
Glu	↑	↑	↑	↓	↑	↑	↑	↑
Gly	↑	↑	↑	↓	↑	↑	↓	↑
Ala	↓	↓	↓	↓	↓	↓	↓	↑
Val	↑	↑	↓	↓	↑	↓	↑	↑
Met	↑	↓	↑	↓	↑	↑	↑	↑
Ile	↑	↑	↓	↓	↑	↑	↓	↑
Leu	↑	↑	↓	↓	↑	↑	↑	↑
Phe	↑	↑	↑	↑	↑	↓	↑	↑
His	↑	↓	↓	↑	↓	↑	↑	↑

Table 17.5 Gas exchange parameters as affected by moisture stress

	A (μmol m^{-2} s^{-1})		g_s (mol m^{-2} s^{-1})		T_{leaf} (°C)		E (mol m^{-2} s^{-1})	
Acc No	Control	Stress	Control	Stress	Control	Stress	Control	Stress
813	2.62		0.02		31.5		0.92	
805	1.18		0.00		33.2		0.19	
910	2.96	1.18	0.02	0.00	33.1	35.7	0.84	0.34
809	2.53		0.02		33.5		1.00	
892	2.69	0.17	0.02	0.00	33.7	36.3	0.83	0.17
1188	3.48		0.02		34.6		0.80	
803	3.82	0.39	0.02	0.00	34.3	36.0	1.08	0.21
1334	2.20		0.01		35.5		0.58	
807	3.62		0.02		35.1		0.98	
1025	3.09		0.02		35.1		0.76	
1108	2.05	1.65	0.01	0.01	35.7	36.2	0.54	0.60
CD at 5 %	NS		NS		3.5		NS	
P value (cont vs stress)	$P < 0.01$		NS		$P < 0.01$		$P < 0.01$	

stabilized in most of the genotypes just after 6 days of stress induction (unpublished, Table 17.5).

17.3.6 Water Stress and Enzyme Activities

Water stress reduced the activity of catalase and acid phosphatase and increased the activities of peroxidase and polyphenol oxidase enzymes. Catalase activity was negatively correlated, while peroxidase activity was positively correlated with membrane damage (Krishnamurthy et al. 2000). Days taken for wilting showed a significant positive correlation with relative water content, significant negative correlation with peroxidase activity and membrane leakage, and no significant correlation with catalase and SOD activities (Fig. 17.1). Acid phosphatase activity decreased during water stress. Lipid peroxidation was high in drought-susceptible cultivars compared to tolerant cultivars (Thankamani et al. 2003).

Glutathione reductase activity increased during stress in general. Tolerant accessions had slightly higher levels of activity but the differences were insignificant. Ascorbate peroxidase activity also followed similar trend as that of peroxidase activity. The activity was highest after 5 days of stress (student's t test, $p = 0.01$ for control vs 5 days after stress and control vs

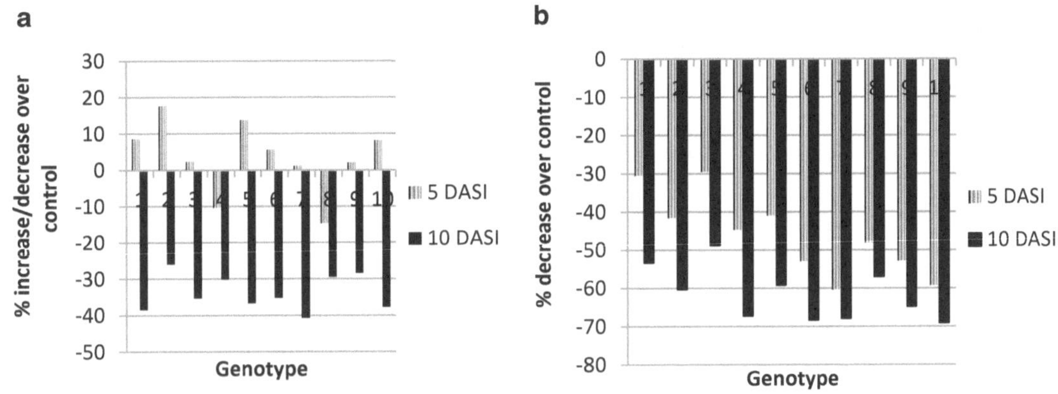

Fig. 17.1 Comparative catalase (**a**) and superoxide dismutase (**b**) activities in tolerant and susceptible accessions. *1–5* Tolerant accessions, *6–10* Susceptible accessions, *DASI* Days after stress induction

10 days after stress) in all the accessions. Like peroxidase activity, here also activity level decreased after 5 days of stress in all the accessions. Activities of all these enzymes have been shown to increase during water stress in coconut and the tolerant varieties maintained higher activities than the susceptible ones (Chempakam et al. 1993). The extent of oxidative stress in a cell is determined by the amounts of superoxide, H_2O_2, and hydroxyl radicals. Therefore, the balance of SOD, ascorbate peroxidase, and catalase activities will be crucial for suppressing toxic reactive oxygen species levels in a cell (Apel and Hirt 2004).

17.3.7 Protein and Isozyme Profiles

SDS-PAGE with Coomassie staining of tolerant and susceptible accessions or control and stress treatments showed similar banding pattern though there was difference in banding intensity. Isozyme profile of catalase showed a single wideband with an Em value from 0.03 to 0.15 and peroxidase showed two bands with Em values of 0.05 and 0.12 in all the accessions both under control and stress conditions, indicating that water stress does not have any influence on isoforms of these enzymes. The results also imply that there is no genotypic difference in the expression of isoforms of these enzymes. SOD showed five isoforms (Em values of 0.19, 0.48, 0.53, and 0.74) under control and six under stress condition in some accessions (Fig. 17.2), indicating that one additional isoform (Em value of 0.60) may be synthesized during water stress in some genotypes, the function of which is not known. There was no difference in banding pattern between tolerant and susceptible accessions. The results indicate that the activity levels of these enzymes may not be influenced by isoforms and that tolerance may not be controlled by polymorphic alleles of these enzymes.

17.3.8 Response of *Piper* Species to Water Stress

Among the species studied, *P. colubrinum* wilted fast and showed high membrane leakage, very high peroxidase activity, and low relative water content during stress period compared to other species, indicating that it is very susceptible, while *Piper chaba* was relatively tolerant (Krishnamurthy and Saji 2006). Acid phosphatase activity decreased during water stress in all the *Piper* species with the exception of *Piper longum* which showed 50 % increase in activity. *Piper colubrinum* showed the highest decrease of 95 %. Total proteins decreased with stress intensity. In general, tolerant varieties had highest proteins under stress compared to susceptible

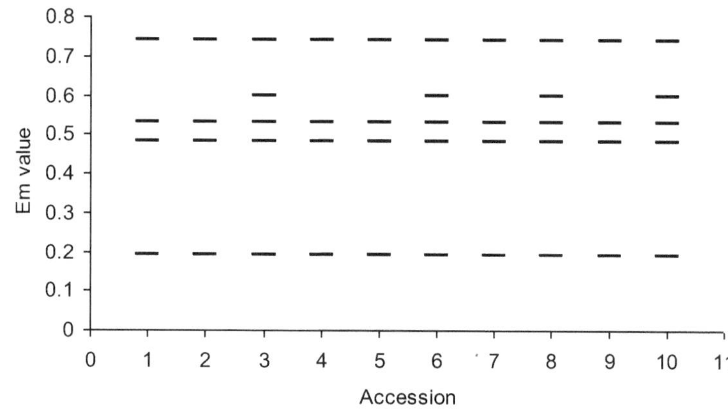

Fig. 17.2 SOD isozyme pattern (*1–5* – control; *6–10* – water stress treatment of 1–5)

accessions. *Piper colubrinum* had the lowest content (both under stress and non-stress), while the cultivated *P. nigrum* varieties Sreekara and Panniyur-1 had the highest levels (Krishnamurthy and Saji 2006).

17.4 Response of Released Varieties of Black Pepper to Moisture Stress

In pot screening, ten released varieties of black pepper were screened for drought tolerance. The varieties used for screening were Panniyur-1, Panniyur-2, Panniyur-3, Panniyur-4, Panniyur-5, Sreekara, Subhakara, Pournami, Panchami, and P-24 (IISR Shakthi). Parameters such as leaf water potential, relative water content, membrane leakage, and enzyme activities such as catalase, peroxidase, etc. were assayed. Among the varieties, cv. Panchami was better than the rest. Leaf water potential reduced drastically in all varieties due to water stress. Leaf water potential reached more than −20 bars after 4 days of stress induction in most of the varieties. The reduction was more pronounced in Pournami and P-24, while Panchami maintained higher leaf water potential than the rest. Panchami also maintained higher relative water content and lesser membrane leakage followed by Panniyur-5 compared to other varieties. Peroxidase activity increased and catalase activity decreased due to water stress in all the varieties (unpublished).

17.5 Differential Proteomics Under Drought

Proteomics-based study was made using tolerant (accession 4226) and susceptible (Subhakara) genotypes for identifying the proteins involved during the drought and also for exogenous application of ABA. Plants were maintained in growth chamber at 30 °C temperature and 60 % relative humidity. After inducing drought one section of plants from both the genotypes were applied with ABA. After 5 days of ABA treatment, leaf proteins were extracted and proteins were subjected to first dimension separated by IEF with defined parameters using IPG strip 4–7 following passive rehydration of strips with 100 ug proteins. The proteins were further separated in second dimension with 12 % continuous SDS-PAGE in Ettan DALT 12 separation system (GE Healthcare). Overall the pattern was different in both genotypes suggesting differential mechanism for drought stress with common reduction in number of protein spots when compared to well-watered plants. Proteins relevant to photosynthesis were downregulated during drought suggesting reduced photooxidative stress and protection of photosynthetic machinery during stress.

17.6 ABA Response (Exogenous ABA) Under Drought Stress

When plants started showing initial wilting symptoms, they were treated with 250 μM ABA (both spray and soil drenching). In susceptible genotype, wilting was delayed by 5–6 days in ABA-treated plants suggesting ABA-mediated drought regulation. New protein spots along with differentially expressed spots could be identified after ABA treatment implicating the role of exogenous ABA in drought tolerance. On the contrary, the tolerant genotype (accession 4226) showed same profile even after ABA application indicating the possible role of endogenous ABA levels in drought tolerance. In this genotype, wilting was very much delayed and ABA treatment had no effect in delaying wilting further.

17.6.1 Promoter Mining for ABA Regulation Against Drought Stress

Important ABA responsive regulatory pathway gene loci (Locus_4364-aldehyde oxidase gene (Fig. 17.3), Locus_30682-NCED gene, Locus_1620-serine threonine protein kinase gene) were identified from the *Piper* transcriptome database. The conserved domain search (InterProScan and PROSITE) for the loci yielded important catalytic domains. These domains had highest match with similar domains from other crop plants, and primers were designed for analysis of these genes to obtain clear understanding of the transcriptional control mechanism of drought and the role of ABA in drought tolerance. Utilizing the sequence information of these identified loci primers were designed targeting cis-trans DNA binding domain in transcriptional regulator elements and regulatory promoter elements toward promoter mining under drought and exogenous application of ABA.

17.7 Gene Expression Analysis in Response to Water-Deficit Stress

Dehydrins are the most prominent proteins with regard to abiotic stress tolerance processes

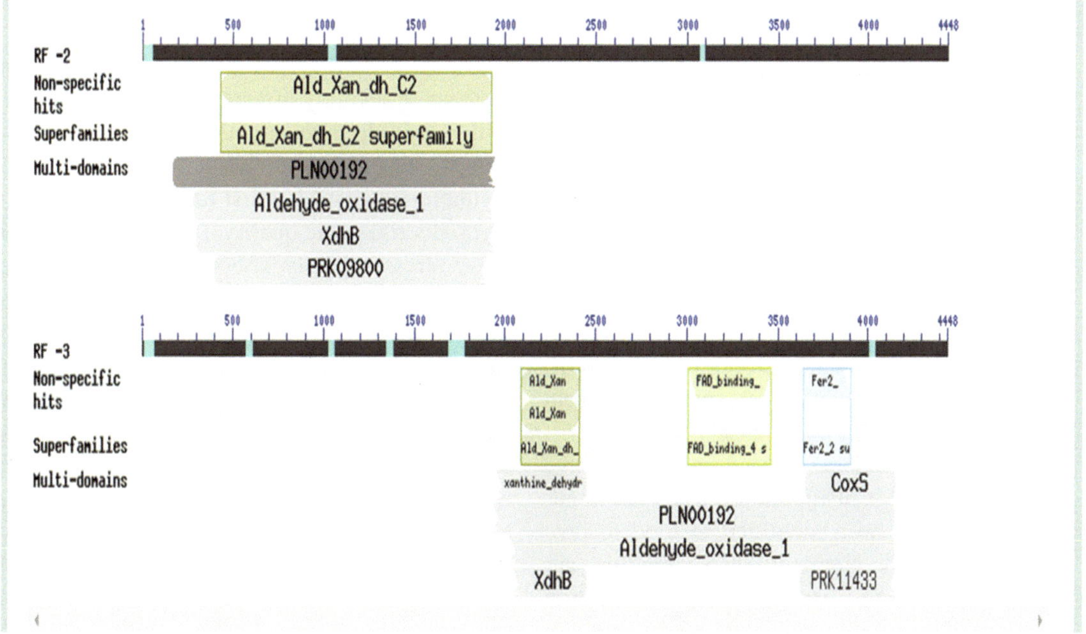

Fig. 17.3 Locus_4364 showing the important domains for aldehyde oxidase gene

(Close 1996). Dehydrins belong to a large family of LEA (late embryogenesis abundant) proteins. These LEA 2 proteins are highly hydrophilic and remain stable under denaturing conditions. They can retain large amounts of water and protect other protein molecules or surfaces of biomembranes from substantial water loss (Tunnacliffe and Wise 2007; Battaglia et al. 2008). The known physical properties of dehydrins suggest roles as stabilizers of nuclear or cytoplasmic macromolecules under water stress conditions (Campbell and Close 1997). The transcript level expression of dehydrin was much high in the drought-tolerant genotype Acc 4216 under water stress (3571-fold) than in the drought-susceptible variety Sreekara (108-fold) when compared to their respective control plants.

Osmotic adjustment (OA) is a key mechanism enabling plants under drought to maintain water absorption and cell turgor pressure, thus contributing to sustained higher photosynthetic rate and expansion growth. Osmotins are believed to be involved in osmotic adjustment of the cells. Osmotin is synthesized and accumulated in cells undergoing gradual osmotic adjustment to either salt or desiccation stress (Singh et al. 1987). Osmotin provides osmotolerance to plants probably by facilitating the compartmentation of solutes (Barthakur et al. 2001) or by being involved in metabolic or structural alterations during osmotic adjustment (Raghothama et al. 1993). Analysis of osmotin transcript levels in well-watered and water-stressed black pepper plants revealed an increase in osmotin expression level under water stress in both the susceptible and tolerant genotypes. Susceptible genotype Subhakara showed 4.2-fold increases in transcript accumulation under water stress over its watered counterpart, while the tolerant accession Acc 4216 showed a higher level of expression (36-fold increase) under water stress over control.

Transcription factors form another attractive target category for manipulation and gene regulation that have been identified to bind to promoter regulatory elements in genes that are regulated by abiotic stresses. The dehydration response element-binding (DREB) factors are one of the most studied transcription factors since they play critical role in plant stress response and tolerance (Novillo et al. 2012). The DREB proteins can bind specifically to the C-repeat (CRT) elements or dehydration responsive elements (DRE) in the promoters of multiple stress-responsive genes and thus activate their expression (Yamaguchi-Shinozaki and Shinozaki 2009). The accumulation of DREB transcript also followed the same expression pattern as that of dehydrin and osmotin in water-stressed and control plants of black pepper. The susceptible variety Sreekara showed only 3.4-fold change with its control when compared to Acc 4216 which showed a 41-fold increase in DREB transcript under drought induction.

Our results suggest that the higher expression of the dehydrin, osmotin, and DREB may impart drought tolerance in Acc 4216 when compared to the Sreekara variety. All the three genes are previously reported to be associated with drought tolerance in plants. The level of expression varied significantly between tolerant and the susceptible genotypes when exposed to water stress. This suggests impending functional roles of these genes in water stress tolerance in black pepper.

17.8 Conclusion

Water stress during flowering to berry development stage severely affects black pepper productivity. Providing fortnightly irrigation (at 50 L vine^{-1}) during March–May months and shade regulation during May first fortnight ensures better productivity and quality. During initial stages, water stress strongly affects leaf area development and leaf elongation rate. Water stress reduces the activities of detoxification enzymes such as catalase and SOD, thus enhancing lipid peroxidation and membrane damage, leading to enhanced solute leakage. Photosynthetic rate and transpiration rate are reduced severely under water stress and 2D gel studies showed that it is to protect photosynthetic machinery under stress. Tolerant genotypes maintained higher relative water content, lower membrane leakage, higher catalase activity and higher amino acid, and reducing sugar levels. Studies on ABA application implicated a role for ABA in drought tolerance. Similarly gene expression studies showed a promising role for dehydrin, DREB, and osmotin genes in stress tolerance. Still there are lots of gaps and concerned efforts are needed to establish the role of ABA in stress tolerance. Promoter mining and protein expression studies may provide some insights on this aspect. Similarly, there is a long way to go with gene expression studies and only a beginning has been made. Identification of QTLs linked to drought tolerance and also identification of candidate genes through SNP approach could also be very useful in developing drought-tolerant genotypes. Now, in the era of climate change, it is essential to develop varieties with better water productivity and also sustainable yield.

References

Ankegowda SJ, Venugopal MN, Krishnamurthy KS, Anandaraj M (2011) Impact of basin irrigation on black pepper production in coffee based cropping system in Kodagu District, Karnataka. Indian J Hortic 68(1):71–74

Apel K, Hirt H (2004) Reactive oxygen species: metabolism, oxidative stress and signal transduction. Annu Rev Plant Biol 55:373–399

Barthakur S, Babu V, Bansal KC (2001) Over expression of osmotin induces proline accumulation and confers tolerance to osmotic stress in transgenic tobacco. J Plant Biochem Biotechnol 10:31–37

Battaglia M, Olvera-Carrillo Y, Garciarrubio A, Campos F, Covarrubias AA (2008) The enigmatic LEA proteins and other hydrophilins. Plant Physiol 148:6–24

Campbell SA, Close T (1997) Dehydrins: genes, proteins, and associations with phenotypic traits. New Phytol 137:61–74

Chempakam B, Kasturi Bai KV, Rajagopal V (1993) Lipid peroxidation in relation to drought tolerance in coconut (*Cocos nucifera* L.). Plant Physiol Biochem 20:5–10

Close TJ (1996) Dehydrins: emergence of a biochemical role of a family of plant dehydration proteins. Plant Physiol 97:795–803

De waard PWF (1969) Foliar diagnosis nutrition and yield stability of black pepper (*Piper nigrum* L.) in Sarawak, Communication No. 58. Dept. of Agrl. Research, Koninklijk, Instituut Voor de Tropen, Amsterdam

Kandiannan K, Krishnamurthy KS, Thankamani CK, Mathew PA (2007) Pattern and variability of black pepper yields in tropical humid climatic conditions. Indian J Hortic 64(3):314–319

Kandiannan K, Parthasarthy U, Krishnamurthy KS, Thankmani CK, Srinivasan V, Aipe KC (2011a) Modeling the association of weather and black pepper yield. Indian J Hortic 68(1):96–102

Kandiannan K, Thankamani CK, Krishnamurthy KS, Mathew PA (2011b) Monthly rainfall trend at high rainfall tract of northern agro-climatic zone in Kerala. In National seminar on recent trends in climate and impact of climate change on South-West India, Department of Physics, St Joseph's College, Devagiri, Calicut, 11 October 2011

Koshy J, Shankar M, Sudhakaran KV (1999) Seasonal climatic influence in pepper production in Idukki district. Spice India 12(12):2–3

Krishnamurthy KS, Saji KV (2006) Response of Piper species to water stress. Indian J Hortic 63(4):433–438

Krishnamurthy KS, Ankegowda SJ, Johnson George K (1998) Impact of water stress on some physiological parameters in black pepper. In: Sadanandan AK, Krishnamurthy KS, Kandiannan K, Korikanthimath VS (eds) Water and nutrient management for sustainable production and quality of spices, Proceedings of the National Seminar. Indian Society for Spices, Calicut, pp 153–157

Krishnamurthy KS, Ankegowda SJ, Saji KV (2000) Water stress effects on membrane damage and activities of catalase, peroxidase and superoxide dismutase enzymes in black pepper (*Piper nigrum* L.). J Plant Biol 27(1):39–42

Krishnamurthy KS, Kandiannan K, Chempakam B, Ankegowda SJ, Anandaraj M (2015) Climate change impact on black pepper and cardamom. In: Choudhary ML, Patel VB, Siddiqui MW, Mahdi SS (eds) Climate dynamics in horticultural science, vol 1, Principles and Applications. Apple Academic Press, Oakville, pp 201–217

Mathai CK, Sastry KSK (1988) Productivity of black pepper vines (*Piper nigrum* L.) as influenced by the light availability during pre-flowering stage. Comput Physiol Ecol 13(3):97–102

Novillo F, Medina J, Rodriguez-Franco M, Neuhaus M, Salinas GJ (2012) Genetic analysis reveals a complex regulatory network modulating CBF gene expression and *Arabidopsis* response to abiotic stress. J Exp Bot 63:293–304

Nybe EV, Prasannakumari amma, Sujatha S, Prabhakaran PV (1999) Survey, evaluation and selection of black pepper cultivars suited for central Kerala. Indian J Arecanut Spices Med Plants 1(2):42–46

Pillay VS, Sasikumar S, Ibrahim KK (1988) Effect of rainfall pattern on the yield of black pepper. In: Rao GSLVP, Nair RR (eds) Agrometeorology of plantation crops. Kerala Agricultural University, Trichur, pp 152–159

Pradeepkumar T, Vasanthakumar, Aipe KC, Kumaran K, George SP, Manmohandas TP, Anith KN (1999) Studies on yielding behaviour of black pepper Cv

Panniyur-I. Indian J Arecanut Spices Med plants 1 (3):88–90

Raghothama KG, Liu D, Nelson DE, Hasegawa PM, Bressan RA (1993) Analysis of an osmotically regulated pathogenesis related osmotin gene promoter. Plant Mol Biol 23:1117–1128

Ramadasan A, Vasantha S (1994) Environmental stress reaction of black pepper. Spice India 7(9):12–30

Ravindran PN, Nirmal Babu K, Sasikumar B, Krishnamurthy KS (2000) Botany and crop improvement of black pepper. In: Ravindran PN (ed) Black pepper. Harwood Academic Publishers, Amsterdam, pp 23–142

Remold M (1997) Pepper prospects in India during 1997–1998 season. Indian Cocoa Arecanut Spices J 21(1):1–5

Ridley HN (1912) Pepper. In: Spices. Macmillan and Co. Ltd., London, pp 239–312

Senanayake YDA, Kirthisinghe JP (1983) Effect of shade and irrigation on black pepper (*Piper nigrum* L.) cuttings. J Plant Crop 11:105–108

Singh NK, Bracker CA, Hasegawa PM, Handa AK, Buckel S, Hermodson MA, Pfankoch E, Regnier FE, Bressan RA (1987) Characterization of osmotin: a thaumatin-like protein associated with osmotic adaptation in plant cells. Plant Physiol 85:529–536

Sivaraman K, Kandiannnan K, Peter KV, Thankamani CK (1999) Agronomy of black pepper (*Piper nigrum* L.). J Spices Aromat Crops 8(1):1–18

Sukumara Pillay V, Sasikumaran S, Venugopalan Nambiar PK (1977) A note on preliminary observation of spike shedding in pepper. Arecanut Spices Bull 8 (4):93–94

Suparman U (1998) The effect of El-Nino & La-Nina on the production of white pepper in Bangka, Indonesia. Intl Pepper News Bull 22:44–45

Thankamani CK, Ashokan PK (2002) Chlorophyll and leaf epicuticular wax contents of black pepper (*Piper nigrum* L.) varieties in response to water stress. J Med Aromat Plant Sci 24(4):943–946

Thankamani CK, Ashokan PK (2004) Solute accumulation in black pepper (*Pipernigrum* L.) varieties in response to water stress. Indian J Hortic 61(1):74–77

Thankamani CK, Chempakam B, Ashokan PK (2003) Water stress induced changes in enzyme activities and lipid peroxidation in black pepper (*Piper nigrum* L.). J Med Aromat Plant Sci 25(3):646–650

Tunnacliffe M, Wise J (2007) The continuing conundrum of the LEA proteins. Naturwissenschaften 94 (10):791–812

Vasantha S, Gopalam A, Ramadasan A (1989) Plastid pigments of black pepper cultivars under heat stress. Indian J Plant Physiol 32(1):78–79

Vasantha S, Varghese Thomas V, Ramadasan A, Zachariah TJ (1990) Drought tolerance in Black pepper (*Piper nigrum* L.) cultivars: an evaluation of physiological parameters. Indian J Plant Physiol 33 (4):363–366

Vijayakumar KR, Mammen G (1990) Effect of contact shading on leaf chlorophyll content and yield performance in black pepper (*Piper nigrum* L.). In: Proceedings of the international congress of plant physiology, 15–20 Feb 1988, vol 2. Society of Plant Physiology and Biochemistry, New Delhi, pp 935–938

Vijayakumar KR, Unni PN, Vamadevan VK (1985) Prevention of photo-induced chlorophyll loss by the use of lime reflectant on the leaves of black pepper (*Piper nigrum* L.). Agric For Meteorol 34(1):17–20

Wahid P, Sitepu D (1987) Current status and future prospects of pepper development in Indonesia. Food and Agricultural Organisation, Regional Office for Asia and Pacific, Bangkok

Yamaguchi-Shinozaki K, Shinozaki K (2009) DREB regulons in abiotic-stress-responsive gene expression in plants. In: Yamada T, Spangenberg G (eds) Molecular breeding of forage and turf. The proceedings of the 5th international symposium on the molecular breeding of forage and turf. Springer, New York, pp 15–28

Oil Palm 18

K. Suresh, R.K. Mathur, and S.K. Behera

Abstract

Oil palm is the highest oil-yielding perennial crop, which produces palm oil and palm kernel oil, being used for cooking and industrial purposes. The palm oil is extracted from the mesocarp of sessile drupe fruit, which contains around 50 % oil. Palm oil is derived from the fleshy mesocarp of the fruit, which contains about 45–55 % of oil. The yield of oil palm varies with genotype, management and environmental factors. Water and fertilizer management holds the key factors in successful management of the crop. The important environmental factors that influence yield are temperature, vapour pressure deficit, relative humidity and rainfall. Breeding for drought and salinity tolerance along with dwarfness and higher oil yield are the key traits for oil palm improvement. As water is the critical factor for successful growth of oil palm, any deficit will act as a signal for repression of female sex expression leading to the production of a large number of male flowers coupled with slow growth leading to poor productivity. Hence a thorough understanding of the morphological, physiological and biochemical basis for drought tolerance is vital for the successful growth of the crop. Drought tolerance phenomenon is complex, since it involves varied mechanisms, which are often antagonistic to each other. These different factors lead to a balance that enables the plant to withstand water stress to varying degrees. Under the Indian context, agronomic techniques for achieving better water reserves at plantation level (through irrigation and erosion control), ensuring more effective distribution of available water to oil palm (by reducing competition with weeds, bare soil cropping, antitranspirants, cropping with other palms) and lessening water consumption of palms during dry seasons through ablation of bunches are effective, but become limited due to technical or economic reasons. The new research options should primarily concentrate

K. Suresh (✉) • R.K. Mathur • S.K. Behera
ICAR-Directorate of Oil Palm Research, Pedavegi, Andhra Pradesh, India
e-mail: sureshkancherla@rediffmail.com

N.K.S. Rao et al. (eds.), *Abiotic Stress Physiology of Horticultural Crops*,
DOI 10.1007/978-81-322-2725-0_18

on the knowledge of responses of adult oil palm and seedlings under water stress. The research on drought tolerance should integrate factors like photosynthetic capacity, assimilate partitioning (source–sink relationship), stomatal regulation and resistance or rigidity of cell membrane integrity in the seedlings and field stages for the same planting materials in order to develop performance tests. The research on the above aspects should lead to new strategies for developing planting materials more closely adapted to the respective agroclimatic conditions.

18.1 Introduction

Oil palm (*Elaeis guineensis* Jacq.) is known to have originated from West Africa. The important oil palm-growing countries in the world are Indonesia, Malaysia, Papua New Guinea (S.E. Asia), Cameroon, Côte d'Ivoire, Congo, Ghana, Liberia, Sierra Leone (Africa), British Guiana, Brazil, Columbia, Costa Rica, Ecuador, Panama, Peru and Venezuela (S. America).

The leading producers of palm oil in the world are Indonesia, Malaysia and Nigeria. Oil palm is the highest oil-yielding perennial crop. It produces palm oil and palm kernel oil. While palm oil is extracted from mesocarp of sessile drupe fruit, palm kernel oil comes from the endocarp of the fruit. Palm kernel oil contains lauric acid. The productivity of palm oil is five times more than that of other known oilseed crops (5 t crude palm oil and 0.5 t palm kernel oil). Breeding for drought and salinity tolerance along with dwarfness and higher oil yield are important traits for oil palm improvement (Cochard et al. 2005).

The yield of oil palm is dependent on genotype, management and environmental factors. Water and fertilizer management are the key factors for better yields. The vital environmental factors affecting yield are temperature, vapour pressure deficit, rainfall and temperature (Henson and Harun 2005). Breeding for drought and salinity tolerance along with dwarfness and higher oil yield are important traits for oil palm improvement. As water is the critical factor for successful growth of oil palm plantations (Henson and Harun 2005), any deficit will act as a signal for female sex suppression (Jones 1997) leading to a large number of male flowers coupled with slow growth leading to poor productivity. Hence a thorough understanding of the morphological, physiological and biochemical basis for drought tolerance is vital for the successful growth of the crop.

18.2 Drought Tolerance in Oil Palm

In tropical regions, rainfall/water is one of the main limiting factors for growing oil palm or at least for obtaining higher yields. Oil palm is best suited for regions having well-distributed rainfall throughout the year. Best yields are obtained when monthly rainfall is around 150 mm (Hemptinne and Ferwerda 1961). Oil palm can also be termed as drought-tolerant crop as it is known to survive even under many months of dry season. When there is moderate water stress, bunch yields get affected. Under extreme water stress, death may occur.

In Benin, death of palms is commonly observed during the dry seasons. Under such condition survival is the key response for drought tolerance (Houssou et al. 1989). The difference in yield with and without drought would be considered under moderate water stress. Though oil palm is being grown in India under irrigated conditions, the palms are often subjected to harsh environments, viz. temperatures as high as 40–45 °C and relative humidity as low as 15–20 % during the summer months, creating atmospheric drought. Under these conditions, it is imperative that a thorough knowledge of the understanding of the basic physiological processes, when the palms are subjected to drought, assumes paramount importance.

Drought tolerance phenomenon is complex, since it involves varied mechanisms, which are often antagonistic to each other. These different factors lead to a balance that enables the plant to withstand water stress to varying degrees. Under the Indian context, agronomic techniques for achieving better water reserves at plantation level (through irrigation and erosion control), ensuring more effective distribution of available water to oil palm (by reducing competition with weeds, bare soil cropping, antitranspirants, cropping with other palms) and lessening water consumption of palms during dry seasons through ablation of bunches are effective, but become limited due to technical or economic reasons. The use of drought-tolerant planting materials will be another option for overcoming drought in the long run. The morphological, physiological and biochemical responses to drought tolerance in oil palm are given below.

18.2.1 Leaf Characters

The most common symptoms of oil palm under drought are occurrence of more unopened leaves, spear leaves, leaf breaking, bending in the middle whorl, yellowing and brownish colour of leaflets in the middle whorl leaves and bunch failure (Umaña and Chinchilla 1989). Villalobos and Rodriguez (1998) reported the occurrence of more number of unopened leaves under water stress. Yellowing of leaves followed by necrosis in leaflets occurs as water stress becomes severe. Some of the leaves in the lowest whorl dry and these tend to bend or break.

Lubis et al. (1993) studied the effect of drought on vegetative growth of oil palm and have divided it into five stages. During the first stage, i.e. when the water deficit is less than 200 mm year^{-1}, the palms do not show any serious problem. The second stage occurs when the water deficit is 200–300 mm year^{-1}. The symptoms like sticking of the frond and immature leaves together and may not open. Also the old fronds become defective. The third stage occurs when the water deficit is 300–400 mm day^{-1}, and the common symptoms are the number of stick and unopened leaves increases to 4–5 and the number of defective old fronds will be seen in 1–1.5 spirals and fronds become dry. Subsequently when the water deficit increases to 500 mm day^{-1}, the young fronds will not open and leaf bud cracks, becomes defective and breaks.

Maillard et al. (1974) coined the term drought sensibility index, SI, which is simply a numerical assessment to the effects of drought on oil palm population. It is calculated as follows: $SI = (10M+5S_3+3S_2+2S_1)/N$, where N is total number of palms, M is number of dead palms, S_3 is palms showing an accumulation of unopened spears, S_2 is palms with four to six leaves broken or collapsed and S_1 is palms with all the leaves withered. During the dry season, less number of unopened leaves was observed in progenies derived from dura population of Angola and teneras of Tanzania. More number of unopened leaves was seen in progenies of Deli duras crossed with Calabar, Ekona and AVROS pisiferas. More tolerant ones were observed in palms belonging to Deli duras crossed with pisiferas of La Me, Yangambi and Tanzania.

Villalobos and Rodriguez (1998) have indicated that the number of dried leaves in the lower whorl ranged from 1 to 21. Less number of desiccated leaves (4–7.2/palm) was seen in crosses of Angola and Bamenda along with wild Malawi populations, while duras belonging to Deli and Tanzania possessed more number of bent and dried leaves. Asemoto et al. (1996) have screened oil palm genotypes in the nursery for stress tolerance using polyethylene glycol 6000 and reported significant differences in the leaf area and leaf elongation.

18.2.2 Root System

Observations on the root system of several oil palm families showed that tolerant crosses have better-developed root system than susceptible crosses. In effect, for an equivalent yield, the cross that can withstand drought is that with the densest root system. Better soil penetration by the roots facilitates water and nutrient uptake, making the plant better able to withstand water stress. However, root density is not the only character to be taken into account, since yields vary considerably between different resistant and moderately resistant families with comparable root densities (1.1 t ha^{-1} for L430T × L404D to 1.8 t/ha for L13T × D8D). However tolerant families manage water reserves more economically, due to better water supply through a more developed root system and to reduced water losses via more efficient stomatal regulation. In the event of the stress, these characteristics make it possible to maintain high turgidity, which leads to better preservation of tissue turgidity and metabolic functioning.

Maillard et al. (1974) developed a methodology for estimating drought tolerance by evaluating seedling root growth under high osmotic pressure using polyethylene glycol. Cornaire et al. (1989) found that better drought tolerance was observed in palms having better root growth. Larger stem volume, efficient root system and better stomatal control are some of the morphological and physiological attributes associated with drought tolerance in oil palm (Villalobos et al. 1991, 1992).

18.2.3 Plant Water Status

Villalobos et al. (1991) studied water status of oil palm in response to drought in Costa Rica and found that during the dry season, nonirrigated palms which were 11 or 17 years old showed leaf water potentials and relative water contents similar to or greater than those in irrigated palms. They also observed a negative correlation between stomatal conductivity and leaf water potential measured at midday in adult palms, which is unusual in cultivated plants and indicates a strong control of stomatal control of leaf water status. Young palms (10 months old) were not able to maintain high leaf water potential status, possibly due to the absence of a voluminous stem and extensive root system. In young palms subjected to 75 days of drought, the relative water content and leaf water potential at midday were lower than in those with a good water supply.

An analysis of the leaf tissue water retention in adult trees has shown that these tissues possessed high apoplastic volume, but did not provide a classification that tallied with drought tolerance of different families. Studies on the relationship between relative water content and water potential could not find differences between resistant and susceptible populations, which suggests that the osmoregulation mechanism is common to all varieties. Similarly, an explanation of the pressure/volume curve showed that the oil palm leaflet tissue characteristics differed significantly from those of seedlings and that neither set of characteristics enables classification of families according to their drought tolerance. However it did show that during water stress, osmotic adjustment mechanisms come into play in all the families. In effect, tissue plasmolysis occurs at lower potentials, when plants are under stress.

18.2.4 Stomatal Regulation

The selection criteria for drought tolerance in oil palm could be membrane breakdown, leaf water potential and stomatal opening (Cornaire et al. 1989). Reduced transpiration due to stomatal closure in response to water deficit well before wilting interferes with photosynthesis, and hence, drought resistance mechanisms based on stomatal sensitivity and reduced transpiration are generally opposed to the maintenance of a higher yield potential. Stomatal activity per se is an imperfect criterion of resistance, unless it is related to both carbon fixation and transpiration under stress. The stomatal regulation in oil palm is very good as it is able to reduce water loss even under drought. However tolerant crosses regulate losses more effectively than susceptible crosses, leading to a greater reduction in stomatal opening during the dry season. At the seedling stage, there are no differences between families with regard to stomatal opening, and it is not therefore possible to predict drought tolerance or susceptibility of the same adult material in the field.

The leaf tissues in oil palm could withstand major falls in water potential with no proportional effects on their relative water contents. Rees (1990) has indicated that midday closure of stomata in oil palm is seen during dry season due to high temperature and low soil moisture. Cornaire et al. (1989) observed that low-yielding drought-tolerant crosses possessed less membrane damage, high leaf water potential and less opened stomata compared to that of high-yielding susceptible crosses. Adiahossou et al. (1984) studied correlations between stomatal resistances of oil palm cultivars and drought tolerance at Benin. Leaf resistance (as a measure of stomatal opening) following suspension of watering differed among the lines, particularly in the period between 10 and 20 days from the beginning of the treatment. The studies on these two lines on the relationship between leaf conductance and photosynthesis suggested that stomatal closure occurs more slowly in high-yielding lines than in low-yielding ones.

Stomatal opening and photosynthetic rate could be selected as physiological markers for drought tolerance as significant variations were observed between populations during dry season in Congo (Smith 1993). In oil palm, the responses of stomata to drought are quick which results in decreased photosynthesis and yields (Suresh and Nagamani 2006). Hence breeding varieties for drought tolerance would not go hand in hand with that of commercial objectives (Villalobos and Rodriguez 1998). Photosynthetic rate along with physiological characters was also studied in oil palm clones under drought (Lamade et al. 1989b).

Suresh et al. (2010, 2012) studied genotypic variations in leaf water potential and gas exchange parameters in five oil palm (*Elaeis guineensis* Jacq.) tenera hybrids at nursery, which indicated that at 24 days of imposition of water stress, the highest leaf water potential was observed in 913 × 1988 hybrid which did not differ significantly with that of 1425 × 2277 and 7418 × 1988 hybrids. Photosynthetic rate and stomatal conductance were significantly decreased in all the hybrids due to water stress. The screening of African dura germplasm based on physiological markers has been completed (Mathur et al. 2001; Suresh and Nagamani 2005; Suresh et al. 2008). Based on these results, superior drought-tolerant duras have been identified. ZS-1 was the most drought-tolerant dura compared to that of other duras, while TS-9 was the most susceptible one. Among Guinea-Bissau duras, GB-25 was the most tolerant one, while GB-21 was the most susceptible. ZS-1 and ZS-2 were the most tolerant and susceptible duras, respectively, among Zambian duras.

18.2.5 Chlorophyll Fluorescence

Suresh et al. (2010, 2012) studied genotypic variations in chlorophyll fluorescence in five oil palm tenera hybrids at nursery, which indicated that there was a reduced apparent electron transport rate in hybrids 7418 × 1988 and 1425 × 2277 at 12 days of rehydration indicating its lesser tolerance to water stress. The highest non-photochemical quenching was observed in 7418 × 1988 hybrid after rehydration, indicating its better adaptation to counter excess light energy produced due to low photosynthesis. The effect of some biotic and abiotic environmental stresses on chlorophyll fluorescence parameters is described in Kalaji et al. (2014).

18.2.6 Sap Flux

To understand plant stress responses in the form of sap flux and transpirational adjustments made by oil palm under Indian conditions, sap flow studies have been undertaken (Suresh et al. 2006; Suresh and Nagamani 2007), and results revealed that sap flux increased gradually from 9:00 AM reaching a peak during 1:00 to 2:00 AM and decreased as the day progressed. Evapotranspiration and vapour pressure deficit also showed similar trend as that of sap flux. Seasonal variations in sap flux indicated higher sap flux during February and March and lower flux during May and June. The lower flux during dry months could be due to closure of stomata after midday as atmospheric vapour pressure deficit increased.

18.2.7 Membrane Structural Resistance

Maintenance of membrane structure and integrity is a major factor in water stress tolerance. Membrane integrity is usually determined by the leakage of solutes (electrolytes, sugars, amino acids, organic acids, hormones, etc.) from cells. The capacity for stem reserve mobilization/translocation appears to be related to drought tolerance or resistance and could be due to accumulation of ABA in response to water stress. However experiments on leaf tissue membrane stability carried out on both seedlings and adult trees showed that an assessment of protoplasmic resistance cannot be used as a sole early test when breeding for drought tolerance. In effect, sometimes the results obtained on seedlings do not tally with adult trees. Hence membrane observations are to be done in comparison with membrane lipid composition and changes in the event of drought. Studies on membrane susceptibility indices (MSI) indicated that ZS-2 recorded the highest value closely followed by TS-9, indicating their better tolerance to drought (Suresh et al. 2008). In another study, 14 dura (Africa) × pisifera crosses were screened for drought tolerance in nursery by undertaking studies on MSI. The highest MSI was recorded in 34CD × 110P closely followed by 124CD × 17P and 66CD × 129P hybrids, indicating its better tolerance to drought. The hybrids 254CD × 14P and 435CD × 14P recorded lower MSI, indicating their poor tolerance to drought (Suresh et al. 2010).

18.2.8 Reserves and Their Mobilization

Although the effect of stomatal regulation on photosynthetic activity has not been quantified, the stomatal closure during water stress occurs to the detriment of carbon nutrition. Hence the palm has to compensate for this deficit. Carbohydrate reserves play a vital role as a source of carbon. It has been shown on seedlings that not only are these reserves larger in resistant families, but mobilization is also better due to the increase in certain enzyme activities (invertase, β-amylase). All the organs are seen to use reserves during water stress, which helps to lessen the demand for carbon that results from stomatal opening and can also play a role in osmoregulation. The leaf starch is hydrolyzed during the dry season and that there is an increase in soluble sugar concentration in oil palm. Initially the soluble sugar levels in the stem fall as the sugars are used for the palm's metabolism and then rise again in a second phase probably due to starch hydrolysis. In adult trees, although starch hydrolysis and carbohydrate accumulation occurred during the dry season, no differences were observed among the families (Cornaire et al. 1994; Adiahossou and Silva 1978).

Another way of solving the oil palm demand for carbon is to regulate the photosynthetic capacity. Results have indicated that the resistant families have higher assimilation rates than susceptible crosses, but the reduction in the event of stress is smaller. The difference can be explained by higher ribulose-1,5-bisphosphate carboxylase/oxygenase (Rubisco) activity that is also less sensitive to stress, in tolerant crosses. Ollagnier (1985) studied ionic reactions along with fertilizer management in relation to drought tolerance in oil palm. Fertilizers acted positively on plant development and yield components such as weight and number of bunches. Drought during the bunch-forming cycle, i.e. 26–28 months, reduces the efficiency of fertilizers.

Accumulation of proline in oil palm has been demonstrated as one of the most evident biochemical indices under water stress (Cha-um et al. 2011). Accumulation of large amounts of proline contributes to osmotic adjustment and serves as a cytoplasmic osmotic balance for potassium accumulation as the main osmoticum in the vacuole. Harun (1997) observed accumulation of proline in leaves of oil palm seedlings under water stress and increased stomatal resistance and reduction in leaf water potential due to increased water deficits. There was an increased concentration of proline due to rise in stomatal resistance and proline content decreased to the normal level after rewatering. In another study, water stress was induced in oil palm by mannitol and PEG, wherein there was a drastic reduction in photosynthetic pigments with increase in osmotic potential of the culture medium (Cha-um et al. 2012).

The proline content increases due to drought and salt stress in oil palm and the mechanism is yet to be unravelled. Nana Yamada et al. (2011) reported an 8319-nucleotide sequence consisting of genomic DNA, cDNA and promoter region of proline transporter gene in oil palm. The transporter gene was having the largest homology with rice *ProT1* and was similar to *Bet/ProT* genes of several plants. There was uniqueness in exon–intron structure of genomic DNA, and more number of *cis*-elements induced by stress was found in the promoter region. There was an increase in the uptake of glycine betaine, choline and proline due to the expression of cDNA *EgProT1* in *Escherichia coli* mutant. The uptake of glycine betaine was more by the roots of oil palm under salt stress compared to that of control, which gives ample evidence that oil palm possesses a unique *Pro/T1* gene.

18.2.9 Effect of Drought on Yield

There exist large variations among oil palm genotypes, and a few of them could be selected for tolerance to stress without forgoing its yield potential (Maillard et al. 1974). The lower carbon exchange rates in oil palm due to stomatal closure caused by any stress could be compensated by aborting few of its inflorescences, resource mobilization and higher annual yield variation (Nouy et al. 1999). Studies by Houssou et al. (1989, 1992) at Benin indicated that high-yielding Deli × La Me crosses had a higher death rate than Deli × Yangambi crosses and yields were not significant in the surviving palms of both the crosses.

Similar results were observed by Nouy et al. (1999) during the evaluation of two crosses at two sites, where annual water deficits ranged from 400 to 700 mm. In general, drought tolerance and yield have to be compromised. However, Houssou et al. (1989) observed some oil palm progenies which possessed high yields and at the same time showed low mortality under stress. Villalobos and Rodriguez (1998) observed that the number of bunches produced during the first 2 years can be considered as an estimate of yield potential of oil palm hybrids under very extreme stress. He found higher number of bunches in the wild palms of Malawi (14.2 bunches/palm) and progenies from Bamenda and Tanzania (11.8 and 9.5 bunches/palm), while Deli dura palms possessed lower number of bunches.

Foo (1998) reported that the moisture stress suppresses female inflorescence formation and increases abortion of female inflorescences. The inadequate water supply to meet such high evapotranspiration demand during the dry period would severely affect sex differentiation and the subsequent inflorescence development process, and this will eventually reduce the ultimate yield. Adequate irrigation and fertilizer application has achieved 59 t FFB ha^{-1} at a palm age of 19 years. An irrigation trial on a semicommercial scale in a dry area has achieved 38.8 t FFB ha^{-1} at the third year of irrigation as compared with 14.1 t FFB ha^{-1} without irrigation. At 8–10 years of age, the irrigated palms could sustain a yield of 30–35 t FFB ha^{-1} as compared with 16–18 t FFB ha^{-1} of the nonirrigated palms. Thus the photosynthetic efficiency and frond and bunch production were also higher during water areas. Villalobos and Rodriguez (1998) observed lesser yields from progenies of Deli dura population, which was severely affected due to water stress. However, progenies of Deli × La Mé and Deli × Yangambi origins possessed moderate yields, which is in concurrence to the findings of

Maillard et al. (1974) and Houssou et al. (1992). The behavioural pattern and yield of Malawi wild palms under stress give an indication that they possess drought-tolerant genes. The most tolerant crosses under water stress were Angola × La Mé and Angola × Tanzania which possessed higher leaf-specific weights and larger leaf water potentials during dawn.

Studies at Benin have classified the oil palm crosses according to their drought tolerance into four main groups. The first contained crosses with high production potential and good drought resistance, as they showed little sign of damage in the event of water stress. The second also contained high-yielding crosses but suffered severe drought damage, even the death of some individuals. The trees in the third group were low yielding and showed little sign of damage, and the last group contained families with a low production potential and highly susceptible to drought (very few families fell into this group).

18.3 High Moisture Stress

Oil palm does not tolerate continuous waterlogging and its roots are unable to respire under submerged conditions. The roots may die under prolonged flooding conditions. Water and nitrogen uptake will be impaired with effects similar to those of water stress due to stomatal closure. The older leaves will become yellow due to the deficiency of nitrogen. Lamade et al. (1989b) observed that waterlogging causes death of roots and decreased photosynthesis and dry matter production due to closure of stomata.

18.4 Low Temperature Stress

Oil palm is very sensitive to low temperature. The effect of low temperature on vegetative growth and physiological processes in leaves of oil palm seedlings revealed lesser height and leaf area under low temperature compared to that of control. Moreover, the seedling growth was stopped due to low temperature stress. These studies indicate that low temperature is a strong inhibitor of oil palm vegetative growth (Tarmizi and Marziah 1995).

Under low temperature stress, there will be an increase in relative conductivity and injury index in oil palm. Relative conductivity is an important index, which reflects cell membrane permeability. When the relative conductivity is higher under low temperature stress, cell membrane permeability will be very high leading to a greater injury to the cell membrane and possibly increase the injury index also. These results confirm the findings that low temperature is the major cause for increased relative conductivity, injury index and cell membrane permeability of plant tissue (Tarmizi and Marziah 1995). The oil palm seedlings subjected to low temperature stress possessed higher contents of MDA and proline compared to that of normal seedlings.

The total cultivated area under oil palm in the world could be tremendously increased by developing oil palm varieties tolerant to cold. Lei et al. (2014) sequenced cold-treated and control (untreated) samples in oil palm which possessed 51,452 unigenes with an average length of 703 bp and were annotated functionally. Large number of genes were induced and repressed from AP2/ERE family, indicating that C-repeat binding factor (CBF) could be involved in cold tolerance of oil palm. The CBFs obtained in the study reached maximum transcript level which was different from the expression pattern of CBFs in rice and *Arabidopsis*. The analysis of promoters of downstream cold-responsive gene (CORs) regulated by CBFs indicated the detection of conservation, mutation and absence of DRE core motif in six CORs. These mutations in DRE motifs indicate that the CORs could not be induced under low temperature stress in oil palm.

18.5 Future Strategies

1. The new research options should primarily concentrate on the knowledge regarding the physiology and biochemistry of adult oil palm and seedling under water stress. These studies should enable the "physiological

characterization" of all the available crosses and parents and provide new tools for genetic improvement programmes.

2. The research on drought tolerance should integrate factors like photosynthetic capacity, assimilate partitioning (source–sink relationship), stomatal regulation and resistance or rigidity of cell membrane integrity in the seedlings as a tool for screening the palms in the field.
3. The research developed on the above aspects should lead to new strategies for developing agroclimatic zone-specific planting materials.

References

Adiahossou F, Silva VD (1978) Soluble glucide and starch contents and resistance to drought in the oil palm. Oleagineux 33:603–604

Adiahossou DF, Louguet P, da Silva JBV (1984) Correlations between stomatal resistance of various crosses of oil palm (*Elaeis guineensis* Jacq.) and drought resistance. Acta Oecol 5:163–178

Asemota O, Isenmila AE, Okwuagwu CO (1996) Screening oil palm genotypes for stress tolerance using polyethylene glycol 6000. Proceedings of the 1996 PORIM International Palm Oil Congress (Agriculture)

Cha-um S, Yamada N, Takabe T, Kirdmanee C (2011) Mannitol-induced water deficit stress in oil palm (*Elaeis guineensis* Jacq.) seedlings. J Oil Palm Res 23:1194–1202

Cha-um S, Takabe T, Kirdmanee C (2012) Physio-biochemical responses of oil palm (*Elaeis guineensis* Jacq.) seedlings to mannitol and polyethylene glycol induced isoosmotic stress. Plant Prod Sci 15:65–72

Cochard B, Amblard P, Durand-Gasselin T (2005) Oil palm genetic improvement and sustainable development. OCL 12:141–147

Cornaire B, Houssou MS, Meunier J (1989) Breeding of drought resistance in oil palm. 2. Kinetics of stomatal opening and protoplasmic resistance. Paper presented at International conference on palms and palm products, 21–25 Nov., Nigerian Inst Oil Palm Res, Benin City

Cornaire B, Daniel C, Zuily-Fodil Y, Lamade E (1994) Oil palm performance under water stress, background to the problem, first results and research approaches. Oleagineux 49:1–11

Foo SF (1998) Impact of moisture on oil palm yield. Kemajuan Penyelidikan 32:5–17

Harun MH (1997) Proline accumulation in the leaves of water stressed oil palm (*Elaeis guineensis* Jacq.) seedlings. Elaeis 9:93–99

Hemptinne J, Ferwerda JD (1961) Influence des précipitations sur les productions du palmier á huile (*Elais guineensis* Jacq.). Oléagineaux 16:431–437

Henson IE, Harun MH (2005) The influence of climatic conditions on gas and energy exchanges above a young oil palm stand in north Kedah, Malaysia. J Oil Palm Res 17:73–91

Houssou M, Omore AS, Meunier J (1989) Breeding of drought resistance in oil palm. 1. Variability of some crosses for their productivity and their mortality. Paper presented at International conference on palms and palm products, 21–25 Nov., Nigerian Inst. Oil Palm Res., Benin City

Houssou M, Cornaire B, Omore A, Adje J (1992) Selection pour la résistance á la sécheresse du palmier á huile. ISOPB, Montpellier

Jones LH (1997) The effect of leaf pruning and other stresses on sex determination in the oil palm and their representation by a computer simulation. J Theor Biol 187:241–260

Kalaji HM, Bosa K, Calatayud A, Guidi L, Schansker G, Ladle RJ, Allakhverdiev SI, Da Browski P, Hogewoning SW, Nebauer SG, Pollastrini M, Sero^dio J, Pancaldi S, Goltsev V, Brestic M, Elsheery NI, Jajoo A, Bussotti F, Ferroni L, Misra AN, Penella C, Romanowska-Duda ZB, Kancherla S, Yanniccari M, Zivcak M, Szulc W, Poli DB, Rutkowska B, Tambussi E (2014) Frequently asked questions about in vivo chlorophyll fluorescence: practical issues. Photosynth Res 122:121–158

Lamade E, Satiyo IE, Muluck CS, Hakim M (1989b) Physiological study of three contrasting clones in Lampung (Indonesia) under drought in 1997. Paper presented at International conference on developments in oil palm plantation industry for the 21st century, 21–22 Sept., Bali

Lei X, Xiao Y, Xia W, Mason AS, Yang Y, Ma Z (2014) Peng M (2014) RNA-seq analysis of oil palm under cold stress reveals a different C-repeat binding factor (CBF) mediated gene expression pattern in *Elaeis guineensis* compared to other species. PLoS One 9 (12):e114482. doi:10.1371/journal.pone.0114482. eCollection 2014

Lubis AU, Syamsuddin E, Pamin K (1993) Effect of long dry season on oil palm yield at some plantations in Indonesia. Paper presented at PORIM International palm oil congress, 20–25 Sept., Kuala Lumpur

Maillard G, Daniel C, Ochs R (1974) Analyse des effets de la sécheresse sur le palmier a huile. Oléagineaux 29(8-9):397–404

Mathur RK, Suresh K, Nair S, Parimala K, Sivaramakrishna VNP (2001) Evaluation of exotic dura germplasm for water use efficiency in oil palm (Elaeis guineensis Jacq.). Indian J Plant Genet Resour 14:257–259

Nouy B, Baudouin L, Djegui N, Omore A (1999) Oil palm under limiting water supply conditions. Plant Rech Dev 6:31–45

Ollagnier M (1985) Ionic reactions and fertilizer management in relation to drought resistance of perennial oil crops (oil palm and coconut). Oleagineux 40:1–10

Rees AR (1990) Mid day closure of stomata in the oil palm. J Exp Bot 12:129–146

Smith BG (1993) Correlations between vegetative and yield characteristics and photosynthetic rate and stomatal conductance in the oil palm (*Elaeis guineensis* Jacq.). Elaeis 5:12–26

Suresh K, Nagamani C (2005) Stomatal characteristics of different African dura germplasm. Paper presented in the national seminar on research and development of oil palm in India at Pedavegi during 19–20th Feb 2005

Suresh K, Nagamani C (2006) Variations in photosynthetic rate and associated parameters with age of oil palm leaves under irrigation. Photosynthesis 44:309–311

Suresh K, Nagamani C (2007) Partitioning of water flux in oil palm plantations – seasonal variations in sap flow under irrigated conditions. Paper published in proceedings of Asiaflux workshop held during 19–21 Oct 2007 at Taoyuan

Suresh K, Nagamani C, Reddy VM (2006) Measurement of transpiration using sap flux probes in Oil Palm grown under irrigated conditions. Paper presented in national seminar held at Hyderabad during 29–31 Jan 2007

Suresh K, Mathur RK, Kochu Babu M (2008) Screening of oil palm duras for drought tolerance – stomatal responses, gas exchange and water relations. J Plant Crop 36:270–275

Suresh K, Nagamani C, Ramachandrudu K, Mathur RK (2010) Gas exchange characteristics, leaf water potential and chlorophyll a fluorescence in oil palm (*Elaeis guineensis* Jacq.) seedlings under water stress and recovery. Photosynthesis 48:430–436

Suresh K, Nagamani C, Kantha DL, Kumar MK (2012) Changes in photosynthetic activity in five common hybrids of oil palm (*Elaeis guineensis* Jacq.) seedlings under water deficit. Photosynthesis 50:549–556

Tarmizi AH, Marziah M (1995) The influence of low temperature treatment on growth and proline accumulation in polyembryogenic cultures of oil palm (*Elaeis guineensis* Jacq.). J Oil Palm Res 7:107–117

Umaña C, Chinchilla C (1989) Sintomatología asociada al déficit hídrico en palma aceitera. Bol Tec OPO-UB 3:50–54

Villalobos E, Rodriguez W (1998) Evaluación de la capacidad de asimilación del carbón, de la eficiencia en el uso del agua y de la resistencia a la sequía, en progenies de palmas aceitera (*Elaeis aguineensis* Jacq.) en Costa Rica. Universidad de Costa Rica, Informe de Proyecto VI-734-97-16, CIGRAS, 1998. p 32

Villalobos E, Chinchilla C, Echandi C, Fernandez O (1991) Short term responses of the oil palm (*Elaeis guineensis* Jacq.) to water deficit in Costa Rica. PORIM Int Conf Kuala Lumpur

Villalobos E, Umana CH, Chinchilla C (1992) Water status of oil palm in response to drought in Costa Rica. Oleagineux 47:1–7

Yamada N, Cha-Um S, Kageyama H, Promden W, Tanaka Y, Kirdmanee C, Takabe T, Peter G (2011) Isolation and characterization of proline/betaine transporter gene from oil palm. Tree Physiol 31:462–468

Tropical Tuber Crops

19

O. Kumari Swadija, Atul Jayapal, and V.B. Padmanabhan

Abstract

Tuber crops are the third most important food crops, after cereals and grain legumes. The important tropical tuber crops are cassava, sweet potato, yams, aroids and a few minor tuber crops. There is a wide difference between potential yield and actual yield of tuber crops obtained in farmers' fields which may be due to genetic as well as soil and environmental factors. Photoperiod, light intensity, temperature and moisture availability are the most important abiotic factors governing the onset of tuber differentiation and rate of tuber bulking. It is necessary to understand the effect of abiotic stress factors on growth and productivity of tuber crops to identify/develop crops/varieties and to evolve appropriate management practices to sustain productivity especially in the context of climate change. Hence, the influence of abiotic stress factors such as solar radiation extremes, temperature extremes, drought, flooding/waterlogging, elevated CO_2 concentration and salinity on the tropical tuber crops is dealt with in this chapter.

19.1 Introduction

Tuber crops are the third most important food crops, after cereals and grain legumes. They form an important staple or subsidiary food for about 500 million of the global population. The important tropical tuber crops are cassava, sweet potato, yams, aroids and few minor tuber crops. Three of these tropical tuber crops, namely, cassava, sweet potato and yams rank among the top 15 crop plants of the world in area under cultivation (Nayar 2014). They are considered as the most efficient converters of solar energy, cassava producing 250×10^3 cal.ha^{-1} day^{-1} and sweet potato 240×10^3 cal.ha^{-1} day^{-1} as compared to 176×10^3 cal.ha^{-1} day^{-1} for rice, 110×10^3 cal.ha^{-1} day^{-1} for wheat and 200×10^3 cal.ha^{-1} day^{-1} for maize. Tuber crops can substitute for cereals due to their higher dry matter production and heavy calorie yield. The tropical tuber crops contribute 3.9 % of human energy for an average consumption of

O.K. Swadija (✉) • A. Jayapal • V.B. Padmanabhan
College of Agriculture, Vellayani, Thiruvananthapuram, Kerala, India
e-mail: okswadija@yahoo.com

N.K.S. Rao et al. (eds.), *Abiotic Stress Physiology of Horticultural Crops*,
DOI 10.1007/978-81-322-2725-0_19

28.6 kg capita^{-1} year^{-1}. It is now proposed that this contribution should be doubled to 8 % by 2030 AD in view of the emerging food crisis and ongoing climate change (Nayar 2014). Besides contributing significantly as source of energy, especially for the weaker sections of the society, these crops are of great industrial value for starch extraction and as animal feed. Often there is a wide difference between potential yield and the actual yield of tuber crops obtained in farmers' fields. This yield difference is due to genetic, soil and environmental factors. Photoperiod, light intensity, temperature and moisture availability are the most important abiotic factors governing the onset of tuber differentiation and rate of tuber bulking. The abiotic stress greatly influences leaf area, leaf production and abscission, leaf photosynthesis, tuber formation and development and dry matter production and partitioning in tuber crops. It is necessary to understand the effect of abiotic stress factors on growth and productivity of tuber crops to identify/develop crops/varieties and to evolve appropriate management practices to sustain productivity especially in the context of climate change. Hence, the influence of abiotic stress factors on tropical tuber crops, particularly, cassava, sweet potato, yams and aroids, is dealt with in this chapter.

19.2 Status of Tropical Tuber Crops

The tropical tuber crops are grown in an area of about 36.2 m ha in the world with a total production of 341.7 mt. Presently they constitute a staple crop in 29 countries/entities of the world and a secondary staple in another 25 countries, out of a total of 236 countries/entities in the world (Nayar 2014). The International Food Policy Research Institute has estimated a projected demand of 37 % increase for roots and tubers globally between 1995 and 2020. The worldwide demand for cassava and other minor roots and tubers is projected to increase by 49 % and for sweet potatoes and yams by 30 %.

19.2.1 Cassava

Cassava (*Manihot esculenta* Crantz) is a woody shrub belonging to the family Euphorbiaceae and native to South America. It is considered as the king of tropical tuber crops as it occupies a dominant role in the global agricultural economy and trade among tuber crops. It is grown today by millions of small-scale farmers in more than 100 countries mostly in tropical and subtropical regions in an area of 20.39 m ha with a total production of 276.76 mt (FAOSTAT 2013). The main producing countries are Nigeria, Brazil, Indonesia, Thailand, Angola, Ghana, Vietnam, India, Paraguay, Columbia and Peru. India ranks first in the productivity of cassava with 36.48 t ha^{-1} when the world average is only 12.84 t ha^{-1}.

19.2.2 Sweet Potato

Sweet potato (*Ipomoea batatas* L.) belonging to the family Convolvulaceae is an important tuber crop widely grown in the tropics and subtropics and warm-temperate regions of the world. Besides being rich in starch, the tubers are good source of vitamin C, B2, B6 and E as well as dietary fibre and minerals like K, Cu, Mn and Fe. The high nutrient content coupled with its anticarcinogenic and cardiovascular disease preventing properties has gained recognition for the crop as a health food. The orange-fleshed sweet potatoes are an important source of β-carotene which is the major provitamin A carotenoid (Chassy et al. 2008), while purple-fleshed sweet potato varieties are rich in anthocyanins and other polyphenolic components (Teow et al. 2007; Steed and Truong 2008). The level of β-carotene and anthocyanin in sweet potato is as high as in carrot juice and pumpkin (Woolfe 1992; Steed and Truong 2008). Sweet potato ranks fifth in economic value, sixth in dry matter production, seventh in energy production and ninth in protein production (Lobenstein 2009).

It is grown in 8.18 m ha with a production of 103.11 mt in the world (FAOSTAT 2013). China ranks first in area and production. The countries leading in the production of sweet potato are China, Uganda, Nigeria, Indonesia, the United Republic of Tanzania, Vietnam, India, the USA, Angola and Burundi (FAOSTAT 2010).

19.2.3 Yams

Yams are monocots belonging to the family Dioscoreaceae of which popular edible species are greater yam (*Dioscorea alata*) and lesser yam (*D. esculenta*) (Asiatic yams) and African yam/white yam (*D. rotundata*). The tubers are rich source of starch (11–39 %), protein (6–13 % on dry weight basis), certain vitamins and high calorific value. Yams are eaten boiled, baked or fried. They are also used as vegetable. These tubers also possess medicinal properties. They are grown in an area of 5.36 m ha in the world with a production of 63.05 mt (FAOSTAT 2013).

19.2.4 Aroids

Aroids, belonging to the monocotyledonous family Araceae, are cultivated and consumed as staple or subsistence food in the tropical climates of Asia, Africa, Latin America and the Pacific (Onwume and Charles 1994). They are rich and cheap source of energy, vitamins and minerals. They are also used in indigenous medicinal preparations for various ailments. Edible aroids are herbaceous plants with enlarged underground storage organs, such as corms and cormels. The major edible aroids are classified into elephant foot yam (*Amorphophallus paeoniifolius* (Dennst.) Nicolson) and cocoyams. The cocoyams include taro (*Colocasia esculenta* (L.) Schott) or old cocoyam and tannia (*Xanthosoma sagittifolium* (L.) Schott) or new cocoyam. In addition, *Amorphophallus konjac* which is cultivated for extraction of glucomannan, konjac flour, konjac gum, etc. and swamp taro (*Colocasia esculenta* var. *stoloniferum*. (L.) are also cultivated to a limited extent. Taro is believed to have been originated in South East Asia, probably India or Malaysia. It is grown in an area of 1.39 m ha in the world with a production of 10.45 mt (FAOSTAT 2013). Cultivation of taro is widespread in India, Burma, China, Japan, Hawaii, Egypt, Africa and the Caribbean. Tannia originated in tropical America (Onwume 1978) and is mainly cultivated by small-scale farmers (Onwume and Charles 1994) in Asia, Africa and Latin America (Wilson 1984). Elephant foot yam has its centre of origin in India and is mainly cultivated in India, the Philippines, Indonesia, Sri Lanka and South East Asia.

19.3 Abiotic Stress in Tropical Tuber Crops

The main abiotic stress factors affecting tuber crops are drought, waterlogging, temperature extremes, solar radiation extremes and nutrient imbalances. Nutrient imbalances can be corrected through soil and plant analysis and adoption of integrated nutrient management practices.

19.3.1 Solar Radiation Extremes

Light is one of the most important environmental factors that influence growth and development of plants. The abiotic stress due to solar radiation may be due to the influence of the photoperiod as well as the light intensity or shade. In general, long days promote vegetative growth in tuber crops and short days favour tuberization. Shade, or reduction of solar radiation, is one of the stress factors, which can cause high-yield losses in crops. A reduction in radiation can be either abiotic due to environmental factors such as cloudy skies during the rainy season or biotic factors such as shading by taller growing adjacent species in the multiple cropping systems.

19.3.1.1 Photoperiod

19.3.1.1.1 Cassava

Photoperiod/day length affects tuberization, photoassimilates' partitioning and flowering in cassava. In the tropics, where day length varies from 10 to 12 h throughout the year, photoperiod may not limit cassava production. Experiments have shown that the optimal light period for cassava is around 12 h with probable varietal differences (Bolhuis 1966). Short day promoted tuber growth and reduced shoot growth, while long day promoted shoot growth and inhibited tuber growth (Lowe et al. 1976; Keating et al. 1985) suggesting an antagonistic relationship between shoot and tuber growth in response to a change in day length. The restrictions regarding cassava distribution outside tropical regions can be due to effects of day length variation on its physiology. The storage root reduction seems to be more related to the change in the distribution patterns of dry matter rather than to a delay in storage root initiation. There are varietal differences in sensitivity to long days (Carvalho and Ezeta 1983; Veltkamp 1985). The increase in shoot dry weight during long days is a result of significant increases in plant height, leaf area per plant, number of apices per plant and number of living leaves per apex. Long photoperiods may increase the growth requirements of the shoots, thereby reducing the excess carbohydrates available for root growth (Veltkamp 1985).

19.3.1.1.2 Sweet Potato

Day length affects both flowering and tuber formation in sweet potato. Porter (1979) observed that 14 h of light treatment increased tuber yield. Day length of 11 h or less promotes flowering, and flowering fails to occur at day length greater than 13.5 h, although yield remains unaffected. Long day inhibits tuber development (McDavid and Alamu 1980) and tended to favour vine development at the expense of tubers. Biswas et al. (1989) reported that 16 h of light stimulated vine length and yield, while number of tubers, tuber yield and total biological yield were promoted by 9 h of light treatment. Short day with low light intensity promotes root development, while excess rainfall and long photoperiod encourage vine growth and reduce tuber yield (Palaniswami et al. 2008).

19.3.1.1.3 Yams

Yam production is also significantly influenced by day length. Allard (1945) reported that a day length of above 12.5 h increased growth of vine, while a day length of less than 12 h promoted production of aerial and the subterranean tubers of *Dioscorea batatas* and *D. alata*. Positive influence of short days on tuberization was demonstrated for yams (*Dioscorea* spp.) by Njoku (1963). Day length greater than 12 h at early stage promotes vine growth, while short photoperiod favours satisfactory tuber production (Onwume 1978).

19.3.1.1.4 Tannia

The required photoperiod for tuberization in tannia is found to be less than 12 h (Ramesh et al. 2007). When plants were given short days immediately after planting, petiole extension retarded, but storage organ (corm) dry weight increased. The increase resulted from a stimulation of cormel initiation which seems to occur within 2 months after planting and also influenced by soil temperature. However, studies indicated that day length had no discernible effect on the number of leaves per plant, leaf size, total leaf area or flower induction (McDavid and Alamu 1980).

19.3.1.2 Shade

19.3.1.2.1 Cassava

Cassava is a sun-loving crop, and its potential yield could be realized only under full sunlight with optimum leaf canopy (Ramanujam et al. 1984a). Cassava requires high solar radiation for efficient photosynthesis (El-Sharkawy et al. 1992; Lebot 2009). In general, crop growth increases as solar radiation increases. Shading decreases radiation interception of cassava which in turn decreases the crop growth rate and root yield (Cock 1978). Shading increases stem elongation and internode length and hence

Table 19.1 Effect of shade on yield characters of cassava

Treatments	Root volume (cm^3 $plant^{-1}$)	Root/shoot ratio	Total dry matter production ($t\ ha^{-1}$)	Tuber yield ($t\ ha^{-1}$)
0 % shade	6.27	0.016	18.55	37.87
25 % shade	5.03	0.010	11.93	15.64
50 % shade	4.34	0.008	8.51	3.20
75 % shade	3.43	0.066	7.03	1.35
SEM±	0.090	0.0015	0.088	0.270
CD (0.05)	0.276	0.0046	0.272	0.830

Source: Geetha et al. (2006)

little carbohydrate is available for root growth. Shading increases plant height, and the leaves tend to become adapted to low-light conditions by increasing leaf area per unit weight (Fukai et al. 1984; Okoli and Wilson 1986; Ramanujam et al. 1984b; Sreekumari et al. 1988). Shading also decreases the leaf life in cassava. Under ideal growing conditions, cassava leaves have the life up to 125 days (Splittstoesser and Tunya 1992). Levels of shade up to about 75 % have very little effect on leaf life, but under 95–100 % shade, leaves abscise within 10 days (Cock et al. 1979). The chlorophyll concentration was higher, while the chlorophyll a:b ratio, carotenoids per unit chlorophyll and weight per unit area of leaf were lower in the shade than in the sun (Ramanujam and Jose 1984; Johnston and Onwume 1998). Reduction in crop growth rate and net assimilation rate was observed in cassava grown under shade (Ramanujam et al. 1984b; Ramanujam and Jose 1984).

When grown under partial shade in intercropping or multiple cropping systems, significant reduction in tuber yield has been reported (Ramanujam et al. 1984a, b; Ghosh et al. 1987, 1988). Artificial shade has also been reported to reduce the root growth rate considerably (Fukai et al. 1984). A considerable delay in tuber initiation and reduction in number of tubers per plant was reported by Ramanujam et al. (1984b). They observed 65–94 % yield reduction due to shade effect depending on the varieties. It was also reported that the dry matter accumulation in the shoots of sun- and shade-grown cassava plants were on par with each other, while marked difference was observed in tuber dry matter accumulation although varietal differences were noticed. Under shaded conditions, cultivars H-97/2, H-165, Cl-590 and H-2304 accumulated more dry matter in tubers than others. Most of the photosynthates of shade-grown cassava plants were utilized for shoot growth affecting tuber growth significantly. Okoli and Wilson (1986) observed 43–80 % reduction in cassava yield when shade levels increased from 20 % to 70 %. Similar reduction in yield with increasing shade was also reported by Kasele et al. (1986) and Sreekumari et al. (1988) in cassava. Geetha et al. (2006) reported that root volume and root/shoot ratio (Table 19.1) showed a decreasing trend with increasing shade, which indicated that the root growth was limited under shade. They also noticed a significant decrease in total dry matter production as well as tuber yield of cassava with increase in shade levels with the highest dry matter production and tuber yield in the open.

19.3.1.2.2 Sweet Potato

Sweet potato requires high levels of solar radiation for optimum growth and tuber yield (Oswald et al. 1985). Among the major tropical tuber crops, sweet potato appears to be the least tolerant to shade (Johnston and Onwume 1998). Sweet potato, when grown in association with annuals or perennials under multiple cropping systems, often records a reduction in tuber yield due to accompanying shade effect. Reduction in flowering, increase in chlorophyll and developing climbing habit are a few of the external manifestations of sweet potato plants under shaded environment (Anil and Palaniswami 2008). Bai and Nair (1982) observed positive influence of shading on vine length in sweet

Table 19.2 Effect of light intensity on growth and yield of sweet potato

Treatments	Fresh weight of tuber (g)		Harvest index	
	105 DAP[a]	140 DAP	105 DAP	140 DAP
0 % shade	205.4	256.5	51.8	51.3
25 % shade	125.2	165.3	45.2	48.3
50 % shade	96.6	133.4	41.3	45.0
75 % shade	70.7	110.2	36.7	42.6
CD (0.05)	5.37	3.41	–	–

Source: Biswas et al. (1990)
[a]*DAP* days after planting

potato. Under shade, sweet potato produced larger leaves and more chlorophyll per leaf (Johnston and Onwume 1998). The chlorophyll a:b ratio, carotenoids per unit chlorophyll and the weight per unit area of leaf were lower in the shade. In sweet potato, relative growth rate tended to decline with increasing shade, and lowest values were recorded in 73 % shade (Roberts-Nkrumah et al. 1986). Decrease in net assimilation rate with shade was reported by Roberts-Nkrumah et al. (1986) in sweet potato.

Shading of the crop will result in reduced tuber yield. Zara et al. (1982) reported 32–92 % yield reduction among cultivars grown under the shade of coconut trees in the Philippines. Under artificial conditions, Martin (1985) observed no significant yield reduction in 20 % shade, whereas 80 % shade resulted in yield reduction as high as 94 %. Shade-tolerant cultivars outyielded standard cultivars under mild shade but were equally inhibited from tuber production under strong shade. The shoot is the storage sink under shaded conditions, and tuber yield is not only reduced by decreased assimilate production as a whole but also by an altered assimilate partitioning (Oswald et al. 1985). It was further observed that decrease in tuber yield was maximum when shade was imposed at the end of the growth season and the least at the beginning. Roberts-Nkrumah et al. (1986) reported that fresh weight of the tuber was not affected in 25 % shade, but in 55 % shade, it was noticeably lower, and in 73 % shade, tuberization was almost completely suppressed in all cultivars. According to Mwanga and Zamora (1989), 31–67 % shading reduced root fresh weight, root bulking ratio, root dry matter content and fresh vine yield by 5–96 %. Differential responses of the varieties were obtained at 0, 31 and 43 % but all responded similarly at 67 % shade. Biswas et al. (1990) recorded an increase in number of tubers, vines and fresh tuber weight and harvest index and a decrease in vine length with an increase in the intensity of light from 25 to 100 % (Table 19.2). Bonsi et al. (1992) found that continuous light did not interrupt the storage root initiation and bulking leading to increased yield. Shading resulted insignificant reduction in root yield and dry matter accumulation and increase in the top/root ratio (Wang et al. 2014). The responses of the photosynthetic parameters and endogenous hormones to shading were closely correlated with the variation in the storage root yield of the different cultivars.

According to Ravi (2000), sweet potato leaf requires at least one third of solar radiation (600 μ mol. m^{-2} s^{-1}) for maximum photosynthetic rate. Compared to plants that grow under 100 % sunlight, mild shade (20–25 % light reduction, LR), moderate (40–55 % LR) and deep shade (60–73 % LR) significantly reduce total dry matter and tuber yield. However, the amount of reduction in total dry matter and tuber yield is greater under deep shade than moderate shade. Moderate shade reduces the number and size of tubers mainly because of suppression of tuber growth but not initiation of tubers. However, deep shade significantly reduces number and size of tubers because of complete suppression of both initiation and growth of tubers.

19.3.1.2.3 Yams

Shade has significant influence on the growth and performance of yams. Shade-grown yam plants exhibited increased plant height and reduced net assimilation rate and crop growth rate. Maximum leaf area index was observed at 50 % shade in greater yam and at 25 % shade in lesser yam (Pushpakumari 1989). Yams compensated for shade by having a large proportional increase in leaf size and more chlorophyll per leaf area and appeared to be moderately tolerant to shade (Johnston and Onwume 1998). The yield of greater yam decreased proportionally as the shade increased and the nutrient requirement under shaded situation was the same as that of open situation. But in the case of lesser yam, the tuber yield under open and 25 % shade level were on par, but under higher shade levels, drastic reduction in yield was observed. Under shaded situation, the nutrient requirement was only 75 % of that of open situation (Pushpakumari and Sasidhar 1992). The total dry matter production of the yams at 25 % shade was superior to all other treatments throughout the growth stages except at harvest stage wherein the open condition recorded maximum dry weight although it was on par with 25 % shade. In greater yam maximum uptake of nitrogen and phosphorus was recorded in open conditions. The uptake of potassium followed a slightly different trend with highest uptake at 50 % shade. Lesser yam had higher uptake of nitrogen, phosphorus and potassium at 25 % shade level followed by open situation but recorded very poor nutrient uptake at intense shade (Pushpakumari and Sasidhar 1996).

19.3.1.2.4 Elephant Foot Yam

Elephant foot yam tolerates shade conditions and is suitable for intercropping between young trees. Under full sunlight and 25 % shade, photosynthetic activity in elephant foot yam reached the maximum of 10 mg CO_2 dm^{-2} ha^{-1} at 20 days after leaf unfolding and then decreased rapidly (Mirua and Osada 1981). Under weak light, the rate was high up to 30 days after leaf unfolding and then gradually decreased. It was also observed that corm dry weight increased under shading. Plant height continued to increase in elephant foot yam grown under shade (Pushpakumari 1989). Higher crop growth rate was recorded at medium and low shade during the first and second phases of growth (Fukai et al. 1984; Pushpakumari 1989). Shading treatments significantly decreased the leaf number and affected the length of petioles and rachis (Santosa et al. 2006). The short lifespan of leaves might enhance the production of new leaves resulting in a large number of leaves under full sunlight. Plants developed the shortest petioles under full sunlight but the longest under 75 % shading. However, maximum leaf area index and total dry matter production were recorded at 25 % shade in elephant foot yam (Pushpakumari 1989). Uptake of nitrogen, phosphorus and potassium by elephant foot yam was maximum in the open and decreased with increasing shade levels (Pushpakumari and Sasidhar 1996). The nutrient dose under partial shade can be reduced to half than that for the open situation. With increase in shade intensity, the yield declined proportionately recording the highest yield under open condition (Pushpakumari and Sasidhar 1992). However, Santosa et al. (2006) reported that the fresh biomass of corms increased with a decrease in light intensity; 75 % shading produced the largest corms and 0 % shading produced the smallest corms. Under full sunlight, necrosis and curling of the edges or the tips of leaflets occurred, causing 25 % crop loss. No damage was observed in the 25, 50 and 75 % shading.

19.3.1.2.5 Cocoyams

Taro (*Colocasia esculenta*) and tannia (*Xanthosoma sagittifolium*) showed enhanced ability to survive stress condition when grown under shade but with low economic yield (Caesar 1980). The cocoyams have been identified as shade tolerant with respect to photosynthetic activity (Schaffer and O'Hair 1987). The photosynthetic rate of leaves of tannia was found to be greater than that of taro. In both taro and tannia leaves, the photosynthetic rate was high during 2nd–4th week after full expansion, and after 4th week the rate declined gradually (Schaffer and O'Hair 1987). Moreover, the photosynthetic rate reached saturation at about 1200 μmol m^{-2} s^{-1}. Leaves of both taro and tannia grown under moderate shade (60 % sunlight) had higher rate of photosynthesis (1300 μmol m^{-2} s^{-1}) than leaves of plants grown under 100 % sunlight. This means that both the crops are adapted to shade conditions (Schaffer and O'Hair 1987).

The plant height of taro grown at full illumination was on par with 25–50 % shade but increased significantly under 75 % shade (Bai 1981). Significantly lower leaf area index was observed at 120 days after sprouting in the full sunlight (open) when compared to shaded treatments. The contents of total chlorophyll and its components in leaves were found to increase up to 75 % shade. A decreasing trend in yield with increasing shade was noticed in taro; however, no marked decrease in yield was observed up to 50 % shade. A declining trend in total dry weight of plant was observed with increase in shade (Bai 1981). Prameela (1990) noticed maximum tuber yield and total dry matter production of taro under 25 % shade followed by a reduction in yield and total dry matter production with further increase in shade intensity. The nitrogen content of taro was the highest under 50 % shade but decreased under high shade. Phosphorus content was more under 25 % shade and it increased with increase in shading. Potassium content also increased with shade (Prameela 1990).

In tannia, maximum leaf area index was recorded at 25 % shade. High crop growth rate was recorded at medium and low shade in the first and second growth phases. Even though net assimilation rate was maximum under open conditions, it was on par with 25 and 50 % shade at second growth phase. There was a significant reduction only at intense shade (Pushpakumari 1989). The highest yield of tannia was recorded under 25 % shade with an almost equal yield under 50 % shade (Pushpakumari and Sasidhar 1992). The total dry matter production at 25 % shade was superior to all other treatments throughout the growth stages except at harvest stage wherein the open treatment recorded maximum dry weight though it was on par with 25 % shade (Pushpakumari and Sasidhar 1996). Higher uptake of nitrogen, phosphorus and potassium was recorded at 25 % shade level followed by 50 % with very poor uptake at intense shade and in the open.

Valenzuela (1990) reported higher leaf area index, crop growth rate, chlorophyll and carotenoid contents and dry weight of shade-grown cocoyams compared to sun grown plants. Taro and tannia are adapted to shade by showing a greater proportional increase in leaf size, chlorophyll and carotenoid contents per leaf and a smaller reduction in leaf weight per unit area (Johnston and Onwueme 1998). Plants grown under less light had greater chlorophyll/N ratios and greater leaf area/leaf dry weight ratios than full sunlight-grown plants. The greater foliage biomass of cocoyam under shaded conditions resulted in increased light interception and photosynthate production. The respiration rate in the leaves of plants grown under shade was found to be lower than plants grown under full sunlight. This might have resulted in an overall increased net biomass accumulation in plants under shade and more partitioning to underground storage organs. The lower leaf temperatures commonly found in canopy under stories also would contribute to decreased respiratory costs than for plants grown under direct sunlight (Ravi 2000). Thus, cocoyam has the potential to grow under moderate shade without sacrificing yields, especially in tropical agroecosystem polycultures.

19.3.2 Drought

Drought or moisture deficit is a major environmental stress condition which limits crop production. Drought stress can be managed by identifying varieties which yield high under drought condition and by giving supplementary irrigation during drought period. Allocation of water for irrigation is a big challenge in the context of surging population and current climate change. The sustainable solution is to develop drought-tolerant/drought-resistant varieties. Variation in the degree of expression of physiological and phenotypic traits depending on severity of drought stress and genotypes (Yang et al. 1991) should be considered for developing drought-tolerant varieties.

19.3.2.1 Cassava

Cassava can be grown in areas ranging from humid (annual rainfall more than 2,000 mm) to semiarid (500–700 mm) conditions. Although it is a drought-tolerant crop, growth and yield decrease under prolonged dry periods. The reduction in storage root yield depends on the duration of water deficit and is determined by the sensitivity of a particular growth stage to water stress. But it is commonly grown in areas receiving <800 mm annual rainfall with a dry season of 4–6 months, where tolerance to water deficit is an important attribute. In cassava, bulk of the roots, 86 % and 96 %, is confined to the upper 10 cm and 30 cm soil, respectively. Majority of the root system laterally spreads up to 45 cm with maximum up to 2 m depth in soil under field conditions (Ravi 2000). Therefore, prolonged water-deficit conditions result in poor establishment of the crop and leaf shedding. Cassava grown under rainfed conditions showed reduction in leaf area index due to drought stress (Ramanujam 1990; DeTafur et al. 1997). The degree of leaf shedding under water-deficit condition varies among cultivars (Ravi and Saravanan 2001). Water stress also reduces leaf chlorophyll content, but inoculation with arbuscular mycorrhizal fungi, *Glomus* sp., was found to enhance the chlorophyll content even under water stress condition. But the effect varied with cassava genotypes and arbuscular mycorrhizal fungi species (Ekanayake et al. 2004).

Significant reduction in leaf area index (18–40 %), light interception (42–70 %), net photosynthetic rate (24–56 %), total dry mass (25–36 %) and tuber yield (28–42 % as evident in Table 19.3) were noticed due to drought stress depending upon the cultivar (Ramanujam 1990). In all the cultivars, leaf respiration closely followed the photosynthetic rate. The high biomass under stress-free conditions could be attributed either to a higher rate of photosynthesis or to a greater interception of solar radiation. However, drought stress affected both the factors significantly resulting in reduction in tuber yield. Under water-deficit condition, photosynthesis decreased by 25–31 % in different cultivars indicating that cassava cultivars maintain considerable rate of photosynthesis even under drought conditions (Ravi and Saravanan 2001). However, reduction in photosynthetic rate and leaf production under water-deficit condition resulted in decreased individual tuber weight as well as tuber number per plant which was ultimately attributed to the reduction in tuber yield (Table 19.4). Field moisture stress led to a decline in plant height by 47 %, stem girth by 15 %, tuber number by 95 % and tuber yield by 87 %. Hence, moisture stress resulted in considerable reduction in both vegetative growth and yield of cassava genotypes (Aina et al. 2007).

The reduction in storage root yield depends on the duration of water deficit and is determined by the sensitivity of a particular growth stage to water stress. The stage of the crop most sensitive to moisture stress is tuberization (Ramanujam 1994). Moisture stress during 3–5½ month of crop growth reduced tuber yield by 32 % (Connor et al. 1981), whereas moisture stress during any time of the initial 6 months period of crop growth reduced tuber yield by 60 % (Oliveira et al. 1982). When there is prolonged moisture deficit during tuber-bulking phase, leaf area index declines and potential productivity is not being realized for want of optimum canopy (Ramanujam and Biradar 1987).

Cassava has a built-in mechanism to avoid drought. Under severe moisture stress, the crop adapts to drought by shedding leaves. Once the crop is relieved off from the moisture stress or remained under drought for a longer period, regeneration of terminal buds with fresh foliage is noticed. This regeneration is initially at the expense of the stored food materials from the tuber. However, the newly formed leaves at the apex are usually smaller in size and are less efficient in synthesis and translocation of photosynthates. Stress occurring later in the season is most detrimental to yield because of reduced ability of old plants to recover leaf area during post-stress period. Water deficit affected the yield of the storage organ but did not alter the pattern of assimilate partitioning. Hence, the harvest index appears to be unaltered. Yield

Table 19.3 Tuber yield (kg m^{-2}) of cassava as affected by drought stress

Cultivar	Irrigated 1986–1987	Rainfed	Mean	Irrigated 1987–1988	Rainfed	Mean	Stress reduction (%)
M-4	4.24	2.79	3.51	3.92	2.81	3.36	31.3
H-165	8.16	4.53	6.34	6.66	4.08	5.87	41.9
H-226	5.30	3.64	4.47	6.25	3.76	5.00	35.9
H-1687	6.21	3.60	4.90	6.07	3.74	4.90	40.2
H-2304	5.99	3.67	4.83	6.22	3.93	5.07	37.7
S-1315	4.91	2.90	3.90	4.48	3.59	4.03	30.8
S-856	6.14	4.25	5.19	5.68	4.31	4.99	27.5
CI-590	5.12	3.08	4.10	5.07	3.18	4.12	38.5
Mean	5.75	3.55	4.65	5.54	3.67	4.60	–
	SE	LSD	SE	LSD			
		$P = 0.05$		($P = 0.05$)			
CV	0.168	0.339	0.193	0.390			
Tr	0.162	0.316	0.063	0.201			
Tr × Cv	0.275	0.667	0.263	0.549			
Cv × Tr	0.237	0.480	0.273	0.552			

Source: Ramanujam (1990)
CV cultivar, *Tr* Treatment

Table 19.4 Tuber yield of six cassava cultivars as affected by water-deficit stress

Cultivars	Single tuber weight (kg)		Tuber number plant^{-1}		Tuber yield (t ha^{-1})		
	No water stress	Water deficit stress	No water stress	Water deficit stress	No water stress	Water deficit stress	Reduction (%)
M-4	0.54	0.41	7.80	6.99	52.25	35.63	31.8
H-1687	0.79	0.61	6.15	5.52	60.90	42.25	30.6
H-165	0.75	0.60	8.65	7.34	81.13	55.50	31.6
H-97	066	0.46	7.19	6.69	59.63	38.13	36.1
H-226	0.66	0.66	8.90	5.69	73.63	47.13	36.0
CI-21	0.61	0.57	6.80	5.85	51.90	41.63	19.79

Source: Ravi and Saravanan (2001)

reduction due to moisture stress, therefore, is due to the reduction in total biomass production.

A close relationship exists between photosynthesis and leaf water potential which is regulated by stomatal movement. Increase in stomatal resistance due to stomatal closure in response to drought stress affected the photosynthetic rate by decreasing the leaf intercellular CO_2 concentration. El-Sharkaway and Cock (1990) observed that only 2 % of the 1,500 cassava cultivars studied had stomata on the upper (adaxial) surface. Cerqueira (1989) also found that only a few stomata are seen on the upper surface and most stomata are located on the lower (abaxial) surface of the cassava leaves. According to Ravi (2000), stomata are distributed only on the lower surface of leaves which is sensitive to air humidity and hence not desirable. Cultivars having stomata on both the surfaces of leaf are considered desirable because of their less sensitivity to air humidity. Reduced rate of leaf production, greater longevity of leaves and distribution of stomata on both sides of leaves (amphistomatous) are the desirable plant characters under drought conditions (Ravi 2000).

Drought causes variations in starch quality, the magnitude of which is influenced by the severity of stress conditions and stage of plant maturity. Despite an increased starch yield, the effect of initial water stress on starch quality is still sustained. In mature plants, starch quality is affected by environmental conditions prior to

Table 19.5 Tuber yield (t ha^{-1}) of sweet potato as affected by water-deficit stress at different stages of crop growth

Treatment	Tuber yield (t ha^{-1})			
	Sree Nandini	H-42	Sree Bhadra	S-108
Water stress at tuber initiation	4.4	4.01	10.0	7.0
Water stress at tuber development	12.4	12.0	17.33	9.0
Water stress at tuber maturity	8.6	5.1	20.16	9.0
Control	8.6	11.0	25.82	11.0

Source: Indira and Kabeerathumma (1988) and Ravi and Indira (1996)

root harvest, especially the onset of rain after a stress period as indicated by a reduced starch paste (Sriroth et al. 2001). Cassava's cyanogenic potential exacerbated during drought (Bograd et al. 2011). Supplementary irrigation increased the starch content and root yield but reduced the protein content.

19.3.2.2 Sweet Potato

Although sweet potato is tolerant to long periods of drought, root yield and quality may be compromised. It requires 500 mm water for 16–20 weeks' growth period (Ravi 2000), and an annual rainfall of 750–1,000 mm is considered the best (Nair 2000). It can tolerate considerable periods of drought, but yields are very much reduced if water shortage occurs at 10–30 days after planting when tuber initiation takes place. Soil moisture deficiency results in physiological and biochemical changes in the leaves and ultimately results in yield reduction. Under water-deficit stress, the leaf total chlorophyll content, relative water content and nitrate reductase activity decrease (Indira 1989; Indira and Kabeerathumma 1988, 1990; Naskar and Chowdhury 1995). The leaf water potential reduction causes increase in stomatal resistance to CO_2 exchange (Indira and Kabeerathumma 1988, 1990) causing reduction in net photosynthetic rate (Ravi and Indira 1996). The water condition in the soil and the plant strongly affects stomatal movement, causing a large change in leaf net photosynthetic rate (Chaves 1991). The effect of ageing on leaf photosynthesis varies with soil water content, and effect of ageing and water content vary with cultivars.

Tuber yield is affected by amount, timing and distribution of water. Tuber yield decreases under water-deficit stress particularly when the available soil moisture decreases below 20 % (Indira and Kabeerathumma 1988; Nair et al. 1996). The tuber initiation period is the most sensitive stage to water-deficit stress due to its effect on tuber number (Indira and Kabeerathumma 1988; Nair et al. 1996; Ravi and Indira 1996) (Table 19.5). The drought stress for 20 days during the critical growing stage decreased yield by 15–39 % (Gong and Wang 1990). Water-deficit stress during tuber initiation period induces lignification of tubers and hampers tuber growth as any lignification of developing roots may impair their potential for the lateral thickening associated with carbohydrate storage (Lewthwaite and Triggs 2012). Lignification and reduction in tuber yield are greater in cultivars with weak sink capacity than those with higher sink capacity (Ravi and Indira 1996). Moisture stress imposed during early phase of growth inhibited tuber formation, while stress imposed during tuber-bulking phase did not affect productivity (Sung 1981). Moreover, insect pests and viral diseases were reported to be very severe in drought conditions (Placide et al. 2013). Insects caused a greater degree of root damage at the water-stressed site, as soil cracking increased insect access to the roots (Lewthwaite and Triggs 2012).

Higher yield and maximum water-use efficiency can be attained by following the irrigation ratio of IW/CPE = 1. Increasing the level of soil moisture beyond the ratio resulted in yield reduction (Indira and Kabeerathumma 1990). According to Ravi (2000), sweet potato yields best under irrigation at 25 % available soil moisture, and there is no increase in yield by maintaining soil moisture >50 %. Sweet potato grown under high rainfall frequently produces vigorous vine but poor tuber yield. The ideal conditions for high yield are good rains during early growth and dry sunny weather during tuber bulking and maturity (Nair 2000).

Sweet potato cultivars vary in both sensitivity and response to drought (Indira 1989; Naskar and Chowdhury 1995; Lewthwaite and Triggs 2012). Of the four sweet potato cultivars having different features in growth and yield, cultivar Koganesengan sustained photosynthetic activity in young as well as aged leaves under drought (Haimeirong and Kubota 2003). This may be due to the stomatal conductance being relatively high in both aged and drought-imposed leaves. Three distinct yield responses, lowered root yield, unmodified yield and increased root yield, were demonstrated under drought conditions (Lewthwaite and Triggs 2012). The cultivar 'TokaToka Gold' demonstrated a 77 % increase in marketable root yield under conditions of water stress, compared with that of the well-watered site. The influence of drought on yield components differed with cultivars, both in the number of storage roots formed and average root weight. Storage root quality was also affected, with the roots of some clones developing fluting along their main axes, while others were unaffected.

Under water-deficit condition, sweet potato plants are known to accumulate proline in the leaf and root tissues (Indira and Kabeerathumma 1988; Ravi and Indira 1997; Ravi and Saravanan 1999). Ravi and Saravanan (1999) observed greater increase in proline content in leaves than in roots under water-deficit stress. Significant difference in proline accumulation was observed among the four cultivars studied under laboratory and field conditions. Since most of the proline accumulation occurs after cessation of growth, it does not seem to influence sweet potato plant growth during water-deficit condition. However, proline accumulation may help the plant to survive moisture stress through osmotic adjustment, strengthening protein stability and building of photosynthetic energy when stomata are closed. Better proline utilization efficiency during post-stress recovery period coupled with high sink capacity may lead to better tuber yield in drought-tolerant sweet potato cultivars (Ravi and Saravanan 1999). Although susceptible cultivars accumulate good amount of proline under water-deficit condition, their poor sink potential results in low yield.

19.3.2.3 Yams

Yams are relatively tolerant to dry conditions but give better tuber yield with the supplementary irrigation. Adequate moisture is required for growth, development and yield of yams which is met from well-distributed rains in the traditional yam-growing areas. Evenly distributed annual rainfall of 100–150 cm spread over 6–-7 months is ideal for white yam. However, lesser yam comes up well in places receiving rainfall of 90–100 cm. The most critical period is the first 5 months after planting (Waitt 1963). Coursey (1967) reported that maintaining optimum moisture regime during 14th–20th week was ideal for better development and tuber yield.

19.3.2.4 Aroids

19.3.2.4.1 Elephant Foot Yam

Elephant foot yam produces large corm and high yield under adequate water supply. Well-distributed rainfall of 1000–1500 mm spread over a period of 6–8 months (George 2000) is optimum for the crop. Supplementary irrigation is necessary for high productivity when the rainy season is shorter than 4 months. Well-distributed rainfall for ample moisture availability during the growth phase of the crop is essential as it promotes proper vegetative development. However, dry condition towards the later phase of crop growth helps in the development and bulking of corm. Soil moisture status does not influence sprouting, but further development of new leaf depends on adequate soil moisture (Ravi et al. 2009). Soil moisture conservation measures like mulching induced higher percentage of early sprouting, greater canopy spread, plant height, greater mean corm weight and corm yield (Mohankumar et al. 1973).

Frequent irrigation at 1-, 3- or 5-day intervals produced large leaves and extended their lifespan compared with less frequent watering (Santosa et al. 2004). A decrease in dry mass of seed corms was more evident with frequent watering, suggesting that reserved carbohydrates in seed corms are not easily metabolized under limited water supply. High ratio of dry mass of daughter corms to that of seed corms under frequent watering treatments could be ascribed to the fact that the soil water availability affects not only the utilization of dry matter in the seed corms but also the production and translocation of photoassimilates into daughter corms (Sugiyama and Santosa 2008). The roots dried earlier than usual when the soil water content decreased to less than 40 % of field capacity (Santosa et al. 2004). The crop tolerates water-deficit stress conditions for about 30–60 days, but prolonged stress may affect corm yield. Infrequent irrigation (at 7- or 15-day intervals) reduced corm yield and forced the corms to enter into dormancy after leaf senescence. The 4–6-month crop growth period is the critical period requiring adequate soil moisture for achieving more corm yield (Ravi et al. 2013b).

19.3.2.4.2 Taro

Taro is primarily adapted to moist environments but can be grown under a wide range of moisture regimes, ranging from lowland paddy culture to upland conditions under irrigation (Wilson 1984). However, under tropical conditions, the crop is often subjected to unfavourable growing situations, due to water-deficit stress. An annual rainfall of approximately 250 cm is considered optimum. But it can be grown in upland areas where the rainfall is about 175 cm provided it is evenly distributed throughout the growing period. It is important to ensure a continuous availability of water in upland cultivation of taro. Wherever rainfall is irregular, irrigation facilities must be provided. High rainfall is needed during the first 20 weeks of growth period corresponding to the maximum leaf development period. Thereafter, drier conditions can be tolerated until harvest (Lebot 2009). Reynold and Netran (1977) observed a close relationship between the rainfall received during 4th, 5th and 6th month period, which coincided with tuber bulking in taro in Western Samoa. Hence, tuber-bulking stage has been considered as the most critical period of water-deficit stress in taro. Decrease in soil moisture drastically reduced the leaf area, leaf longevity, stomatal opening and dry matter production (Ravi and Chowdhury 1991, 1993) and corm yield in taro (Ravi and Chowdhury 1996) (Table 19.6). Hence, water is critical input throughout the growth period of taro.

Among the two types of taro, the eddoe type is considered to be tolerant to drought conditions (Kay 1973). Ravi and Chowdhury (1991) observed that 25 % of soil moisture saturation adversely affected the growth and corm yield of eddoe taro. Dasheen type grown under erratic moisture conditions showed peculiar dumb-bell-like-shaped tubers reflecting constrictions in growth during dry periods. Under water-deficit stress, eddoe-type taro produced few cormels (Fujimoto 2009).

Compared to taro plants grown under 100 and 50 % soil moisture saturation, plants grown under 25 % soil moisture saturation showed greater reduction in leaf relative water content, total chlorophyll, sugar content and membrane stability. Accumulation of greater amount of proline in leaves of taro plants grown under 25 % soil moisture than those grown under 50 and 100 % soil moisture saturation also indicates that plants grown under 25 % have undergone maximum stress than those grown under 50 and 100 % soil moisture saturation. Taro plants, grown under 25 % soil moisture, showed greater reduction in growth and yield as well as greater stomatal closure than those grown under 50 and 100 % soil moisture (Ravi and Chowdhury 1991; 1993b). Therefore, 50 % soil moisture saturation can be considered as the critical level below which taro crop may experience water-deficit stress and yield reduction.

19.3.2.4.3 Tannia

Among the aroids, tannia is relatively more tolerant to dry conditions but performs better under irrigation, especially when rainfall is irregular (Onwume and Charles 1994). Tannia are plants of tropical rain forest region which require

Table 19.6 Tuber yield of taro genotypes as affected by water stress

	Mean yield of rainfed crop		Mean yield under drought		Yield reduction
Taro cultivar	g plant^{-1}	t ha^{-1}	g plant^{-1}	t ha^{-1}	%
CI-106	131.8	4.9	75.0	2.8	42
C-266	127.9	4.7	64.7	2.4	49
N-82	238.6	8.8	56.5	2.1	76
CI-154	214.0	7.9	62.0	2.3	71
Jangdi local	207.25	7.7	83.1	3.1	59
Aiginia local	180.8	6.7	84.0	3.1	53
Thelia local	216.84	8.0	77.4	2.9	64
N-129	225.0	8.3	30.0	2.2	73
N-149	142.0	5.3	33.0	2.4	54
Chiruli local	165.2	6.1	28.0	2.0	66

Source: Ravi and Chowdhury (1996)

copious rainfall and adequate soil moisture (Lopez et al. 1995). The crop is suited to high-rainfall areas receiving an annual precipitation of 140–200 cm, but it can also grow well with an annual rainfall as low as 100 cm, provided it is evenly distributed (Ramesh et al. 2007).

Tannia can be grown under upland conditions with irrigation, and certain early maturing cultivars can be grown without irrigation in comparatively dry situations. Tannia produces the highest yield under shade and adequate water supply. Under water-deficit stress, only the main corms grew, while the growth of cormels was negligible (Caesar 1980). As a result of the increased development of above-ground plant organs under shade, the plant can withstand water-deficit stress but with a low edible tuber yield.

In Puerto Rico, in a study involving different depths of free water tables, cormel yields of 9.9, 11.7 and 16.6 t ha^{-1} at water table levels of 15, 30 and 45 cm below the soil surface, respectively, were obtained. According to Silva and Irizarry (1980), the free water table must be kept at 45 cm below soil surface to get high yields.

19.3.3 Flood/Waterlogging

Maintenance of crop productivity in waterlogged environment is a challenging task. The waterlogged environment imposes several difficulties in traditional cultivation practices keeping the land either fallow or sub-productive. In general, tuber crops like cassava, sweet potato, yams, tannia and elephant foot yam cannot tolerate waterlogging and require well-drained soil for their successful cultivation.

Cassava is very susceptible to flooding and most plants die if flooded for more than 4–5 days. Cassava shows adaptive ability on peat with different depths of water table (Tan and Ambak 1989). When water table was high, the storage roots concentrated near the soil surface. When water table was low, the root growth extended downwards towards the water. The effects of flooding for 30 days on the root and leaf carbohydrate levels of cassava were investigated by Onwugbuta-Envi and Kpekot (2008). The levels of starch in the flooded roots decreased rapidly, indicating that root starch was readily mobilized under flooded condition. Marked accumulation of starch in the leaf was observed in the flooded plant suggesting that phloem transport was reduced. Sucrose, glucose and fructose levels in the flooded roots increased several folds with initial flooding and were maintained at a high level than those of non-flooded control during the flooding period. These results show that starch accumulation in the leaves does not result in deficiency of sugar in flooded roots, and root sugar level is not critical for flooding tolerance.

Though grown in areas with relatively high rainfall, sweet potato cannot withstand waterlogging. Tuber formation or development is affected under inundation. Inhibition of tuber development in sweet potato under flooded environment is attributed to high levels of ethylene production under such conditions (Patterson et al. 1979). Since sweet potato cannot withstand waterlogging, it is grown in ridges and mounds (Gomes 1999). In waterlogged conditions, roots of sweet potato may die out. Even if the storage roots are formed, they may develop enlarged lenticels which give tuber a rough unattractive appearance.

Proper control of flood and soil salinity is critically important for successful yam production in tidal lands of Korea (Sohn et al. 2011). In yam seedlings raised from aerial bulblets grown in ridges, flooding injury very seriously led to plant death during the rainy summer season. The main reason for flooding injury included the decreased rainfall acceptable capacity of the soil after the rising of water table and slowdown of water infiltration rate after the formation of an impermeable soil crust in the furrow bottom with continuous and heavy downpour during the rainy summer season.

One of the simple but effective strategies to increase productivity of waterlogged area is cultivation of crops which have natural adaptation to such conditions. Although elephant foot yam can tolerate temporary flooding, anaerobic waterlogging causes corm rot (Ravi et al. 2011). Apart from rice and lotus, taro is one of the few crops in the world that can be grown under flooded conditions. The large air spaces in the petiole permit the submerged parts to maintain gaseous exchange with the atmosphere (Sunell and Arditti 1983). Also, it is important that the water in which taro grows remains cool and continuously flowing, so that it can have maximum dissolved oxygen. Swamp taro (*Colocasia esculenta* var. *stoloniferum*. (L.) Schott.) locally known as 'panikachulati' or 'kachulati' (Saud and Baruah 2000; Sen and Goswami 2000) also has the ability to grow under swampy, waterlogged conditions and hence the name. All plant parts like stolons or runners, petioles and leaves are popularly consumed as green vegetable (Chowdhury et al. 2008). The crop showed light utilization strategies for sustained growth in waterlogged environment (Chowdhury et al. 2009). It can grow in swampy areas prone to submergence and frequently subjected to excess water situation and partial or even total submergence (Chowdhury et al. 2010). However, varietal variation was observed in the performance of the crop when grown under swampy conditions in low-lying areas and subjected to brief submergence. Maintenance of greater leaf area, leaf chlorophyll and root nitrate reductase activity in post-submergence period and higher runner yield suggested the suitability of the variety, BCST 15 for cultivation in low-lying flood prone areas (Chowdhury et al. 2010, 2011). Thus, tuber crops like taro and swamp taro can be recommended for cultivation in regions prone to brief submergence by flood.

19.3.4 Temperature Extremes

As with all crops, both high and low temperatures negatively affect the production of tuber crops. It is predicted that mean global surface air temperature may increase by 3–5 °C from present levels (Schneider 1989). Heat stress is one of the most important factors affecting crop yield (Wrigley et al. 1994). All plant processes are sensitive to and can be irreversibly damaged by heat. Heat stress induces oxidative injury by forming reactive oxygen, which damage cell membranes besides denaturing native proteins (Dhindsa et al. 1981), causing a reduction in loss of chlorophyll (Kyle 1987) and number of chloroplast per cell (Hurkman and Tanaka 1987). Elevated temperatures accelerate senescence, reduce the duration of viable leaf and diminish photosynthesis (Harding et al. 1990). Cultivation of suitable tuber crops under partial shade and maintaining optimum water status in plant will alleviate heat stress. Growth and tuber production are affected when the tropical tuber crops are cultivated in areas with prolonged and extreme winter condition due to low soil temperature. Change in the pattern of rainfall with rise in temperature coupled with frequent occurrences of drought and flood will ultimately affect the crop production.

19.3.4.1 Cassava

Temperature has considerable effects on cassava growth as it affects sprouting, leaf formation, leaf size, storage root formation and, ultimately, the tuber yield. The optimum temperature favourable for the growth of cassava is 25–29 °C. But it can tolerate temperatures from 16 to 38 °C (Cock 1984). However, it cannot withstand frost since growth ceases at temperatures below 10 °C. At low temperatures (16 °C), sprouting of stem cuttings is considerably delayed, and the rate of leaf production, total and storage root dry weight decrease (Cock and Rosas 1975). Sprouting is hastened when the temperature increases up to 30 °C, but it slows down above 37 °C (Keating and Evenson 1979). At lower temperatures leaf area development becomes slow because the size of individual leaves is smaller, and fewer leaves are produced at each apex although leaf life is increased (Irikura et al. 1979). At temperatures between 15 and 24 °C, the leaf life is approximately 200 days (Irikura et al. 1979), while at higher temperatures, leaf life is only 120 days (Splittstoesser and Tunya 1992). At 24 °C, the leaf life varies according to the genotype and plant age and is approximately 60–80 days during the first 4 months of growth (El-Sharkawy 2006). Different varieties showed differential response (Table 19.7) to temperature rise (Irikura et al. 1979). Significant increase in tuber yield of four varieties was observed between 20 and 24 °C. No significant increase in tuber yield was observed between 24 and 28 °C. There is considerable genotype x temperature interaction indicating that natural selection is highly significant on varietal adaptation to local environments.

The main effect of temperature is on biological production, as dry matter partitioning does not change much when cassava is cultivated under different temperatures (Cock and Rosas 1975). Higher temperatures are associated with a greater crop growth rate and high photosynthetic rate. Photosynthetic rate of cassava leaves has been reported to be maximum (90–100 %) between 30 and 35 °C (El-Sharkawy and Cock 1990; El-Sharkawy et al. 1992) (Table 19.8). Under controlled growth chamber conditions, the highest net photosynthetic (P_N) rate was noticed at 25 °C, and thereafter it declined (Mahon et al. 1977). Similar trend was observed in studies conducted under field conditions in India (Ravi et al. 2008). Leaves of cassava plants grown at 29/24 °C day/night temperature regime had higher P_N (19.0 mg CO_2 $dm^{-2}h^{-1}$) than plants grown at 24/19 °C (12.0 mg CO_2 $dm^{-2}h^{-1}$) showing an increase in P_N at 1.4 mg CO_2 $dm^{-2}h^{-1}$ per 1 °C rise in temperature (Mahon et al. 1977). Cassava plants grown at 20–22/12–15 °C day/night temperature regimes showed an increase in P_N at 0.2–0.6 μmol CO_2 $m^{-2}s^{-1}$ between 25 and 35 °C (El-Sharkawy et al. 1992). On evaluation of the potential photosynthesis of three cultivars from contrasting habits under different growing environments, El-Sharkawy et al. (1992) concluded that P_N increased with increase in temperature, reaching its maximum at 30–40 °C. In all the cultivars, P_N was substantially low in the leaves that had developed in the cool climate than in those from the warm climate. The high sensitivity of P_N to temperature points to the need of genotypes more tolerant to low temperature, which could be used in the highland tropics and subtropics.

Although cassava sustains vegetative growth and biomass at high temperatures (40 °C) under adequate soil moisture, sucrose synthesis and export from the leaves and starch synthesis in tubers are affected. In cassava, tuber starch

Table 19.7 Tuber yield of four cassava varieties as affected by temperature

	Fresh tuber yield (t ha^{-1})		
Variety	20 °C	24 °C	28 °C
M Col 22	9.3	22.7	39.4
M Mex 59	22.8	38.8	30.4
M Col 113	24.2	26.1	23.9
Popayan	28.9	15.7	9.4
Mean	21.3	25.83	25.78
[a]		1.13	0

Source: (Irikura et al. 1979)
[a]Yield increase due to 1 °C rise in temperature between 20 and 24 °C

Table 19.8 Photosynthetic rate with increase in temperature in cassava

Leaf temperature (°C)	Photosynthetic rate (P_N) (μmol CO_2 m^{-2} s^{-1})	Increase (+) or decrease (−) in P_N per 1 °C	Average increase in P_N due to 1 °C rise between 15 and 35 °C	Average increase in P_N due to 1 °C rise between 35 and 45 °C
15–20	15–18	+0.6		
20–25	18–20	+0.4		
25–30	20–25	+1.0		
30–35	25–28	+0.6	0.65	
35–40	28–25	−0.6		
40–45	25–14	−2.2		1.4

Source: El-Sharkawy and Cock (1990)

content is greater by 3–5 % at cooler climate in high altitude than warmer conditions in plain (Ravi et al. 2008). Higher temperatures and drought also can divert photosynthates to lignification (Amthore 2003) and can significantly reduce the tuber yield and starch content of cassava.

19.3.4.2 Sweet Potato

Sweet potato is a warm weather crop. A temperature range of 20–27 °C is best suited for vine growth and tuberization. The ideal temperature is 20–25 °C (Palaniswami et al. 2008). The areas with an average day temperature of more than 24 °C and an average rainfall of 750 mm or more are suitable for cultivation. Enzymes of photosynthesis (*Rubisco* activase) as well as electron transport in sweet potato leaves are sensitive to temperatures more than 35 °C (Cen and Sage 2005) (Table 19.9). No tuber production takes place at very high temperature (35–40 °C) and also temperatures below 15 °C. At temperatures below 10 °C, the plants are damaged. Hence, sweet potato can be grown throughout the tropics, but in temperate regions, the aerial portion dries out during winter (<10 °C). At air temperatures more than 30 °C, an increase in IAA oxidase activity causes reduction in tuber formation and growth, while increases in gibberellic acid promote shoot growth. Under field conditions, tuber yields were greater in winter (*rabi*) season than in rainy (*kharif*) season, and sweet potato yield decreased by 2.37 t ha^{-1} due to 1 °C rise in temperature from 23.5 to 25.9 °C (Nawale and Salvi 1983).

Warm sunny days and cool nights are very much favourable for storage root formation in sweet potato. Root formation requires cool temperature, while the weather should be warm for root bulking. Night air temperature seems to be the most critical factor for tuber growth which may be due to greater translocation of sugar from the shoot to roots during this time. Night temperatures between 15 and 25 °C promote tuber formation and growth (Singh and Mandal 1976). Soil temperatures between 20 and 30 °C favour tuber formation, while soil temperatures more than 30 °C promote shoot growth at the expense of tuber growth. Sajjapongse (1989) developed a yield prediction equation, and it was concluded that temperature was more important yield determinant than precipitation and solar radiation. The changes in yield of sweet potato with increase in temperature (Table 19.10) were simulated using SPOTCOMS model (Mithra and Somasundaram 2008).

19.3.4.3 Yams

Yams are tropical crops and hence thrive well under warm sunny weather with plenty of rains. Warm climate with temperature of 25–30 °C and adequate moisture are ideal for vegetative growth and tuber bulking. Yams cannot tolerate frost and growth is affected below 20 °C.

19.3.4.4 Aroids

Taro is best suited in swampy tropical climate with temperature range of 25–30 °C. It is grown well in flooded conditions in Hawaii. In moist environment, it can be grown at temperature of 20–21 °C. A hot and humid condition with

Table 19.9 Photosynthetic rate as a function of intercellular CO_2 and temperature in sweet potato leaves

Temperature (°C)	Photosynthetic rate (CO_2 μmol $m^{-2}s^{-1}$) Intercellular CO_2 (ppm)					Increase in photosynthetic rate (CO_2 μmol $m^{-2}s^{-1}$) due to increase in CO_2 between 250 and 560 ppm
	250	280	420	500	560	
20	14.0	12.5	15.0	19.0	15.0	1.0
25	15.0	16.0	21.0	22.0	21.0	6.0
34	16.0	18.0	27.0	25.5	34.5	18.5
38	13.0	17.0	24.0	24.0	28.0	15.0
[a]	0.14	0.39	0.86	0.46	1.36	

Source: Cen and sage (2005)

[a]Increase in photosynthetic rate (CO_2 μmol $m^{-2}s^{-1}$) due to 1 °C rise between 20 and 34 °C

Table 19.10 Sweet potato yield changes due to 1 °C rise in temperature based on SPOTCOMS model

Variety	Seasonal mean temperature (°C)		Increase in yield (t ha^{-1}) due to 1 °C rise in mean temperature between 19.24 and 20.24 °C
	19.24	20.24	
Sree Arun	5.01	6.23	1.26
Sree Bhadra	4.17	6.45	
Sree Rethna	7.97	8.25	
Mean	5.72	6.98	
Variety	Seasonal mean temperature (°C)		Decrease in yield (t ha^{-1}) due to 1 °C rise in mean temperature between 28.24 and 29.24 °C
	28.24	29.24	
Sree Arun	8.2	8.36	0.12
Sree Bhadra	7.76	7.49	
Sree Rethna	10.24	9.98	
Mean	8.73	8.61	

Source: Mithra and Somasundaram (2008)

temperature range of 25–30 °C is ideal for the cultivation of tannia. Elephant foot yam requires warm humid climate with an average temperature of 30–35 °C and well-distributed rainfall of 1,000–1,500 mm to promote better shoot emergence and vegetative growth, while dry condition with comparatively low temperatures of 20–25 °C is favourable for corm development.

19.3.5 Rise in Atmospheric CO_2 Level

The concentration of CO_2 in the atmosphere has increased from 280 to 387 ppm from 1750 to 2000 AD and continues to increase at 1.5 ppm every year (IPCC 2007). Hence, research is being conducted to study the response of tuber crops to increased CO_2 concentration.

19.3.5.1 Cassava

Cassava being a C_3 plant (Veltkamp 1985; Edwards et al. 1990; Ueno and Agarie 1997) possesses a greater advantage to respond to elevated CO_2 concentrations. Mahon et al. (1977) observed an increase in net photosynthetic rate up to 500 ppm CO_2, and thereafter the increase in P_N rate was negligible. Under controlled conditions, tuber weight of cassava increased significantly with increase in CO_2 from 350 to 700 ppm, while the increase was greater at day/night temperature 33/21 °C than at 28/21 °C (Imai et al. 1984). The water-use efficiency of cassava plants grew at high CO_2 also increased to 10.5–17.1 mg CO_2 g H_2O^{-1} than control plants (3.3–4.5 mg CO_2 g H_2O^{-1}). Fernandez et al. (2002) reported significant increase in leaf photosynthesis of cassava grown in elevated CO_2 (680 cm^3 m^{-3}) above ambient CO_2 (300 cm^3m^{-3} CO_2).

Table 19.11 Sweet potato yield increase due to high CO_2

CO_2 (ppm)	Tuber yield (g $plant^{-1}$)	Tuber yield (t ha^{-1})[a]	Yield increase due to increase in CO_2 from 363 to 514/666 ppm
363	346.9	28.91	3.76 t ha^{-1}
514	388.2	32.35	
666	395.9	32.99	

Source: Bhattacharya et al. (1992)
[a]Tuber yield ha^{-1} estimated from yield $plant^{-1}$ at 83333 plants ha^{-1}

Based on several physiological, anatomical and biochemical leaf traits, cassava can be considered a C_3–C_4 intermediate species (Angelov et al. 1993; El-Sharkawy 2006). Cassava leaves possess elevated PEP case activity that reaches 15–25 % of those in C_4 tropical crops such as maize and sorghum and much greater than activities observed in typical C_3 species such as common beans. The PEP case activity of cassava increased by 13 %, whereas the RUBP case activity decreased by 42 % under water-deficit stress and heat stress conditions (El-Sharkawy 2006). The effects of high CO_2 on cassava tuber yield and starch enhancement appear to be strongly temperature dependent (Ravi et al. 2008).

19.3.5.2 Sweet Potato

Photosynthetic rate of sweet potato leaves increased with increase in CO_2 up to 560 ppm at temperature up to 34 °C. Beyond 34 °C, no increase in photosynthetic rate was observed with increase in CO_2. Photosynthetic rate, specific leaf weight, biomass and tuber yield increased (Table 19.11) under high CO_2 (514, 666 ppm) (Bhattacharya et al. 1990b, 1992; Biswas et al. 1986). High CO_2 ameliorated the effect of drought with increase in root/shoot ratio (Bhattacharya et al. 1990b). Carotene, starch and glucose contents in tuber increased under high CO_2 (675 ppm) (Bhattacharya et al. 1989, 1990a).

Since the response of tuber crops to elevated CO_2 concentration is temperature dependent, a detailed investigation is warranted regarding the interactive effect of high temperature (>35 °C), elevated CO_2 concentration and water availability on the growth, yield and quality of tuber crops.

19.3.6 Salinity

Global area of salt-affected soils including saline and sodic soils is about 831 million hectares (Martinez-Beltran and Manzur 2005), out of which approximately 50 % is accounted by saline soils. Soil salinity is caused by the presence of chlorides and sulphates of sodium, calcium and magnesium. In the arid and semiarid regions, the problem of salinity is further aggravated due to loss of water by evaporation and transpiration under the prevailing hot dry climate. Poor drainage in agricultural fields also increases salinity by raising the water table near the surface. Salinity inhibits plant growth and interferes with uptake and transport of essential nutrients (Greenway and Munns 1980). Reclamation, drainage and irrigation management can minimize the extent and spread of soil salinity. Identification and breeding of salt-tolerant crops/varieties is a promising strategy that could be integrated with water and land management strategies.

Tuber crops are also sensitive to salinity, and in general, soils with 5.5–7 pH are ideal for cultivation. Cassava can be grown successfully in acid soils where few other crops will yield (Mohankumar 2008). In cassava, decay of the cortical parenchyma and formation of tyloses in the vessels are the major anatomical changes noticed in the roots in relation to the toxic effect of salinity (Indira 1978). It is seen that arbuscular mycorrhizal fungus colonization provides a biological mechanism by which cassava clones increased plant biomass and salt tolerance. Inoculation with the arbuscular mycorrhizal fungus, *Glomus intraradices*, improved salt tolerance and promoted the plant development in all the studied clones of cassava, irrespective of salinity level (Carretero et al. 2008).

Despite their calorie yield, the productivity of sweet potato is affected to a greater extent due to salinity (Horton 1989). Root growth is much more sensitive to salinity than vine growth resulting in low productivity (Greig and Smith 1962). Salinity significantly reduces growth of stems and roots and results in lateral rolling of leaf lobules, reduction in leaf size and necrosis of older leaves (Villafane 1997). Foliar analysis of salt-stressed plants revealed the accumulation of Na^+ and Cl^- ions to >400 % as compared to plants irrigated with normal water, suggesting the lack of extrusion mechanism (Nair 2000). Richardson and Caligari (2014) also noticed severe growth reduction in sweet potato under salinity stress. Visual symptoms observed in NaCl-stressed plants included the shedding of leaves and necrotic lesions in the roots.

Development of salt-tolerant sweet potato varieties would help to supplement food to the poor people living in coastal areas. Besides being rich in carbohydrates and minerals, the orange- and purple-fleshed tubers are cheap sources of β-carotene and anthocyanin, respectively. A regular intake of 100 g per day of orange-fleshed sweet potato tubers can provide the recommended daily amount of vitamin A for children and adults, which prevents night blindness. Sodium chloride tolerance in sweet potato genotypes has been reported by Mukherjee (1999, 2001, 2002) under both in vivo and in vitro conditions. The orange-fleshed genotypes like CIP 3, CIP 12 and JO 14 are found tolerant to high NaCl stress and could be included in the varietal improvement programme to overcome salt stress conditions (Dasgupta et al. 2006). The genotypes, ST-14, SB-198/115, CIPSWA-2, CIP-420027, CIP-440127 and ST-13 showed tolerance to soil salinity (6–8 dS m^{-1}) with high tuber yield (16.69–22.49) anthocyanin content (85–90 mg 100 g^{-1}) in purple-fleshed genotype (Mukherjee and Naskar 2012).

The ability to transport Ca^{2+} to the shoot during salt stress is considered by La Haye and Epstein (1971) an indication of salt tolerance. Increasing levels of Ca^{2+} in the external medium alleviates the adverse effects of NaCl salinity on plant growth (Bilski et al. 1988). Richardson and Caligari (2014) observed differences in ion uptake of sweet potato plants subjected to NaCl salinity and low and high concentrations of $CaCl_2$. Salinity tolerance of sweet potato appears to be associated with its ability to control rates of Na^+, K^+ and Ca^+ ion uptake and transport, in order to maintain ionic adjustments within the plant tissues during salt stress. There appears to be a salt-tolerance mechanism operating in which, during salt stress, ionic adjustment within the plant tissues is maintained by controlling the ion uptake and transport rates.

19.4 Tuber Crops: Climate-Resilient Crops

Climate change is among the major global issues of the twenty-first century. Global climate change mainly involves the rise in atmospheric carbon dioxide (CO_2) up to 700 ppm, temperature up to 6 °C and frequent soil drought (Ravi et al. 2008). Variations in rainfall and temperature due to climate change affect the production of major food crops like rice and wheat, thereby food security. In cereals, the flowering and grain-filling stages are highly sensitive to environmental stresses, and the occurrence of stresses at these critical stages leads to complete crop loss (Naskar and Ravi 2011). However, in tuber crops, both tuber and shoot growth occurs simultaneously during favourable conditions. These crops cease tuber development as well as vegetative growth and become dormant during unfavourable conditions such as drought. During favourable conditions, tuber growth is resumed and complete crop failure is avoided. Hence, they are considered as climate-resilient crops.

Cassava is known to be a very drought-tolerant and water-efficient crop, while it is also exceptionally tolerant to high soil acidity (Howeler 2004). Cassava maintains nearly 50 % of its photosynthetic rate under drought conditions (Ravi and Saravanan 2001). Cassava is sometimes referred to as the drought, war and famine crop of the developing world, and reliance on this crop is expected to increase in the coming years under global climate change conditions (Burns et al. 2010). Studies have shown that large areas under major food crops including maize, sorghum, millet, beans, potato and bananas mainly in Africa will become too dry and would be negatively affected by the impact of climate change, while cassava would be the least affected. Cassava with high optimum temperature requirement for photosynthesis and growth and positive response to elevated CO_2 levels would be a potential food, feed and energy crop in tropical and subtropical zones which are likely to be adversely affected by climate change (Jarvis et al. 2012). Ezekiel et al. (2012) opined that cassava production does not vary with the current levels of climate change, although there are many other factors having impact on cassava productivity, like variety, soil type, pests and disease and other socio-economic constraints. Tuber crops like yams can also tolerate prolonged drought conditions. Cassava and yams are presently among the major crops that show the highest rates of increase in area under cultivation. This may be attributed to their resilience to climate change.

Development of improved drought-tolerant orange-fleshed sweet potato will increase its production in arid and semiarid lands where seasonal drought is a problem. It will also have a positive impact on the livelihood and health of the people especially in sub-Saharan Africa. Research is under way to develop saline-tolerant varieties of sweet potato which can be used for the reclamation of waste lands and coastal areas. Aroids like taro, tannia and elephant foot yam are well adapted to shade conditions and highly suitable for intercropping. Taro is a versatile crop as it can be grown on a variety of soil conditions from the waterlogged swampy condition to the upland rainfed conditions (Naskar and Ravi 2011; Ravi et al. 2013a). Tannia is one of the most shade-tolerant food crops (Ramesh et al. 2007). These features offer tropical tuber crops a better choice in the future under climate change conditions. The ability to yield reasonably well under changing climatic conditions makes them the 'future crops'.

19.5 Conclusion

Tuber crops are the energy reservoirs of nature, mainly in the tropical regions. Due to high starch content and calorie value, these crops have a major role in meeting the food security of our ever-increasing population. Tuber crops have wide adaptability to diverse agroclimatic conditions. They are known as crops for adversity since they have resilience to adverse weather conditions. Hence, tuber crops have high relevance to combat abiotic stresses in the context of climate change and play a prominent role in feeding the world in the coming decades. As with all crops, challenges of abiotic stress in tuber crops need to be addressed by developing stress-tolerant/stress-resistant varieties and also by adopting proper management practices.

References

Aina OO, Dixan AGO, Akinrinde EA (2007) Effect of soil moisture stress on growth and yield of cassava. Pak J Biol Sci 10(18):3085–3090

Allard HA (1945) Some behaviours of yam (Dioscorea). Castanea 10:8–13

Amthore JS (2003) Efficiency of lignin biosynthesis: a quantitative analysis. Ann Bot 91:673–695

Angelov MN, Sun J, Byrd GT, Brown RH, Black CC (1993) Novel characteristics of cassava, *Manihot esculenta* Crantz, a reputed C_3–C_4 intermediate photosynthesis species. Photosynth Res 38:61–72

Anil SR, Palaniswami MS (2008) Botany, physiology and biodiversity. In: Palaniswami MS, Peter KV (eds) Tuber and root crops. New India Publishing Agency, New Delhi, pp 41–84

Bai EKL (1981) Shade response of common rainfed intercrops of coconut. MSc (Ag) thesis, Kerala Agricultural University, Thrissur, Kerala, p 161

Bai EKL, Nair RV (1982) Shade response of some common rainfed intercrops. Indian Society for Plantation Crops, Kasargod, p 410

Bhattacharya NC, Bhattacharya S, Strain BR, Biswas PK (1989) Biochemical changes in carbohydrates and proteins of sweet potato plants. J Plant Physiol 135 (3):261–266

Bhattacharya NC, Hileman DR, Ghosh PP, Musser RL, Bhattacharya S, Biswas PK (1990a) Interaction of enriched CO_2 and water stress on the physiology and biomass production in sweet potatoes grown in open top chambers. Plant Cell Environ 13:933–940

Bhattacharya S, Bhattacharya NC, Tolbert MEM (1990b) Characterization of carotene in sweet potato (Ipomoea batatas) grown at CO_2 enriched atmosphere under field conditions. In: Hodgson RH (ed) Proceedings of the Plant Growth Regulator Society of America 17th annual meeting. Plant Growth Regulator Society, Ithaca, pp 123–126

Bhattacharya NC, Ghosh PP, Hileman DR, Alemayehu M, Huluka G, Biswas PK (1992) Growth and yield of sweet potato under different carbon dioxide concentrations. In: Hill WA, Bonsi CK, Loretan PA (eds) Sweet potato technology for the 21st century. Tuskegee University, Tuskegee, pp 333–336

Bilski JJ, Nelson DC, Conlon RL (1988) The response of four potato cultivars to chlorine salinity, sulphate salinity and calcium in pot experiments. Am Potato J 65(2):85–90

Biswas PK, Hilean DR, Bhattacharya NC, Ghosh PP, Bhattacharya S, Johnson JH, Mbikai NT (1986) Response of vegetation to carbon dioxide: growth, yield and plant water relationships in sweet potatoes in response to carbon dioxide enrichment. Rep. 30.U. S. Dept. Energy. Carbondioxide Res. Div., Office of Energy Res, Washington, DC

Biswas J, Sen H, Mukhopadhyay SK (1989) Effect of light hours on growth and tuber yield in sweet potato. J Roots Crops 15(2):123–125

Biswas J, Sen H, Mukhopadhyay SK (1990) Effect of light intensity on shoot growth and tuber development in sweet potato. J Roots Crops 16(1):1–3

Bograd JH, Sarobol E, Rojanaridpiched C, Sriroth K (2011) Effect of supplemental irrigation on reducing cyanide content of cassava variety. Kasetsart J (Nat Sci) 45:985–994

Bolhuis GG (1966) Influence of length of illumination period on root formation in cassava (*Manihot utilissima* Pohl). Neth J Agric Sci 14:251–254

Bonsi CK, Loretan PA, Hill WA, Morley DG (1992) Response of sweet potatoes to continuous light. Hortic Sci 27(5):471

Burns A, Gleadow R, Cliff J, Zacarias A, Cavagnao T (2010) Cassava: the drought, war and famine crop in a changing world. Sustainability 2:3572–3607

Caesar K (1980) Growth and development of Xanthosoma and Colocasia under different light and water supply conditions. Field Crop Res 3:235–244

Carretero CL, Cantos M, Garc'ıa JL, Azc'on R, Troncoso A (2008) Arbuscular-mycorrhizal contributes to alleviation of salt damage in cassava clones. J Plant Nutr 31:959–971

Carvalho P de CL, Ezeta FN (1983) Efeito do fotoperiodo sober a 'tuberizacao' da mandioca. Rev Bras Mandioca 2:51–54

Cen YP, Sage RF (2005) The regulation of Rubisco activity in response to variation in temperature and atmospheric CO_2 partial pressure in sweet potato. Plant Physiol 139:979–990

Cerqueira YM (1989) Efeito da deficiencia de agua na anatomia foliar de cultivares de mandioca (*Manihot esculenta* Crantz). MSc thesis, Universidade Federal da Bahia, cruz das Almas, Brazil

Chassy B, Egnin M, Gao Y, Kleter G, Nestel P, Newell-McGloughlin M, Phipps R, Shillito R (2008) Nutritional and safety assessments of foods and feeds nutritionally improved through biotechnology. Compr Rev Food Sci Food Saf 7(1):50–113

Chaves MM (1991) Effect of net water deficit on carbon assimilation. J Exp Bot 42:1–16

Chowdhary SR, Brahmanand PS, Kumar A (2011) Changes in leaf water potential, chlorophyll content and runner yield of swamp taro (*Colocasia esculenta*) grown in waterlogged environment. J Roots Crops 37 (1):32–36

Chowdhury SR, Thakur AK, Kumar A (2008) Comparative study of utilization of incident radiation and chlorophyll fluorescence in swamp taro (*Colocasia esculenta*), Alocasia and tannia (*Xanthosoma* sp.) leaves. J Roots Crops 34:5–9

Chowdhury SR, Kumar A, Sahoo N (2009) Diurnal changes in chlorophyll fluorescence and light utilization in Colocasia esculenta leaves grown in marshy waterlogged area. Biol Plants 53:167–170

Chowdhury SR, Anand PSB, Kundu DK, Kumar A (2010) Performance of swamp taro cultivars (Colocasia esculenta (L.) Schott) as affected by brief submergence. J Roots Crops 36(2):204–209

Cock JH (1978) Adaptability of cassava in cassava production. CIAT, Cali

Cock JH (1984) Cassava. In: Goldsworthy PR, Fisher NM (eds) The physiology of tropical field crops. Wiley, Chichester, pp 529–549

Cock JH, Rosas S (1975) Ecophysiology of cassava. In: Symposium in ecophysiology of tropical crops. Communications Divisions of Ceplac, Ilheus, Bahia, Brazil, pp 1–4

Connor DJ, Cock JH, Parra GE (1981) Response of cassava to water shortage. I. Growth and yield. Field Crop Res 4:181–200

Coursey DG (1967) Yams. An account of the nature, origin, cultivation and utilization of some useful members of the Dioscoreaceae. Longmans, London, p 230

Dasgupta M, Mukherjee A, Sahoo MR, Naskar SK, Kole PC (2006) Response of orange fleshed sweet potato (*Ipomoea batatas* L.) genotypes to sodium chloride stress conditions. J Roots Crops 32(1):53–58

DeTafur SM, El-Sharkawy MA, Cadavid LF (1997) Response of cassava (*Manihot esculenta* Crantz) to water stress and fertilization. Photosynthetica 34:233–239

Dhindsa RS, Plumb-Dhindsa P, Thorpe TA (1981) Leaf senescence: correlated with increased levels of membrane permeability and lipid peroxidation and decreased levels of superoxide dismutase and catalase. J Exp Bot 32:93–101

Edwards GE, Sheta E, Moore B, Dai Z, Franceschi VR, Cheng SH, Lin CH, Ku MSB (1990) Photosynthetic characterization of cassava (*Manihot esculenta* Crantz), a C_3 species with chlorenchymatous bundle sheath cells. Plant Cell Physiol 31:1199–1206

Ekanayake JI, Oyetunji OJ, Osonubi O, Layasse O (2004) The effects of arbuscular mycorrhizal fungi and water stress on leaf chlorophyll production of cassava (*Manihot esculenta* Crantz). Food Agric Environ 2 (2):190–196

El-Sharkawy MA (2006) International research on cassava photosynthesis, productivity, eco-physiology and responses to environmental stresses in the tropics. Photosynthetica 44:481–512

El-Sharkawy MA, Cock JH (1990) Photosynthesis of cassava (*Manihot esculenta*). Exp Agric 26:325–340

El-Sharkawy MA, De Tafur SM, Cadavid LF (1992) Potential photosynthesis of cassava as affected by growth conditions. Crop Sci 32:1336–1342

Ezekiel AA, Olawuyi SO, Ganiyu MO, Ojedokun IK, Adeyemo SA (2012) Effects of climate change on cassava productivity in Ilesa – East Local Government area, Osun State, Nigeria. Br J Art Soc Sci 10 (2):153–162

FAOSTAT (2010) Food and Agriculture Organization (FAO) FAOSTAT. http://faostat.fao.org/site/612/default.aspx#ancor

FAOSTAT (2013) Food and Agriculture Organization (FAO) FAOSTAT. http://faostat.fao.org/site/612/default.aspx#ancor

Fernández MD, Tezara MDW, Rengifo E, Herrera A (2002) Lack of down regulation of photosynthesis in a tropical root crop, cassava, grown under elevated CO_2 concentration. Funct Plant Biol 29(7):805–814

Fujimoto T (2009) Taro (*Colocasia esculenta* (L.) Schott) cultivation in vertical wet dry environments: farmers techniques and cultivation diversity in South Western Ethiopia. Econ Bot 63(2):152–166

Fukai S, Alcoy AB, Lamela AB, Patterson RD (1984) Effect of solar radiation on growth of cassava (*Manihot esculenta* Crantz). Field Crop Res 9:347–360

Geetha V, Pushpakumari R, Krishnaprasad BT (2006) Influence of light intensities on yield, root characters and nutrient uptake of cassava. J Roots Crops 3 (2):130–133

George J (2000) Cultural and manurial requirements of yams and aroids. In: Mohankumar CR, Nair GM, George J, Ravindran CS, Ravi V (eds) Production technology of tuber crops. Central Tuber Crops Research Institute, Thiruvananthapuram, pp 102–108

Ghosh SP, Nair GM, Pillai NG, Ramanujam T, Mohankumar B, Lekshmi KR (1987) Growth, productivity and nutrient uptake by cassava in association with perennials. Trop Agric Trinidad 64:233–236

Ghosh SP, Ramanujam T, Jose JS, Moorthy SN, Nair RG (1988) Tuber crops. Oxford and IBH Publishing Co, New Delhi

Gomes G (1999) Sweet potato growth characteristics, 1st edn. Academic, New York, pp 4–7

Gong Y, Wang G (1990) An investigation on the effect of drought stress on growth of sweet potato and measures to improve drought resistance and stabilize yields. Zhejiang Acad Agric Sci 1:26–29

Greenway H, Munns R (1980) Mechanisms of salt tolerance in non-halophytes. Annu Rev Plant Biol 31:149–190

Greig JK, Smith FW (1962) Salinity effects on sweet potato growth. Agron J 54:309–313

Haimeirong, Kubota F (2003) The effects of drought stress and leaf ageing on leaf photosynthesis and electron transport in photosystem 2 in sweet potato (*Ipomoea batatas* Lam.) cultivars. Photosynthetica 41(2):253–258

Harding SA, Guikema JA, Paulsen GM (1990) Photosynthetic decline from high temperature stress during maturation of wheat. Plant Physiol 92:648–653

Horton D (1989) Constraints to sweet potato production and use. In: Improvement of sweet potato in Asia. CIP, Lima, Peru, pp 219–223

Howeler R (2004) Cassava in Asia- present situation and its future potential in agro-industry. J Roots Crops 30 (2):81–92

Hurkman WJ, Tanaka CK (1987) The effects of salt on the pattern of protein synthesis in barley roots. Plant Physiol 83:517–524

Imai K, Coleman DF, Yanagisawa T (1984) Elevated atmospheric partial pressure of carbon dioxide and dry matter production of cassava (*Manihot esculenta* Crantz). Japan J Crop Sci 53:479–485

Indira (1978) Photoperiodic effect on flowering in cassava. J Roots Crops 4(1):65–66

Indira P (1989) Drought tolerant traits in sweet potato genotypes. J Roots Crops 15(2):139–144

Indira P, Kabeerathumma S (1988) Physiological response of sweet potato under water stress: 1. Effect of water stress during different phases of tuberization. J Roots Crops 14(1):37–40

Indira P, Kabeerathumma S (1990) Physiometabolic changes in sweet potato grown under different levels of soil moisture. J Roots Crops 16:28–32

IPCC (Integrated panel on climate change) (2007) Climate Change (2007) The physical science basis. Cambridge University Press, Cambridge, p 176

Irikura Y, Cock JH, Kawano K (1979) The physiological basis of genotype temperature interactions in cassava. Field Crop Res 2:227–239

Jarvis A, Ramirez-Villegas J, Campo BVH, Navarro-Racines C (2012) Is cassava the answer to African climate change adaptation. Trop Plant Biol 5 (1):9–29

Johnston M, Onwume IC (1998) Effect of shade on photosynthetic pigments in the tropical root crops: yam, taro, tannia, cassava and sweet potato. Exp Agric 34 (3):301–312

Kasele IN, Hahn SK, Oputa CO (1986) Effect of shade, nitrogen and potassium on cassava. Potash Res 12:22–25

Kay DE (1973) Crop and product digest 2. Root crops. Tropical Products Institute, London, p 245

Keatin BA, Evenson JP (1979) Effect of soil temperature on sprouting and sprout elongation of stem cuttings of cassava. Field Crop Res 2:241–252

Keating BA, Wilson GL, Evenson JP (1985) Effect of photoperiod on growth and development of cassava (*Manihot esculenta* Crantz). Aust J Plant Physiol 12:621–630

Kyle DJ (1987) The biochemical basis for photoinhibition of photosystem II. In: Kyle DJ et al (eds) Photoinhibition. Elsevier, Amsterdam, pp 197–226

LaHaye PA, Epstein E (1971) Calcium and salt toleration by bean plants. Physiol Plant 25:213–218

Lebot V (2009) Tropical root and tuber crops: cassava, sweet potato, yams and aroids. CIRAD/CAB International, Paris/Wallingford, p 413

Lewthwaite SL, Triggs CM (2012) Sweet potato cultivar response to prolonged drought. Agron N Z 42:1–10

Lobenstein G (2009) Origin, distribution and economic importance. In: Lobenstein G, Thottappilly G (eds) The sweet potato. Springer Science Publishers, Business Media, BV, Berlin pp 9–12

Lopez M, Vasquez E, Lopez F (1995) Raices y tuberculos. Publo Educacion. Universidad Central de Las Villas, Santa Clara, p 312

Lowe SB, Mohan JD, Hunt LA (1976) The effect of day length on shoot growth and formation of root tubers in young plants of cassava (*Manihot esculenta* Crantz). Plant Sci Lett 6:57–62

Mahon JD, Lowe SB, Hunt LA, Thiagaraja M (1977) Environmental effects on photosynthesis and transpiration in attached leaves of cassava (*Manihot esculenta* Crantz). Photosynthetica 11:121–130

Martin FW (1985) Differences among sweet potatoes in response to shading. Trop Agric 62(2):161–165

Martinez- Beltran J, Manzur CL (2005) Overview of salinity problem in the world and FAO strategies to address the problem. Proceedings of the International Salinity Forum, Riverside, pp 311–313

McDavid CR, Alamu S (1980) The effect of day length on the growth and development of whole plants and rooted leaves of sweet potato. Trop Agric 57:113–119

Mirua K, Osada A (1981) Effect of shading on photosynthesis, respiration, leaf area and corm weight in konjak plants. Japan J Crop Sci 50:553–559

Mitra S, Somasundaram K (2008) A model to stimulate sweet potato growth. World Appl Sci J 4(4):568–577

Mohankumar CR (2008) Cultural and manurial requirement of cassava. In: Mohankumar CR, Nair GM, George J, Ravindran CS, Ravi V (eds) Production technology of tuber crops. Central Tuber Crops Research Institute, Thiruvananthapuram, pp 7–36

Mohankumar CR, Mandal RC, Singh KD (1973) Effect of mulching and plant density on growth, yield and quality of Amorphophallus. Indian J Agron 18:62–66

Mukherjee A (1999) Tuber crops. In: Ghosh SP (ed) Biotechnology and its application in horticulture. Narosha Publishing House, New Delhi, pp 267–294

Mukherjee A (2001) Effect of NaCl on axillary shoot proliferation in sweet potato. Ann Trop Res 23:1–10

Mukherjee A (2002) Effect of NaCl on in-vitro propagation of sweet potato. Appl Biochem Biotechnol 102:431–441

Mukherjee A, Naskar SK (2012) Performance of orange and purple fleshed sweet potato genotypes in coastal locations of Odisha. J Roots Crops 38:26–31

Mwanga ROM, Zamora OB (1989) Agronomic responses of sweet potato (*Ipomoea batatas* (L.) Lam.) to varying levels of shade. Philipp J Crop Sci 14(2):83–90

Nair GM (2000) Cultural and manurial requirements of sweet potato. In: Mohankumar CR, Nair GM, George J, Ravindran CS, Ravi V (eds) Production technology of tuber crops. Central Tuber Crops Research Institute, Thiruvananthapuram, pp 44–64

Nair GM, Nair VM, Sreedharan C (1996) Response of sweet potato to phasic stress irrigation in summer rice fallows. J Roots Crops 22:45–49

Naskar SK, Chowdhary SR (1995) Genotypic variation in relative water content and nitrate reductase activity in sweet potato under field drought condition. J Roots Crops 21(1):46–49

Naskar SK, Ravi V (2011) Tropical root and tuber crops for food security under changing climatic conditions in India. In: Sajeev MS, Anantharaman M, Padmaja G, Unnikrishnan M, Ravi V, Suja G, Hegde V (eds) Climate change and food security: challenges and opportunities for tuber crops. Proceedings of a national seminar, Central Tuber Crops Research Institute, Sreekariyam, Thiruvananthapuram, pp 35–40

Nawale RN, Salvi MJ (1983) Effects of season on yield of sweet potato. J Roots Crops 9:55–58

Nayar NM (2014) The contribution of tropical tuber crops towards food security. J Roots Crops 40(2):3–14

Njoku E (1963) The propagation of yams (*Dioscorea* spp.) by vine cuttings. J West Afr Sci Assoc 8:29–32

Okoli PSO, Wilson GF (1986) Response of cassava (*Manihot esculenta* Crantz) to shade under field conditions. Field Crop Res 14:349–359

Oliveira SC, Macedom MC, Portos MCM (1982) Effect of water stress on cassava root production. Resqui Agrop Brasileira 17:121–124

Onwugbuta-Envi JA, Kpekot KA (2008) Flooding tolerance and sugar level in cassava (*Manihot esculenta* Crantz). Acta Agronómica Niger 8(1):7–12

Onwume IC (1978) The tropical tuber crops. Wiley, England, p 234

Onwume IC, Charles WB (1994) Cultivation of Cocoyam. In: Tropical Root and Tuber Crops, production, Perspectives and Future Prospects. FAO Plant Production and Protection Paper 126, Rome, pp 228

Oswald A, Alkamper J, Midmore DJ (1985) The effect of different shade levels on growth and tuber yield of sweet potato. II. Tuber yield. J Agron Crop Sci 175 (1):29–40

Palaniswami MS, Anil SR, Singh PK (2008) Agronomy and production technologies. In: Palaniswami MS, Peter KV (eds) Tuber and root crops. New India Publishing Agency, New Delhi, pp 147–178

Patterson DR, Earhart DR, Fugue MC (1979) Effect of flooding level on storage root formation, ethylene production and growth of sweet potato. Hortic Sci 77:452–457

Placide R, Shimelis H, Laing M, Gahakwa D (2013) Physiological mechanism and conventional breeding of sweet potato (*Ipomoea batatas* (L.) Lam) to drought tolerance. Afr J Agric Res 8(18):1837–1846

Porter WC (1979) Sweet potato growth as affected by photoperiod. Hortic Sci 14:124

Prameela P (1990) Screening different morphotypes of colocasia (*Colocasia esculenta*) for shade tolerance. M. Sc. (Ag) thesis, Kerala Agricultural University, Thrissur, pp 62

Pushpakumari R (1989) Fertilizer management of minor tuber crops in coconut based cropping system. Ph.D thesis, Kerala Agricultural University, Thrissur, pp 278

Pushpakumari R, Sasidhar VK (1992) Fertilizer management of yams and aroids in coconut based cropping system. J Roots Crops 18(2):99–102

Pushpakumari R, Sasidhar VK (1996) Dry matter production and uptake of nutrients by yams and aroids as influenced by shade intensities. In: Kurup GT, Palaniswami MS, Potty VP, Padmaja G, Kabeerathumma S, Pillai SV (eds) Tropical tuber crops – problems, prospects and future strategies. Oxford & IBH Publishing Co. Pvt. Ltd, New Delhi, pp 274–279

Ramanujam T (1990) Effect of moisture stress on photosynthesis and productivity of cassava. Photosynthetica 24:217–224

Ramanujam T (1994) Water management in cassava. In: Chadha KL, Nayar G (eds) Advances in horticulture, vol 8, Tuber Crops. Malhotra Pub. House, New Delhi, pp 333–342

Ramanujam T, Biradar RS (1987) Growth analysis in cassava (*Manihot esculenta* Crantz). Indian J Plant Physiol 30:144–153

Ramanujam T, Jose JS (1984) Influence of light intensity on chlorophyll distribution and anatomical characters of cassava leaves. Turrialba 34:467–471

Ramanujam T, Indira P, Nair GM (1984a) Productivity of cassava under shade. Indian Farm 33:39–42

Ramanujam T, Nair GM, Indira P (1984b) Growth and development of cassava (*Manihot esculenta* Crantz) genotypes under shade in a coconut garden. Turrialba 34:367–374

Ramesh V, John KS, Ravindran CS, Edison S (2007) Agrotechniques and plant nutrition of tannia (*Xanthosoma* sp.): an overview. J Roots Crops 33 (1):1–11

Ravi V (2000) Physiological aspects of tuber crops. In: Mohankumar CR, Nair GM, George J, Ravindran CS, Ravi V (eds) Production technology of tuber crops. Central Tuber Crops Research Institute, Sreekariyam, pp 133–168

Ravi V, Chowdhury SR (1991) Growth and yield of taro to different moisture regimes. J Roots Crops 17:129–133

Ravi V, Chowdhury SR (1993) Stomatal movements in the leaves of taro grown under different soil moisture regimes. J Roots Crops 19:1–7

Ravi V, Chowdhury SR (1996) Growth and yield response of colocasia accessions to drought stress. In: Kurup GT, Palaniswami MS, Potty VP, Padmaja G, Kabeerathumma S, Pillai SV (eds) Tropical tuber crops – problems, prospects and future strategies. Oxford & IBH publishing Co. Pvt Ltd, New Delhi, pp 289–297

Ravi V, Indira P (1996) Anatomical studies on tuberization in sweet potato under water deficit stress and stress free conditions. J Roots Crops 22:105–111

Ravi V, Indira P (1997) Investigation on physiological factors limiting yield potential in sweet potato under drought stress. In: Annual Report. Central Tuber Crops Research Institute, Trivandrum, Kerala, India

Ravi V, Saravanan R (1999) Proline metabolism and its relation to drought tolerance in sweet potato. J Roots Crops 25(2):135–142

Ravi V, Saravanan R (2001) Photosynthesis and productivity of cassava under water deficit stress and stress free conditions. J Roots Crops 27(1):214–218

Ravi V, Ravindran CS, Ramesh V (2008) The impact of climate change on photosynthesis and productivity of cassava and sweet potato: effect of rise in temperature, CO_2 and UV-B radiation: an overview. J Roots Crops 34(2):95–107

Ravi V, Ravindran CS, Suja G (2009) Growth and productivity of elephant foot yam (*Amorphophallus paeoniifolius* (Dennst. Nicolson): an overview. J Roots Crops 35(2):131–142

Ravi V, Ravindran CS, Suja G, George J, Neduncheziyan M, Byju G, Naskar SK (2011) Crop physiology of elephant foot yam (*Amorphophallus paeoniifolius* (Dennst.) Nicolson). Adv Hortic Sci 25 (1):51–63

Ravi V, Chakrabarti SK, Saravanan R (2013a) Mitigating climate change – reaping bumper harvest of tubers in adverse conditions. Indian Hort 58(3):34–36

Ravi V, Suja G, George J, Nedunchezhiyan M, Byju G (2013b) Critical period of water requirement for elephant foot yam (*Amorphophallus paeoniifolius*). In: International Conference on Tropical Roots and Tubers for Sustainable Livelihood under Changing Agro-climate. Abstract of papers. July 9-12, 2013. Indian Society for Root Crops, Thiruvananthapuram, Kerala, India, pp 208–209

Reynold SG, Netran P (1977)The effect of rainfall and time of planting on yield of taro in Western Samoa. In: Leakey CLA (ed) Proceedings of the 3rd international symposium on tropical tuber crops. International Society for Tropical Root Crops, International Institute for Tropical Agriculture, Nigeria, pp 374–376

Richardson KVA, Caligari PDS (2014) Calcium chloride and salinity stress effects on growth, dry matter allocation and ion uptake of sweet potato. J Roots Crops 40(1):56–65

Roberts-Nkrumah LB, Ferguson TU, Wilson LA (1986) Response of four sweet potato cultivars to levels of shade 2. Tuberisation Trop Agric 63(4):265–270

Sajjapongse A (1989) The effect of climatic factors on the yield of sweet potato. IBSRAM Newsl 5(13):7

Santosa E, Sugiyamma N, Sulistyono E, Sopandie D (2004) Effect of watering frequency on the growth of elephant foot yams. Japan J Trop Agric 48:235–239

Santosa E, Sugiyamma N, Nakata M, Lee ON (2006) Growth and corm production of Amorphophallus at different shading levels in Indonesia. Japan J Trop Agric 50:87–91

Saud BK, Baruah RKSM (2000) 'Pani-Kachu' – a special taro cultivation in southern Assam. Intensiv Agric 38:26–27

Schaffer B, O'Hair SK (1987) Net CO_2 assimilation of taro and cocoyam as affected by shading and leaf age. Photosynth Res 11:245–251

Schneider SH (1989) The greenhouse effect: science and policy. Sciences 243:771–781

Sen H, Goswami SB(2000) Swamp taro cultivation in Teri climate- a case study. Paper presented in the international symposium of tropical root and tuber crops. January 19–22, 2000. Thiruvananthapuram, Kerala, India

Silva S, Irizarry H (1980) Effect of depth of water table on yield of tannia. J Agric Univ P R 64(2):241–242

Singh KD, Mandal RC (1976) Performance of coleus and sweet potato in relation to seasonal variations (time of planting). J Roots Crops 2(2):17–22

Sohn Y, Song J, Jeon G, Kim D, Park M (2011) Effect of flooding and soil salinity on the growth of yam (Dioscorea batatas) transplanted by seedling of aerial bulblet in Saemangeum reclaimed tidal land. Korean J Soil Sci Fert 44(1):8–14

Splittstoesser WE, Tunya GO (1992) Crop physiology of cassava. Hortic Rev 13:105–129

Sreekumari MT, Abraham K, Ramanujam T (1988) The performance of cassava under shade. J Roots Crops 14:43–52

Sriroth K, Piyachomkwan K, Santisopasri V, Oates CG (2001) Environmental conditions during root development: drought constraint on cassava starch quality. Euphytica 120:95–101

Steed L, Truong VD (2008) Anthocyanin content, antioxidant activity, and selected physical properties of flowable purple fleshed sweet potato purees. J Food Sci 735(5):215–221

Sugiyamma N, Santosa E (2008) Edible amorphophallus in Indonesia-potential crops in agroforestry. Gadjah Mada University Press, Yogyakarta, p 125

Sunell LA, Arditti JC (1983) Physiology and phytochemistry. In: Wang JK (ed) Taro, a review of Colocasia esculenta and its potentials. University of Hawaii Research, Honululu, pp 34–139

Sung FJM (1981) The effect of leaf water status on stomatal activity, transpiration and nitrate reductase of sweet potato. Agric Water Manag 4:465–470

Tan SK, Ambak K (1989) A lysimeter study on the effect of watertable on cassava grown on peat. MARDI Res J 17(1):37–142

Teow CC, Truong VD, McFeeters RF, Thompson RL, Pecota KV, Yencho GC (2007) Antioxidant activities, phenolic and beta-carotene contents of sweet potato genotypes with varying flesh colours. Food Chem 103 (3):829–883

Ueno O, Agarie S (1997) The intercellular distribution of glycine decarboxylase in leaves of cassava in relation to the photosynthetic mode and leaf anatomy. Japan J Crop Sci 66:268–278

Valenzuela HR (1990) Shade tolerance and photosynthesis of cocoyams (*Xanthosoma sagittifolium*). Sci Eng 52:173

Veltkamp HJ (1985) Growth, total dry matter yield and its partitioning in cassava at different day lengths. Agric Univ Wagening Pap 85:73–86

Villafane R (1997) Effect of soil salinity on growth of sweet potatoes. Agron Trop (Maracay) 47(2):131–139

Waitt AW (1963) Yams- dioscorea species. Field Crop Abstr 16:145–157

Wang Q, Hou F, Dong S, Xie B, Li A, Zhang H, Zhang L (2014) Effects of shading on the photosynthetic capacity, endogenous hormones, and root yield in purple-fleshed sweet potato (*Ipomoea batatas* (L.) Lam). Plant Growth Regul 72:113–122

Wilson JE (1984) Cocoyam. In: Goldsworthy PR, Fisher NM (eds) The physiology of tropical field crops. Wiley, New York, pp 589–605

Woolfe JA (1992) Sweet potato: an untapped food resource. Cambridge University Press, New York

Wrigley CW, Blumenthal C, Gras PW, Barlow EWR (1994) Temperature variation during grain filling and changes in wheat grain quality. Aust J Plant Physiol 21:875–885

Yang RC, Jana S, Clarke J (1991) Phenotypic diversity and associations of some potentially drought-responsive characters in durum wheat. Crop Sci 31 (6):1484–1491

Zara DL, Cueves SE, Carlos IJ Jr (1982) Performance of sweet potato under coconuts. In: Villareal RL, Griggs TD (eds) Sweet potato. Asian Vegetable Research and Development Centre, Taiwan, pp 233–242